T0141173

Membrane Technology

Sustainable Solutions in Water, Health, Energy and Environmental Sectors

Membrane Technology

Sustainable Solutions in Water, Health, Energy and Environmental Sectors

Edited by
Sundergopal Sridhar

CRC Press
Taylor & Francis Group
Boca Raton London New York

CRC Press is an imprint of the
Taylor & Francis Group, an **informa** business

CRC Press
Taylor & Francis Group
6000 Broken Sound Parkway NW, Suite 300
Boca Raton, FL 33487-2742

First issued in paperback 2020

© 2019 by Taylor & Francis Group, LLC
CRC Press is an imprint of Taylor & Francis Group, an Informa business

No claim to original U.S. Government works

ISBN 13: 978-0-367-57125-2 (pbk)
ISBN 13: 978-1-138-09542-7 (hbk)

This book contains information obtained from authentic and highly regarded sources. Reasonable efforts have been made to publish reliable data and information, but the author and publisher cannot assume responsibility for the validity of all materials or the consequences of their use. The authors and publishers have attempted to trace the copyright holders of all material reproduced in this publication and apologize to copyright holders if permission to publish in this form has not been obtained. If any copyright material has not been acknowledged please write and let us know so we may rectify in any future reprint.

Except as permitted under U.S. Copyright Law, no part of this book may be reprinted, reproduced, transmitted, or utilized in any form by any electronic, mechanical, or other means, now known or hereafter invented, including photocopying, microfilming, and recording, or in any information storage or retrieval system, without written permission from the publishers.

For permission to photocopy or use material electronically from this work, please access www.copyright.com (http://www.copyright.com/) or contact the Copyright Clearance Center, Inc. (CCC), 222 Rosewood Drive, Danvers, MA 01923, 978-750-8400. CCC is a not-for-profit organization that provides licenses and registration for a variety of users. For organizations that have been granted a photocopy license by the CCC, a separate system of payment has been arranged.

Trademark Notice: Product or corporate names may be trademarks or registered trademarks, and are used only for identification and explanation without intent to infringe.

Library of Congress Cataloging-in-Publication Data

Names: Sridhar, Sundergopal, editor.
Title: Membrane technology : sustainable solutions in water, health, energy and environmental sectors / Sundergopal Sridhar, editor.
Description: Boca Raton : Taylor & Francis, 2018. | Includes bibliographical references and index.
Identifiers: LCCN 2018007554| ISBN 9781138095427 (hardback : acid-free paper) | ISBN 9781315105666 (ebook)
Subjects: LCSH: Membranes (Technology)
Classification: LCC TP159.M4 M4724 2018 | DDC 660/.28424--dc23
LC record available at https://lccn.loc.gov/2018007554

Visit the Taylor & Francis Web site at
http://www.taylorandfrancis.com

and the CRC Press Web site at
http://www.crcpress.com

Dedicated to my mother, S. Kannammal, my twin daughters, Srishti

and Santushti, my wife, Santoshi, and all my well-wishers.

Contents

Section III   Health

Section IV   Membrane Process Design

Section V   Energy

Section VI Environment

Preface

Membrane technology has undergone rampant growth in western countries besides acquiring immense popularity in developing and underdeveloped nations for applications that appeal to the common man, which include drinking water purification, wastewater treatment and hemodialysis. The ability of membranes to remove most of the impurities in a single step is what makes them so special in the vast spectrum of separation processes. The intervention of membrane technology in several unit operations in chemical and allied industries has not just enabled water reclamation but also recovery of valuable products. This book makes an attempt to reveal maximum possible data, even patentable ones, for the benefit of the reader, especially young aspiring scientists. While most patents do not reveal critical information, this work attempts to break away from that genre, so that the revealed information provides motivation to researchers, academicians and industrialists alike, to take calculated risks and solve problems pertaining to the four major sectors of water, health, energy and environment. Apart from important chapters from leading Indian institutes, such as IITs, BARC, CSIR Laboratories, NITs and other reputed academia, the book also presents data from two major industries, viz., Permionics Membrane Pvt. Ld. based in Vadodara, India, which specializes in manufacturing low fouling, high-flux membranes for pressure driven processes, and Porifera Inc., United States, which is the world leader in forward osmosis membranes and systems manufacture. The Editor's group at IICT comes under the CSIR umbrella of 39 national laboratories and has been engaged in most of the first-generation membrane activities over the past 20 years, except liquid membrane technology. This work reflects their efforts in bringing forth a stage-wise depiction of how membranes that exhibit optimum performance on a laboratory level can be harnessed to pilot plant and commercial system application, to serve both the industry and the society. The future of membrane technology looks rosy what with modular configurations, such as hollow fiber and spiral wound available in different dimensions with large effective surface areas, which provide flexibility in building systems of different capacities for various applications. Moreover, second-generation membrane techniques such as membrane crystallization, membrane adsorption, membrane contactors and catalytic membrane reactors are already in vogue. The combined effect of concentration polarization and membrane fouling constitute a gray area of membrane research and development. These challenges have been addressed to a large extent by manipulating the hydrodynamic scenario in the membrane module and employing effective membrane cleaning and storage protocols, as described in this book. The major challenge yet to be addressed is to make large-scale manufacture of pervaporation, gas separation, liquid membrane, fuel cell and forward osmosis membranes more simple and affordable, through design of tailor-made casting machines and procedures, such that solvent recovery, air purification, desalination, energy generation and several other unit operations become economically viable, to benefit the common man in the globe.

Editor

Dr. Sundergopal Sridhar is a Chemical Engineer from University College of Technology, Osmania University, Hyderabad, who has been working as a scientist in Membrane Separation Processes at CSIR-IICT, Hyderabad for the past 20 years. He has developed several membrane technologies for chemical and allied industries, in addition to contributing immensely to rural welfare through water purification projects and academic growth via extensive Human Resources and laboratory development in several educational institutions. Major highlights of Dr. Sridhar's career include: (i) Commissioning of several membrane pilot plants based on Electrodialysis, Nanofiltration and Gas Permeation of capacities varying from 500–5000 L/h to facilitate solvent recovery, effluent treatment and gas purification in pharmaceutical, steel, textile and petrochemical industries, and (ii) Design and installation of 15 model defluoridation plants of 600–4000 L/h capacity and 25 highly compact low cost systems of 100–200 L/h capacity for purification of ground and surface water for a population of more than 5 million affected by fluorosis, gastroenteritis, jaundice, typhoid and other waterborne diseases in the villages of Telangana, Andhra Pradesh, Karnataka and Tamil Nadu, which have been widely appreciated by the press, masses and His Excellency, the Governor of Telangana and Andhra Pradesh States, along with union ministers of science and technology. A free water camp based on compact system design was established by Dr. Sridhar on Uppal Road in Hyderabad city and has been providing healthy water to the urban population including pedestrians, drivers of buses, cars and two-wheelers, totally free of cost, since April 2016. Similar camps at the All India Industrial Exhibition in January–February 2017 & 2018, and the CSIR Science Exhibition have benefited a population of three lakhs, including school children. Dr. Sridhar has published 130 research papers in reputed international journals such as *Journal of Membrane Science* and *Macromolecules*, which are widely cited by peers, more than 6000 times in high impact journals with an h-index of 38. He has 12 foreign patents to his credit, as well as 24 book chapters and 200 proceedings of various symposia/conferences. Dr. Sridhar has trained 350 B.Tech./M.Tech and M.Sc. students from different universities and institutes for project work apart from guiding 6 scholars on their PhD track. His students have been bestowed with 20 prizes for meritorious work or best oral paper/poster presentation made under his guidance in various conferences and symposia. Dr. Sridhar credits part of his success as a researcher to his all-round sports activities that keep him physically and mentally fit, with him having represented his State, University and All India CSIR Teams in Cricket, Lawn Tennis and Table Tennis with distinction, both as a player and captain. Dr. Sridhar is a recipient of 30 prestigious Science Awards including 15 National Awards and 3 State Awards such as the IIChE Amar Dye-Chem Award 2003, CSIR Young Scientist Award 2007, Engineer of the Year Award from A.P. State Govt. in 2009, Scopus Young Scientist Award 2011, NASI–Reliance Industries Platinum Jubilee Award 2013, VNMM award from IIT-Roorkee 2015, CIPET national awards for 2016 and 2017 and the Nina Saxena Excellence in Technology Award from IIT-Kharagpur in 2017.

Contributors

P. Anand
Membrane Separations Laboratory
Chemical Engineering Division
CSIR-Indian Institute of Chemical
   Technology
Hyderabad, India

Usha K. Aravind
Advanced Centre of Environmental
   Studies and Sustainable Development
and
Centre for Environment Education and
   Technology (CEET)
Mahatma Gandhi University
Kerala, India

Charuvila T. Aravindakumar
School of Environmental Sciences
Mahatma Gandhi University
Kerala, India

Sibdas Bandyopadhyay
Ceramic Membrane Division
CSIR-Central Glass & Ceramic Research
   Institute
Kolkata, India
and
CSIRES, Department of Environmental
   Studies
Visva-Bharati Santiniketan
West Bengal, India

Suresh K. Bhargava
Royal Melbourne Institute of Technology
   (RMIT)
Melbourne, Australia

Santoshkumar D. Bhat
CSIR-Central Electrochemical Research
   Institute-Madras Unit,
CSIR Madras Complex
Chennai, India

Chiranjib Bhattacharjee
Chemical Engineering Department
Jadavpur University
Kolkata, India

E. Bhuvanesh
Polymer and Process Engineering
I.I.T. Roorkee, Saharanpur Campus
Uttar Pradesh, India

Anusha Chandra
Polymer and Process Engineering
I.I.T. Roorkee, Saharanpur Campus
Uttar Pradesh, India

S.S. Chandrasekhar
Membrane Separations Laboratory
Chemical Engineering Division
CSIR-Indian Institute of Chemical
   Technology
Hyderabad, India

Somak Chatterjee
Chemical Engineering Department
Indian Institute of Technology (IIT)
Kharagpur, West Bengal, India

Sujay Chattopadhyay
Polymer and Process Engineering
I.I.T. Roorkee, Saharanpur Campus
Uttar Pradesh, India

Satyendra P. Chaurasia
Department of Chemical Engineering
Malaviya National Institute of Technology
   Jaipur
Jaipur, India

and

Catalysis and Chemical Engineering
   Laboratories
Department of Chemical and Biological
   Engineering
University of Saskatchewan
Saskatoon, Canada

Ravi Dhabhai
Catalysis and Chemical Engineering
    Laboratories
Department of Chemical and Biological
    Engineering
University of Saskatchewan
Saskatoon, Canada

Ajay K. Dalai
Catalysis and Chemical Engineering
    Laboratories
Department of Chemical and Biological
    Engineering
University of Saskatchewan
Saskatoon, Canada

and

Department of Chemical Engineering
Banasthali University
Rajasthan, India

Chandan Das
Department of Chemical Engineering
Indian Institute of Technology Guwahati
Guwahati, India

Ranjana Das
Chemical Engineering Department
Jadavpur University
Kolkata, India

Sirshendu De
Chemical Engineering Department
Indian Institute of Technology (IIT)
Kharagpur, West Bengal, India

N. L. Gayatri
Membrane Separations Laboratory
Chemical Engineering Division
CSIR-Indian Institute of Chemical
    Technology
Hyderabad, India

B. Govardhan
Membrane Separations Laboratory,
    Chemical Engineering Division
CSIR-Indian Institute of Chemical
    Technology
Hyderabad, India

Rehana Anjum Haldi
Membrane Separations Laboratory
Chemical Engineering Division
CSIR-Indian Institute of Chemical
    Technology
Hyderabad, India

Anjali Jain
Department of Chemical Engineering
Banasthali University
Rajasthan, India

and

Catalysis and Chemical Engineering
    Laboratories
Department of Chemical and Biological
    Engineering
University of Saskatchewan
Saskatoon, Canada

F. Dileep Kumar
Membrane Separations Laboratory
Chemical Engineering Division
CSIR-Indian Institute of Chemical
    Technology
Hyderabad, India

M. Madhumala
Membrane Separations Laboratory,
    Chemical Engineering Division
CSIR-Indian Institute of Chemical Technology
Hyderabad, India

Mainak Majumder
Monash University
Melbourne, Australia

and

Ceramic Membrane Division
CSIR-Central Glass & Ceramic Research
    Institute
Kolkata, India

Mary Lidiya Mathew
Advanced Centre of Environmental
    Studies and Sustainable Development
Mahatma Gandhi University
Kerala, India

Satya Jai Mayor
Permionics Membranes Pvt. Ltd.
Gujarat, India

Arijit Mondal
Chemical Engineering Department
Jadavpur University
Kolkata, India

Kaustubha Mohanty
Department of Chemical Engineering
Indian Institute of Technology Guwahati
Guwahati, India

Siddhartha Moulik
Membrane Separations Laboratory
Chemical Engineering Division
CSIR-Indian Institute of Chemical
    Technology
and
Academy of Scientific and Innovative
    Research (AcSIR)
Hyderabad, Telangana, India

A. Muthumeenal
PG & Research Department of Chemistry
Polymeric Materials Research Lab
Alagappa Government Arts College
Karaikudi, India

Harsha Nagar
Membrane Separation Laboratory
Chemical Engineering Division
CSIR-Indian Institute of Chemical
    Technology
Hyderabad, India

A. Nagendran
PG & Research Department of Chemistry
Polymeric Materials Research Lab
Alagappa Government Arts College
Karaikudi, India

Barun Kumar Nandi
Department of Fuel and Mineral
    Engineering
Indian Institute of Technology (Indian
    School of Mines)
Dhanbad, Jharkhand, India

Kaushik Nath
New Separation Laboratory, Department of
    Chemical Engineering
G H Patel College of Engineering &
    Technology
Gujarat, India

Shaik Nazia
Membrane Separations Laboratory,
    Chemical Engineering Division
CSIR-Indian Institute of Chemical Technology
Hyderabad, India

and

Royal Melbourne Institute of Technology
    (RMIT)
Melbourne, Australia

Anil Kumar Pabby
Nuclear Recycle Board
Bhabha Atomic Research Centre
Tarapur, India
and
Homi Bhabha National Institute
Mumbai, India

Tejal M. Patel
New Separation Laboratory
Department of Chemical Engineering
G H Patel College of Engineering &
    Technology
Gujarat, India

M. Praveen
Membrane Separations Laboratory
Chemical Engineering Division
CSIR-Indian Institute of Chemical
    Technology
Hyderabad, India

Mihir Kumar Purkait
Department of Chemical Engineering
Indian Institute of Technology Guwahati
Guwahati, Assam, India

Mehabub Rahaman
Department of Chemical Engineering
Jadavpur University
Kolkata, India

Gutru Rambabu
CSIR-Central Electrochemical Research
    Institute-Madras Unit,
CSIR Madras Complex
Chennai, India

D. Rana
Department of Chemical and Biological
    Engineering
Industrial Membrane Research Institute
University of Ottawa
Ottawa, ON, Canada

Y.V.L. Ravikumar
Membrane Separations Laboratory,
    Chemical Engineering Division
CSIR-Indian Institute of Chemical
    Technology
Hyderabad, India

Ravindra Revanur
Porifera Inc.
San Leandro, California

Kulbhushan Samal
Department of Chemical Engineering
Indian Institute of Technology Guwahati
Guwahati, India

Subha Sasi
Advanced Centre of Environmental
    Studies and Sustainable Development
Mahatma Gandhi University
Kerala, India

Nivedita Sahu
Membrane Separations Laboratory
Chemical Engineering Division
CSIR-Indian Institute of Chemical
    Technology
Hyderabad, India

R. Saranya
Membrane Separations Laboratory
Chemical Engineering Division
CSIR-Indian Institute of Chemical
    Technology
Hyderabad, India

M. Sri Abirami Saraswathi
PG & Research Department of Chemistry
Polymeric Materials Research Lab
Alagappa Government Arts College
Karaikudi, India

Rosilda Selvin
Department of Chemistry
School of Science
Sandip University
Maharashtra, India

N. Shiva Prasad
Membrane Separations Laboratory,
    Chemical Engineering Division
CSIR-Indian Institute of Chemical
    Technology
Hyderabad, India

K.K. Singh
Chemical Engineering Division
Bhabha Atomic Research Centre
Mumbai, India
and
Homi Bhabha National Institute
Mumbai, India

Randeep Singh
Department of Chemical Engineering
Indian Institute of Technology Guwahati
Guwahati, Assam, India

S. Sridhar
Membrane Separations Laboratory
Chemical Engineering Division
CSIR-Indian Institute of Chemical
    Technology
Hyderabad, India

C. Sumana
Process Simulation and Control Group
CSIR-Indian Institute of Chemical
    Technology
Hyderabad, India

Sankaracharya M. Sutar
Membrane Separations Laboratory,
  Chemical Engineering Division
CSIR-Indian Institute of Chemical
  Technology
Hyderabad, India

Biswajit Swain
Nuclear Recycle Board
Bhabha Atomic Research Centre
Tarapur, India
and
Homi Bhabha National Institute
Mumbai, India

Jogi Ganesh Dattatreya Tadimeti
Chemical Engineering Department
Parala Maharaja Engineering College
Berhampur, Odisha, India

Pavani Vadthya
Chemical Engineering Division
CSIR-Indian Institute of Chemical
  Technology
and
Academy of Scientific and Innovative
Research (AcSIR)
Hyderabad, Telangana, India

Bukke Vani
Membrane Separations Laboratory,
  Chemical Engineering Division
CSIR-Indian Institute of Chemical
  Technology
Hyderabad, India

Section I

Membrane Technology
for Sustainable Development

1

Processing of Complex Industrial Effluents and Gaseous Mixtures through Innovative Membrane Technology

Sundergopal Sridhar

CONTENTS

1.1 Introduction

1.1.1 Impact of Industrial Effluents and Off-Gases on the Environment

Almost all chemical and allied industries generate wastewater and off-gases during various processing steps and cleaning protocols. Industrial effluents contain a wide range of impurities including color, turbidity, chemical oxygen demand (COD), biochemical oxygen demand (BOD), total dissolved solids (TDS), total suspended solids (TSS), chlorides, sulfides, phosphates, phenols, ammoniacal nitrogen, heavy metals such as cadmium, lead, chromium, arsenic, iron and mercury, besides oil and grease (Sugasini et al., 2015; Chowdhury et al., 2013; Sharma et al., 2013; Tikariha et al., 2014; Lin et al., 1994). Detailed compositions of effluents from pharmaceutical, distillery, textile, tanning, oil, dairy and fermentation industries, besides wastewater generated by agricultural and domestic activities, are reported in detail in the literature (Ghaly et al., 2014; Noukeu et al., 2016; Al-Jasser et al., 2011; and Abdallh et al., 2016). Steel industry generates off-gases from blast furnaces, coke ovens and LD converters that possess calorific value for being rich in inflammable gases like CO, H_2 and CH_4, which need to be separated from pollutants like SO_2, H_2S, NH_3, greenhouse gases such as CO_2, inert N_2 and moisture (Diez et al., 2002; Chao et al., 1978). Similarly, power plants emanate huge quantities of CO_2 (Bram et al., 2011). Process gases, such as refinery gas mixture (RFG), are rich in propylene or hydrogen (present in off-gas that is usually flared), which require isolation from other hydrocarbons for making polypropylene or fuel (Heinz, 2008). LPG has a considerable amount of H_2S impurity while natural gas and biogas contain CO_2 and H_2O apart from H_2S (Suzuki et al., 2001; Ahn et al., 2008). When let out untreated into water bodies, the industrial effluents ruin the ecosystem and make surface water unfit for drinking. Recalcitrant compounds present in pharmaceutical, textile and steel industrial effluents destroy aquatic life and cause sustained damage to the environment for several years (Praneeth et al., 2014). On the other hand, when treated in ETPs, the costs involved are extremely high, because most of the methods focus on destroying the chemicals rather than their recovery and recycle. Final treatment by multiple effect evaporation (MEE) involves a cost of US$ 0.008 to 0.016 (Rs. 0.53-1.07). Industrial off-gases cause air pollution and global warming if vented out without treatment.

Therefore, the focus must quickly shift toward recovery of valuable constituents such as solvents and water from effluents, isolation of gases with calorific value like H_2 or CO from industrial exhausts or implementation of technologies for enrichment of CH_4 from biogas, landfill gas or natural gas (Sridhar et al., 2007; Sridhar et al., 2007; Madhumala et al., 2014; Ravikumar et al., 2013; Praneeth et al., 2014). Similarly, reclamation of water from municipal sewage should be combined with the production of bio manure and biogas.

Membrane process described in this study have not only reduced the enormous costs involved in activated sludge process (aerobic digestion) utilized in ETPs and STPs, but have also brought down raw material costs owing to the recycle of valuable chemicals and reclamation of water.

1.1.2 Mechanisms of Mass Transfer in Hydrostatic Pressure-Driven Membrane Process and Gas Permeation

Hydrostatic pressure is the driving force in pressure driven membrane processes such as reverse osmosis (RO), nanofiltration (NF), ultrafiltration (UF) and microfiltration (MF), which are widely employed in treatment of industrial and municipal wastewater, while partial pressure gradient is responsible for mass transfer in gas permeation (Sridhar et al., 2007). Four different mechanisms viz., Knudsen's diffusion, Viscous flow, Capillary condensation and Molecular sieving can happen in porous membranes used in NF, UF and MF processes, whereas for nonporous (dense) membranes used in gas separation, pervaporation and RO, solution-diffusion is the predominant mechanism, which involves preferential interaction of selective component for sorption, diffusion and desorption through the membrane (Moulik et al., 2016). Figure 1.1 (a) illustrates the molecular sieving effect in a porous membrane where the smaller molecule (say water) permeates faster to

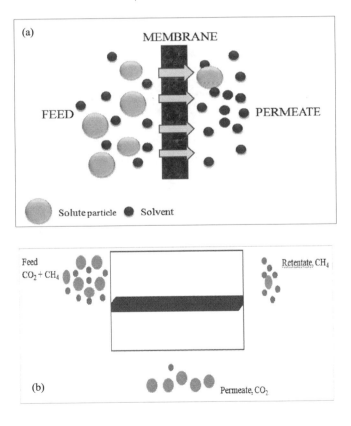

FIGURE 1.1
(a) Separation by molecular sieving in porous membranes and (b) separation through affinity between membrane and feed components in nonporous membranes.

the other side relative to larger molecules, like salts. On the other hand, affinity-dominated separation of the quadrupolar acid gas CO_2 from relatively inert and saturated CH_4 can be possible in spite of CO_2 being a bigger molecule (Figure 1.1 (b)). Affinity-based mass transfer explains why only partial rejection can be achieved with respect to say urea or organic molecules that constitute the COD in the effluent in comparison to the nearly complete rejection of inorganic salts through RO membranes. In such cases membranes need to be retrofitted into other processes to constitute hybrid techniques that are more effective in removal of most of the impurities.

1.1.3 Objectives of the Work and Scope for Membrane Technology to Meet Challenges

The present study focuses on the development of various cost-effective membrane technologies for recovery of water and value-added organics from industrial effluents and separation of useful gases from off-gases and process gas mixtures emanating from steel industry and petroleum refineries. Apart from zero liquid discharge (ZLD) and prevention of air pollution, the aim is to economize productivity through the recycling of valuable solvents, gases, catalysts and to reduce the need for fresh water resources in domestic and industrial applications. Membranes possess a wide scope for treating several of these industrial effluents and off-gas streams, especially since many of the impurities can be removed in a single step. Moreover, the membrane can perform bulk separation followed by final cleanup using conventional methods to achieve desired purity levels. NF can be used to separate monovalent salts and solvents from bi- and multivalent impurities as well as bigger organic molecules. Electrodialysis (ED) can separate ionizable chemicals like salts, acids and alkalis from non-ionizable compounds, such as organic solvents and active pharmaceutical ingredients. Pervaporation and membrane distillation can separate a volatile molecule from a non-volatile species. Gas permeation can permeate polar gas molecules with retention of non-polar moieties. Thus, differences in molecular sizes, physical and chemical properties can be taken advantage of to retrofit an appropriate membrane technique into the existing framework and bring about the recovery of valuable chemicals, gases, and water from industrial wastes. Membrane capabilities in this context are demonstrated with the help of a few interesting case studies.

1.2 Case Study on Chloride Separation from Coke Oven Wastewater in Steel Industry

1.2.1 Source of Chloride Effluent

During steel manufacture, there are various pollutants including chloride and cyanide which contaminate the process water, making it unfit for disposal or reuse. In the manufacturing process of steel, coke is an important ingredient. It is a solid carbon source used to melt and reduce iron ore. Coke production begins with pulverization of bituminous coal, which is fed into a coke oven and heated to very high temperatures (Diaz et al., 2002).

TABLE 1.1

Characteristics of Coke Oven Wastewater in Steel Industry

S. No.	Parameter	Value
1.	pH	7.6
2.	Total Dissolved Solids	1832–3000 mg/L
3.	Conductivity	3.68 mS/cm
4.	Aluminum (as Al)	0.087 mg/L
5.	Calcium (as Ca)	53.3 mg/L
6.	Chloride (as Cl)[a]	942–2000 mg/L
7.	Iron (as Fe)	0.43 mg/L
8.	Magnesium (as Mg)	18.5 mg/L
9.	Manganese (as Mn)	0.175 mg/L
10.	Sulphate (as SO_4)	50 mg/L
11.	Total Hardness	209 mg/L
12.	Carbonate (CO_3^{-2})	Nil
13.	Silica (as SiO2)	8.54 mg/L
14.	Sodium (as Na)	697.8 mg/L

[a] Major impurity present in the wastewater.

After the coke is finished, it is moved to a quenching tower where it is cooled by sprayed water. Once cooled, the coke is moved directly to an iron melting furnace for steel production. Chloride also comes into the effluent stream during the production of metal in the blast furnace. Steel effluent contains 500 to 10,000 ppm of suspended particles. This effluent stream is sent for sedimentation followed by a clarification step to remove suspended particles. After this treatment, the water still contains a large amount of dissolved solids and various ions. The typical composition of quenching tower effluent from a steel industry is provided in Table 1.1. The effluent stream is collected from this stage for further purification, with particular emphasis on the removal of chloride whereas cyanide does not corrode the coke oven. However, cyanide can be highly hazardous in case the effluent is disposed without treatment and hence needs to be separated along with excess chloride (> 800 ppm). Conventional processes, such as multiple effect evaporation and bio-sorption, etc., are found to be highly energy intensive or have low throughput. In order to overcome these problems, it is necessary to employ a less energy-consuming, high-throughput alternative.

Since RO gives low water recovery due to limitations posed by high osmotic pressure, NF could be an ideal alternative to enable higher water recovery at a low pressure in view of the less stringent limits of chloride in reusable water. The reject stream rich in chloride could be treated with a coagulant to precipitate most of the chloride and bring back the TDS to a value equal to the original effluent to enable its reprocessing.

1.2.2 Synthesis of Nanofiltration Membrane

15% w/v of polyethersulfone (PES) was mixed with 3% propionic acid in *N,N*-dimethylformamide (DMF) solvent and the resultant polymer solution was de-aerated to remove air bubbles. Subsequently, the clear solution was cast on a nonwoven polyester fabric support using a doctor's blade to achieve the desired thickness followed by immersion

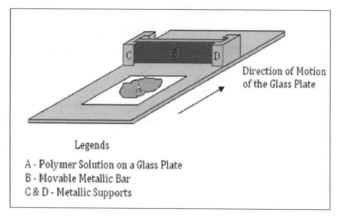

FIGURE 1.2
Membrane casting technique using doctor's blade.

in an ice-cold water bath (Figure 1.2). The PES substrate prepared by the phase inversion method was ultraporous in nature with a MWCO of 50 kDa (Venkata Swamy et al., 2013). To prepare polyamide NF membrane by interfacial polymerization, the PES substrate was soaked in 1% aqueous solution of piperazine for 1 min. Excess water was drained off and the substrate was then immersed in hexane bath containing 0.1 % Trimesoyl chloride (TMC) for 30 sec. The membrane was then heated in an oven at 110°C for 5 min. to obtain thermally cross-linked NF membrane of 400 MWCO. The NF membrane was functionalized by dip coating it in 1.5 wt% polyvinyl alcohol (PVA) solution, which was in turn prepared in deionized water at 90°C containing glutaraldehyde (GA) crosslinker in similar proportion with PVA, to finally obtain hydrophilized polyamide (HPA) NF membrane of a four-layered structure with a MWCO of 100 Da.

1.2.3 Nanofiltration Process for Treatment of TATA Steel Industrial Effluent of 5 m³/h Capacity

Experimental trials were performed in the laboratory for a feed capacity of 60 L using spiral wound hydrophilized polyamide (HPA) based nanofiltration membrane of 100 Da MWCO. The membrane exhibited a permeate flux of 150 L/h with 85% chloride rejection and 90% water recovery at a feed pressure of 21 kg/cm². Upon successful demonstration of the technology at pilot scale, a NF plant of 5–8 m³/h of permeate capacity was designed and installed in a steel industry in West Bengal for separation of cyanide and excess chloride from wastewater. The Piping & Instrumentation Diagram (P&ID) and the actual photograph of the NF plant of 5000 L/h capacity, installed for quenching tower effluent treatment at a steel manufacturing unit, are shown in Figure 1.3 (a, b). The membrane exhibited 75% chloride rejection with 80% water recovery at a feed pressure of 7 kg/cm². The reject of ≤ 20% rich in chloride was treated by the conventional coagulation process using a unique flocculating reagent called sodium aluminate. The commercial plant has completed 18 months of successful operation already without requiring any significant troubleshooting.

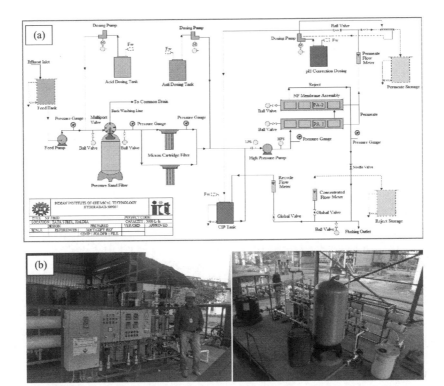

FIGURE 1.3
(a) Process and instrumentation diagram and (b) actual photograph of 5000 L/h nanofiltration plant for quenching tower effluent treatment at a steel industry.

1.3 Recovery of Dimethyl Sulfoxide Solvent from Pharmaceutical Effluent

1.3.1 Source of Pharmaceutical Effluent

Most effluents from pharmaceutical industries may contain valuable chemicals (solvents) which need to be recovered along with recalcitrant compounds (Ravikumar et al., 2014; Larsson et al., 2007). One such aqueous effluent from Astrix Laboratories Ltd. (ALL), Medak, contains 12–15% of dimethylsulfoxide (DMSO) solvent along with 2–3% salts comprising of the highly hazardous and explosive sodium azide (NaN_3) and the corrosive ammonium chloride (NH_4Cl). DMSO is the reaction medium for production of the antiretroviral drug Zidovudine used to prevent and treat HIV/AIDS while NaN_3 is the catalyst. The salts need to be removed completely to recover the DMSO solvent, which is widely used in bulk drug manufacture. The effluent was being sent to ETP for disposal, resulting in heavy loss of the expensive DMSO solvent at the rate of approximately 1000–1500 kg/day, which also contributes to excess COD that poses effluent treatment problems and cumbersome methods for neutralization of the azide. Therefore, an appropriate separation scheme needed to be identified or developed for the removal of NaN_3 to facilitate recovery of DMSO.

Electrodialysis (ED) is a membrane-based process capable of removing dissolved salts like NaN_3 and NH_4Cl from aqueous solutions without significantly changing the composition of the non-ionic constituents such as DMSO. The objective is to remove all the NaN_3 and NH_4Cl present in the effluent by ED and optimize operating parameters to achieve the best membrane performance. The desalted feed could then be safely distilled to obtain pure DMSO.

1.3.2 Design and Commissioning of Electrodialysis Pilot Plant of 7,500 L/batch/day

a. Laboratory Trials

The prototype ED system at IICT consists of a membrane stack resembling a filter press with a number of cell pairs. Each cell pair comprises of one cation transfer membrane and one anion transfer membrane. The total area of 10 cell pairs in the stack was 0.525 m². The other major components of the electrodialyzer were three tanks of 10 L capacity each and magnetically coupled centrifugal pumps of 6 L/min flow rate (rating: 0.37 kW) and a DC power supply (110 V, 30 A). A batch of 10 L of effluent was charged at a time as feed and subjected to ED operation. Analysis of the original feed (effluent), diluate (desalted liquor), concentrate and rinse solutions were carried out by conductivity measurements and gas chromatography. The conductivity of the sample was monitored at regular intervals and the termination of the ED operation was done when the feed conductivity (15–21 mS/cm) attained a value less than 0.08 mS/cm. The losses of the DMSO solvent from the feed tank to the concentrate tank was found to be negligible. The experiments were repeated with fresh batches of the effluent from Astrix to collect at least 40–60 liters of desalted liquor which contained < 60 ppm of NaN_3 + NH_4Cl and about 14% DMSO. This solution, which was nearly free from the salts, was subjected to vacuum distillation at 20–30 mmHg. Under these conditions, the reboiler temperature was found to range from ambient (35°C) to 70°C during removal of water in the first step. The second distillation step involved the recovery of pure, colorless DMSO from the color-imparting, high-boiling organic impurities under a similar vacuum of 20–30 mmHg, wherein the reboiler temperature was found to vary from 90 to 130°C.

The process was successfully demonstrated by the IICT team to the pharmaceutical industry using over nine electrodialysis batches of 10 L each—six batch distillations to remove water and three batch distillations to recover pure DMSO from the mother liquor supplied by Astrix. The residues of the three distillations were again combined and distilled to assess the overall DMSO recovery that was possible. A total of 25 kg of pure DMSO was recovered from a total desalted effluent quantity of 180 kg. The four distillations to recover colorless DMSO yielded purities of 99.25%, 99.49%, 98.78% and 99.43%, respectively.

b. Commissioning of Commercial Plant

Electrodialysis for Salt Separation. The ED unit would be operated in batch mode with continuous recirculation of the diluate (effluent feed), concentrate and electrode rinse solutions. The PFD of the ED pilot plant of 7500 L/batch/day is shown in Figure 1.4 (a), whereas the actual photograph of the installed pilot plant is shown in Figure 1.4 (b). 7500 L of the pharmaceutical effluent is initially stored in ST-101 and then pumped at a rate of 1000 L/h through the pre-filtration assembly comprising of a bag filter, BF-101, and a micron filter, MF-101, to remove suspended solids and

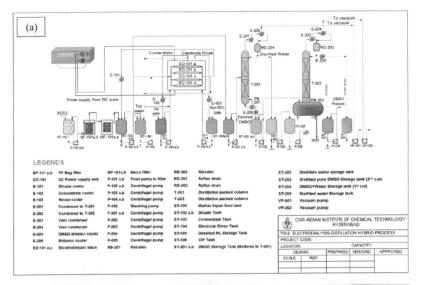

FIGURE 1.4
(a) Process flow diagram of ED + Distillation pilot plant for pharmaceutical effluent (b) actual Photograph of ED + Distillation pilot plant.

turbidity. The pre-filtration operation is expected to last about 8 h after which the effluent is stored in diluate tank ST-102. Tap water of quantity equivalent to half the diluate volume is filled in the concentrate tank ST-103 to facilitate transfer and collection of NH_4Cl and the hazardous NaN_3. 500 L of 1% w/v aqueous sodium bisulphate is charged into ST-104 for rinsing the electrodes. The solutions filled in tanks ST-102, ST-103 and ST-104 are then pumped through the ED stack at controlled flow rates of at least 100 mL/sec each using centrifugal pumps P-102, P-103 and P-104, respectively at equal flow rates to ensure uniform pressure drop across the stack. After stabilizing the flow rates, electrical potential (50–70 V) is applied across the ED stack through the DC power supply DC-101 to attain a specific current density to enable ionizable compounds ($NaN_3 + NH_4Cl$) to migrate from the diluate to the concentrate. Samples of the outlet streams are collected at regular intervals to determine the conductivity of all three solutions using online digital conductivity meters. Once the conductivity of the diluate solution falls below 0.1 mS/cm the operation is terminated and the desalted diluate is pumped to storage tank ST-105 for subsequent separation of DMSO and water by vacuum distillation. During the recirculation of streams, the temperatures of the diluate, concentrate and rinse streams could increase and hence coolers E-101, E-102 and E-103 are provided to cool the respective streams. The ED operation is expected to run for duration of 20 h.

To retain membrane efficiency after each ED batch, the system is washed with tap water followed by cleaning with 2% v/v HCl solution and subsequently another tap water wash. For this, about 200 L of tap water is charged into the CIP tank ST-106 and is circulated for about 15 min. through the diluate and concentrate compartments. A similar volume of 2% v/v aqueous HCl is charged into the CIP tank and circulated through the diluate and concentrate compartments for about 30 min. to remove mineral scales and metal precipitates. This step is followed by a similar water wash. During the washing operation, the electrode compartments are circulated with tap water. Once a week, an extra washing step for 15 min. is provided with 1.5% NaOH + 0.5% tetra sodium EDTA + 0.1% sodium lauryl sulfate containing aqueous solution which helps to remove stubborn organic foulants from the membrane surface. The weekly cleaning operation requires incorporation of an extra water washing step. The total membrane cleaning process is expected to carry on for a period of 1 h everyday and about 1.5 h during the seventh day when EDTA cleaning is given.

DMSO Recovery by Distillation. Pure, colorless DMSO is isolated by vacuum distillation in Section 200 using two distillation columns T-201 (continuous) and T-202 (batch) (Figure 1.4 (a)). Desalted DMSO-water mixture is pumped from ST-105 to T-201 at a specific flow rate of 450–500 L/h. Distillation is carried out T-202 under a vacuum of about 70–80 mmHg absolute. By maintaining the required reflux ratio as well as reboiler and condenser duties, water of high purity is collected as the top product in ST-202, whereas DMSO is collected as the bottom product in ST-201 after cooling in E-206. Water collected in ST-205 is pumped by P-205 for reuse in preparation of CIP, rinse or membrane cleaning solutions. Apart from condenser E-201 to T-201, a vent condenser E-203 is also provided to ensure complete condensation. The vacuum pump VP-201 provides the low pressure required in the continuous distillation setup. The DMSO collected in ST-201 has traces of water and light brown color due to the presence of unknown heavy organic impurities. The desired DMSO product which is pure (> 99.5%), colorless and free from traces of water is obtained by simple batch distillation in T-202 under the vacuum of about 70–80 mmHg absolute. Crude DMSO is pumped from ST-201 to the reboiler still RB-202. The top product from the column which comprises mostly of DMSO and traces of water is initially collected in ST-204 as the first cut. Subsequently, pure DMSO is collected as the top product in ST-203 after cooling in E-205. The residue in the still is then drained for incineration resulting in loss of DMSO to the tune of ≤ 10% of the total solvent present in the effluent. The vacuum pump VP-202 provides the necessary evacuation for the batch distillation setup.

Once the ED operation with the first batch is commenced using ST-102 (a) as diluate tank, the pre-filtration step for the second batch of another 7500 L effluent volume can be begin simultaneously to fill the standby tank ST-102 (b). Similarly, once the distillation of desalted batch 1 has started, the ED operation for the second batch can commence. Hence the pre-filtration unit, ED set-up and distillation columns could be operated simultaneously and independently from the second batch onward to ensure processing of at least 7500 L each day to recover about 1000 kg of pure DMSO assuming 90% recovery.

1.4 Decolorization of Aqueous Sodium Thiocyanate Solution in Acrylic Fiber Industry

1.4.1 Origin of Textile Effluent

In acrylic fiber industry, acrylonitrile along with methyl acrylate and sodium methyl sulphonate are used as co-monomers in the polymerization process to manufacture polyacrylonitrile (Sridhar, 2014). The reaction is initiated by sodium chlorate. The chain transferring agent is sodium metabisulfite thioglycol. Termination of the reaction is accomplished by adjusting the pH to 4. Uncontrolled pH conditions lead to the formation of unwanted low molecular weight compounds. The unconverted monomer is distilled off under vacuum. However, the polymer contains monomer traces and low molecular weight compounds as impurities at this stage. During the dope making that involves dissolution of polymer in sodium thiocyanate (NaSCN) solution with the addition of a small amount of sodium metabisulfite as an oxygen scavenger, traces of monomer present in the polymer reacts with sodium metabisulfite to form organic impurities, βSPA (β Sulpho Propionic Acid) and βSPN (β Sulpho Propio Nitrile). Acrylic fiber manufacturers use the aqueous solution rich in NaSCN (58%) for spinning the fiber. During coagulation in spinning, NaSCN gets released and is sent to solvent recovery. The spent NaSCN picks up impurities like sodium sulfate, iron, calcium, sodium chloride, low molecular weight compounds and some other unknown impurities, which degrade fiber quality. The method employed for purifying this solution is a multiple step process comprising of activated carbon column, leaf filter, gel filter and rhodonate filter. Gel filtration column (GFC) requires expensive imported gel, high operating costs and can only be operated at low capacities. The quantity of the water required for the gel filtration column is four times the feed solution, which is another drawback. The rate of purification is also slow and only about 7 m^3 per day of solution containing 4% NaSCN is obtained.

1.4.2 Depiction of Nanofiltration Process

NF was conceptualized as a viable and economical alternative for purifying this process stream for the erstwhile Consolidated Fibers & Chemicals Ltd. (CFCL) located at Haldia, West Bengal, since NasCN is a monovalent salt while impurities are mostly bivalent salts (Na_2SO_4, salts of Fe and Ca) or larger compounds (β-SPA, β-SPN, low mol. wt polymer). NF could offer advantages like faster rate of purification (production capacity can be enhanced), lower applied pressure and water consumption, removal of impurities and color in a single step, and higher NaSCN concentration in the purified stream. Thus, there exists a strong possibility to extract the monovalent NaSCN along with water using a NF membrane.

A novel hydrophilized polyamide (HPA) membrane of 250 MWCO was used for the extraction of pure NaSCN from an effluent containing 10–12% of sodium thiocyanate, 3–5% of impurities such as color, sodium sulfate, Fe, Ca, β-SPA, β-SPN and low molecular weight polymer in water. Pilot-scale NF experiments were carried out with a spirally wound module of HPA-250 module (scaled up with the help of Permionics Membranes

TABLE 1.2

Performance of Hydrophilized Polyamide (HPA-250) Nanofiltration Membrane for Recovery of Sodium thiocyanate in Three Stages with Intermittent Dilution

Quality Parameters	Stage I (No dilution)			Stage II (Dilution 1: 0.5)			Stage III (Dilution 1: 0.75)		
	Feed	Permeate	Reject	Feed	Permeate	Reject	Feed	Permeate	Reject
NaSCN (Conc. %)	11.63	11.81	11.24	8.02	8.62	7.57	4.65	5.31	4.01
Total Impurities (%)	2.35	0.59	4.77	3.41	0.67	5.83	3.65	0.65	7.3
Color (APHA Units)	285	39	499	363	41	650	373	24	778

Note: Avg. Flux: 27 L/m^{2}h; % Rejection (Impurity): 80%; % Rejection (Color): 90%.

Pvt. Ltd., Vadodara, India) having an effective separation area of 2.5 m^2 at an operating pressure of 21 bar (300 psi) to process a feed volume of 100 L in batch mode with retentate recycle. The membrane gave optimum performance in terms of color and total impurity rejection (85–95%), flux (25–40 L/m^{2}hr) and sodium thiocyanate recovery (> 99%). Table 1.2 depicts the performance of hydrophilized polyamide (HPA-250) membrane for extraction of impurity free sodium thiocyanate over three NF stages with intermittent dilution.

1.4.3 Design and Installation of Pilot Nanofiltration Plant of 6,000 L/batch/day

A commercial plant employing a six-stage process with intermittent dilution was designed to treat 6 m^3/day of the effluent. The plant incurred a capital investment of US$ 21,500 (Rs. 14 Lakhs) and an operating cost of < US$ 2.00 (Rs. 120/-) per cubic meter, which is half the cost incurred by GFC column in generating 4% impurity-free sodium thiocyanate aqueous solution, which is further evaporated to 58% for spinning the fiber. The commercial plant was successfully commissioned by the IICT team during January 21–23, 2006 at CFCL, Haldia to process 6 m^3/day of the effluent. The PFD and the photographs of 6 m^3/day pilot NF plant during erection and after commissioning are shown in Figure 1.5 (a) and (b). A photograph depicting yellow-colored effluent and colorless permeate is shown as an inset in Figure 1.5 (b).

An excel program (Figure 1.6) was developed for optimizing the process parameters during the five-stage operation considering a feed capacity of 6000 kg/day of 12% of NaSCN + 5% of impurities including color. Variables such as the quantity of initial feed or DM water (W), initial concentrations of NaSCN (Z1) and impurity (Z2), enrichment factors (K), stage cut and fraction of solvent recycle (f1) can be given as inputs (in yellow boxes) for determining the quantity of fresh water required for dilution from cycle 2 onward. The excel program can even determine the quantity of product recovery and impurity rejected at each stage of operation. The permeates collected from the first three stages are mixed together to obtain a NaSCN product concentration of > 10%, which is sent for evaporation and spinning, whereas the permeate streams from Stages 4 and 5 containing < 1% of NaSCN are recycled back for dilution which thereby reduces the quantity of demineralized water consumed during the overall process. The total feed in each stage after dilution is kept constant at 9000 kg, whereas the quantity of permeate and reject streams collected in each stage were 5400 kg and 3600 kg, respectively. The block diagram of overall material balance shows that the NF process enabled recovery of 99.05% of NaSCN by rejecting 94.2% of impurities. The flux and membrane area requirement for each stage of operation were kept constant at around 21 kg/m^{2}h and 250 m^2 which means that 7 modules of 8″ diameter × 40″ length spiral membranes are required. An operating pressure in the range 15–25 bar is maintained at each stage.

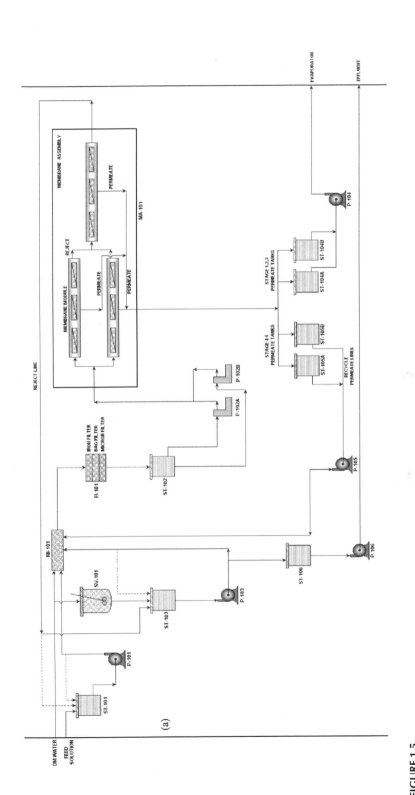

FIGURE 1.5
(a) Process flow diagram.

(Continued)

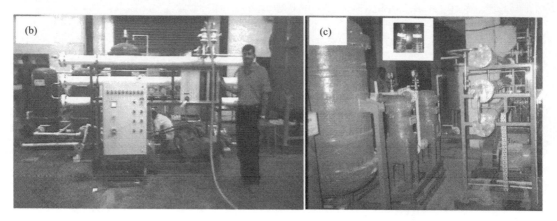

FIGURE 1.5 (CONTINUED)
(b) Actual photograph of 6 m³/day Pilot nanofiltration plant for acrylic fiber industry (c) (inset) actual photograph of feed and purified wastewater.

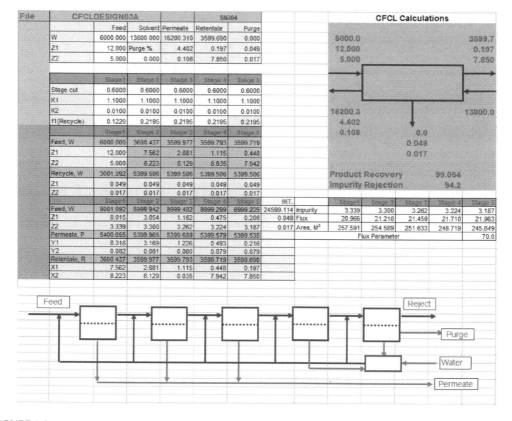

FIGURE 1.6
An Excel program for optimizing the process parameters during a five stage NF operation considering feed capacity of 6000 kg/day of 12% of NaSCN and 5% of impurities.

1.5 Effluent Treatment for Chloralkali Industry

Wastewater generated from chloralkali and metal industries contains important inorganic acids, alkalis, metal salts and other toxic chemicals. Hydrochloric acid (HCl) is an inorganic reagent that finds application in the acid pickling process for removing oxides and corrosion products from metal surfaces, as a catalyst in several important chemical reactions, and for various purposes in industries including textile dyeing, petroleum, chemical and electroplating etc. (Rosocka, 2010). Several technologies including evaporation, ion exchange, distillation and adsorption, etc., investigated for recovery of volatile inorganic acids have been considered as disadvantageous due to the involvement of high capital and operating costs and complex handling. An efficient methodology has been developed for economical recovery of HCl from the effluent sample containing 32.8% (w/v) aqueous HCl along with impurities such as Fe salt (10–20 ppm) and heavier hydrocarbons (C_9–C_{14}) (50 ppm) (Madhumala et al., 2014). Bench-scale experiments were carried out using indigenous vacuum membrane distillation (VMD) set-up incorporated with Polytetrafluoroethylene (PTFE) membrane of 0.22 µm pore size. Effect of operating parameters such as feed composition, permeate pressure and feed temperature on VMD performance was evaluated. Increase in feed acid concentration resulted in enhanced flux which can be attributed to the volatile gaseous nature of HCl combined with the negligible vapor pressure of impurities, whereas increasing permeate pressure reduced the flux due to a lower rate of desorption (Figure 1.7 (a, b)). Raising the feed temperature enhanced HCl flux without compromising on separation performance due to large differences between volatility of the desirable acid component and undesirable hydrocarbons and Fe salts (Figure 1.7 (c)). Cost estimation that the microporous PTFE membrane exhibited commercial potential for recovery of impurity-free HCl (33 wt%) at a low operating cost of only US$ .035 (Rs. 2.28/–) per L (Figure 1.7 (d)).

1.6 Application of Ultrafiltration in Wastewater Treatment

Ultrafiltration (UF) process has wide application in food industry. Figure 1.8 (a) shows the process flow diagram for the application of submerged UF process for wastewater treatment from a food industry of 500 L/h product capacity. The wastewater generated from the process plant comes to a bag filter and then goes to the effluent equalization tank (EET). The EET consists of a micron filter cartridge and a submerged UF system. The initial turbidity of the feed water is measured as 500 NTU. The permeate turbidity from the process is found to be 54 NTU, with a 5-log reduction in bacteria. The operating cost of the process was found to be < US$ 0.05 per L (Rs. 0.03).

Figure 1.8 (b) depicts the schematic of UF system integrated with an extended aeration process for treatment of municipal wastewater. In this process, municipal wastewater will be initially treated in an aerobic moving bed bioreactor (MBBR) for reducing organic content. Only the overflow from the MBBR is sent as a side stream to the UF system for clarification and disinfection of the treated wastewater. UF permeate can be directly utilized for agriculture or further treated using UV or ozonation for reuse in process industry. The major advantage of this scheme is that mixed liquor suspended solids (MLSS) will not be recirculated through the UF membranes, which minimizes fouling. The sludge

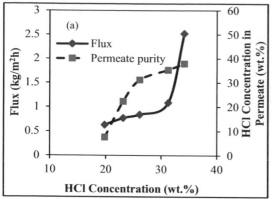

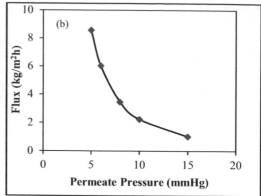

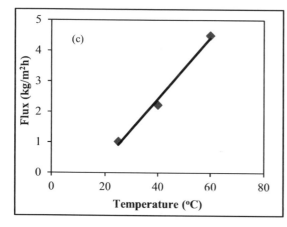

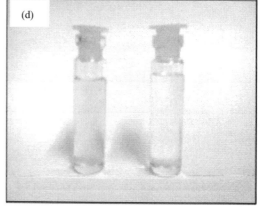

FIGURE 1.7

Effect of (a) Feed HCl concentration composition (b) permeate pressure (c) feed temperature on VMD separation performance and (d) actual photograph of (left) feed HCl effluent and (right) MD permeate.

from MBBR unit will be further treated by an anaerobic digester for production of biogas. Aerobes such as Alcaligenes sp., Pseudomonas sp., P. aeruginosa, P. putida, P. fluorescens, P. syringae, Bacillus sp., Veilonella sp., Chloroflexi sp can be used for the activated sludge process while microbes such as *Clostridium botulinum, Staphylococcus aureus, Escherichia sp. Methanogenic bacteria, Desulfitabacterium chlororespirans, Lactobacillus, Bifidobacterium, Bacteroides* are generally used in the anaerobic MBBR process (Shchegolkova et al., 2016; Wagner et al., 2002; Bedoya et al., 2016).

Figure 1.8 (c) shows a photograph of a pilot membrane bioreactor (MBR) system installed at IICT for domestic wastewater treatment. The feed tank of 2000 L capacity is installed for effluent storage and submerged polyvinyl chloride (PVC) flat sheet membranes are arranged in plate and frame configuration to treat the wastewater in either aerobic or anaerobic condition. The reject and permeate streams are recycled back to the MBR until the permeate quality reaches the desired COD and BOD levels. The permeate is then collected in a storage tank for reuse. In the MBR scheme, the membrane is directly exposed to the MLSS due to submersion, as in the case of Figure 1.8 (c), or recirculation of a side stream. Hence, the chances of fouling are greater in the MBR mode of operation as compared to MBBR illustrated in Figure 1.8 (b).

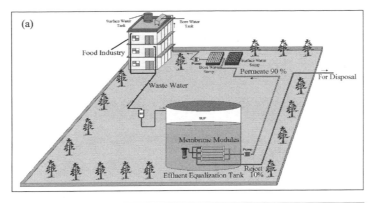

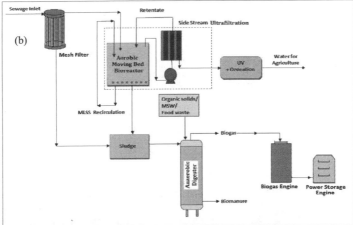

FIGURE 1.8
(a) Process flow diagram of submerged ultrafiltration process for treatment of wastewater from food industry (b) schematic of ultrafiltration system integrated with moving bed bioreactor (MBBR) for treatment of municipal wastewater and (c) treatment of domestic wastewater by integration of biological treatment method with membrane filtration in membrane bioreactor (MBR) configuration.

1.7 Separation of Industrial Off-Gases and Process Gas Mixtures

1.7.1 Membranes for Gas Separation

Dense membranes used in the present study were prepared by solution casting and solvent evaporation technique, while hollow fiber modules were procured from commercial suppliers.

1.7.1.1 Poly(ether-block-amide) Membrane

4% (w/v) dope solution was prepared by adding Pebax pellets to a solvent mixture of 70% ethanol and 30% water. The polymer was dissolved at 90°C with vigorous stirring over a period of 2 h. The bubble-free solution was then cast on to a PES substrate to the desired thickness and finally dried by controlled evaporation of the solvent to obtain a nonporous layer at the top of the composite.

1.7.1.2 Silver tetrafluoroborate Loaded Pebax Membrane

$AgBF_4$ in proportion of 35, 40 and 50% of the polymer weight was added to 4% Pebax solution and kept under stirring. A small quantity of an oxidizing agent, commonly hydrogen peroxide, is added to stabilize the solution and prevent poisoning and loss of silver (Murali et al., 2015). The bubble-free solution was cast on a PES substrate to the desired thickness and dried by removing solvent by a controlled evaporation technique followed by vacuum drying for 24 h to remove residual solvent. The physical structure of dense $AgBF_4$ loaded Pebax membrane is represented in Figure 1.9 (a), whereas Figure 1.9 (b) and 1.9 (c) show actual photographs of Pebax and Ag-Pebax membranes prepared in this study.

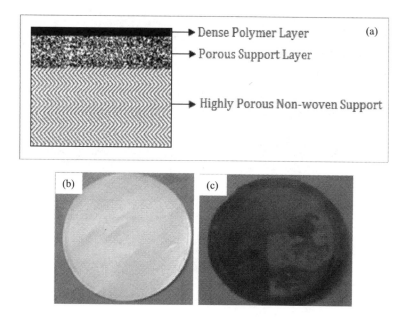

FIGURE 1.9
(a) Physical structure of dense $AgBF_4$ loaded Pebax membrane and actual photograph of (b) Pebax and (c) Ag-Pebax membranes.

1.7.1.3 Cobalt(II) phthalocyanine Incorporated Pebax Membrane

Cobalt(II) phthalocyanine (CoPc) is added to the polymer solution in concentrations of 0.01, 0.1, 0.5 and 1 wt% of the Pebax polymer weight. The solutions were kept under constant stirring at 60–70°C for 10 h and subsequently subjected to sonication for 5 h to uniformly disperse the metal complex. The bubble-free solution was cast on the ultraporous PES substrate to the desired thickness and dried to remove entire solvent resulting in CoPc incorporated Pebax (CoPc-Pebax).

1.7.1.4 Matrimid Hollow Fiber Module

The Matrimid hollow fiber module used in the study was procured from the University of Twente, Netherlands. As per details obtained from the supplier, the Matrimid polymer was dissolved in a mixture of *N*-Methyl-2-pyrrolidone and a methanol solution for spinning the hollow fibers by a phase inversion process after partial evaporation (Figure 1.10 (a)). The fibers were shredded to the module size and heated in an oven at 200°C for 5 min. before assembling in SS 316 housing. The housing consisted of a cylindrical tube of 30 cm length, 6 cm OD with end caps at both the ends. Provision was made for the permeate to pass through the module. Fibers were assembled inside the module using epoxy adhesive at both ends.

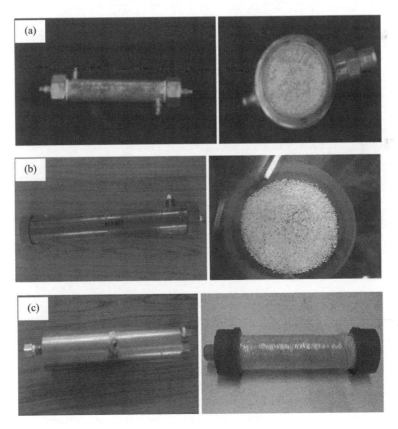

FIGURE 1.10
Actual photographs of (a) matrimid hollow fiber module (b) polysulfone hollow fiber module and (c) poly (ether ether ketone) hollow fiber module.

1.7.1.5 Polysulfone Hollow Fiber Module

Solvay's sulfone polymers have been used to make membranes for a wide variety of applications including UF and gas separation (Figure 1.10 (b)). Due to properties of high-temperature resistance, high fiber strength, toughness, rigidity and hydrolytic stability, Udel polysulfone is found to be the most cost-effective amongst the sulfone polymers and has become an industrial standard. Asymmetric polysulfone fibers prepared by a phase inversion technique have excellent hydrolytic stability and are compatible in pH ranging from 2 to 13. A PVC tube with 45 mm OD and a length of 55.5 cm was used as housing to assemble fibers. The potting of hollow fibers was done by inserting a bundle of fibers in the housing and applying epoxy glue on both ends for perfect sealing. After complete drying of the glue, fibers were cut carefully. The module was fitted with end caps having provision for gas inlet and outlet.

1.7.1.6 Poly(ether ether ketone) Hollow Fiber Module

Poly(ether ether ketone) PEEK-SEP™ hollow fiber membranes find their application in specialty gaseous mixture separation and high-temperature gas filtration processes. Figure 1.10 (c) shows the actual photograph of the PEEK membrane module. PEEK fibers are potted in stainless steel (SS) housing with an effective membrane area of 0.5 m². Melt extrusion method is used for fabricating porous PEEK hollow fibers.

1.7.1.7 Palladium Coated Ceramic Tubular Membrane

Palladium (Pd) coated tubular membrane is procured from Media and Process Technology Inc., PA, United States. Pd and Pd alloy membranes are commercially available for hydrogen separation and purification involving relatively clean streams requiring an ultrapure re product of > 99.99% purity. Pd and Pd alloy membranes are composed of dense metallic thin film supported on a macroporous α-Al_2O_3 commercial tubular ceramic substrate, in contrast to existing Pd membranes supported on porous SS. Pd membrane with ceramic substrate is considerably more economical than its SS supported counterpart. Figure 1.11 shows the actual Pd coated ceramic membrane. An electrode-free plating method is used for the preparation of Pd coating on the tubular membrane. In order to plate Pd onto the support, the surface initially undergoes activation followed by dissolution in a solution containing dissolved Pd ions & EDTA, to stabilize the amine complex. Hydrazine is used as the major reducing agent and is added to cause the Pd ions to get deposited out of the solution onto the support, specifically onto the areas that were activated before.

1.7.2 Experimental Description of Laboratory Gas Separation Unit

The gas permeation tests were conducted using an indigenously designed and fabricated permeation test cell made of stainless steel 316 (Figure 1.12). The gas inlet and outlet ports were provided in the test cell for the transport of feed, permeate and retentate streams. The test cell contained a circular perforated plate affixed with a mesh to support a membrane of effective area 42 cm². Neoprene rubber 'O' ring and vacuum grease were used to provide a leak-tight arrangement. Feed, permeate and retentate lines were made of ¼-inch SS 316 tubes. Nut and ferrule compression fittings were used in the manifold to transport the gas streams to the manifold without any leaks. The vacuum line consisted of a network of high vacuum rubber and glass valve connections capable of providing a

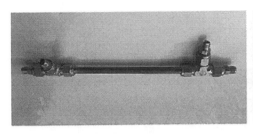

FIGURE 1.11
Palladium coated ceramic tubular membrane for hydrogen separation.

FIGURE 1.12
Schematic of high pressure gas separation manifold.

pressure of ≤ 0.5 mmHg. Needle valves of ¼-inch SS 316 were used to regulate the flow of inlet and outlet streams. The experimental manifold was flexible for studies with both pure gases and mixtures.

1.7.3 Membrane Performance for Gas Separation

1.7.3.1 Separation of H_2 and N_2

Palladium (Pd) based membranes for hydrogen separation has been the focus of many studies because of their affinity for H_2, and as a potential consequence for hydrogen separation in membrane reactors or catalytic dehydration. An additional benefit of using a Pd-based membrane is that it does not oxidize as easily as the other hydrogen-permeable metals such as vanadium, tantalum, and niobium. In the present study, a Pd-Ceramic membrane composed of dense metallic thin film supported on the commercial tubular ceramic substrate was tested rather than the existing expensive Pd membranes, which are usually supported on porous stainless steel (SS) tubes. The membrane adsorbs the molecular hydrogen which dissociates to atomic hydrogen and further diffuses through the dense membrane before recombining to molecular hydrogen prior to getting desorbed (Figure 1.13 (a)). The effect of feed pressure on the Pd-coated ceramic membrane is shown in Figure 1.13 (b). An increase in pressure from 1–5 kg/cm² resulted in an enhancement

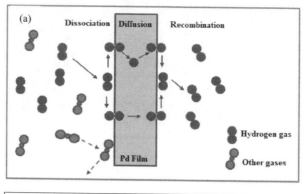

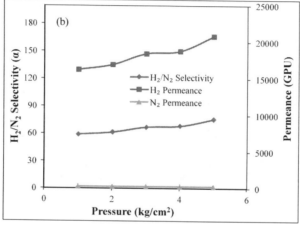

FIGURE 1.13

(a) Facilitated transport mechanism for preferential H_2 permeation through palladium metallic film and (b) performance of palladium-ceramic composite tubular membrane module for H_2/N_2 separation.

in permeance of H_2 gas from 16256.8 to 20833.3 GPU, whereas the permeance of N_2 was observed to be constant at 273.2 GPU, and the H_2/N_2 selectivity was found to rise from 59.5 to 76.25 for the system.

1.7.3.2 Separation of CO_2 and N_2

The PEEK hollow fiber module has been used for separation of CO_2 and N_2 gases. Figure 1.14 represents the effect of feed pressure in the range 0–5 kg/cm² on gas permeation properties and CO_2/N_2 selectivity. At an applied pressure of 1 kg/cm², the permeance of CO_2 and N_2 were found to be 93.4 and 4.7 GPU, respectively. With an increase in pressure to 5 kg/cm², the permeance of CO_2 showed gradual improvement to 249.5 GPU, owing to greater sorption of the preferentially permeating quadrupolar gas, whereas the more or less inert N_2 gas exhibited a marginal improvement to 7.09 GPU. Higher partial pressure gradient across the membrane resulted in higher trans-membrane flux.

1.7.3.3 Separation of O_2 and N_2

The separation of O_2 and N_2 gases was studied using CoPc incorporated Pebax, matrimid and polysulfone membranes. The results pertaining to permeance of O_2 are shown in

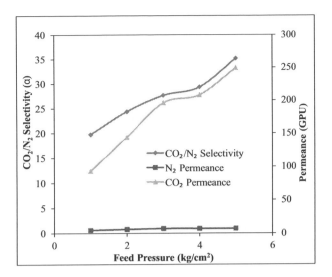

FIGURE 1.14
Effect of feed pressure on CO_2/N_2 separation by PEEK hollow fiber membrane module.

Figure 1.15 (a). The permeance of O_2 is high and steeply increased with a reduction in upstream feed pressure for matrimid and polysulfone membranes. If the state of CoPc in the membrane is inactive, the permeance does not get enhanced, whereas O_2 permeance through a barrier containing active CoPc sites significantly increases (Nagar et al., 2015). One observes that O_2 permeance at low pressures is significantly enhanced in membranes possessing a higher fraction of active CoPc. These results indicate that the fixed CoPc carrier in the membrane interacts with O_2 and facilitates its transport. In other words, O_2 permeance decreases rapidly with increasing feed pressure because the CoPc carrier is saturated with oxygen at high pressures due to Langmuir-type, oxygen-binding equilibrium and therefore hardly contributes to O_2 transport (Nagar et al., 2015; Ferraz et al., 2007). The hypothetical facilitated transfer of O_2 from the feed to the permeate side of the barrier is shown in Figure 1.15 (b). A reduction in permeance with increasing pressure might also be due to dual mode sorption of O_2 in the membrane. With increasing pressure from 2–8 kg/cm^2, the permeance of O_2 drops from 0.38 to 0.35 GPU for 1 wt% CoPc incorporation. Figure 1.15 (c) shows the effect of pressure on O_2/N_2 selectivity which reduces for all the membranes. Permeance of N_2 appears to be independent of feed pressure, whereas that of O_2 decreases resulting in a fall in selectivity. 1 wt% CoPc-Pebax exhibits commercial potential by displaying a maximum selectivity of 9.28 at 2 kg/cm^2 which, however lowers to 5.5 at 8 kg/cm^2 upstream pressure.

1.7.3.4 Recovery of Propylene from Refinery Fuel Gas Mixture

Spiral and hollow fiber membranes were tested for recovering propylene from the heavier hydrocarbons constituting the refinery fuel gas (RFG) mixture. Amongst the membranes tested for propylene recovery, the Matrimid hollow fiber module was found to exhibit a permeate concentration of 38.1% propylene from the RFG feed mixture cylinder containing approximately 20% of the gas. The RFG mixture was supplied by Inox Air Products, Mumbai, at a low pressure of 2.0–2.5 kg/cm^2 and hence feasibility of recovering propylene from the mixture needs to be established at higher pressures (7–8 kg/cm^2). C_4 components

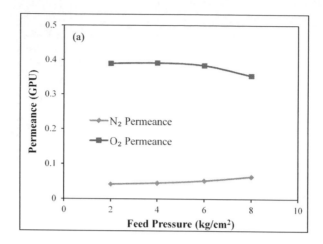

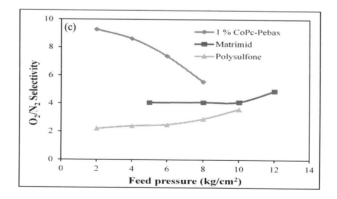

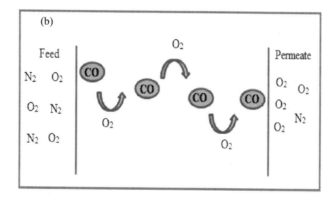

FIGURE 1.15
(a) Permeance of O_2 and N_2 through 1% CoPc loaded Pebax membrane (b) mechanism of facilitated transport for air separation by cobalt thiocyanine and (c) comparison of selectivities of matrimid, polysulfone hollow fiber and CoPc-Pebax membranes for air separation.

(butanes and butenes) interfered in the separation of propylene from the multi-component RFG mixture by permeating significantly along with other gas components, due to the transfer of momentum resulting in the dragging of saturated hydrocarbons, as well.

1.7.3.5 Separation of Propane/Propylene Binary Gas Mixture

Propane is very similar to propylene in molecular size and physical properties. Hence their separation factor is critical. The C_3 fraction could be processed using a lower number of membrane stages compared to distillation, which has to overcome the hurdle of very low relative volatility. The study on separation of propylene from its binary mixture with propane using Ag-Pebax or Matrimid membranes is described in this section. Ag-Pebax membranes showed higher selectivity (25.5) towards propylene due to the facilitated transport phenomena, whereas the Matrimid polyimide membrane module showed greater permeance (2.4 GPU) towards propylene due to the lower thickness (< 0.5 μm) of the skin layer in the hollow fibers.

Figure 1.16 (a) shows the effect of feed pressure on gas permeance through 40% $AgBF_4$-Pebax 1657 membrane. As the pressure is increased from 1–4 kg/cm², the flux of propylene is increased from 4.16 to 6.19 L/m²h and the propylene selectivity was found to decrease from 0.51 to 0.30 for the system. Figure 1.16 (b) represents the effect of $AgBF_4$ wt% on permeance and propylene selectivity. With increasing silver loading from 0 to 50 wt%, the propylene permeance was found to increase from 3.81 to 123.25 GPU and the propylene selectivity also increased from 2.11 to 20.38.

Figure 1.16 (c) shows hypothetical facilitated transport phenomena of the olefin (C_3H_6)/ paraffin (C_3H_8) gas mixture occurring through Ag-loaded dense membrane. Facilitated transport through dense membrane involves the reaction of the olefin with a complexing agent or a carrier alongside the regular solution-diffusion mechanism. The penetrant initially dissolves in the membrane and reacts with the complexing agent prior to subsequent downward diffusion due to a concentration gradient. Propane being a saturated molecule does not react with the complexing agent and thus, the separation of propylene from propane is facilitated by the reversible carrier-gas complexation. Silver has a special affinity to bind with olefins and has been used successfully by several researchers for facilitated transport through liquid membranes. A unique characteristic of *d*-block transition elements, like silver, is the presence of partially-filled *d*-orbitals and the ease of electron exchange due to the low energy gap with σ-orbitals in forming hybrid orbitals. The unpaired *d*-electrons readily participate in the formation of complexes or bonds with certain electron donating groups (ligands) and sometimes, reversibly with molecules having double bonds such as propylene. The reversible bonding involves (i) the overlap of π-electrons of feed molecules with σ-type acceptor orbitals on the transition metal atom and (ii) a back bond resulting from the flow of electron density from the filled d_{xz} or other $d\pi$-$p\pi$ hybrid orbitals into antibonding orbitals of the C_3H_6 molecule. However, the long-term stability of the silver-loaded membrane needs to be established.

1.7.3.6 Natural Gas Sweetening

The worldwide increase in energy consumption by burning fossil and petroleum fuels has led to the emission of large amounts of CO_2 and other greenhouse gases, which has resulted in global warming affecting the ecosystem and human life to a large extent. Hence, there is a need for capturing carbon dioxide and other pollutant gases like SO_x and NO_x. Due to its properties of low carbon intensity and high compressibility, natural gas is found

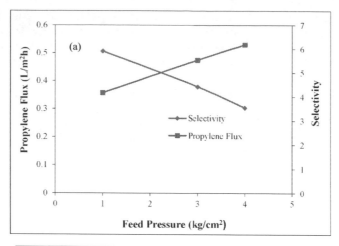

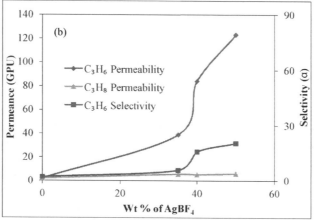

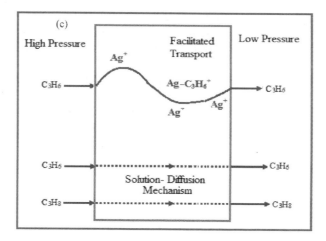

FIGURE 1.16
(a) Performance of 40% AgBF$_4$-Pebax1657 membrane module (b) effect of silver loading in Ag-Pebax membranes on permeability and (c) facilitated transport phenomena for C$_3$H$_6$/C$_3$H$_8$ separation through Ag-Pebax membrane.

to be an energy efficient source for meeting future fuel demands. The presence of CO_2 and H_2S in natural gas mixture results in pipeline corrosion, decreases calorific value and reduces the overall efficiency of the process. Conventional methods like cryogenic distillation, chemical absorption and pressure swing adsorption are available for the capture and storage of greenhouse gases. In recent years, membrane technology has fascinated several researchers for natural gas separation (Kesting & Fritzsche, 1993). CSIR-IICT has successfully designed, developed and commissioned the first gas separation pilot plant in India of 100 NM³/h capacity at Oil and Natural Gas Corporation (ONGC), Hazira, in collaboration with Engineers India Limited (EIL), Gurgaon, Haryana, India for natural gas sweetening (Figure 1.17 (a)). The indigenously-developed, thin film composite Pebax membrane enabled the effective removal of water and acid gases (H_2S and CO_2) from natural gas, which thereby increased its calorific value and reduced polluting amine solvent circulation rates (Sridhar et al., 2007). Figure 1.17 (b) shows the Excel program for a two-stage gas separation pilot plant to process 100 M³/h of natural gas feed, assuming it to be a binary

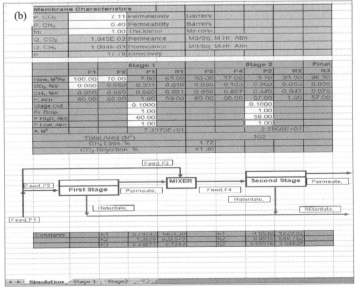

FIGURE 1.17
(a) Actual photograph of pilot plant gas separation unit for natural gas sweetening and (b) an Excel program for optimization of a two stage process scheme for CO_2 separation from natural gas (Methane).

mixture of 95 mol% of major hydrocarbon CH_4 + 5 mol% of major impurity CO_2, using an effective membrane area of 102 m^2, to minimize hydrocarbon losses. A fraction of the feed was bypassed before entering the Stage 1 membrane and mixed with the retentate of Stage 1 to constitute the feed for Stage 2. At an operating pressure of 60 atm and a temperature of 25°C, the membrane exhibited a CO_2 permeate concentration of 33% in Stage 1 and 56% in Stage 2 with a minimum CH_4 loss of 1.72%. A cascade arrangement is avoided to reduce operating expenditure incurred through recompression of the permeate.

1.8 Conclusions and Future Perspectives

In recent years, membrane technology has proven to be advantageous for various industrial applications including effluent treatment, water purification, solvent recovery and gas separation. Among pressure driven membrane based processes, nanofiltration (NF) has achieved significant results during the processing of steel, pharmaceutical and textile industrial effluents. The NF process enabled an effective separation of chloride from steel industrial effluent, which facilitated the prevention of pipeline corrosion and the achievement of zero liquid discharge. An integrated process of electrodialysis and distillation has proven to be a cost effective and efficient technology for the recovery of dimethyl sulfoxide, a polar aprotic solvent from pharmaceutical effluent, which can be used for the manufacture of antiretroviral drugs with simultaneous reclamation of utility water. The five-stage NF process recovered pure sodium thiocyanate which can be used for the manufacture of high-quality fiber in the textile industry. The Polytetrafluoroethylene membrane exhibited a potential for the recovery and resale of aqueous HCl (32–34%) from industrial effluent using the vacuum membrane distillation technique. The application of the ultrafiltration process in combination with the biological treatment method could prove successful for the treatment of municipal wastewater. Novel composite membranes based on Pebax, Matrimid, PEEK, Polysulfone and Pd-coated ceramic membranes were investigated for the separation of binary gas pairs like H_2/N_2, CO_2/N_2, O_2/N_2 and C_3H_8/C_3H_6, as well as multicomponent gas streams like natural gas and were found to exhibit commercial potential, owing to their high separation performance. The scale-up of gas membranes, especially in India, remains a challenge due to lack of membrane casting infrastructure, which must be overcome through a technically efficient and economical approach. The membrane technologies developed during this study could revolutionize the process and environmental safety of chemical, petroleum, pharmaceutical, textile and food and other allied industries.

References

Abdallh, M.N., W.S. Abdelhalim, and H.S. Abdelhalim. 2016. Biological treatment of leather tanneries wastewater effluent bench scale modeling. *International Journal of Engineering Science and Computing* 6(9): 2271–2286.

Ahn, J., W.J. Chung, I. Pinnau, and M.D. Guiver. 2008. Polysulfone/silica nanoparticle mixed-matrix membranes for gas separation. *Journal of Membrane Science* 314: 123–133.

Al-Jasser, A.O. 2011. Saudi wastewater reuse standards for agricultural irrigation: Riyadh treatment plants effluent compliance. *Journal of King Saud University-Engineering Sciences* 23(1): 1–8.

Archana, A. and K.K. Sahu. 2009. An overview of the recovery of acid from spent acidic solutions from steel and electroplating industries. *Journal of Hazardous Materials* 171: 61–75.

Bedoya, L.M.S., M.S. Sánchez-Pinzón, G.E. Cadavid-Restrepo, and C.X. Moreno-Herrera. 2016. Bacterial community analysis of an industrial wastewater treatment plant in Colombia with screening for lipid-degrading microorganisms. *Microbiological Research* 192: 313–325.

Bram, M., K. Brands, T. Demeusy, L. Zhao, W.A. Meulenberg, J. Pauls, G. Gottlicher, K.V. Peinemann, S. Smart, H.P. Buchkremer, and D. Stove. 2011. Testing of nanostructured gas separation membranes in the flue gas of a post-combustion power plant. *International Journal of Greenhouse Gas Control* 5: 37–48.

Chao, J.T., P.J. Dugdale, D.R. Morris, and F.R. Steward. 1978. Gas composition, temperature and pressure measurements in a lead blast furnace. *Metallurgical Transactions B* 9(2): 293–300.

Chowdhury, M., M.G. Mostafa, T.K. Biswas, and A.K. Saha. 2013. Treatment of leather industrial effluents by filtration and coagulation processes. *Water Resources and Industry* 3: 11–22.

Díez, M.A., R. Alvarez, and C. Barriocanal. 2002. Coal for metallurgical coke production: Predictions of coke quality and future requirements for coke making. *International Journal of Coal Geology* 50(1–4): 389–412.

Ferraz, H.C., L.T. Duarte, M. Di Luccio, T.L.M. Alves, A.C. Habert, and C.P. Borges. 2007. Recent achievements in facilitated transport membranes for separation processes. *Brazilian Journal of Chemical Engineering* 24: 101–118.

Ghaly, A.E., R. Ananthashankar, M. Alhattab and V.V. Ramakrishnan. 2014. Production, characterization and treatment of textile effluents: A critical review. *Chemical Engineering & Process Technology* 5: 1–18.

Heinz, W.H. 2008. *Industrial Gases Processing*. Weinheim: John Wiley & Sons.

Kesting, R.E. and A.K. Fritzsche. 1993. *Polymeric gas separation membranes*. New York: Wiley.

Larsson, D.G.J., C.D. Pedro, and N. Paxeus. 2007. Effluent from drug manufactures contains extremely high levels of pharmaceuticals. *Journal of Hazardous Materials* 148(3): 751–755.

Lin, S.H. and C.F. Peng. 1994. Treatment of Textile Wastewater by Electrochemical Method. *Water Research* 28(2): 277–282.

Madhumala, M., D. Madhavi, T. Sankarshana, and S. Sridhar. 2014. Recovery of hydrochloric acid and glycerol from aqueous solutions in chloralkali and chemical process industries by membrane distillation technique. *Journal of the Taiwan Institute of Chemical Engineers* 45(4): 1249–1259.

Moulik, S., S. Nazia, B. Vani, and S. Sridhar. 2016. Pervaporation separation of acetic acid/water mixtures through sodium alginate/polyaniline polyion complex membrane. *Separation and Purification Technology* 170: 30–39.

Nagar, H., P. Vadthya, N. Shiva Prasad, and S. Sridhar. 2015. Air separation by facilitated transport of oxygen through a Pebax membrane incorporated with a cobalt complex. *RSC Advances* 5: 76190–76201.

Noukeu, N.A., R.J. Gouado, D. Priso, V.D. Ndongo, S.D. Taffouo, G. Dibong, and E. Ekodeck. 2016. Characterization of effluent from food processing industries and stillage treatment trial with Eichhornia crassipes (Mart.) and Panicum maximum (Jacq.). *Water Resources and Industry* 16: 1–18.

Praneeth, K., D. Manjunath, S.K. Bhargava, J. Tardio, and S. Sridhar. 2014. Economical treatment of reverse osmosis reject of textile industry effluent by electrodialysis–evaporation integrated process. *Desalination* 333(1): 82–91.

Praneeth, K., S. Moulik, P. Vadthya, S.K. Bhargava, J. Tardio, and S. Sridhar. 2014. Performance assessment and hydrodynamic analysis of a submerged membrane bioreactor for treating dairy industrial effluent. *Journal of Hazardous Materials* 274: 300–313.

Ravikumar, Y.V.L., S. Kalyani, S.V. Satyanarayana, and S. Sridhar. 2014. Processing of pharmaceutical effluent condensate by nanofiltration and reverse osmosis membrane techniques. *Journal of the Taiwan Institute of Chemical Engineers* 45(1): 50–56.

Ravikumar, Y.V.L., S. Sridhar, and S.V. Satyanarayana. 2013. Development of an electrodialysis–distillation integrated process for separation of hazardous sodium azide to recover valuable DMSO solvent from pharmaceutical effluent. *Separation and Purification Technology* 110: 20–30.

Rohit, K.C. and P. Ponmurugan. 2013. Physico-Chemical analysis of textile, automobile and pharmaceutical industrial effluents. *International Journal of Latest Research in Science and Technology* 2: 115–117.

Rosocka, M.R. 2010. A review on methods of regeneration of spent pickling solutions from steel processing. *Journal of Hazardous Materials* 177: 57–69.

Sharma, N., S. Chatterjee, and P. Bhatnagar. 2013. An evaluation of physicochemical properties to assess quality of treated effluents from Jaipur. *International Journal of Chemical, Environmental and Pharmaceutical Research Pharmaceutical Research* 4: 54–58.

Shchegolkova, N.M., G.S. Krasnov, A.A. Belova, A.A. Dmitriev, S.L. Kharitonov, K.M. Klimina, N.V. Melnikova, and A.V. Kudryavtseva. 2016. Microbial community structure of activated sludge in treatment plants with different wastewater compositions. *Frontiers in Microbiology* 7: 90–97.

Sridhar, S. 2014. Forays in membrane technology to promote industrial development, rural welfare and academic progress. Chemeca 2014: Processing excellence, Powering our future, Australia: 1547–1563.

Sridhar, S., B. Smitha, and T.M. Aminabhavi. 2007. Separation of carbon dioxide from natural gas mixtures through polymeric membranes—A review. *Separation & Purification Reviews* 36(2): 113–174.

Sridhar, S., R. Suryamurali, B. Smitha, and T.M. Aminabhavi. 2007. Development of crosslinked poly (ether-block-amide) membrane for CO_2/CH_4 separation. *Colloids and Surfaces A: Physicochemical and Engineering Aspects* 297: 267–274.

Sugasini, A., and K. Rajagopal. 2015. Characterization of physicochemical parameters and heavy metal analysis of tannery effluent. *International Journal of Current Microbiology and Applied Sciences* 4(9): 349–359.

Suzuki, T., H. Iwanami, O. Iwamoto, and T. Kitahara. 2001. Pre-reforming of liquefied petroleum gas on supported ruthenium catalyst. *International Journal of Hydrogen Energy* 26(9): 935–940.

Tikariha, A., and O. Sahu. 2014. Study of characteristics and treatments of dairy industry waste water. *Journal of Applied & Environmental Microbiology* 2(1): 16–22.

Venkata Swamy, B., M. Madhumala, R.S. Prakasham, and S. Sridhar. 2013. Nanofiltration of bulk drug industrial effluent using indigenously developed functionalized polyamide membrane. *Chemical Engineering Journal* 233: 193–200.

Wagner, M. and A. Loy. 2002. Bacterial community composition and function in sewage treatment systems. *Current Opinion in Biotechnology* 13: 218–227.

2

Comprehensive Process Solutions for Chemical and Allied Industries Using Membranes

Satya Jai Mayor and Sundergopal Sridhar

CONTENTS

2.1 Introduction

Membrane separation processes and its applications are important in many scientific disciplines such as chemical engineering, environmental engineering, pharmacology, textile dying, bulk drug manufacturing and biotechnology (Feng & Huang, 1997). Figure 2.1 depicts the wide spectrum of membrane technology applications in process industry. Through sustained research and development, these processes have been utilized in drinking water purification, wastewater reclamation, hazardous industrial waste treatment, food and beverage production, gas and vapor separation, energy conversion and storage, air pollution control, hemodialysis and protein concentration, etc. (Strathmann, 1981). Therefore, a comprehensive understanding of the separation phenomenon that can be achieved through membranes is necessary. A membrane acts as permselective barrier through which selective components pass more readily than the others by differences in one or more properties of the components. Transport of the components through the membrane is affected by convection and depends on the nature and extent of driving force applied like concentration, pressure, temperature or electric potential gradient. These membranes can be either homo- or heterogeneous, symmetric or asymmetric in structure, solid or liquid, which may carry either positive or negative charges, or those that are neutral or bipolar in nature. In recent years, the scope of membrane separation processes is widening, stimulated by the developments of novel or improved membrane

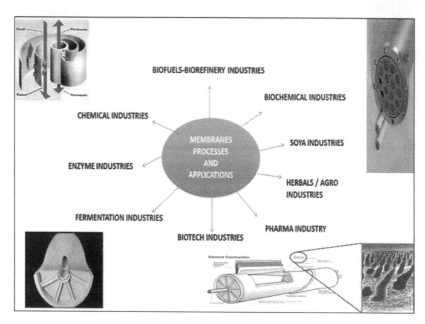

FIGURE 2.1
Spectrum of membrane applications in process industry.

materials with better chemical, thermal and mechanical properties alongside enhanced permeability and selectivity characteristics resulting in reduction of capital and operating costs for commercial success (Moulik et al., 2015). Stringent environmental regulations and norms for cost effectiveness have prompted chemical engineers to develop better and alternative separation methods than the presently-available conventional techniques. However, the selection of an appropriate process depends upon major factors like yield, efficiency of the process, economic feasibility, environmental hazards and its durability. The traditional molecular separation processes, like distillation, evaporation, adsorption, extraction and crystallization, have become uneconomical particularly when operated in small-scale production units. On the other hand, membrane separation processes are energy efficient, eco-friendly natured and easy to scale up when compared to conventional processes. Membranes used in these systems are compact and modular. In addition, membranes processes can sometimes produce novel results. Based on the main driving force, which is applied to accomplish the separation, membrane processes can be distinguished from one another. Pressure-driven membrane processes have become effective techniques for the removal of environmentally undesirable and hazardous contaminants from aqueous systems such as sewage, ground and polluted surface waters, etc. (Praneeth et al., 2014). On the other hand, the integration of membrane separation processes with the conventional process has been explored over the last couple of years for drinking water systems throughout the world. These integrated processes can achieve higher water recoveries of more than 90%, which brings us closer to the goal of zero liquid discharge (ZLD) specified by pollution control boards. The overall scope of membrane processes in health and biotechnology, energy, water treatment, food and beverage industry is depicted in Figure 2.2.

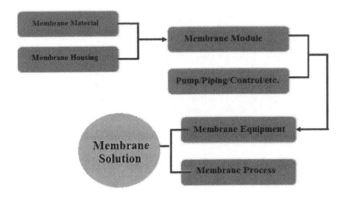

FIGURE 2.2
Step-wise attainment of process solutions and their commercialization.

This chapter provides an entire overlook on several aspects of membrane technologies starting from membrane synthesis route to choice of membrane process and subsequent retrofitting of membrane equipment into the existing plant layout.

2.2 Milestones on Industrial Applications of Membrane Technology in India

At the very beginning (1975–1980), Permionics started working on the synthesis and development of cellulose acetate-based reverse osmosis membranes for seawater desalination and concentration of streptomycin antibiotic solution. Since then, its membrane and element manufacturing facility at Vadodara, India, has undergone vast expansion as seen from Table 2.1.

In 1981, the world's first membrane system for papain enzyme concentration and India's first plant for metal recovery in electroplating industry were being developed. From 1986–1990, polysulfone-based ultrafiltration membrane had been developed for surface water purification and double pass RO unit for ultrapure water production and peritoneal dialysis. In the years 1991–1995, India's first system for the concentration of gelation and its ultrafiltration unit for concentration of cheese whey were developed by Permionics. A membrane system of 4000 m³/day for the production of boiler feed water from brackish water was developed in the year 1997. From 1998–2000, many membrane-based processes were being developed for oil-water separation from jet spinning wastewater, brine recovery for the sugar processing industry, hydroblast water treatment, etc. The years 2006–2011 saw the development of the process for coconut water clarification and concentration and a membrane system for the concentration of herbal extracts and acid recovery from etching bath. For the last five years, Permionics has been developing high temperature and pH tolerant spiral wound elements and membranes, solvent stable membranes, methanol and isopropanol recovery from mother liquor, triglyceride concentration from extracting

TABLE 2.1

Expansion of Membrane and Element Manufacturing Facility at Permionics Membranes Private Limited, Vadodara, India

Time Frame of Five-Year Plan	Process Employed	Application
1975–1980	RO	• Seawater desalination using cellulose acetate
	UF	• RO membrane System for concentration of streptomycin antibiotic solution
1981–1985	UF	• World's first membrane system for Papain enzyme concentration
	RO	• Commissioning of India's first RO system for drought affected village by seawater desalination
	RO	• India's first plant for metal recovery in electroplating industry
1986–1990	UF	• Polysulfone membrane for UF applications
	Q RO	• Ultrapure water production
	Double pass RO	• Peritoneal dialysis
1991–1995	UF	• India's first system for concentration of Gelation
	RO	• India's first system for distillery effluents treatment
	RO	• Indigenous spiral wound TFC polyamide membrane
	RO	• India's largest plant for pyrogen free water production
	NF	• Chlorine tolerant membrane
	NF	• India's first plant for dyes desalting and concentration
	UF	• India's first system for concentration of cheese whey
1996–2000	RO	• Production of boiler feed water from brackish water (4000 m³/day)
	NF	• Concentration of tea extract
	MF	• Fermentation broth clarification
	NF	• India's first plant for recovery and recycle of water from textile dye effluent
2001–2005	NF	• Decolorizing and recovery of dye bath effluent
	UF	• Oil water separation from jet spinning wastewater
		• Brine recovery for sugar processing industry
		• Concentration and purification of food dyes
		• Hydroblast water treatment
2006–2011		• Concentration of coconut water
	RO	• RO system for waste water recycling for laundries, UK
		• Membrane system for concentration of herbal extracts
	NF	• Acid recovery from etching bath
2012–2017	RO, NF	• Introduced high temperature and pH tolerant spiral wound elements and membranes
		• First fish protein recovery (stick water concentration)
	RO, NF	• Introduced solvent stable membranes
		• First membrane system for product and solvent recovery – methanol and isopropanol recovery from mother liquor
		• Triglyceride concentration from extract of fish oil
	UF	• Process for oily wastewater treatment
	Two pass RO + EID	• High purity water

FIGURE 2.3
Glimpse of production facility for membrane systems (Permionics Membranes Private Limited, Vadodara).

fish oil and a process for oily wastewater treatment. Figure 2.3 provides a glimpse of a production facility for the membrane systems at Permionics Membranes Private Limited, Vadodara.

2.3 Membrane Development and Scale-Up

Details of membrane synthesis and characterization techniques are mentioned in other chapters of this book. An overview of the different kinds of membranes produced and commercialized by Permionics is provided in Table 2.2. The typical configuration of a spiral wound membrane module in a partly unwound state is depicted in Figure 2.4. The geometry of the module with a spacer was constructed with a large number of flow cells and filaments. In spiral wound module, flat sheet membrane leaves are sandwiched between layers of feed and permeate spacers and folded around a sheet of feed spacers. The membrane backing is in contact with the permeate spacer, which is attached to the permeate tube. Large-scale water supply systems run on surface water sources, whereas smaller systems tend to rely on groundwater. Water supplied for public use, including drinking purposes, should be potable. However, the raw water available from surface water sources is unfit for drinking purpose. Therefore, the scope of the present work is to produce safe and potable drinking water by treating ground, surface and wastewater using membrane-based separation processes. A complete membrane separation laboratory has been designed, fabricated and installed to provide safe and purified drinking water. A truck mounted mobile water treatment plant has been developed and commissioned for the Defense Research & Development Organization (DRDO) and is shown in Figure 2.5 (a) whereas, Figure 2.5 (b) depicts a containerized system for easy installation in remote areas. The system does not require a shed for installation and is can be easily shifted.

TABLE 2.2

Gist of Membranes Produced and Commercialized by Indian Industry (PMPL)

Membrane Type	MWCO (Da)	Process	Properties	Resistance to Fouling and Chemical Stability			
				Low Fouling	BOD/COD Tolerance	Cl_2 Tolerance	Acid/Alkali Resistance
Modified Polyether urea	NA 150 300	RO NF NF	Stable in pure lower alcohols	√	√	Limited	×
Hydrophilised Polyamide	NA 100–1000	RO NF	For aqueous solutions	√	√	Limited	×
Crosslinked Polyethersulfone	400–600	NF	Chlorine and alkali resistant	√	√	√	√
Hydrophilised Polyethersulfone	5000 30000 50000	UF UF UF	pH range 2–11 Stable in aqueous lower alcohols	√	√	√	×
Polyethersulfone	5000 10000 30000	UF UF UF	Sharper cut-offs	√	√	√	Limited
Ceramic	0.05–0.8 µm	MF[a]	Channel diameter: 2–6 mm For broth, slurry and biomass clarification	×	√	√	√

Note: NA, not applicable.

[a] MF ceramic membranes are imported.

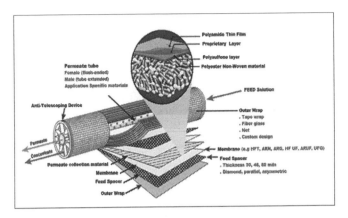

FIGURE 2.4
Typical configuration of spiral wound membrane module in partly unwound state.

(a)

(b)

FIGURE 2.5
(a) Mobile plant for outreach to remote villages and strategic purpose in defense, (b) containerized systems for easy installation in remote areas.

2.4 Industrial Process Solutions

Membranes technology has a variety of applications, like drinking water purification, effluent treatment, bulk drug manufacturing, dairy, pharma, textile dyeing, and generally within the biotech and chemicals manufacturing sectors (Figure 2.6). Membrane-based technologies for drinking water purification are discussed in detail in other chapters of this book. In this section, we will focus on the applications of membrane technology in various disciplines such as health, biotechnology, food and dairy, energy and industrial effluent treatment.

2.4.1 Health

Wide ranges of membranes are available in different makes and cut-off suits best for the processing of pharmaceutical bulk drugs, active pharmaceutical ingredients (APIs) and intermediates. Membrane processes are integrated in pharma industries for the cost-effective concentration, purification, and separation of compounds, enabling a high-product yield. Specialized membranes can successfully replace centrifugation, extraction, distillation and chromatography wherever possible to provide the most economical solution. In many pharmaceutical processes, where the traditional methods for purifying API's such as extraction and chromatography do not work, membrane intervention may help by taking advantage of size difference between the API and the impurities present. Integrating membranes for purifying API adds value as membranes offer better savings compared to all other conventional purification technologies. These membranes are compatible with different hydrophilic and hydrophobic solvents to replace the

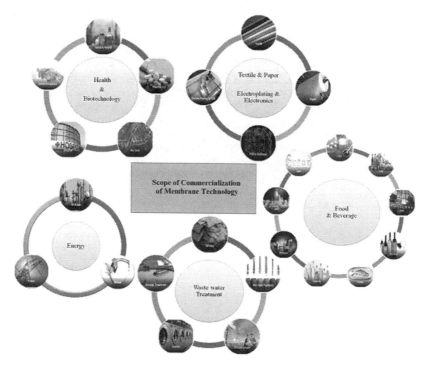

FIGURE 2.6
Scope of commercialization of membrane technology.

conventional distillation—an energy-consuming process. Membranes can concentrate the dilute product/API and recover the solvents that can be recycled back, thus improving the economics of the process (Peeva et al., 2014). Membrane processes are also capable of processing herbals, natural products, phytochemical extracts and products which can help to reduce the cost of extraction, separation, concentration and purification by replacing the conventional methods, and thus offer the best solution. Membrane processes can also help to concentrate dilute extracts from herbs or natural products and reduce the volume of the extract by 70–80%. Membrane processes can be directly used for the concentration of extracts even after chromatography and other purification methods to concentrate the purified herbal/natural product. Peshev et al. (2011) developed a solvent-resistant nanofiltration process for the concentration of antioxidant extracts of rosemary. Table 2.3 shows the data for rejection and permeate flux obtained during experimental trials with 200, 300 and 500 Da molecular weight cut-off (MWCO) membranes. The Rosmarinic acid rejection of the 200 Da MWCO NF membrane is almost complete (> 99%), but this is at the expense of lower flux (approximately 37%), compared to the 300 Da MWCO NF membrane.

Membrane processes are also used for the separation/purification of herbal and natural products where there is enough difference in the sizes of the product of interest and impurities to be removed. Thus, the cases in which a conventional purification method such as extraction and chromatography does not work, membranes offer a dual advantage, i.e., product concentration as well as its purification. Most herbal extraction processes require the removal of salts from the extract which can be achieved using membranes. Various herbal and natural product streams have provided active ingredients for manufacturing products such as Aloe Vera, Garcinia, Methi, Capsaicin, Beetroot extracts. Nayak and Rastogi (2010) studied the potential of the forward osmosis process for the concentration of anthocyanin from Garcinia indica Choisy. During a lab scale trial, the osmotic agent flow rate and concentration during the experiments were maintained at 100 mL/min and 6.0 M, respectively. With an increase in feed flow rate from 50 to 125 mL/min, the transmembrane flux and anthocyanin concentration was found to increase from 4.2 to 4.7 L/m²h and 50.9 to 57.8 mg/L, respectively (Figure 2.7).

Membrane processes are also applied for the downstream processing of Serratiopeptidase. Serratiopeptidase (serrapeptase) is a proteolytic enzyme that has fibrinolytic, anti-edemic, and anti-inflammatory activity and has been used successfully for pain and inflammation due to arthritis, trauma, surgery, sinusitis, bronchitis, etc. Serratiopeptidase is derived from bacteria belonging to the genus *Serratia* by a fermentation process. The typical broth is composed of 25–30% of bacterial cells while soluble components in the broth include

TABLE 2.3

Performance of Duramem™ series with Solution of Caffeic and Rosmarinic Acid

Operating Pressure, (bar)	Membrane	% Rejection		Permeate Flux, (L m⁻² h⁻¹)
		Caffeic Acid	Rosmarinic Acid	
20	Duramem™ 200	93.9	99.7	15.4
	Duramem™ 300	87.8	94.0	24.4
	Duramem™ 500	82.0	94.5	21.4
40		96.1	99.5	25.7
		90.7	94.7	41.1
		91.5	97.4	31.3

Source: Peshev, D., L.G. Peeva, G. Peev, I.I.R. Baptista, and A.T. Boam, *Chemical Engineering Research and Design*, 89(3): 318–327, 2011.

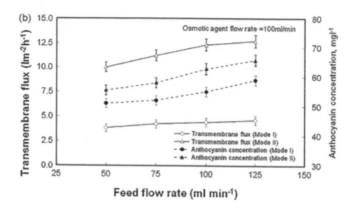

FIGURE 2.7

Effect of feed flow rate on transmembrane flux and anthocyanin concentration. (From Carrère, H., F. Blaszkow, and H.R. Balmann, *Journal of Membrane Science*, 186(2): 219–230, 2001.)

dilute serratiopeptidase (52 kDa) along with micronutrients (inorganics) and macronutrients. Thus, the need was to have an efficient separation process that can remove the bacterial cells as well as concentrate, enrich and purify the serratiopeptidase until the desired activity was achieved. Permionics has successfully provided integrated MF-UF systems that are able to meet the said requirements. The microfiltration process can remove the bacterial cell and provide cell free broth containing the enzyme of interest, along with other soluble broth components, that are further concentrated and purified using UF process. The UF process helps to concentrate the broth by reducing volumes and enriching the serratiopeptidase. The UF process completely retains the serratiopeptidase and helps to remove the other undesired components from the broth.

2.4.2 Biotechnology

Concentration and separation are the two unit processes in which membranes play a vital role. Membrane separation processes help to purify products, enhance yield, reduce cost and provide selective separations. Antibiotics, proteins, enzymes, etc. can be simultaneously concentrated and deashed/purified by selecting a suitable membrane (Charcosset, 2006). Laboratory and pilot plants for antibiotic concentration is shown in Figure 2.8. Membrane processes are widely used for different application involving separation,

FIGURE 2.8

Laboratory and pilot plants for antibiotic concentration.

concentration or purification of various bio-based products made by fermentation or bio-catalysis process (Teplyakov et al., 1996). The recovery of product from the fermentation broth is an important step after any fermentation processes that is done using bacterial, yeast or fungal cells. The cross-flow MF membrane system that is used for clarification of the broth and subsequently to wash out the remaining product from the cells (called diafiltration) achieves complete recovery of the product (Carrère et al., 2001). Membrane systems have been developed with capability to handle cell densities ranging from 2–40% (w/v) in fermentation broths, recovery of macromolecules like proteins post fermentation, separation of nutrients and other media components from the product to facilitate their recycle to the fermentation process. Membrane-based processes have been developed for the recovery of biocatalysts like *cellulases, lipases, peroxidases,* etc. from reaction media for reuse, which is important for improving economics. One of the major applications of membrane technology in biotech industry is gelatin concentration, de-ashing and recovery. Gelatin is a water soluble, high molecular weight protein (20–200 KDa) prepared by the thermal denaturation of collagen, isolated from animal skin and bones with very dilute acid/alkali, even extracted from fish skins. Gelatin broths after the extraction process have a typical solids content of 2–6% (w/v). A Polysulfone (PES)-based 5 KDa MWCO ultrafiltration membrane was synthesized by dissolving 13 wt% PES in dimethylformamide solvent and applied to concentrate the gelatin to 10–18% (w/v). This process not only concentrates gelatin but also purifies it by removing ash and undesired components from the extract. Integrating membranes prior to the evaporator improves economics as the evaporator load can be reduced by 70–80% thus adding value to the process. The capital cost of the ultrafiltration process for gelatin concentration, which can remove water at the rate of 7000 L/h, is US$ 250.00 (INR 16272) and the operating cost per hour is US$ 9.50 (INR 619), whereas the operating cost per hour of a conventional five-effect evaporator is US$ 27.25 (INR 1774). Figure 2.9 shows the commercial ultrafiltration unit for gelatin concentration.

2.4.3 Food and Dairy

Membrane technology has a wide range of applications in food processing industry for skimmed milk concentration, de-ashing of skimmed milk, protein enrichment/

FIGURE 2.9
Pilot ultrafiltration system for gelatin concentration and sugar cane juice clarification.

standardization, whey processing, edible oil degumming and impurity removal. The developed nanofiltration technology can effectively concentrate skimmed milk by removing 30–40% of its water. Integrating NF will add value and save costs by reducing the volume and operational cost of an evaporator in making skimmed milk powder. The specification of skimmed milk powder has a limit for ash content, typically ~ 7% w/w. However, the ash content in skimmed milk has been higher at times and it becomes difficult to meet the desired specification of final ash content in the powder. To address this challenge, a specialized, high flux NF membrane system is developed that can partially remove the ash without losing any proteins and lactose. The system offers a dual advantage as it will remove some ash along with water, thereby reducing the evaporator load and improving economics. Due to adulteration in milk, the lowered protein content does not meet the specification of SMP, which needs 36–37% w/w protein content. The ultrafiltration process can increase the protein content in milk by removing lactose. The integration of ultrafiltration with diafiltration can concentrate whey protein by up to 35–80%. For whey concentration and partial demineralization, the nanofiltration system can be a good alternative where membrane is selective towards most of the monovalent ions, organic acids, and some of the lactose. Atra et al. (2005) investigated the potential of ultra- and nanofiltration for the utilization of whey protein and lactose. The lactose content was found to be increased with the operating temperature (Figure 2.10). An increase in temperature helps to reduce the solution viscosity, which assists permeate flow rate, and to increase diffusivity by dispersion of the polarized layer. The nanofiltration plant for lactose concentration developed by Pemionics is presented in Figure 2.11.

In a lab scale study, indigenous sulfonated poly(ether ether ketone) (SPEEK) ultrafiltration (UF) membranes were prepared by the phase inversion technique and used in the clarification of sugar cane juice. The pilot ultrafiltration system for sugar cane juice clarification is shown in Figure 2.9. At an applied pressure of 10 bar, UF flux was found to be as high as 12.82 L/m²h (Figure 2.12). A photograph of UF clarified sugar cane juice is shown in Figure 2.13. The clarified cane juice was further treated using indigenously synthesized functionalized polyamide (FPA) nanofiltration membranes of 150 molecular weight cut off (MWCO) for the production of a concentrated sugar solution. The pilot nanofiltration unit for the concentration of sugar cane juice is depicted in Figure 2.14. Sucrose rejection was found to be 92.31 for FPA-150 nanofiltration membranes at a feed pressure of 10 bar

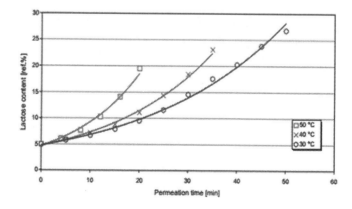

FIGURE 2.10

Influence of temperature on lactose content in the retentate during nanofiltration at feed pressure of 20 bar and 200 l/h flow rate. (From Atra, R., G. Vatai, E.B. Molnar, and A. Balint, *Journal of Food Engineering*, 67(3): 325–332, 2005.)

FIGURE 2.11
Nanofiltration plant for lactose concentration.

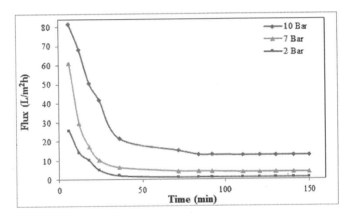

FIGURE 2.12
Variation of flux with time for SPEEK UF membrane.

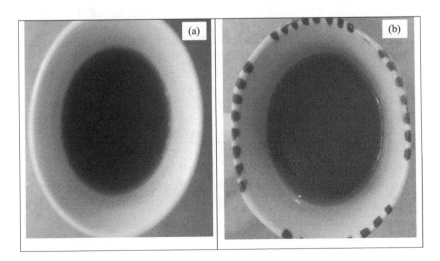

FIGURE 2.13
(a) Color of feed sugar cane juice (b) color of UF permeate.

FIGURE 2.14
Nanofiltration pilot plant for sugar cane juice concentration.

(Figure 2.15 (a, b)). The UF/NF integration exhibits promise for commercialization owing to high flux combined with minimum sugar loss.

Membrane processes are widely used for edible oil degumming and impurity removal (Moura et al., 2005). Crude vegetable oils consist mainly of triglycerides or neutral oil with fat-soluble and suspended impurities. The nontriglyceride fraction contains variable quantities of impurities such as free fatty acid, phospholipids (gums), color pigments, metal complexes, sterols, waxes, carbohydrates, proteins, water, and dirt. Most of these impurities are detrimental to the quality of the finished product and thus the required purification of crude vegetable oil has gained more attention. Membrane processes can help to remove these impurities by providing a good quality of oil at much cheaper cost compared to the traditional methods. The removal of phospholipids ("degumming") is the first step of the crude vegetable oil refining process. Conventional methods use water degumming whereby crude oil is treated with water, salt solutions, or dilute acid to remove phospholipids. This process changes phosphatides into hydrated gums, which are insoluble in oil and are separated from the oil by filtering, settling, or centrifugal action. In this procedure, a considerable loss of neutral oil occurs, a large amount of wastewater is produced, and the energy consumption is fairly large. To overcome this, membrane process for degumming of crude oil/hexane mixture will offer better savings compared to all of the traditional methods. Apart from triglycerides, crude oil contains different impurities such as fatty acid, phospholipids (gums), color pigments, metal complexes, sterols, waxes, carbohydrates, proteins, water, and dirt. Colored materials, some free fatty acids and other impurities are trapped by the reverse micelles and thus can be removed by ultrafiltration membrane processes (Gzara & Dhahbi, 2001). Membrane processes can also be used for the separation of free fatty acids (FFA) and triglycerides and thus help to reduce the FFA content in the crude oil. The advantages of this technology are that it is simple, it can be done at ambient temperature if required, and it demands a small amount of energy. Further, because there are no chemicals used, there is no wastewater generation.

Another more rapidly growing and commercially attractive area in soybean processing involves the production of protein concentrates and isolates, products containing 70–95% protein. Membrane processes can add value to the soya meals or flours by providing a complete solution for the concentration and purification of soya proteins. Membrane

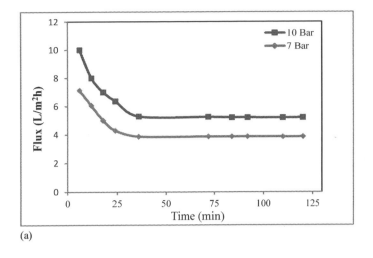

(a)

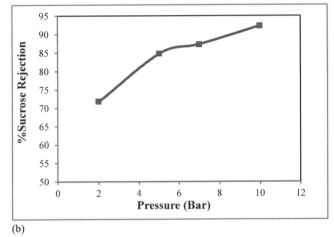

(b)

FIGURE 2.15
(a) Variation of flux with time, (b) variation of % rejection with feed pressure for HPA-150 nanofiltration membrane.

separation processes consume less energy when compared to other concentrating techniques, such as freeze-drying, spray-drying or evaporation. Another advantage is that membrane processes can be tailored to operate at low, ambient, and high temperatures depending on the nature of the application and the solids to be concentrated. After the digesting of soya meal/flour, we have membrane processes that can concentrate as well as purify proteins by removing sugars and minerals. Integrating membranes will also reduce the spray-drying load and offer significant economic improvements. Membrane processes are used for the concentration of sugars and recovery of alcohols that are used in the processes for producing soya protein concentrates (SPC) from the soya meal/flour. Integrating membranes for the recovery of alcohol in the SPC process will fare better economically compared to the conventional distillation process. The removal of ash to improve the quality of soya peptide is a major concern, and membrane processes can achieve this in the most economically efficient way compared to the conventional technologies.

2.4.4 Energy

The production of biofuels from lignocellulosic biomass (LBM) would require mega scale plants and thus would require less energy-intensive, high throughput, continuous and scalable unit operations. Membrane filtration technologies such as microfiltration (MF), ultrafiltration (UF), nanofiltration (NF) and reverse osmosis (RO) are therefore poised to prove cost-effective solutions for separations, concentrations and recovery processes in the entire technology (Wei et al., 2014). Permionics has expertise of integrating membranes in different stages of biofuel production, viz., alkali pre-treatment, acid hydrolysis, enzymatic hydrolysis, sugar concentrations and water recovery. The company provides membranes along with complete systems for biofuel technology with better recoveries/separations/ concentrations compared to conventional technologies, thus reducing the overall capital and operating expenses. Membrane technology can be integrated with many unit operations during the production of biofuel such as acids/alkali recovery, hydrolysis and enzyme recovery, sugar concentration, fermentation, etc. The recovery of acid or alkali that is used for the pre-treatment or fractionation of lignocellulosic biomass in biofuel technology is a challenging task. In processes that use acid/alkali treatments, the extracts are composed of hemicelluloses or lignin depending on the pretreatment deployed, and membrane processes can be used to recover and concentrate these streams, adding value. Acidic treatment for biomass is used in the case that converts biomass to dilute soluble sugars. Membrane processes can be applied for the concentration of these dilute sugars and the recovery of acids that can be recycled back. Recovered enzymes used in the hydrolysis process can be reused multiple times using membrane technology, which improves the process's overall economy.

2.4.5 Wastewater Treatment

With the expansion of urbanization, various industries have been installed, which, apart from providing us with various useful products, also contribute to various types of pollution (Pendergasta & Hoek, 2011). This is one of the primary reasons lying behind the depletion of clean water resources as water is the chief raw material in any industry. Hence, the invention of several wastewater treatment technologies have become imperative as the temporary solutions. Now, apart from its purpose for drinking, there are many such areas where the ultrapure quality of water is not required at all. Wastewater can be divided into two categories based on its toxicity, harmful effects, total dissolved solids, and those are known as black water and gray water. Membrane technology has a huge potential for directly treating the effluent through membrane systems to reclaim the water. Special low fouling hydrophilic nanofiltration and ultrafiltration membranes are having high COD and BOD tolerances. These membranes are able to handle high TDS and SS loads and are able to take in TDS and COD of about 40,000 ppm directly. The smooth surface and its near neutral to negative charge give the added advantage of it being resistant to biofilm formation and fouling. Membrane systems clubbed with an evaporator will make an industry an almost zero discharge industry thereby reducing pollution-related problems and making the company eco-friendly. Membrane-based effluent treatment systems have been installed for treating effluents from textile, dyeing, biotech and pharma, PCB manufacturing, vegetable oil manufacturing, pesticides industries, oil and gas, etc. The supplied staged systems enable recoveries of 80–85% of the wastewater. Figure 2.16 represents state-of-the-art treatment of industrial wastewater using indigenously synthesized membranes

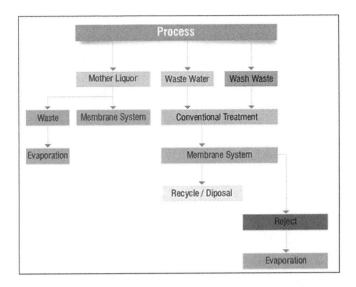

FIGURE 2.16
Flow sheet depicting state-of-the-art treatment of industrial wastewater.

stable in acids, bases and solvents which are available in different molecular weight cut-offs and can cater to multiple applications. Some of the applications are listed below:

- Decolorization/clarification/demineralization of spent acid and caustic solutions for recycle
- Concentration and demineralization of dyes
- Concentration/desalination/softening of in-process and waste streams
- Product concentration and solvent recovery
- COD/BOD reduction in waste streams from pulp and paper industry

The spent caustic solution is generated from the various CIP processes in dairy industry such as cleaning and sterilization of heat exchangers, evaporators, surfaces of silos and pipes. The spent caustic is contaminated with dispersed and soluble organics. The unique alkaline stable NF membrane system is efficient for the removal of impurities from the spent caustic. The pilot robust reverse osmosis and nanofiltration plant for industrial effluent treatment is depicted in Figure 2.17 whereas Figure 2.18 represents the ultrafiltration membrane system for tertiary treatment.

2.4.6 Ultrapure Water Production

For products where there is a concern about pyrogens, it is expected that high purity water will be used. This applies to products of both pharma and semiconductor industries. This also applies to the final washing of components and equipment used in their manufacture. For this, Permionics provides pre-filtration systems like MGF + Softener + UF and Reverse Osmosis (RO) along with EDI + UV with a Post UF system, which is the only acceptable method listed in the USP for producing water for injection. However, in the bulk of the

FIGURE 2.17
Reverse osmosis and nanofiltration based industrial effluent treatment plant.

pharmaceutical and biotechnology industries and some companies, ultrafiltration (UF) is employed to minimize endotoxins in drug substances that are administered parenterally. Another consideration is the temperature of the system. It is recognized that hot systems are self-sanitizing. While the cost of other systems may be less expensive for a company, the cost of maintenance, testing and potential problems may be greater than the cost of energy saved. In the distribution of the USP water system, we will treat the PW Distribution & WFI Distribution, PSG Distribution Streams using Sanitary Design Equipment. Figure 2.19 shows the membrane-based electro deionizer unit for the production of ultrapure water containing zero ppm dissolved solids for use as medical grade water or high purity water in caustic soda plants and biotechnological industries.

FIGURE 2.18
Ultrafiltration based STP as tertiary treatment.

2.5 Conclusions

Several newer membranes are synthesized by researchers on a laboratory scale every day, as are processes based on membranes in small batches. In order to make membranes a viable alternative for niche applications, process scale-up is an essential criterion. However, only a few membranes/processes reach pilot plant or commercial level due to lack of proper infrastructure or design of machines that can fabricate different types of membranes on a large scale. Permionics has shown how, many of the membranes and processes developed in India could be taken from lab to land through scale-up using highly precise casting and coating machines besides appropriate design of membrane systems with the requisite peripherals. Systems of large capacities have been successfully commissioned for applications in the chemical industry and most of its allied industries. Challenges remain in the form of pervaporation and gas separation membranes which do not get produced on a large scale and hence cannot be applied for important applications including the production of absolute alcohol or the sweetening of natural gas and biogas. Forward osmosis is another such area of concern. However, there certainly exists a scope for tailoring of membrane casting machines to suit the particular membrane type and fabricate it into modular geometries in an economical manner.

FIGURE 2.19
Ultrapure water production through (a) EDI integrated RO and (b) double pass RO with mixed bed resin column.

References

Atra, R., G. Vatai, E.B. Molnar, and A. Balint. 2005. Investigation of ultra- and nanofiltration for utilization of whey protein and lactose. *Journal of Food Engineering* 67(3): 325–332.

Carrère, H., F. Blaszkow, and H.R. Balmann. 2001. Modelling the clarification of lactic acid fermentation broths by cross-flow microfiltration. *Journal of Membrane Science* 186(2): 219–230.

Charcosset, C. 2006. Membrane processes in biotechnology: An overview. *Biotechnology Advances* 24(5): 482–492.

Feng, X., and R.Y.M. Huang. 1997. Liquid separation by membrane pervaporation: A review. *Industrial & Engineering Chemistry Research* 36(4): 1048–1066.

Gzara, L., and M. Dhahbi. 2001. Removal of chromate anions by micellar-enhanced ultrafiltration using cationic surfactants. *Desalination* 137 (1–3): 241–250.

Moura, J.M.L.N., L.A.G. Gonçalves, J.C.C. Petrus, and L.A. Viotto. 2005. Degumming of vegetable oil by microporous membrane. *Journal of Food Engineering* 70(4): 473–478.

Moulik, S., P. Vadthya, K.Y. Rani, S. Chenna, and S. Sridhar. 2015. Production of fructose sugar from aqueous solutions: Nanofiltration performance and hydrodynamic analysis. *Journal of Cleaner Production* 92: 44–53.

Nayak, C.A. and N.K. Rastogi. 2010. Forward osmosis for the concentration of anthocyanin from Garcinia indica Choisy. *Separation and Purification Technology* 71(2): 144–151.

Peeva, L., J.S. Burgal, I. Valtcheva, and A.G. Livingston. 2014. Continuous purification of active pharmaceutical ingredients using multistage organic solvent nanofiltration membrane cascade. *Chemical Engineering Science* 116: 183–194.

Pendergasta, M.T.M. and E.M.V. Hoek. 2011. A review of water treatment membrane nanotechnologies. *Energy & Environmental Science* 4: 1946–1971.

Peshev, D., L.G. Peeva, G. Peev, I.I.R. Baptista, and A.T. Boam. 2011. Application of organic solvent nanofiltration for concentration of antioxidant extracts of rosemary (Rosmarinus officiallis L.). *Chemical Engineering Research and Design* 89(3): 318–327.

Praneeth, K., S. Moulik, P. Vadthya, S.K. Bhargava, J. Tardio, and S. Sridhar. 2014. Performance assessment and hydrodynamic analysis of a submerged membrane bioreactor for treating dairy industrial effluent. *Journal of Hazardous Materials* 274: 300–313.

Strathmann, H. 1981. Membrane separation processes. *Journal of Membrane Science* 9(1–2): 121–189.

Teplyakov, V., E. Sostina, I. Beckman, and A. Netrusov. 1996. Integrated membrane systems for gas separation in biotechnology: Potential and prospects. *World Journal of Microbiology and Biotechnology* 12(5): 477–485.

Wei, P., L.H. Cheng, L. Zhang, X.H. Xu, H. Chen, and C. Gao. 2014. A review of membrane technology for bioethanol production. *Renewable and Sustainable Energy Reviews* 30: 388–400.

3

An Insight into Various Approaches toward Flux Enhancement and Fouling Mitigation of Membranes during Nano and Ultrafiltration

Kaushik Nath and Tejal M. Patel

CONTENTS

3.1 Introduction

Membrane-based separation processes are broadly acknowledged as efficient and viable 'cleaner technologies' for the treatment of wastewater as well as the concentration and isolation of useful solutes from aqueous streams from a large spectrum of chemical industries encompassing textile, leather, paint, paper and pulp, petrochemicals, pharmaceuticals, food processing and so on. In recent years, membrane technologies have been developing by leaps and bounds and their cost is continuing to reduce while the application possibilities are ever-expanding. Performance of separation without phase change, less energy consumption, operability at ambient temperature, small footprint requirement, and a lower cost of installation have given an edge to membrane processes over their traditional counterparts. Based on the main driving force, which is applied to accomplish the separation, most of the membrane processes can be distinguished. These are pressure-driven (microfiltration, ultrafiltration, nanofiltration, reverse osmosis) concentration gradient driven (dialysis), temperature gradient driven (membrane distillation) and electrical potential driven (electrodialysis).

Nanofiltration (NF) and ultrafiltration (UF), as subsets of pressure-driven liquid membrane processes, have evolved from a novel approach into reliable and techno-economically attractive standard unit operations in the recent past. The lower applied pressures compared to reverse osmosis (RO), and unique selectivity of the membranes render them less energy-intensive and eco-friendly downstream operations of immense future options. NF finds its applications in concentrating, clarifying/fractionating and/or purifying various edible or inedible products from the dilute streams, thereby enhancing process efficiency to a great extent. It can significantly reduce levels of dissolved solids, colors, organics, hardness, turbidity, divalent and multivalent ions and facilitate the required desalting of permeate streams (Van der Bruggen et al., 2008). On the other hand, UF, the immediate elder sibling of NF in terms of membrane pore size, can be applied not only for the removal of high molecular-weight substances, colloidal materials, and organic and inorganic polymeric molecules in food and beverage, pharmaceutical and biotechnological industries, but also for the effective pretreatment of water for NF or RO (Mohammad et al., 2012). Both NF and UF have proven their potentials as stand-alone separation processes, or in combination with one or more conventional as well as membrane processes in the reclamation of a plethora of wastewater streams and recovery of value-added solutes.

3.2 Principle and Mechanism

NF employs a pressure gradient to selectively transport solvent and certain solutes across a membrane. It spans the gap in particle size between RO and UF. An NF membrane may be aptly considered a very 'tight' UF or a very 'loose' RO membrane. The size of the solutes excluded in this process is of the order of 1 nm. The filtration process takes place on a selective separation layer formed by an organic semipermeable membrane. The driving force of the separation process is the pressure difference between the feed (retentate) and the filtrate (permeate) side at the separation layer of the membrane. However, because of its selectivity, one or several components of a dissolved mixture are retained by the membrane despite the driving force, while water and substances with a molecular weight < 200 Da are able to permeate the semipermeable separation layer. Moreover, NF membranes have a slightly charged surface that originates from the dissociation of ionizable groups present therein and from within the membrane pore structure (Hilal et al., 2004). These groups may be acidic or basic in nature or indeed a combination of both depending on the specific materials used during the fabrication process.

The mechanism of transport and rejection characteristics of the NF membrane can be broadly explained by two major theories—Sourirajan's "sorption surface-capillary flow" approach, and the "solution-diffusion" theory. The sorption surface-capillary flow theory describes the preferential sorption of water molecules in the membrane and the desorption of multivalent ions (by dielectric forces) causing the exclusion of charged solutes, even smaller than the membrane pores, from movement into the membranes (Donnan exclusion). Effective charge density, pore radii and the ionic strength determine the rejection of monovalent ions. The phenomenon of dielectric exclusion is based on two different hypotheses. These are the so-called "image forces" phenomenon (Yaroshchuk, 1998) and the "solvation energy barrier" mechanism (Bowen & Welfoot, 2002). Both exclusion mechanisms operate due to extreme spatial confinement and nano-length scales that are plausibly present in NF membrane separations and are effectively charge-based exclusion phenomena. These interactions have been extensively reviewed by Oatley et al. (2012). Solution-diffusion theory describes the membrane as a porous film into which both water and solutes (ion) dissolve. The solute moves in the membrane mainly under concentration gradient forces, while the water transport is dependent on the hydraulic pressure gradient. The transport of the solute through the membrane depends on hindered diffusion and convection. The transportation of an uncharged solute through an NF membrane is considered to be determined by a steric exclusion mechanism. Steric exclusion applies to NF membranes as well as UF and MF membranes. A separation between two different uncharged solutes is determined predominantly by the difference in their sizes and shapes. For charged solutes, Donnan exclusion and dielectric exclusion theories are also recognized. Due to the slightly charged nature of the membrane, solutes with an opposite charge compared to the membrane (counter-ions) are attracted, while solutes with a similar charge (co-ions) are repelled. At the membrane surface, a distribution of co- and counter-ions takes place, thereby causing an additional separation. According to the dielectric exclusion theory, water molecules show a polarization in the pore due to the charge of the membrane and the dipole moment of water. This results in a decrease of the dielectric constant inside the pore, thereby making it less favorable for a charged solute to penetrate.

On the other hand, the primary removal mechanism in UF is size exclusion, although the electrical charge and surface chemistry of the particles or membrane may affect the purification efficiency. UF pore ratings range from approximately 1,000 to 500,000 Da, thereby making UF more permeable than NF (2,001,000 Da). Since only high-molecular weight species are removed in UF, the osmotic pressure differential across the membrane surface is negligible. Low applied pressures are therefore sufficient to achieve high flux rates from an UF membrane. In conventional UF configurations, the process solution is pressurized, typically between 10 to 70 psi, while in contact with a supported semipermeable membrane. Solutes smaller than the molecular weight cut-off (MWCO), emerge as ultrafiltrate and retained molecules are concentrated on the pressurized side of the membrane.

3.3 Core Issues: Concentration Polarization and Membrane Fouling

While these membrane processes have frequently shown great promise, their wider use at a commercial scale is often marred by the problem of gradual drop in solvent flux. Decline in permeate flux or throughput over the period of operation is a major and inevitable drawback of all the pressure-driven membrane processes including NF and UF. This is primarily attributed to the phenomena of concentration polarization and membrane fouling, which pose a serious problem in membrane process design and operation. The combined effect of concentration polarization and membrane fouling constitute a gray area of membrane research and development. Before we proceed further, it would not be out of context to shed some light on these two core issues. Localized concentration of solutes builds up at the solution-membrane interface resulting in the formation of a relatively stable boundary layer adjacent to the membrane. As osmotic pressure π increases with the increase in boundary layer concentration, the overall driving force $(\Delta P - \Delta\Pi)$ decreases. Eventually, water flux drops considerably. Hence, often ΔP must be increased to compensate, which incurs a higher power cost. Concentration polarization is governed by solute properties, membrane properties and hydrodynamics. In essence, the consequences of concentration polarization are the increased solute transport, increased film resistance to mass transfer and considerable decrease of flux. However, concentration polarization is reversible.

A further complication of the situation occurs when solute membrane interaction increases due to several reasons such as adsorption, polymerization, microbial interaction and so on, leading to irreversible membrane fouling. Membrane fouling indicates plugging or blocking of membrane pores as a result of deposition of particles and colloidal matter on the membrane surface. A pore size to foulant size ratio is important in the determination of the mechanisms involved in rejection and fouling. Fouling in pressure driven membrane process can be categorized into four types. These are organic fouling as a result of deposition of colloidal organic matter such as natural organic matter (NOM), proteins, polysaccharides, etc., colloidal or particulate fouling due to the accumulation of colloidal or particulate matters, clay minerals, organic colloids; inorganic fouling or scaling which is the precipitation of inorganic salts like calcium sulfate, calcium carbonate, barium sulphate and biofouling due to adhesion and growth of microorganisms accompanied by agglomeration of extracellular materials on the membrane surface (Antony et al., 2011; Mohammad et al., 2015).

3.4 Feed Pretreatment

Feed pretreatment is one of the popular options to mitigate membrane fouling and probably to reduce so much reliance on the end-of-pipe solutions such as membrane cleaning or even the premature replacement of a membrane element. Prior to feed pretreatment the characteristics of the feed needs to be analyzed in terms of various physico-chemical parameters such as pH, suspended and dissolved solids, total hardness, dissolved oxygen, conductivity, specific solute concentration, COD, BOD, etc. to ascertain the selection of one or more suitable pretreatment methods. Amongst various processes of feed treatment coagulation/flocculation, acidification, ion-exchange softening, adsorption or even MF or UF in case of RO or NF feed are widely reported in literature. Another important trend towards reducing the load on the membrane for water treatment is the integration of two or more pretreatment strategies to improve the performance of NF and UF (Huang et al., 2009).

3.4.1 Coagulation/Flocculation

Coagulation is an efficient pretreatment practiced to primarily remove particulate and colloidal matters, both prior to NF and UF. Coagulants help reduce and eliminate the internal clogging of the membrane by forming relatively large aggregates, thereby facilitating higher water flux. Important mechanisms for coagulation process include double layer compression, interparticle bridging, enmeshment in a precipitate and charge neutralization. Zahrim et al. (2011) presented a comprehensive review of coagulation with polymers for NF pretreatment of highly concentrated dye. According to the authors, cationic, anionic and natural polymers could be effectively used as flocculants for dye removal. However, the optimization of a suitable type of metal coagulant-polymer dose as well as mixing conditions could enhance dye removal at a wider range of operating pH and reduce sludge production. Literature cites a number of inorganic coagulants with organic polymers as flocculating aids for pretreatment of feed for NF and UF. Some of these inorganic coagulants include ferric-based, such as ferric chloride, ferrous sulphate, polyferric chloride; aluminium based such as aluminium sulphate, aluminium oxide, polyaluminium chloride, potassium-aluminium sulphate docedahydrate, magnesium chloride along with various cationic, anionic and natural polymers (Suksaroj et al., 2005; Gao et al., 2007; Joo et al., 2007; Choi et al., 2013).

A study reported flux improvement to the tune of 20% during the treatment of an artificial dyeing wastewater using a combination of alum, color removing agent and anionic polymer as pre-treatment for the polyamide NF membrane. The improvement of flux was most probably due to the decrease in the content of dye and COD (Mo et al., 2008). Pretreatment involving coagulation-sedimentation of a textile wastewater using four different commercial polymer-based coagulants revealed that Zetag 7103—a cationic polymer at 1000 mg/L concentration could ensure 55% of COD reduction at pH 9. With NF, the COD and conductivity were reduced to 100 mg/l and 1000 mS/cm respectively rendering the permeate suitable enough for re-use (Bes-Pia et al., 2005). The same group of scientists (Bis-Pia et al., 2002) also reported the application of $FeCl_3$ (200 mg/l) with anionic polymer (1 mg/l) as pretreatment for NF of a textile wastewater. The COD and conductivity removals were reported to be 100% and 85% respectively. It was also observed that UF was not able to remove COD as well as conductivity. Riera-Torres et al. (2010) studied the pretreatment of five reactive dye solutions using $FeCl_3$ as a coagulant agent along with two different resins: melamine-urea-formaldehyde and polyamine resin prior to NF. NF with coagulation-flocculation could achieve more than 85% color removal as compared to stand-alone NF, which could remove only in the order of 40–80%. Different pretreatment strategies combining microfiltration and coagulation-flocculation using alum and non-ionic polymer were explored in the reclamation of print dyeing wastewater from a carpet manufacturing industry using NF (Capar et al., 2007). Coagulation with alum was reported to be the best pretreatment, among others, based on the highest color and turbidity removal, which was more than 90%. Recently, Choo et al. (2007) reported an integrated coagulation-UF integrated process for textile wastewater treatment using various types of coagulating chemicals such as polyamine, alum, polyaluminium chloride, and ferric salts. The polymeric coagulant at a concentration greater than 4 mg/l was found to increase membrane fouling, while the inorganic coagulants helped mitigating. Furthermore, polyaluminum turned out to be quite effective amongst the other coagulants tested, although ferric salt was better than alum in controlling fouling. The structure of flocs and membrane cakes could be useful in optimizing feed pretreatment in terms of the tradeoff between the mass of the cake, its permeability and responsiveness to hydraulic and chemical cleaning (Amjad et al., 2015). The coagulation-UF hybrid process was carried out to increase natural organic matter (NOM) removal for drinking water treatment using alum as a coagulating agent. The results showed excellent water quality well beyond the Malaysian and WHO drinking water quality regulations with removal of more than 96% of color, 99% of suspended solids and 99% of turbidity (Zularisam et al., 2009). A study on the effects of aluminum hydrolysis products and natural organic matter on NF fouling with polyaluminium chloride coagulation pretreatment by Choi et al. (2013) reveals that proper control of coagulation conditions is necessary to avoid residual polymeric Al or aluminate ion. In order to reduce the footprint of the entire NF or UF filtration facility, coagulants may be applied "in-line" in which the coagulated water will directly enter the membrane filtration system (Choi & Dempsey, 2004). A similar "in-line" coagulation-UF hybrid process was investigated using three different coagulants, viz., $FeCl_3$, $Fe_2(SO_4)_3$ and $Al_2(SO_4)_3$ for the treatment of surface water from the Czarna Przemsza river (Silesia region, Poland) (Konieczny et al., 2009). The highest efficiency of the process was achieved when the $Al_2(SO_4)_3$ was used as coagulant at a dose of 2.9 mg/dm^3. The water quality improved substantially. Taken together, the observations from different studies indicate that the right selection of coagulating/flocculating agents play an important role in the pretreatment prior to either NF or UF. However, since there is a generation of sludge or concentrated stream, its utilization also needs to be addressed adequately to make the coagulation-membrane hybrid technologies become more practical. It should not result in the generation of additional unwanted solid waste.

3.4.2 Adsorption

Adsorption onto different types of adsorbents have been used by several researchers as a pretreatment step in low pressure membrane processes such as NF and UF for the alleviation of fouling. These adsorbents, having relatively large specific surface areas due to their high dispersity or porosity, can bind small contaminants present in feed solution. Meier et al. (2002) proposed a new process combination for the treatment of severely contaminated wastewater where powdered adsorbent was injected into the feed of a NF unit and was subsequently removed from the concentrate by a thickener. The process combination resulted in higher recovery rate, lower operating pressure and energy consumption. In another study, Chakraborty et al. (2005) combined adsorption with saw dust and NF for the treatment of a textile dye-house effluent containing a mixture of two reactive dyes. The permeate flux for the integrated method was reported to be about twice that for the direct NF method. The percentage removal of COD was greater than 99%, and the salt recovery was of the order of 90%, both of which being significantly higher compared to adsorption. A number of adsorbents have been synthesized for the purpose of fouling control and tested at bench scale membrane filtration processes. Borbély and Nagy (2009) reported the combination of complexation-membrane filtration processes to remove zinc and nickel ions from industrial wastewater using polyethylenimine and polyacrylic acid (PAA) as the organic adsorbent. The study demonstrated that the use of a polyethersulfone (PES) membrane with 10 kDa cut-off and of PAA as complexation agent at pH 8 allowed the removal of Zn(II) and Ni(II) from polluted solutions satisfactorily. Another study conducted with poly(dimethylamine-coepichlorohydrin-coethylene-diamine) and assisted by UF permitted an almost complete removal of Cu(II) (Molinari et al., 2004).

3.4.3 Advanced Oxidation

Advanced oxidation processes (AOPs) generally aim at reducing membrane fouling by oxidizing NOM—particularly humic acids and larger organic molecules (Real et al., 2012) and other micropollutants that could otherwise block membrane surface/pores. Ozonation and photolysis with or without H_2O_2 are the only AOPs that seem viable as a pretreatment stage for membrane filtration because others like Fenton oxidation, photocatalysis and EAOPs are not viable for large volume water systems with low pollutant concentrations, as commonly encountered under real conditions. The high concentration of micropollutants in the concentrates provides an enabling condition for an effective performance of the process since most AOPs are highly efficient at elevated pollutants concentration. Ganiyu et al. (2015) critically reviewed various advanced oxidation processes (AOP) such as ozonation, peroxone (O_3/H_2O_2), UV/H_2O_2, photo-fenton, photocatalysis and electro-chemical oxidation as the pretreatments for membrane separation used in treatment of pharmaceuticals wastewater with special reference to NF. Real et al. (2012) comparatively investigated the efficiency of combined chemical oxidation (O_3, Cl_2, O_3/H_2O_2, UV or UV/H_2O_2) and membrane filtration techniques (UF or NF) for removal of five selected pharmaceuticals namely amoxicillin, hydrochlorothiazide, metoprolol, naproxen and phenacetin from water systems. Ozonation (initial dose of 2.25 mg L^{-1}) followed by NF showed high removal efficiency of over 70% in the permeate stream generated in experiments carried out with natural waters. The study also examined ozonation as a post treatment using a sequence of NF or UF followed by ozonation. A pilot plant study demonstrated that a combined pretreatment effect of 0.5 mg/L of permanganate and 1 mg/L of chlorine could markedly reduce the fouling propensity of a UF membrane in an integrated coagulation-UF system during the treatment of a reservoir water impaired by algae (Heng et al., 2008). Probably the synergic action of permanganate and chlorine was effective in the deactivation of the algal biomass. Schlichter et al. (2004) reported a hybrid process, integrating ozonation and UF for drinking water production from river water using a ceramic membrane. UF with prior ozonation could achieve as high as 99% of yield without reducing membrane permeability. It was also observed that permeate flux increased with the increase in ozone dose, however, up to certain limits. This underscores the need of optimizing the dosage of the oxidizing agent for pretreatment. A hybrid ozonation-ceramic-UF system was introduced by Kim et al. (2008) for treating natural water. Ozone introduction into the upstream feed water of the module resulted in substantial enhancement of the permeate flux over a wide range of operational conditions. Moreover, at higher ozone dosages, the effects of cross-flow velocity and TMP on the extent of fouling were more pronounced compared to lower ozone dosage. Karnik et al. (2005) investigated an integrated ozonation-UF system using a commercial membrane (5 kDa MWCO) coated with iron oxide nanoparticles for the treatment of model wastewater containing tricholoromethane and haloacetic acid. The reduction of trichloromethanes and haloacetic acid were reported to be 90% and 85%, respectively, using the combined system. From the performance of various studies reported in literature, it can be inferred that a combined membrane separation—AOPs treatment—could be a much more promising technology than that of a stand-alone UF or NF. The process integration can not only eliminate the drawbacks of both membrane filtration and AOPs but also complement the advantages of each other. Nonetheless, due to inadequate data regarding the predictability of membrane fouling, further rigorous experimental studies using pilot plants with parametric analysis assume importance to obtain additional insights of the pretreatment processes.

In addition to the methods described above, there are a few other pretreatment strategies applied for some specific processes or feed types. Heat-treatment followed by settling is used to remove immunoglobulins and fats from cheese whey prior to its UF for protein separation. Chemical processes include feed pH adjustment so that molecular or colloidal foulants become far from their isoelectric point, thereby reducing their tendency to form a gel layer. Divalent ions from the feed solution can be removed using the ion-exchange process. The use of proprietary chemicals, such as anti-scalants like polyacrylic acid-based, organo-phosphonate-based, sodium hexametaphosphates or disinfectants, like different biocides, is also practiced as feed pretreatment (Vrouwenvelder et al., 2000).

3.5 Imparting Fluid Instabilities

The use of high membrane shear rates by imparting fluid instabilities has been attempted by a large body of literature as a means towards alleviation of flux decline by minimizing concentration polarization and fouling. As against the traditional ways of increasing the cross-flow velocity along the membrane and reducing tube diameter or channel thickness generating large axial pressure, the focus is now shifted toward dynamic or shear enhanced membrane filtration. In this method, shear rate at the membrane surface is augmented by incorporating a variety of techniques, such as rotating the disk membrane, vibrating the membrane either longitudinally, turbulent promoter, ultrasonic vibration and so on. The traditional methods of promoting turbulence not only incurs high-energy expenditure, but also results in the decrease of transmembrane pressure (TMP) along the membrane leading to non-optimal membrane utilization. Jaffrin (2008), in a recent review, surveyed extensively the application of rotating disks, rotating membranes and vibrating systems in pressure-driven membrane processes. The review also highlights the merits of various designs in the light of fluid mechanics and energy considerations. Several experimental as well as simulation studies of the rotating disk module have also been reported over the last decade. In another study, Al-Bastaki and Abbas (2001) reviewed some of the key methods of generating flow instabilities implemented in the membrane separation processes by various workers during the decade, 1990–2001. In the following subsections, we shall attempt to examine some of the widely reported means and ways of dynamic filtration with particular reference to various NF and UF processes.

3.5.1 Rotating Disk Module

According to the basic configuration of the rotating disk membrane module, feed solution is introduced into the module and is then passed through the gaps between the rotary disks and the membrane element. The permeate passes through the filter elements and exits the module through a permeate outlet, while the retentate or reject exits the module through a retentate outlet. The relative movement of the rotary disks and the membrane elements ensures feed solution in the gaps between the rotary disks and the membrane elements, to sweep the membrane surface thereby preventing solute accumulation on the surface and subsequent fouling or clogging of the membrane is minimized. This extends the useful life of the filter elements. Some common types of rotating disk modules include cylindrical rotating membranes involving couette flow (Vigo & Uliana, 1986; Belfort et al., 1993), disks or blades rotating near a fixed membrane (Mänttäri et al., 2006), rotating flat circular membranes (Murkes & Carlsson, 1988), multi-shaft systems with overlapping rotating ceramic membranes (He et al., 2006; Kaiser, 2004) and vibrating systems with toroidal membrane oscillations around an axis, or vibrating hollow fibers cartridges. Frappart et al. (2006) demonstrated the rotating disk module equipped with a Desal 5 DK membrane in the NF of a model of dairy process waters in the cheese making industry. The system was reported to be very efficient in terms of higher permeate flux than those obtained with spiral wound modules made with the same membrane. Luo et al. (2012) reported that the permeate flux could be maintained stable for long time during NF of model dairy wastewater in a rotating disk module. Several other studies also report the use the rotating disk module for the NF of skimmed milk and dairy waste water (Luo et al., 2010; Luo and Ding 2011). Applications of rotating disk systems in NF are scarce compared to UF and MF as few rotating disk systems can sustain the high pressures required in NF. Table 3.1 summarizes various rotating disk modules used in NF and UF from literature.

Nuortila-Jokinen and Nyström (1996) presented a comparison of a cross-rotational CR 500 filter pilot plant in UF of paper mill process waters at 710 and 50 kDa cut-off with PCI tubular polymeric and inorganic membranes of same cut-off. According to the study, an acidic pH could increase membrane fouling as compared to neutral pH. Four- to five-fold

TABLE 3.1

Summary of Various Rotating Disk Modules Used in NF and UF from Literature

Sr. No	Membrane Process	System Used	Module and Membrane	Flux Performance	Reference
1	UF	Simulated solution of bovine serum albumin (BSA) and glucose	Rotating disk membrane module, made up of SS316 with two different blade angle 45° and 90°. Polyether sulphone (PES) membranes of 30 kg mol^{-1} MWCO; TMP: 019-0.29 MPa	45° blade angle vane produced maximum permeate flux at low TMP and membrane speed for even high concentration of the feed solution	Sen et al. (2010)
2	UF	Kraft Black liquor	q-stirred rotating disk module (capable of also using a fixed disk module) using a cellulose triacetate membrane, MWCO: 5 kDa, rotating at 300 rpm and 600 rpm at a fixed TMP of 5 kg/cm^2 and stirrer speed of 500 rpm.	Flux enhancement of the order of 60% after a period of 1 h of UF experimentation using rotating disk module as compared to the fixed disk one at a TMP of 7 kg/cm^2.	Bhattacharyya et al. (2006)
3	UF	Casein whey	Rotating disk system with 30 kD and followed by 5 kD polyethersulfone membrane TMP: 392-980 kPa, and membrane speeds of zero and 300 rpm; T: 30±2 °C.	The membrane rotation was found to reduce the concentration polarization and subsequent fouling of the membrane in a significant amount.	Sarkar et al. (2009)
4	NF	Model dairy wastewater from commercial UHT skim milk	The rotating disk module, one Disk mounted on a single shaft and rotating near a fixed circular membrane, Speed: 2500 rpm NF270 a thin-film polyamide NF membrane with polysulfone support; TMP: 0.4 – 2.0 MPa; T: 25°C	Flux remained stable and the threshold flux can reach 6.1×10^{-5} ms^{-1} for diluted milk, much higher than the usual critical flux in NF	Luo et al. (2012)
5	NF	Commercial UHT skim milk (Carrefour, France) diluted with demineralized water	Rotating disk system with eight 6 mm vanes, speed: 500-2500 rpm single polyamide circular flat membrane Desal 5 DK (Osmonics, USA) with a cut-off between 150 and 300 g mol^{-1} and an area of 188 cm^2 TMP: 200-400kPa; T: 45°C	At initial concentration, the permeate fluxes at a TMP of 4000 kPa and 45°C ranged from 130 L h^{-1} m^{-2} for a smooth disk at 1000 rpm to 230 L h^{-1} m^{-2} using a disk with vanes at 2000 rpm.	Frappart et al. (2006)

(Continued)

TABLE 3.1 (CONTINUED)

Summary of Various Rotating Disk Modules Used in NF and UF from Literature

Sr. No	Membrane Process	System Used	Module and Membrane	Flux Performance	Reference
6	NF	Model dairy wastewater from commercial UHT skim milk	Disk equipped with 6-mm high vanes can rotate at adjustable speeds, ranging from 500 to 2500 rpm, NF270 (Dow-Filmtec) membrane TMP: 10–40 bar T: 25°C	Reduced concentration polarization, permeate flux could increase continuously when TMP rose from 10 to 40 bar.	Luo et al. (2010)
7	NF	A model dairy effluent from commercial UHT skim milk	RDM module, rotating speed: 2000 rpm, Polyamide membrane NF270 (Dow-Filmtec), MWCO: 150–200 Da, TMP: 41 bar maximum, T: 45°C	A dairy wastewater with pH of 7–8 was most suitable to be treated by the NF-RDM module under extreme hydraulic conditions because of a good compromise between permeate flux and membrane fouling as well as a satisfactory permeate quality.	Luo and Ding (2011)
8	UF	Commercial UHT Žsterilized at ultra high temperature. skim milk	Cylindrical housing of an inner radius 0.0775 m with an aluminum disk of radius 0.0725 m rotating at adjustable speed up to 3,000 rpm around a horizontal shaft equipped with eight radial vanes. A 190 cm² polyethersulfone ŽPES. 50 kDa fixed membrane, supported by a 0.3 mm thick polypropylene grid, was mounted on the flat front end of the housing	The large module yielded a stabilized permeate flux of 276 L h⁻¹m⁻² at 45°C and a TMP of 1,000 kPa when fitted with a 2-mm vanes disk rotating at 1,500 rpm, confirming the reduction of concentration polarization by high shear rate.	Ding et al. (2003)
9	UF	Cutting oil (Doall No. 407) emulsions with tap water	Centrifugal unit, model ST-IIL., rotor speed: 1800 rpm; Ultrason 6010 polyethersulfone (PES) and commercial 50 kDa MWCO hydrophilic PES membranes; TMP: 69–310 kPa; T: 25°C	Module could operate at extremely high shear rates (>10⁵ s⁻¹) which in the pressure controlled regime above 600 rpm with TMP up to 345 kPa.; concentration polarization or gel formation was minimal	Dal-Cin et al. (1998)

(Continued)

TABLE 3.1 (CONTINUED)

Summary of Various Rotating Disk Modules Used in NF and UF from Literature

Sr. No	Membrane Process	System Used	Module and Membrane	Flux Performance	Reference
10	UF	Bovine Serum Albumin	Spinning basket membrane module: a hollow shaft-mounted spinning basket with four radial arms, PVDF membrane (Membrane type: HFM™-100, asymmetric, MWCO: 50 kDa) TMP: 210–830 kPa; Temperature 65.5°C.	Permeate flux decline could be restricted within 10% of its initial value after 5 h of continuous run.	Sarkar et al. (2012)
11	UF and NF	Mineral suspension of silica and effluent from detergent industry	Vibrating shear-enhanced filtration system (VSEP) equipped with Polyethersulfone Nadir membranes of 150, 50, 20 and 10 kD cut-off were used in UF, and an Osmonics Desal DS5 DL membrane for NF at high frequency of 60.75 Hz,	Permeate fluxes obtained with the rotating disk fitted with vanes were consistently higher than those of the VSEP, up to 45% in NF at high pressures.	Bouzerar et al. (2003)
12	UF	Black liquor from alkaline sulfite pulping industry	Stirred rotating disk module with cellulose triacetate membrane of 5000 MWCO (Millipore Corporation, Bedford, MA, USA). TMP: 3–8 kg cm⁻² stirrer speeds: 500–1000 rpm	32% enhancement of initial flux as compared to fixed disk at 600 rpm speed, and 22% increase in initial flux as compared to fixed disk at 300 rpm at constant stirrer speed (1000 rpm).	Bhattacharjee and Bhattacharya (2006)
13	UF	Aqueous solution of ethylene glycol	RDM, multi-shaft disk membrane module Moist 'Spectra-Por C5' asymmetric cellulose acetate complex membrane (cut-off size: 5000), Spectrum Medical Industries (USA).	Analytical expression of back transport flux generated due to rotation-induced shear field was evaluated	Sarkar and Bhattacharjee (2008)
14	UF	Solution of sodium dodecyl benzne sulphonate	Rotating disk system with Polyethersulfone, Rotating speed: 500–2750 rpm; MWCO 10 kDa, Microdyn- Nadir GmbH); T: 25°C; TMP: 500–2100 kPa	Above 1000 rpm, the Permeate flux kept rising without reaching a plateau up to a TMP of 2100 kPa. system was proved to be very efficient for decreasing concentration polarization, due to high shear rates	Tu et al., 2009

rise of permeate flux was achieved with the CR filter rotating at 470 rpm and producing a peripheral rotor speed of 12ms^{-1} than tubular membranes at cross-flow velocities of 2.1–2.5 ms^{-1}. After an initial decline, a quasi-steady state flux could be sustained close to 100 h. The separation of anionic surfactant sodium dodecyl benzene sulfonate in aqueous solutions was carried by Moulai-Mostefa et al. (2007) using dynamic ultrafiltration system with PES membrane of 10–50 kDa. The maximum surfactant rejection was observed to be 92% with a 10 kDa membrane, against 58% at 50 kDa. At lower pressure (900 kPa), no increase of flux was observed above 500 rpm. This was plausibly due to strong interactions between surfactant molecules. Al-Akoum et al. (2006) reported the treatment of aqueous ionic surfactant solutions by dynamic UF by rotating disk module for the concentration soy milk proteins while recovering trypsin inhibitors in the permeate by UF at 50 kDa. Various other rotating disk membranes are described in Table 3.1. Dynamic mode of membrane filtration using rotating disk stands out to be an effective process, in terms of sustenance of the permeate flux and achieving high membrane selectivity. As the use of a rotating disk reduces concentration polarization remarkably, the concentration of rejected solutes at the membrane is lowered. This in turn minimizes the concentration gradient and diffusive solute transfer through the membrane thereby increasing the solute rejection rate. However, the conventional dynamic mode of membrane filtration suffers from certain obvious disadvantages. Firstly, they are very difficult to clean without completely disassembling the set-up. Cracks and crevices are common problems encountered by the membrane housing, filter unit, and rotational unit. Moreover, the presence of stagnant regions or regions of low-flow velocity within the filter as well as the rotor unit aggravates the cleaning problems. Yet another disadvantage with conventional dynamic filter assemblies is that the rotation of the disks within the housing causes the process fluid to heat up. The heat transferred to the feed solution from the rotating disk can harm thermolabile valuable components of the feed mixture.

3.5.2 Turbulence Promoters and Secondary Flow

The application of static turbulence promoters and imparting secondary flow can improve hydrodynamic conditions in the membrane module without significantly increasing energy and investment costs. Turbulence promoters are expected to disrupt the boundary layer at the feed-membrane interface, thereby reducing concentration polarization and subsequent fouling. The concept includes not only the increased local velocity in the vicinity of the membrane, but also the changes in the flow fields appearing as secondary flows. Moreover, certain geometry of turbulent promoters and use of shear enhancing accessories are important to increase the cross-flow velocity. High membrane shear arrests the growth of the polarized layer and therefore enhances the permeate flux. So far, different types of cross-flow modules with varied geometry are reported in membranes literature. Auddy et al. (2005) reported the use of a turbulent promoter in the NF of crystal violet using a polyamide thin film composite membrane, with a polysulfone support layer. Thin wires of a diameter of 0.19 mm were placed laterally (along the width of the channel) in between the gasket and the membrane as turbulent promoters. The gel-type layer resistance decreased up to 68% due to the presence of the turbulent promoters leading to an enhancement of the permeate flux by about 109%. Four geometrical types of turbulence promoters in the form of cylindrical, column cross-section, winding and helical inserts, made of stainless steel were used in the UF of oil field wastewater (Xiang-hua et al., 2006).

The insertion of turbulence promoters resulted in a large improvement in permeate flux to the tune of 83–164% and the winding inserts with 20 mm ditches could cause the largest improvement of the permeate flux with the least energy consumption among the four kinds of turbulence promoters. In another study, helical baffles in the form of wound stems in circular helices suspensions helped in enhancing the permeate flux during the UF of a suspension from the settler of an activated sludge plant. Around 30% flux enhancement was reported to be achieved in the process with a substantial reduction of cake deposition (Ghaffour et al., 2004). Zakrzewska-Trznadel et al. (2009) reported the application of helical Couette Taylor flow (CTF) for the alleviation of membrane fouling in the treatment of simulated radioactive wastewater using the ultrafiltration-complexation hybrid process. According to the authors, two-phase CTF could be endowed with a number of advantages such as high values of mass transfer coefficients, low axial dispersion and a large interfacial area. A near three-fold increase in permeate flux was achieved in comparison with the static filtration. Modification of the hydrodynamic area above the membrane can be accomplished in a plate module by inducing a rotational shear in addition to the normal axial shear (Kamifiski & Stawczyk, 1997). Generation of two streams tangentially flowing over the membrane is expected to cause transfer of momentum. As a result, two complex flows—an axial laminar flow and a rotational flow around the channel axis—are formed. The presence of the additional flow component in the channel could substantially reduce the thickness of the concentration polarization level. The authors reported a 110% increase of permeate flux. Schwinge et al. (2004) developed a three-layers spacer (A3LS) with superior mass transfer characteristics and less fouling propensity as compared to conventional two-layer spacers at both identical mesh length and identical hydraulic diameter for spiral wound module. The three-layer spacer improves flux without covering additional membrane area by filaments adjacent to the membrane wall under both fouling and non-fouling conditions. Even though the pressure loss of A3LS is increased due to the additional flow resistance of the additional transverse filament. Schwinge (2000) investigated a zigzag spacer-filled cross-flow channel for feed solutions containing dextran, silica and reconstituted whey protein concentrate. For cross-flow velocities between 0.1 and 0.8 m/s, the spacer-filled channel achieved flux enhancements over the empty channel of up to three times for dextran and silica UF. Flux enhancements of up to two times were measured for reconstituted whey protein concentrate.

3.5.3 Ultrasonic Irradiation

In ultrasonic irradiation, ultrasound is propagated via a series of compression and rarefaction (decompression) waves induced in the molecules of the medium through which it passes. As a result, cavitation of gas- and/or vapor-filled bubbles takes place at frequencies of roughly 20–100 kHz and up to 1,500 W power. Ultrasound dislodges the adherent foulants by a combined cavitation and microstream mechanism. The growth and implosion of the cavitation bubble has significant mechanical and chemical effects, which tend to alter the interfacial phenomena of the solid phase. This in turn leads to reduction of viscosity of the liquid phase in a suspension making room for improved dewatering (Tarleton & Wakeman, 1992). Muthukumaran et al. (2005) put forward four specific effects of ultrasonic irradiation for pressure-driven membrane processes. These are an agglomeration of fine particles by sonication, thereby reducing pore blockage, generation of mechanical vibrational energy to keep particles partly suspended, cavitating bubbles to scour surfaces and acoustic streaming, causing turbulence. A combined effect of all these results in bulk water movement toward and away from the membrane cake layer. There are two different modes of use of ultrasonic irradiation for membrane processes. In one mode, it is used ex-situ, as the cleaning of fouled membrane while not in-use in the module. The ultrasonically cleaned membrane could be able to recover the water flux considerably. The other mode of operation involves the in-situ attachment of the ultrasonic bath or coupling an irradiation device with the membrane housing while in operation. Although literature cites both types of operation in large numbers, here we shall briefly focus on the latter mode of in-line ultrasonic irradiation used in NF and UF processes. A summary of ultrasonic irradiation including the mode of operation, frequency and flux performances from literature are presented in Table 3.2.

Significant improvements in permeate flux were observed for both online continuous and offline pulse sonication. However, optimization on frequency and power intensity for sonication may be needed to ensure the structural integrity of the membranes. Shahraki et al. (2014) investigated the effect of four different irradiation modes of sonication such as continuous, pulsed, sweeping and degassing on flux enhancement during the UF of skimmed milk. The study revealed that the permeation flux increased with decreasing ultrasound frequencies and 37 kHz irradiation in pulsed mode provided the best condition. Physicochemical properties of the colloidal feed suspension as a result of structural features arising out of differences in particle size and interparticle interactions affect ultrasound efficiency. Detailed analysis of deposit micro-structure upon application of ultrasound gives further insight for potential application (Hengl et al., 2014).

Although there are quite a good number of studies on ultrasonication for flux improvement in the UF process, the use of the same in NF is scarce. Probably in case of the UF membranes, a pronounced deposition of foulants render them to be effectively dislodged by acoustic irradiation unlike NF processes, where solute deposition and associated flux decline mostly occur by the concentration polarization phenomenon. Patel and Nath (2013) demonstrated that the flux decline during the NF of a ternary mixture of dyes and salt could be effectively mitigated using low-frequency ultrasonic irradiation of about 34 ± 3 kHz. Among the continuous and intermittent mode of irradiation, the former resulted in 3–4% increase in permeate flux when compared with the conventional unassisted NF. Estimation of various filtration resistances using a resistance-in-series model revealed that concentration polarization resistance could be reduced by ultrasonic irradiation. Figure 3.1 presents a schematic as well as a photograph of NF in the presence of low frequency acoustic irradiation for the treatment of dye wastewater carried out by the authors. A perusal of the figure indicates that the flat sheet module has been immersed into the ultrasonic bath.

3.5.4 Flow Reversal and Pulsating Flow

Conceptually, flow reversal indicates the change of the hydrodynamics of membrane systems by periodically reversing the direction of the flow of the feed stream to the membrane. Periodic reversal of the direction of flow of the feed stream in the membrane module, while maintaining the cross-flow, has been attempted by several researchers to keep the system in a hydrodynamically transient state, thereby preventing the formation of an undesirable stable boundary layer at the membrane surface. Hargrove et al. (2003) introduced the concept of periodic reversal of feed flow in a cross-flow UF operation for flux enhancement in a hollow-fiber membrane module using bovine serum albumin (BSA) solution as feed. The results suggest that by flow reversal, significant enhancement

TABLE 3.2

Summary of Ultrasonic Irradiation Used in NF and UF from Literature

Sr. No	Membrane Process	Feed Solution	Membrane & Operating Condition	Mode of Irradiation and Frequency Used	Flux Performance	Reference
1	UF	A clay solution made from surface soil and deionized (DI) water	Polsulfone hollow-fiber membrane (pore size of 10,000 Da)	Online continuous and offline pulse sonication, Frequency: 40, 68 & 170 kHz	Significant improvements of the permeate flux for both online continuous and offline pulse sonication.	Li et al., 2011
2	UF	Laponite XLG dispersion in DM water	A flat polyether sulfone UF membrane 100 kD,	A thin titanium vibrating blade embedded in the feed compartment Frequency: 20 kHz	An immediate increase of permeate fluxes, and steady states could be reached almost immediately after every operation below 1.1 × 10^5 Pa.	Jin et al., 2014
3	UF	Colloidal suspensions of (i) Natural Wyoming–Na Montmorillonite clay	Flat commercial Polyethersulfone membrane MWCO: 100 kDa TMP: 0 to 1.5 bar	Metallic blade specially shaped (4 mm ×110 mm ×50 mm) connected to an ultrasonic transducer linked to a generator Frequency: 20 kHz ultrasonic bath	Permeation flux for Montmorillonite Wyoming–Na clay suspension increased by a factor 7.1, from 13.6 L $m^{-2}h^{-1}$ without ultrasound to 97 $Lm^{-2}h^{-1}$ with ultrasound	Hengl et al., 2014
4	UF	Aqueous solution of dextran	Polyethersulfone UF membrane MWCO of 30 kDa	Frequency: 28, 45 and 100 kHz	At TMP of 0.4 bar the flux improvement by US at 28 kHz was significantly stronger than that at 45 kHz	Cai et al., 2010
5	UF	Collidal silica solution	Whatman (Clifton, NJ) Anodisc™ γ-Al_2O_3 ceramic membranes with a polypropylene support ring T: 22.4°C, TMP: 6.9 to 55.2 kPa	Ultrasonic probe with 20 kHz frequency. Continuous and pulsed sonication	For short pulse interval (i.e., 0.1 s), the relative permeate flux improvement was 73 ± 4%, which was similar to continued sonication (75 ± 7%).	Chen et al., 2006

(Continued)

TABLE 3.2 (CONTINUED)

Summary of Ultrasonic Irradiation Used in NF and UF from Literature

Sr. No	Membrane Process	Feed Solution	Membrane & Operating Condition	Mode of Irradiation and Frequency Used	Flux Performance	Reference
6	UF	Real extract of a natural product, *Radix astragalus*	Polyethersulfone flat sheet UF membrane with MWCO 10,000 Da and effective membrane area of 41.8 cm²	Ultrasonic transducer plates at frequencies of 28, 45 and 100 kHz, and variable output power ranging 0–200, 0–300 and 0–600 W, respectively	An enhancement of 12–15% in permeate flux at effective power of 10W and frequency of 28 or 45 kHz;At US power of 120W, the fluxes increased dramatically with the increased ultrasonic intensity, nearly 70% higher in flux than those with stirring only.	Cai et al., 2009
7	UF	Wastewater from paper industry	Alumina-based ceramic membranes with mean pore sizes of 0.12, 0.19 and 0.25 μm, and a commercial polymeric PES-50H membrane	Transducers operated with frequencies of 27, 40 or 200 kHz. 27 and 40 kHz piezoceramic transducer elements were built up using four Langewin type sandwich transducers side by side.	Significant flux improvement when using a frequency of 27 kHz but only minor when using 200 kHz.	Kyllonen et al., 2006
8	UF	Dextran with average molecular weight of 260 Da	UF membranes (PTTK type, Millipore, Bedford, MA) with MWCO 30 kDa and an effective filtration area of 39.6 cm²	Low-frequency US was produced by a classical probe system (3 mm diameter) operating at 20 kHz	Intermittent use of US less effective than continuous use. Although the permeate flux started to increase as soon as the pulse is applied	Simon et al., 2000
9	UF	*Radix astragalus* extracts (natural product)	Polysulfone Hollow fibre UF membrane MWCO 10 kDa and an effective area of 150 cm². TMP: 0.2 to 1.0 bar	The ultrasonic plate has 45 kHz frequency and output power ranged from 0 to 300 W in an interval of 30 W.	The optimum conditions of the process have been found to be ultrasonic power of 120 W, ultrasonic irradiation mode of continuous, TMP of 0.6 bar and temperature of 20 C with fouling degree of 38.5–43% and process duration of 53–58 min.	Cai et al., 2012

(Continued)

TABLE 3.2 (CONTINUED)

Summary of Ultrasonic Irradiation Used in NF and UF from Literature

Sr. No	Membrane Process	Feed Solution	Membrane & Operating Condition	Mode of Irradiation and Frequency Used	Flux Performance	Reference
10	UF	Lake water from Lake Washington at adock adjacent to the University of Washigton in Seattle.	Single capillary membrane fibre made of polysulfone, MWCO: 100 kDa TMP: 0.5–7 psi; T: 25°C	Membrane assembly immersed in an Ultrasonic bath, at frequency: 45 kHz.	Ultrasound could limit reversible fouling leading to an increase in filtration rate.	Naddeo et al., 2015
11	UF	Solution of skimmed milk powder with 1% solid content	Flat sheet polyether sulfone membrane with 10 kD MWCO with effective membrane area 112 cm² T: 20 ± 2°C	Frequencies: 37, 80 kHz and tandem. Sonication mode: Continuous, pulsed, sweeping and degassing.	Pulsed irradiation mode at 37 kHz frequency could significantly increase the permeation flux than other modes.	Shahraki et al., 2014
12	UF	Solution of spray dried non-hygroscopic whey powder	Flat sheet polysulfone UF membranes with 30,000 MWCO; an effective membrane area of 30 cm²	Ultrasonic bath, at frequency: 50 kHz, power 300 W, Spacers were used as turbulent promoters	Significant enhancement of the permeate flux with an enhancement factor of between 1.2 and 1.7.	Muthukumaran et al., 2005.
13	UF	Binary BSA and lysozyme (Ly) mixture	Flat sheet polyether sulfone membrane (MWCO, 30,000)	Ultrasonic bath, at frequency: 25 kHz, power: 240 W	The use of ultrasound at 25 kHz and 240 W resulted in an increase of UF flux by 135% and 120%	Teng et al., (2006)
14	NF	Mixture of Reactive Black 5 and reactive yellow 160	Flat sheet polyamide membrane, MWCO 400, TMP: 294, 490, 686 and 980 kPa Temp: 35–55°C	Ultrasonic bath, at frequency: 34 ± 3 kHz, both continuous and intermittent irradiation	Permeate flux increased up to 16.5% for intermittent irradiation. For continuous mode the flux could be maintained over the whole concentration and TMP range tested.	Patel and Nath 2013

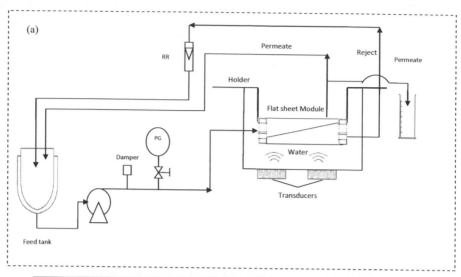

FIGURE 3.1

(a) Schematic of experimental set-up of nanofiltration of reactive dye and salt mixture assisted by low frequency acoustic irradiation (PG: pressure gauge; RR: reject rotameter), (b) A photograph of nanofiltration in presence of low frequency acoustic irradiation for the treatment of dye waste water carried out by the authors. The flat sheet module has been immersed into the ultrasonic bath.

of flux could be accomplished as compared to conventional unidirectional flow. The flux improvements were also found out to be very high with increasing feed concentration and operating TMP. On the other hand, in back pulsing, the permeate stream is periodically forced back through the membrane module permeate outlet under the influence of an induced pressure gradient. Literature cites two different types of back pulsing namely rapid water back pulsing and gas back pulsing. In water back pulsing, liquid phase permeate is forced back through the membrane by reverse TMP. In gas back pulsing, different inert gases can be used for their periodic pulsing through the membrane from the permeate side. It has been demonstrated that both types are effective for the enhancement of permeate flux in UF processes (Ma et al., 2001). The incorporation of pulsating flow in order to produce oscillations into the feed, filtrate and permeate channels has been reported to be effective in improving the flux. A study reports an approximately 300% increase in flux improvements in case of periodically-spaced, doughnut-shaped baffles in UF tubes and pulse flows (Finnigan & Howell, 1989). The effects of varying the pulse amplitude and duration on permeate flux and solute throughput has been studied for a BSA solution in cross-flow transmembrane pressure pulsed UF using a polysulfone membrane (Wilharm & Rodgers, 1996). The average duration of pulse was kept between 0.13 and 0.34 s of backpressure per second of operation. The mass flux of BSA was enhanced due to pulsing by as much as a factor of three. Strategies to modify fluid hydrodynamics of the bulk stream have been investigated widely by membrane researchers. Each of them has its merits and demerits and hence it cannot give a conclusive solution of the fouling problem. The advantage of using pulsation overcomes the disadvantage of increased power consumption. However, the problems of energy dissipation and reduced cross-flow, which results in a lower net filtering capacity, remain. Additionally, fluid instabilities due to flow in curved ducts, known as Taylor and Dean flows, have been investigated to mitigate the flux-limiting effects of concentration and fouling. The problems associated with this and other similar external devices are the high energy required to operate the devices and the difficulty to repair or scale them up.

3.6 Air Sparging and Gas Slug

Gas or air sparging usually consists of generating an intermittent gas-liquid two-phase flow by injecting gas such as nitrogen, argon, carbon dioxide or air directly into the concentrate compartment. Gas injection is one of the most effective ways to ameliorate membrane fouling by enhancing shear force, which in turn increases turbulence. Two-phase flow is very useful in delivering improved permeate flux over time by creating hydrodynamic instabilities in the flow channel. The basic mechanism of flux enhancement by gas sparging in membrane filtration can be ascribed to the secondary flow induced by air bubbles, which facilitates local mixing, thereby reducing the thickness of mass transfer boundary layer (Patel & Nath, 2014). The important governing parameters are cross-flow velocity, gas flow rate, size and spatial distribution of the gas bubbles, flow patterns, TMP and feed concentration. Wang et al. (2004) reported different types of gas (compressed air and *n*-hexadecane) in a UF membrane made of regenerated cellulose for *n*-hexadecane/water and air/water systems for which flux enhancements were 25% and 17%, respectively, compared to the single-phase flow. Cui et al. (2003) published a thorough review of the use of gas bubbles to enhance membrane processes. Since then, a vast body of literature on the use of two-phase flow in membrane processes has appeared. Yet, one more excellent comprehensive review on the gas sparged membrane processes was presented by Wibisono et al. (2014). The review encompasses the basic concepts of the two-phase flow process, including flow patterns in tubular channels, analysis of normalized data from the database and provides a perspective on two-phase flow in membrane processes until 2013. A perusal of information and data provided in the review indicates that gas/liquid ratio is one of the key parameters having direct influences on membrane process performance. The same gas/liquid ratio can give different results when the channel geometries vary as flow patterns are largely responsible for generating wall shear stress. Slug flow is known as the best flow pattern for improving the permeate flux in non-submerged systems. Smaller bubble flow, however, was reported to have better performance in submerged systems (Wibisono et al., 2014). It is also worth mentioning that gas sparging at relatively high-pressure applications of NF processes is not entirely trouble-free. Back pressure of dissolved gas to the permeate side and high energy requirements for pumps and compressors to achieve the necessary gas-to-liquid volume ratio continue to be grey areas of a gas sparged operation requiring further research.

Here, we would focus on a few other recent research projects on gas sparged NF and UF processes that were carried out in last four to five years. Sriniworn et al. (2015) studied the effect of gas sparging on flux enhancement and fouling resistance during the UF separation of protein and oligosaccharide from tofu whey. Gas sparging resulted in the increase in shear stress, which could minimize the effect of concentration polarization and fouling and improve permeate flux to approximately 45.1%. Charoenphun and Youravong (2016) reported the separation of angiotensin-I converting enzyme (ACE) inhibitory peptides from tilapia protein hydrolysate by UF using a polysulfone hollow fiber membrane under gas sparged conditions. The injection of gas into the UF system resulted in the increase of the shear stress number. As a result, the permeate flux improved substantially with a marked reduction of concentration polarization and fouling. According to the authors, the bubble-induced secondary flow played a major role by promoting local mixing in the bubble wake to minimize solute accumulation on the membrane surface. The effect of gas bubbling including slug and bubble flows on enhancing shear force in the UF process has been investigated experimentally by image processing and numerical simulation (Javid et al., 2017). Nitrogen was used as the gas phase and whey protein concentrate (WPC) solution with a 1% wt solid content concentration was used as the liquid feed for the UF process employing a polyethersulfone flat sheet membrane (Sepro, United States) of 10 kDa MWCO with an effective membrane area of 300 cm^2. The permeate flux for the slug and bubble flows was increased by 78% and 30%, respectively, compared to the case with no gas bubbling.

Gas sparging has been observed to significantly affect the performance of the NF treatment of molasses wastewater using a hydrophilized polyamide membrane in a flat sheet module (Nath & Patel, 2014). Shear stress number increased with increasing cross-flow velocity and this was more pronounced in presence of gas sparging. An almost 1.5–2-fold increase in shear stress number was observed using cross-flow velocity of 0.8 ms^{-1} with gas sparging, compared to the case of "no sparging." Moreover, the mass transfer coefficient also increased with increasing cross-flow velocity for both the conditions of "gas sparging" and "no sparging," but its value became considerably higher with gas sparging. The same group of authors also predicted a semi-empirical resistance-in-series model by combining solution-diffusion and film theory to predict the permeate flux and mass transfer resistances during the gas sparged cross-flow NF of molasses wastewater in a flat-sheet module (Patel & Nath, 2014). The authors considered three resistances, namely membrane hydraulic resistance, osmotic pressure resistance and the concentration polarization resistance in series. The concentration polarization resistance was correlated to the feed concentration, TMP and the cross-flow velocity for a set of selected experiments. There was an appreciable reduction of the concentration polarization resistance in the presence of gas sparging, which ultimately resulted in the alleviation of flux decline over a period of time. Figure 3.2 presents a schematic as well as a photograph of the nitrogen gas sparged NF set-up used for the treatment of molasses wastewater by the authors in their laboratory.

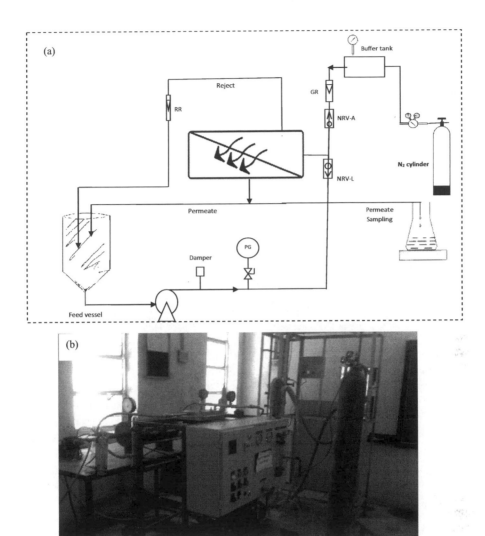

FIGURE 3.2
(a) Schematic of experimental set-up of gas-sparged nanofiltration for molasses wastewater (PG: pressure gauge; RR: reject rotameter, GR: gas rotameter; NRV-A: non return valve for air; NRV-L: non return valve for liquid), (b) A photograph of nitrogen gas sparged NF set-up used for the treatment of molasses waste water by the authors in their laboratory.

3.7 Membrane Surface Modification

Since membrane separation is essentially a surface phenomenon where the skin layer or top surface layer plays the vital role, it is quite rational to modify the membrane surface for reducing the fouling propensity. Apart from the strategies discussed so far, there is one more important approach towards flux improvement by means of membrane surface modification. Both physical and chemical modifications of membrane surface have been attempted by various research scientists to reduce irreversible fouling. But it has been observed that surface modification yields satisfactory results mostly in the case of adhesive fouling, which is caused due to adsorptive phenomena. In a comprehensive review,

Rana and Matsuura (2010) surveyed various methods for membrane surface modification. The emergence of a variety of spectroscopy particularly atomic force microscopy (AFM) facilitates the correlation of the surface roughness and fouling by enabling the measurement of surface roughness in nanoscale. One of the basic methods of surface modification is to increase surface hydrophilicity by treating the base polymer materials with different surfactants via surface coating, dip coating, grafting, cross linking by covalent attachment, UV treatment, plasma treatment, ion beam irradiation, incorporating nanoparticles and so on. Coating and grafting are most widely reported methods. Coating involves the use of a solution containing the component with antifouling character, while in grafting covalent immobilization of hydrophilic component is done on to the membrane (Hilal et al., 2005). Photo-induced grafting is another new technique for membrane surface functionalization. But the coating and grafting techniques suffer from the drawbacks such as easy erosion of the coating layer, increased cost due to the extensive use of organic solvents, and unsatisfactory reliability and durability of the modified surface. The flux of the nanoparticle incorporated membranes becomes very high due to the larger effective surface area of the membrane caused by the nodular shapes having ridges and valleys. Yan et al. (2016) modified a polyamide-thin film composite (PA-TFC) membrane by surface grafting with triethanolamine (TEOA) through esterification between hydroxyl group of TEOA and the residual acyl chloride group. The as-prepared membrane became more hydrophilic rendering higher water permeability. Ample amount of research has been devoted to improving the antifouling characteristics of the PA-TFC NF membranes through suitable surface modifications encompassing physical and/or chemical treatments to mitigate the deposition of foulants on a membrane surface (Ingole et al., 2014; Saenz de Jubera et al., 2013; Kochkodan et al., 2015; Zhu et al., 2015). However, many of these techniques cannot be applicable at large scale applications. The reduction of fouling can also be accomplished by increasing the surface charge, both positive as well as negative, by grafting with carboxylic acid and sulfonic acid, amination, grafting monomers with trimethyl/ethyl ammonium etc. Immobilization of biomacromolecules plays a significant role in preventing biofouling. This is one of the latest attempts to reduce membrane fouling. The formation of a biomimetic surface is considered to be one of the most effective ways to prevent biofouling. One of the attractive options of surface treatment is to modify the commercial membrane surfaces with bacteristatic components to prevent biofouling initiated by microbial growth (Hilal et al., 2003).

3.8 Conclusion and Future Outlook

Nanofiltration and ultrafiltration have been reigning the kingdom of industrial separation for almost three decades. Notwithstanding their numerous successful applications in the reclamation of wastewater and recovery of valuable solutes, their widespread commercial acceptance has not attained the desired level of theoretical expectation of high throughput coupled with improved selectivity. Growing academic and research interest in this field, as evidenced by an increasingly large body of literature, are not commensurate with the pace of its commercialization. Various strategies to mitigate fouling have been critically discussed in the present review and each of them has their own advantages and limitations. Hence, it is most unlikely that any single strategy could become a panacea for flux decline and the fouling problem. It would be judicious to adopt a strategy either as stand-alone or in synergic combination with another one depending on the feed quality, permeate requirement, membrane chemistry and morphology, and other physico-chemical process parameters. On top of this, cost and energy consumption also need to be given due consideration. The optimization of hydrodynamic conditions, selection of suitable turbulence promoters, gas sparging and assisted filtration using acoustic irradiation along with additional force field are the most sought-after techniques for the alleviation of flux decline. Besides the development of multifunctional membrane materials offering higher permeabilities, smart nanomaterials for doping and sophisticated analytical tools have undeniably made membrane research more focused toward commercial acceptance. Here comes the paramount importance of further research in process modification, membrane materials, surface chemistry, cleaning of fouled membrane and pretreatment processes. The issue of significant reductions in both capital investment and operating costs should also be addressed. In essence, it can be predicted that the use of NF and UF will continue to dominate downstream operations in days to come.

References

Al-Akoum, O., D. Richfield, M.Y. Jaffrin, L.H. Ding, and P. Swart. 2006. Recovery of trypsin inhibitor and soy milk concentration by dynamic filtration. *Journal of Membrane Science* 279: 291–300.

A1-Bastaki, N. and A. Abbas. 2001. Use of fluid instabilities to enhance membrane performance: A review. *Desalination* 136: 255–262.

Amjad, H., Z. Khan, and V.V. Tarabara. 2015. Fractal structure and permeability of membrane cake layers: Effect of coagulation–flocculation and settling as pretreatment steps. *Separation and Purification Technology* 143: 40–51.

Auddy, K., S. De, and S. DasGupta. 2005. Performance prediction of turbulent promoter enhanced nanofiltration of a dye solution. *Separation and Purification Technology* 43: 85–94.

Belfort, G., P. Mikulasek, J.M. Pimbley, and K.Y. Chune. 1993. Diagnosis of membrane fouling using a rotating annular filter. 2. Dilute particle suspensions of known particle size. *Journal of Membrane Science* 77: 23–39.

Bes-Pia, A., M-I. Iborra-Clar, A. Iborra-Clar, J.A. Mendoza-Roca, B. Cuartas-Uribe, and M.I. Alcaina-Miranda. 2005. Nanofiltration of textile industry wastewater using a physicochemical process as a pre-treatment. *Desalination* 178: 343–349.

Bes-Pia, A., J.A. Mendoza-Roca, M.I. Alcaina-Miranda, A. Iborra-Clar, and M.I. Iborra-Clar. 2002. Reuse of wastewater of the textile industry after its treatment with a combination of physico-chemical treatment and membrane technologies. *Desalination* 149: 169–174.

Bhattacharjee, S., S. Datta, and C. Bhattacharjee. 2006. Performance study during ultrafiltration of Kraft black liquor using rotating disk membrane module. *Journal of Cleaner Production* 14: 497–504.

Bhattacharjee, C. and P.K. Bhattacharya. 2006. Ultrafiltration of black liquor using rotating disk membrane module. *Separation and Purification Technology* 49: 281–290.

Bowen, W.R. and J.S. Welfoot. 2002. Modelling the performance of membrane nanofiltration—Critical assessment and model development. *Chemical Engineering Science* 57: 1121–1137.

Borbély, G. and E. Nagy. 2009. Removal of zinc and nickel ions by complexation–membrane filtration process from industrial wastewater. *Desalination* 240: 218–226.

Bouzerar, R., P. Paullier, and M.Y. Jaffrin. 2003. Concentration of mineral suspensions and industrial effluent using a rotating disk dynamic filtration module. *Desalination* 158: 79–85.

Cai, M., S. Wang, and H. Liang. 2012. Optimization of ultrasound-assisted ultrafiltration of Radix astragalus extracts with hollow fiber membrane using response surface methodology. *Separation and Purification Technology* 100: 74–81.

Cai, M., S. Wang, Y. Zheng, and H. Liang. 2009. Effects of ultrasound on ultrafiltration of *Radix astragalus* extract and cleaning, of fouled membrane. *Separation and Purification Technology* 68: 351–356.

Cai, M., S. Zhao, and H. Liang. 2010. Mechanisms for the enhancement of ultrafiltration and membrane cleaning by different ultrasonic frequencies. *Desalination* 263: 133–138.

Capar, G., U. Yetis, and L. Yilmaz. 2007. Most effective pre-treatment to nanofiltration for the recovery of print dyeing wastewaters. *Desalination* 212: 103–113.

Charoenphun, N. and W. Youravong. 2017. Influence of gas–liquid two-phase flow on angiotensin-I converting enzyme inhibitory peptides separation by ultra-filtration. *Journal of the Science of Food and Agriculture* 97(1): 309–316.

Choi, Y.H., J.A. Nason, and J.H. Kweon. 2013. Effects of aluminum hydrolysis products and natural organic matter on nanofiltration fouling with PACl coagulation pretreatment. *Separation and Purification Technology* 120: 78–85.

Choi, K.Y.J. and B.A. Dempsey. 2004. In-line coagulation with low pressure membrane filtration. *Water Research* 38(19): 4271–4281.

Chakraborty, S., S. De, J.K. Basu, and S. DasGupta. 2005. Treatment of a textile effluent: Application of a combination method involving adsorption and nanofiltration. *Desalination* 174: 73–85.

Chai, X., T. Kobayashi, and N. Fujii. 1998. Ultrasound effect on crossflow filtration of polyacrylonitrile ultrafiltration membranes. *Journal of Membrane Science* 148: 129–135.

Chen, D., L.K. Weavers, and H.W. Walker. 2006. Ultrasonic control of ceramic membrane fouling by particles: Effect of ultrasonic factors. *Ultrasonics Sonochemistry* 13: 379–387.

Choo, K.H., S.J. Choi, and E.D. Hwang. 2007. Effect of coagulant types on textile wastewater reclamation in a combined coagulation/ultrafiltration system. *Desalination* 202: 262–270.

Cui, Z.F., S. Chang, and A.G. Fane. 2003. The use of gas bubbling to enhance membrane processes. *Journal of Membrane Science* 221(1–2): 1–35.

Dal-Cin, M.M., C.N. Lick, A. Kumar, and S. Lealess. 1998. Dispersed phase back transport during ultrafltration of cutting oil emulsions with a spinning membrane disc geometry. *Journal of Membrane Science* 14: 165–181.

Ding, L-H., O. Akoum, A. Abraham, and M.Y. Jaffrin. 2003. High shear skim milk ultrafiltration using rotating disk filtration systems, *AIChE Journal* 49(9): 2433–2441.

Finnigan, S.A. and J.A. Howell. 1989. The effect of pulsetile flow on ultrafiltration fluxes in a baffled tubular membrane system. *Chemical Engineering Research and Design* 67: 278–282.

Frappart, M., O. Akoum, L.H. Ding, and M.Y. Jaffrin. 2006. Treatment of dairy process waters modelled by diluted milk using dynamic nanofiltration with a rotating disk module. *Journal of Membrane Science* 282: 465–472.

Ganiyu, S.O., E.D. van Hullebusch, M. Cretin, G. Esposito, and M.A. Oturan. 2015. Coupling of membrane filtration and advanced oxidation processes for removal of pharmaceutical residues: A critical review. *Separation and Purification Technology* 156(3): 891–914.

Gao, B.Y., Y. Wang, Q.Y. Yue, J.C. Wei, and Q. Li. 2007. Color removal from simulated dye water and actual textile wastewater using a composite coagulant prepared by polyferric chloride and polydimethyldiallylammonium chloride. *Separation and Purification Technology* 54: 157–163.

Ghaffour, N., R. Jassim, and T. Khir. 2004. Flux enhancement by using helical baffles in ultrafiltration of suspended solids. *Desalination* 167: 201–207.

Hargrove, S.C., H. Parthasarathy, and S. Ilias. 2003. Flux enhancement in cross-Flow Membrane filtration by flow reversal: A case study on ultrafiltration of BSA. *Separation Science and Tecnology* 38: 3133–3144.

He, G., L.H. Ding, P. Paullier, and M.Y. Jaffrin. 2006. Experimental study of a dynamic filtration system with overlapping ceramic membranes and non-permeating disks rotating at independent speeds. *Journal of Membrane Science.* 276: 232–240.

Heng, L., Y. Yanling, G. Weijia, L. Xing, and L. Guibai. 2008. Effect of pretreatment by permanganate/chlorine on algae fouling control for ultrafiltration (UF) membrane system. *Desalination* 222 (1–3): 74–80.

Hengl, N., Y. Jin, F. Pignon, S. Baup, R. Mollard, N. Gondrexon, A. Magnin, L. Michot, and E. Paineau. 2014. A new way to apply ultrasound in cross-flow ultrafiltration: Application to colloidal suspensions. *Ultrasonics Sonochemistry* 21: 1018–1025.

Hilal, N., H. Al-Zoubi, N.A. Darwish, A.W. Mohamma, and M. Abu Arabi. 2004. A comprehensive review of nanofiltration membranes: Treatment, pretreatment, modelling, and atomic force microscopy. *Desalination* 170: 281–308.

Hilal, N., L. Al-Khatib, B.P. Atkin, V. Kochkodan, and N. Potapchenko. 2003. Photochemical modification of membrane surfaces for (bio) fouling reduction: A nano-scale study using AFM. *Desalination* 158: 65–72.

Hilal, N., O.O. Ogunbiyi, N.J. Miles, and R. Nigamatullin. 2005. Methods employed for control of fouling in MF and UF membranes: A comprehensive review. *Separation Science and Technology* 40: 10, 1957–2005.

Huang, H., K. Schwab, and J.G. Jacangelo. 2009. Pretreatment for Low Pressure Membranes in Water Treatment: A review. *Environmental Science and Technology* 43(9): 3011–3019.

Ingole, P.G., W. Choi, K.H. Kim, C.H. Park, H.K. Choi, and H.K. Lee. 2014. Synthesis, characterization and surface modification of PES hollow fiber membrane support with polydopamine and thin film composite for energy generation. *Chemical Engineering Journal* 243: 137–146.

Javid, S.M., M. Passandideh-Fard, A. Faezian, and M. Goharimanesh. 2017. Slug and bubble flows in a flat sheet ultrafiltration module: Experiments and numerical simulation. *International Journal of Multiphase Flow* 91: 39–50.

Joo, D.J., W.S. Shin, J.H. Choi, S.J. Choi, M.C. Kim, M.H. Han, T.W. Ha, and Y.H. Kim. 2007. Decolorization of reactive dyes using inorganic coagulants and synthetic polymer. *Dyes and Pigments* 73: 59–64.

Jaffrin, M.Y. 2008. Dynamic shear-enhanced membrane filtration: A review of rotating disks, rotating membranes and vibrating systems. *Journal of Membrane Science* 324: 7–25.

Jin, Y., N. Hengl, S. Baup, F. Pignon, N. Gondrexon, A. Magnin, M. Sztucki, T. Narayanan, L. Michot, and B. Cabane. 2014. Effects of ultrasound on colloidal organization at nanometer length scale during cross-flow ultrafiltration probed by in-situ SAXS. *Journal of Membrane Science* 453: 624–635.

Kaiser, B. 2004. Dynamic crossflow filtration with rotating membranes, in: Proceedings of Euromembrane Congress, Hamburg, September 28–October 1: 620.

Karnik, B.S., S.H. Davies, M.J. Baumann, and S.J. Masten. 2005. Fabrication of catalytic membranes for the treatment of drinking water using combined ozonation and ultrafiltration. *Environmental Science and Technology* 39(19): 7656–7661.

Kim, J., S.H.R. Davies, M.J. Baumann, V.V. Tarabara, and S.J. Masten. 2008. Effect of ozone dosage and hydrodynamic conditions on the permeate flux in a hybrid ozonation-ceramic ultrafiltration system treating natural waters. *Journal of Membrane Science* 311(1–2): 165–172.

Koniecznya, K., D. Sąkolb, J. Płonkaa, M. Rajcaa, and M. Bodzek. 2009. Coagulationultrafiltration system for river water treatment. *Desalination* 240: 151–159.

Kochkodan, V., and N. Hilal. 2015. A comprehensive review on surface modified polymer membranes for biofouling mitigation. *Desalination* 356: 187–207.

Kyllonen, H., P. Pirkonen, M. Nystrom, J. Nuortila-Jokinen, and A. Gronroos. 2006. Experimental aspects of ultrasonically enhanced cross-flow membrane filtration of industrial wastewater. *Ultrasonics Sonochemistry* 13: 295–302.

Li , X., J. Yu, and A.G. Agwu Nnanna. 2011. Fouling mitigation for hollow-fiber UF membrane by sonication. *Desalination* 281: 23–29.

Luo, J., L. Ding, Y. Wan, and M.Y. Jaffrin. 2012. Threshold flux for shear-enhanced nanofiltration: Experimental observation in dairy wastewater treatment. *Journal of Membrane Science* (409–410): 276–284.

Luo, J., L. Ding, Y. Wan, P. Paullier, M.Y. Jaffrin. 2010. Application of NF-RDM (nanofiltration rotating disk membrane) module under extreme hydraulic conditions for the treatment of dairy wastewater. *Chemical Engineering Journal* 163: 307–316.

Luo, J., and L. Ding. 2011. Influence of pH on treatment of dairy wastewater by nanofiltration using shear-enhanced filtration system. *Desalination* 278: 150–156.

Ma, H.M., L.F. Hakim, and C.N. Bowman. 2001. Factors affecting membrane fouling reduction by surface modification and backpulsing. *Journal of Membrane Science* 189: 255–270.

Mänttäri, M., L.Vitikko, and M. Nyström. 2006. Nanofiltration of biologically treated effluents from the pulp and paper industry. *Journal of Membrane Science* 272: 152–160.

Meier, J., T. Melin, L.H. Eilers. 2002. Nanofiltration and adsorption on powdered adsorbent as process combination for the treatment of severely contaminated waste water. *Desalination* 146: 361–366.

Moulai-Mostefa, N., L.H. Ding, M. Frappart, M.Y. Jaffrin. 2007. Treatment of aqueous ionic surfactant solutions by dynamic ultrafiltration. *Separation Science and Technology* 42: 2583–2594.

Molinari, R., P. Argurio, and T. Poerio. 2004. Comparison of polyethylenimine, polyacrylic acid and poly(dimethylamine-co-epichlorohydrin-co-ethylenediamme) in Cu^{2+} removal from wastewaters by polymer-assisted ultrafiltration. *Desalination* 162: 217–228.

Mohammad, A.W., C.Y. Ng, Y.P. Lim, and G.H. Ng. 2012. Ultrafiltration in food processing industry: Review on application, membrane fouling, and fouling control. *Food Bioprocess Technology* 5: 1143–1156

Mohammad, A.W., Y.H. Teow, W.L. Ang, Y.T. Chung, D.L. Oatley-Radcliffe, and N. Hilal. 2015. Nanofiltration membranes review: Recent advances and future prospects. *Desalination* 356: 226–254.

Mo, J.H., Y.H. Lee, J. Kim, J.Y. Jeong, and J. Jegal. 2008. Treatment of dye aqueous solutions using nanofiltration polyamide composite membranes for the dye wastewater reuse. *Dyes and Pigments* 76: 429–434.

Muthukumaran, S., S.E. Kentish, M. Ashokkumar, and G.W. Stevens. 2005. Mechanisms for the ultrasonic enhancement of dairy whey ultrafiltration. *Journal of Membrane Science* 258: 106–114.

Murkes, J. and C.G. Carlsson. 1988. *Cross Flow Filtration: Theory and Practice*. New York: John Wiley & Sons.

Naddeo, V., V. Belgiorno, L. Borea, M.F.N Secondes, and F. Ballesteros, Jr. 2015. Control of fouling formation in membrane ultrafiltration by ultrasound irradiation. *Environmental Technology* 36: 1299–1307.

Nath, K. and T.M. Patel. 2014. Mitigation of flux decline in the cross flow nanofiltration of molasses wastewater under the effect of gas sparging. *Separation Science and Technology* 49: 1479–1489.

Nuortila-Jokinen, J. and M. Nyström. 1996. Comparison of membrane separation processes in the internal purification of paper mill water. *Journal of Membrane Science* 119: 99–115.

Oatley, D.L., L. Llenas, R. Pérez, P.M. Williams, X. Martínez-Lladó, and M. Rovira. 2012. Review of the dielectric properties of nanofiltrationmembranes and verification of the single oriented layer approximation. *Advances in Colloid and Interface Science* 173: 1–11.

Patel, T.M. and K. Nath. 2013. Alleviation of flux decline in cross flow nanofiltration of two-component dye and salt mixture by low frequency ultrasonic irradiation. *Desalination* 317: 132–141.

Patel, T.M. and K. Nath. 2014. Modeling of permeate flux and mass transfer resistances in the reclamation of molasses wastewater by a novel gas-sparged nanofiltration. *Korean Journal of Chemical Engineering* 31(10): 1865–1876.

Rana, D. and T. Matsuura. 2010. Surface modifications for antifouling membranes. *Chem. Rev.* 110: 2448–2471.

Real, F.J., F.J. Benitez, J.L. Acero, and G. Roldan. 2012. Combined chemical oxidation and membrane filtration techniques applied to the removal of some selected pharmaceuticals from water systems. *Journal of Environmental Science Health Part A*: 47: 522–533.

Riera-Torres, M., C. Gutierrez-Bouzan, and M. Crespi. 2010. Combination of coagulation–flocculation and nanofiltration techniques for dye removal and water reuse in textile effluents. *Desalination* 252: 53–59.

Schlichter, B., V. Mavrov, and H. Chmiel. 2004. Study of a hybrid process combining ozonation and microfiltration/ultrafiltration for drinking water production from surface water. *Desalination* 168(1–3): 307–317.

Suksaroj, C., M. Heran, C. Allegre, and F. Persin. 2005. Treatment of textile plant effluent by nanofiltration and/or reverse osmosis for water reuse. *Desalination* 178: 333–341.

Sen, D., W. Roy, L. Das, C. Sadhu, and C. Bhattacharjee. 2010. Ultrafiltration of macromolecules using rotating disc membrane module (RDMM) equipped with vanes: Effects of turbulence promoter. *Journal of Membrane Science* 360: 40–47.

Sarkar, P., S. Ghosh, S. Dutta, D. Sen, C. Bhattacharjee. 2009. Effect of different operating parameters on the recovery of proteins from casein whey using a rotating disc membrane ultrafiltration cell, *Desalination* 249: 5–11.

Sarkar, A., S. Moulik, D. Sarkar, A. Roy, and C. Bhattacharjee. 2012. Performance characterization and CFD analysis of a novel shear enhanced membrane module in ultrafiltration of Bovine Serum Albumin (BSA). *Desalination* 292: 53–63.

Sarkar, D. and C. Bhattacharjee. 2008. Modeling and analytical simulation of rotating disk ultrafiltration module. *Journal of Membrane Science* 320: 344–355.

Saenz de Jubera, A.M., J.H. Herbison, Y. Komaki, M.J. Plewa, J.S. Moore, D.G. Cahill, and B.J. Mariñas. 2013. Development and performance characterization of a polyamide nanofiltration membrane modified with covalently bonded aramide dendrimers. *Environmental Science & Technology* 47: 8642–8649.

Schwinge, J., D.E. Wiley, and A.G. Fane. 2004. Novel spacer design improves observed flux, *Journal of Membrane Science* 229: 53–61.

Schwinge, J., D.E. Wiley, A.G. Fane, and R. Guenther. 2000. Characterization of a zigzag spacer for ultrafiltration. *Journal of Membrane Science* 172: 19–31.

Shahraki, M.H., A. Maskooki, and A. Faezian. 2014. Effect of various sonication modes on permeation flux in cross flow ultrafiltration membrane. *Journal of Environmental Chemical Engineering* 2: 2289–2294.

Simon, A., N. Gondrexon, S. Taha, J. Cabon, G. Dorange. 2000. Low-Frequency ultrasound to improve dead-end ultrafiltration performance. *Separation Science and Technology* 35(16): 2619–2637.

Sriniworn, P., W. Youravong, S. Wichienchot. 2015.Permeate flux enhancement in ultrafiltration of tofu whey using pH-shifting and gas- liquid two-phase flow. *Separation Science and Technology* 50(15).

Tarleton, E.S. and T.J. Wakesman. 1992. Electro-acoustic cross flow microfiltration. *Filtration & Separation* 29: 425–432.

Teng, M.-Y., S.-H. Lin, and R.-S. Juang. 2006. Effect of ultrasound on the separation of binary protein mixtures by cross-flow ultrafiltration. *Desalination* 200(1–3): 280–282.

Tu, Z., L. Ding, M. Frappart, and M.Y. Jaffrin. 2009. Studies on treatment of sodium dodecyl benzene sulfonate solution by high shear ultrafiltration system. *Desalination* 240: 251–256.

Van der Bruggen, B., M. Manttari, and M. Nystrom. (2008). Drawbacks of applying nanofiltration and how to avoid them: A review. *Separation and Purification Technology* 63: 251–263.

Vigo F., and C. Uliana. 1986. Influence of the vorticity at the membrane surface on the performances of the ultrafiltration rotating module. *Separation Science and Technology* 21(4): 367–381.

Vrouwenvelder, J.S., S.A. Manolarakis, H.R. Veenendaal, and D. van der Kooij. 2000. Biofouling potential of chemicals used for scale control in RO and NF membranes. *Desalination* 132: 1–10.

Wang, H.-M., C.-Y. Li, S.-J. Chen, T.-W., Cheng, and T.-L. Chen. 2004. Abatement of con- centration polarization in ultrafiltration using n-hexadecane/water two-phase flow. *Journal of Membrane Science* 238: 1–7.

Wilharm, C. and V.G.J. Rodgers. 1996. Significance of duration and amplitude in transmembrane pressure pulsed ultrafiltration of binary protein mixtures. *Journal of Membrane Science* 121: 217–228.

Wibisono, Y., E. Cornelissen, A. Kemperman, W. Van Der Meer, and K. Nijmeijer. 2014. Two-phase flow in membrane processes: a technology with a future. *Journal of Membrane Science* 453: 566–602.

Xiang-hua, Z., Y.U. Shui-li, W. Bei-h, and Z. Hai-feng. 2006. Flux enhancement during ultrafiltration of produced water using turbulence promoter. *Journal of Environmental Sciences* 18(6): 1077–1081.

Yan, F., H. Chen, Y. Lü, Z. Lü, S. Yu, M. Liu, and C. Gao. 2016. Improving the water permeability and antifouling property of thinfilm composite polyamide nanofiltration membrane bymodifying the active layer with triethanolamine. *Journal of Membrane Science* 513: 108–116.

Yaroshchuk, A.E. 1998. Rejection mechanisms of NF membranes. Serono. Sym.: 9–12.

Zakrzewska-Trznadel, G., M. Harasimowicz, A. Miskiewicz, A. Jaworska, E. Dłuska, and S. Wronski. 2009. Reducing fouling and boundary-layer by application of helical flow in ultrafiltration module employed for radioactive wastes processing. *Desalination* 240: 108–116.

Zhu, W.P., J. Gao, S.P. Sun, S. Zhang, and T.S. Chung. 2015. Poly(amidoamine) dendrimer (PAMAM) grafted on thin film composite (TFC) nanofiltration (NF) hollow fiber membranes for heavy metal removal. *Journal of Membrane Science* 487: 117–126.

Zahrim, A.Y., C. Tizaoui, and N. Hilal. 2011. Coagulation with polymers for nanofiltration pretreatment of highly concentrated dyes: A review. *Desalination* 266: 1–16.

Zularisam, A.W., A.F. Ismail, M.R. Salim, M. Sakinah, and T. Matsuura. 2009. Application of coagulation–ultrafiltration hybrid process for drinking water treatment: Optimization of operating conditions using experimental design. *Separation and Purification Technology* 65: 193–210.

Section II

Water

4

Fabrication and Applications of Functionalized Membranes in Drinking Water Treatment

Somak Chatterjee and Sirshendu De

CONTENTS

4.1 Introduction

4.1.1 Different Classes of Membranes

Membranes are a special class of separation devices, allowing transport of some selective ions while hindering the permeation of others (Mulder, 2012). Depending upon the pore size, membranes operating on applied hydrostatic pressure can be classified as microfiltration (MF), ultrafiltration (UF), nanofiltration (NF) and reverse osmosis (RO). These membranes have different applications in water treatment; MF membranes are used for removal of bacteria and suspended solids; UF membranes, for removal of proteins, polymers, viruses and humic acid; NF membranes are for separation of salts, synthetic dyes and polyphenols; while RO membranes are employed for large-scale desalination.

4.1.2 Utilities of Different Membranes in Water Treatment

The most common application of membranes in water treatment is the large-scale desalination of sea and brackish water, which have a high salt concentration in the range of 3000 mg/L and 35000 mg/L, respectively. RO provides efficient removal of over 90%

of monovalent salts, with a flux of 60–67 m³/day at 5 kWh/m³ energy consumption (Shimokawa, 2009). Filtered water is used for the power industry, semiconductors, in addition to household purposes. Membranes are also used in removal of hardness from water, which is due to calcium, iron and magnesium salts that often form scales on the walls of process vessels, like heat exchangers. Scaling also downplays the performance of detergents in domestic usage. In this case, the NF-70 nanofiltration membrane (synthesized by Dow Filmtec) removes 70% calcium and magnesium salts at a pressure of 0.5 MPa (Bodzek & Konieczny, 2011). Anionic micropollutants like nitrate, fluoride and arsenic can also be separated from a contaminated stream by the use of RO and NF membranes. Toxic metals like cadmium, lead, copper, zinc and chromium can be removed by electrodialysis (ED), which utilizes a charged electric interface for transporting ionic contaminants through the membrane (Pedersen et al., 2003). Some pollutants are organic, like natural organic matter (NOM), micro-pollutants, like polycyclic aromatic hydrocarbons (PAH), pharmaceutical active compounds (PAC) and endocrine disrupting compounds (EDC). NOMs, such as fulvic acid and humic acid that cause extensive coloration of water streams are removed by MF and UF membranes. For PAHs, the NF membrane NF-MQ16 exhibits a rejection as high as 93.3 % for a feed containing 70 ng/L of benzo-perylene (Yoon et al., 2006). In case of EDCs, NF membranes are efficient in removing compounds having molecular weights higher than 190 Dalton (Taylor & Wiesner, 2000). For PACs, higher efficiency is attained when RO is used in series with UF as a pre-filtration step. For example, a common non-steroidal anti-inflammatory drug, Ibuprofen is present in pharmaceutical wastewater at 87 ng/L concentration. Pretreatment with UF membranes leads to 12% removal followed by 80% removal after first stage RO, and 98% removal after the second stage (Heberer & Feldmann, 2008). Removal efficiency can be magnified using a hybrid system. For example, boron is removed from seawater feed through uptake by ion exchange resins (IER), followed by filtration from treated stream by MF membranes (Bryjak et al., 2008). Metal removal from water is attained by the integration of oxidation and the UF membrane, like in the Zeeweed method (Bodzek et al., 2011).

4.1.3 Organic and Inorganic Membranes

Membranes can also be classified as organic and inorganic based on synthetic materials (Nath, 2017). Organic membranes made of natural and synthetic polymers are used in industrial and household water purification. Natural polymeric membranes are derived from pure cellulose, while synthetic polymeric membranes are made from polyvinylidene fluoride (PVDF), polysulfone (PSF) and polytetrafluoroethylene (PTFE), etc. These membranes have relatively low chemical stability and are thermally unstable. Inorganic membranes include metallic, ceramic and zeolitic membranes, which are synthesized by sintering a metal/metal oxide followed by subsequent deposition on a porous substrate. They are inert at extreme pH and can be used for catalysis, hydrogen adsorption, etc. However, the major limitations are surface poisoning, membrane cracking at high temperature and low flux.

4.1.4 Organic-Inorganic Mixed Matrix Membranes (MMMs)

A special class of membranes can be synthesized using inorganic particles like alumina, iron-oxide or titanium nanoparticles in a polymeric matrix. Such membranes behave as a hybrid unit of adsorption and filtration in situ. They possess more open pores and requires much less power consumption, unlike RO and NF membranes (Jun & Deng, 2015).

Primarily they are used in gas separation by impregnating zeolites in polymers. They also find utility in fuel cells, lithium ion (Li-ion) batteries, sensors, pervaporation and organic solvent nanofiltration (OSN) (Aroon et al., 2010). Doping inorganic additives not only improves the selectivity of the membrane but also improves permeability, hydrophilicity and surface roughness for better performance. In the case of water treatment, they are suitable for the removal of ionic contaminants like fluoride (Chatterjee & De, 2014), arsenic (Chatterjee & De, 2017), heavy metals (Mukherjee et al., 2016), nitrate (Mukherjee & De, 2014) and organics like phenols (Mukherjee & De, 2014), dyes (Daraei et al., 2013) and other dissolved particulates (Panda et al., 2015).

4.2 Synthesis and Fabrication of MMMs

Organic-inorganic mixed matrix membranes (MMMs) can be synthesized by dissolving polymers like cellulose acetate phthalate (CAP), polysulfone (PSF), polyethersulfone (PES) or polyacrylonitrile (PAN) in an organic solvent like dimethyl formamide (DMF) or dimethyl acetamide (DMAC) (Yin and Deng, 2015). The choice of the polymer entirely depends upon its thermal and mechanical strength. Other determining factors are biocompatibility, biodegradability, hydrophilicity and the antifouling nature of the membrane (Singh & Hankins, 2016). The chemical structure of the polymer plays a significant role in determining various properties of the synthesized membrane. For example, nitrile groups present in PAN polymers impart a negative charge on the membrane surface thereby enabling repulsion assisted separation of anions (Wang et al., 2007). Chemical treatment of the base polymer sometimes increases its selectivity. For example, arsenic removal efficiency of the PAN membrane is magnified when it is treated with sodium hydroxide (Lohokare et al., 2008). Their properties are also improved by insitu-polymerization or by cross-linking. For example, a graft polymerization PAN membranes surface is performed by immersion in a benzophenone-methanol solution, followed by a soaking in acrylic acid under ultraviolet (UV) radiation (Nouzaki et al., 2002). This significantly improves water flux and dextran rejection capability of the membrane. Similarly, montmorillonite impregnated-PVDF nanocomposite UF membrane is grafted for enhanced separation proteins (Wang et al., 2012). Following the dissolution of the polymer in organic solvent, the inorganic material can be added to the solution and then allowed to homogenize for 16–18 h. Organic additives like polyvinyl-pyrollidone (PVP) or polyethylene-glycol (PEG) can be added to increase the mechanical strength and the porosity of the membrane (Zheng et al., 2006; Han & Nam, 2002). Membrane performance also depends upon its method of preparation. There are four different types of preparation methods, by which a polymer solution can be cast onto a porous support. The most common method of preparation is non-solvent (like water) induced phase separation of the polymer (Ulbricht, 2006). Membranes are also prepared by evaporation, vapor or the thermally-induced phase separation method. Preparation parameters affect membrane morphology. Increased phase inversion time (36–40 hours), use of organic solvents (acetone or ethanol) and elevated temperature in gelation bath can directly influence membrane performance by controlling its structure. Prepared membranes can be fabricated as flat sheet (FS) or hollow fibers (HF). Conventional preparation method of flat sheet membrane (Norman et al., 2011) and schematic of hollow fiber spinneret (Bhandari et al., 2016) are shown in Figures 4.1 and 4.2, respectively.

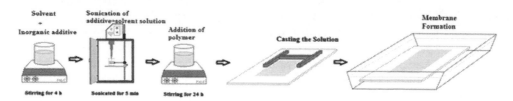

FIGURE 4.1
Schematic representation for preparation of an organic inorganic casting solution and subsequent casting.

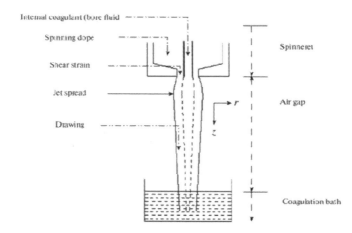

FIGURE 4.2
Conventional Spinneret assembly.

4.3 Application of Organic-Inorganic Mixed Matrix Membranes for Water Purification

4.3.1 Literature Survey

Table 4.1 illustrates the applications of different MMMs. Morphological, mineralogical, surface and inherent membrane properties were improved with addition of these inorganics. These membranes are used in removal of fluoride (Chatterjee & De, 2014), arsenic (Chatterjee & De, 2017), heavy metals (Mukherjee et al., 2016), nitrate salts (Mukherjee & De, 2014), phenol (Mukherjee & De, 2014), dye (Daraei et al., 2013), humic acid (Panda et al., 2015) and protein filtration (Wang et al., 2012). Different MMMs have been synthesized for removal of arsenate (Ganesh et al., 2013). An enhancement of antifouling behavior and flux enhancement with UV radiation has been reported (Zinadini et al., 2014; Vatanpour et al., 2012). PAN-treated laterite and PSF-iron manganese binary oxide (FMBO) particles have been used for removal of arsenate and arsenite species, respectively (Chatterjee & De, 2015; Gohari et al., 2013). Titanium dioxide (TiO_2) and multiwalled carbon nanotubes (MWCNTs) have been impregnated in PVDF and PES membranes, respectively, for improvement in self-cleaning properties and protein removal (Safarpour et al., 2014). Polyethylenimine (PEI) soft nanoparticles in PES polymer matrix (quaternized by bromoethane) have been synthesized for dye removal (Celik et al., 2011). Hydrous ferrous oxide (HFO) impregnated PSF membranes possess lead removal ability (Zhu et al., 2015). Enhancement in hydrophilicity,

TABLE 4.1

Different Mixed Matrix Membrane and Their Utilities in Water Treatment

Polymer	Inorganic	Applications in Water Treatment	Membrane Inherent Properties			Reference
			Permeability/ Pure Water Flux	Porosity	MWCO/ Pore Radius	
CAP	activated alumina	Fluoride removal	PP: 3×10^{-11} m/Pa.s MMM: 1.4×10^{-11} m/Pa.s	PP: 65% MMM: 40%	PP: 110 kDa MMM: 24 kDa	Chatterjee and De (2014)
PSF	iron ore slime	Total Arsenic removal	PP: 6×10^{-11} m/Pa.s MMM: 2.6×10^{-11} m/Pa.s	PP: 75% MMM: 46%	PP: 75 kDa MMM: 45 kDa	Chatterjee and De (2017)
PSF	GO	Heavy metals (cadmium, chromium, lead, copper) removal	PP: 3.3×10^{-11} m/Pa.s MMM: 4.2×10^{-11} m/Pa.s	PP: 49% MMM: 58%	PP: 70 kDa MMM: 150 kDa	Mukherjee et al. (2016)
PAN	Alumina nanoparticles	Nitrate salts	PP: 1.8×10^{-10} m/Pa.s MMM: 2.3×10^{-10} m/Pa.s	Not reported	PP: 79 Å MMM: 100 Å	Mukherjee and De (2014)
CAP	Alumina nanoparticles	Phenol removal	PP: 1.15×10^{-11} m/Pa.s MMM: 1.25×10^{-11} m/Pa.s	PP: 42% MMM: 53%	PP: 82 kDa MMM: 103 kDa	Mukherjee and De (2014)
PVDF+Chitosan	nanoclay	Dye removal	PP: 599 kg/m²h MMM: 20 kg/m²h	Not reported	Not reported	Daraei et al. (2013)
Chitosan coated PAN	Iron oxide	Humic acid removal	PP: 6×10^{-11} m/Pa.s MMM: 2.86×10^{-11} m/Pa.s	Not reported	PP: 95 kDa MMM: 44 kDa	Panda et al. (2015)
PVDF+PVP	MMT	BSA protein filtration	PP: 18 L/m²h.Bar MMM: 58 L/m²h.Bar	PP: 55.3% MMM: 81%	PP: 14.4 nm MMM: 14.7 nm	Wang et al. (2012)
PSF	GO	Arsenate removal	PP: not reported MMM: 5 L/m²h.Bar	PP: Not reported MMM: 75%	Not reported	Rezaee et al. (2015)
PES	GO	Enhancement of antifouling behavior	PP: 8.2 kg/m²h MMM: 14 kg/m²h	PP: 73% MMM: 79%	PP: 3.2 nm MMM: 3.8 nm	(Zinadini et al. (2014)
PES	TiO₂	Enhancement of flux and antifouling behavior in presence of UV radiation	PP: 3.5 L/m²h MMM: 7.2 L/m²h	PP: 70% MMM: 73%	PP: 2 nm MMM: 3 nm	Vatanpour et al. (2012)

(Continued)

TABLE 4.1 (CONTINUED)

Different Mixed Matrix Membrane and Their Utilities in Water Treatment

Polymer	Inorganic	Applications in Water Treatment	Membrane Inherent Properties			Reference
			Permeability/Pure Water Flux	Porosity	MWCO/Pore Radius	
PAN	Treated laterite	Arsenate removal	PP:10.2 × 10^{-10} m/Pa.s MMM: 3.4 × 10^{-10} m/Pa.s	PP: 65% MMM: 46%	PP: 110 kDa MMM: 48 kDa	Chatterjee and De (2015)
PSF	FMBO	Arsenite removal	PP: 39 L/m²h MMM: 94 L/m²h	PP: 80.5% MMM: 83.9%	Not reported	Gohari et al. (2013)
PVDF	GO/TiO_2	Enhancement of self-cleaning and antibacterial properties	PP: 125 L/m²h MMM: 225 L/m²h	PP: 73% MMM: 83.1%	PP: 55.1 nm MMM: 72.3 nm	Safarpour et al. (2014)
PES	MWCNT's	Protein removal	PP: 12 L/m²h MMM: 58 L/m²h	PP: 42% MMM: 38%	PP: 26.5 kDa MMM: 24.7 kDa	Celik et al. (2011)
PES	PEI	Dye removal	PP: 12 L/m²h.Mpa MMM: 112.5 L/m²h.Bar	Not reported	PP: 482 Da MMM: 781 Da	Zhu et al. (2015)
PSF	HFO	Lead removal	PP: 246 L/m²h.Bar MMM: 942 L/m²h.Bar	PP: 48.7% MMM: 88.8%	PP: 166.8 nm MMM: 77.2 nm	Abdullah et al. (2016)
PSF	GO	Enhanced hydrophilicity and salt rejection	PP: 19.7 L/m²h MMM: 46.4 L/m²h	PP: 48.3% MMM: 82.1%	PP: 6.9 nm MMM: 8.7 nm	Rezaee et al. (2015)
PES	Boehmite nanofillers	Enhancement of antifouling nature	PP: 3.9 kg/m²h MMM: 4.1 kg/m²h	PP: 68% MMM: 38.1%	PP: 2.36 nm MMM: 2.41 nm	Vatanpour et al. (2012)
CA	Silver nanoparticles	Enhancement in antibacterial properties	PP: 5.6 L/m²h Atm MMM: 6.1 L/m²h Atm	Not reported	Not reported	Chou et al. (2005)
PES	TiO_2	Coupling inorganics with UV radiation to decrease membrane fouling	PP: 35 kg/m²h MMM: 30 kg/m²h	Not reported	Not reported	Rahimpour et al. (2008)
PVDF	TiO_2	Self-cleaning, antibacterial and photo-catalytic activity	PP: 5.2 L/m²h. kPa MMM: 3.25 L/m²h. kPa	Not reported	PP: 0.5 µm MMM: 0.4 µm	Damodar et al. (2009)
PVDF	Hydrous zirconia	Arsenate removal	PP: 84.3 L/m²h MMM: 177.6 L/m²h	PP: 86.8% MMM: 88.9%	Not reported	Zheng et al. (2011)

**PP- Pure polymeric; MMM- Mixed matrix membrane with highest additive concentration.

antifouling nature and antibacterial properties have been observed by using graphene oxide (GO)-PSF, Boehmite nanofillers-PES and silver nanoparticles-CA membranes, respectively (Rezaee et al., 2015; Vatanpour et al., 2012; Chou et al., 2005). TiO_2 in PES and PVDF matrix reduces fouling tendencies in the presence of UV radiation (Rahimpour et al., 2008; Damodar et al., 2009). The application of the hydrous zirconia-PVDF membrane in water filtration has also been explored (Zheng et al., 2011). The effect of additives in the properties of pure polymeric and MMM has also been shown in Table 4.1. Detailed discussions are made in the subsequent sections.

4.3.2 Morphological, Mineralogical and Surface Roughness Variation

Two types of structures, i.e., micropore network and macrovoids are observed along the cross-section of a membrane. The formation of these structures is proportional to the solvent-non-solvent exchange rate (demixing) across the membrane interface. Micropore network with denser structures is formed by delayed demixing, while macrovoids result from instantaneous demixing (Chatterjee & De, 2014). Inorganic additives are hydrophilic in nature and they draw in more non-solvent, i.e., water toward the interface. The increased flow of water toward the membrane occurs with a rise in additive concentration, resulting in instantaneous demixing. Hence, formation of macrovoids occurs with the increasing concentration of additives in the casting solution. Additionally, the viscosity of the casting solution also determines membrane morphology. Viscosity increases with the addition of these inorganics, hindering the exchange rate between the solvent and non-solvent. This leads to a delayed demixing process, thereby resulting in the formation of a micropore network along the cross-section of the membranes. Higher viscosity and more stable thermodynamic properties would induce delayed phase separation or contribute to the decrease of instantaneous phase separation rate. Formation of macrovoids or micropores often results from a tradeoff between the hydrophilicity of the additive and its rising concentration in the casting solution. The effect of additives on membrane cross-section is shown in Figure 4.3, which depicts the formation of macrovoids when iron ore slime is added to PSF solution (Chatterjee & De, 2017). The size of the macrovoids increases with the concentration of the additive in the casting solution. In this case, the faster exchange rate due to the addition of hydrophilic additive dominates the effect caused by increased viscosity, resulting in the formation of macrovoids. A similar phenomenon is observed when alumina is added to CAP polymer (Chatterjee & De, 2014). But, increasing viscosity dominates during the laterite addition to PAN MMM, resulting in the formation of microstructures (Chatterjee & De, 2015). Sometimes macrovoids become narrow when GO is added in the PSF and PES matrix (Ganesh et al., 2013; Zinadini et al., 2014). They also become elongated and reduced in size, as in the case of the addition of TiO_2 nanoparticles in the PES matrix (Vatanpour et al., 2012). The membrane top surface shows a clear morphological difference due to the addition of inorganics. For example, pores at the top surface get blocked with an increasing concentration of alumina in the CAP membrane (Chatterjee & De, 2014). In another instance, pore density increases when FMBO particles are added to the PSF membrane, though the pore size gets smaller (Gohari et al., 2013).

Mineralogical phase changes also occur from the incorporation of inorganics, as observed from Figure 4.3. Peaks corresponding to the polymer and inorganic components are relevant in the diffraction pattern of the MMM. For example, peaks corresponding to PSF and PVP are present in the pure polymeric membrane (Chatterjee & De, 2017). Peaks corresponding to goethite, kaolinite, hematite, ferrihydrite and silica are present in the inorganic additive of treated slime of iron. Prepared MMM shows diffraction peaks

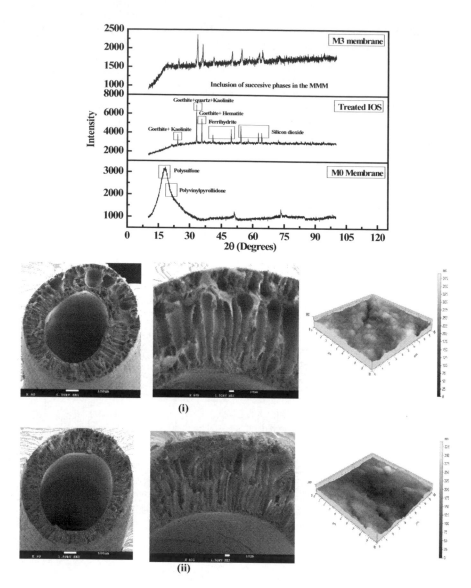

FIGURE 4.3
Morphological, mineralogical and surface roughness variation of a typical treated iron ore slime-PSF mixed matrix membrane. *(Continued)*

corresponding to mineral phases of iron ore slime, PSF and PVP. In some cases, polymer-inorganic interaction shifts the characteristic peaks owing to changes in the crystal structure of the additive caused by the polymer substrate during membrane formation, e.g., shift in TiO_2 peaks due to presence of PVDF in the matrix (Safarpour et al., 2014). Surface roughness is considerably increased by the addition of these inorganics, which affects the roughness factor (Gohari et al., 2013). The agglomeration or even the distribution of the fillers on the membrane's surface may be the reason for this increase. For example, the addition of oxidized MWCNT's in PES nanocomposite membrane increases surface roughness due to agglomeration of carbon nanotubes (Celik et al., 2011). Some researchers pointed out that nodules of polymeric spheres are formed on the membrane surface during phase

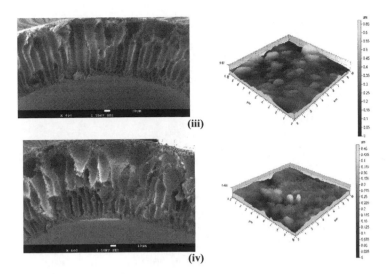

(iii)

(iv)

FIGURE 4.3 (CONTINUED)
Morphological, mineralogical and surface roughness variation of a typical treated iron ore slime-PSF mixed matrix membrane.

inversion, resulting in increases in roughness and porosity. The increase in surface roughness improves the uptake capacity of the contaminants on the MMM surface. For example, the surface roughness of the prepared iron ore slime-PSF MMM increased from 40 nm to 97 nm (Chatterjee & De, 2017). Due to this, the arsenic removal percentage increased from 15% to 90% (Chatterjee & De, 2017).

4.3.3 Improvement in Membrane Inherent Properties

Critical membrane properties include permeability, porosity, molecular weight cut-off (MWCO) and hydrophilicity. Permeability of a membrane is the measure of its throughput and calculated as,

$$J_w = L_p(\Delta P - \Delta \pi) \tag{4.1}$$

where, J_w is the pure water flux, L_p is the membrane permeability, ΔP and $\Delta \pi$ are the transmembrane pressure drop and osmotic pressure drop, respectively. It is observed from Table 4.1 that the permeability is improved with the addition of inorganics in the polymeric matrix. For example, with the addition of alumina nanoparticles, the permeability increased from 1.8×10^{-11} m/Pa.s to 2.3×10^{-11} m/Pa.s (Mukherjee & De, 2014). For HFO-PSF MMM, the permeability increased almost four times from 246 L/m²h.Bar to 942 L/m²h.Bar (Abdullah et al., 2016). The increase in permeability is due to (a) an increase in pore density and (b) an increase in the pore size, though the number density remains same with respect to pure polymeric membrane. However, in some cases, the permeability is decreased due to blockage of pores. For example, the permeability of alumina-CAP MMM, the permeability decreased from 3×10^{-11} m/Pa.s to 1.4×10^{-11} m/Pa.s (Chatterjee & De, 2014). Similar behavior is noticed in the case of iron ore slime-PSF, nanoclay-chitosan coated PVDF, laterite-PAN MMM (Chatterjee & De, 2017; Daraei et al., 2013; Chatterjee & De, 2015).

Another important feature of the membrane performance is determined by its MWCO, which also directly signifies the pore size of the membrane (Bodzek et al., 2011; De, 2010). MWCO is measured by the calculation of rejection for different molecular weight PEGs as,

$$R = \left(1 - \frac{C_p}{C_0}\right)\%$$ (4.2)

where, R is the rejection percentage, C_p and C_o are the concentrations of solute in permeate and feed, respectively. As the cut-off increases, pore size also increases, resulting in higher permeability. For example, the addition of GO nanoparticles increases the cut-off for PSF from 70 kDa to 150 kDa (Mukherjee et al., 2016). Another example includes in-situ synthesized polyethylenimine (PEI) soft nanoparticles in PES polymer matrix quaternized by bromoethane, where the MWCO increases from 482 Da to 781 kDa (Zhu et al., 2015). MWCO of the inorganic-organic MMMs generally increases due to combined effect caused due to free volume increment and reduction in polymer chain segmental mobility, which is caused by inorganic material present in the casting solution (Wara et al., 1995). However, the opposite behavior was observed in specific cases where the cut-off decreased due to an inorganic addition in the polymer matrix. For example, the addition of iron oxide in a chitosan-coated PAN membrane decreased the cut-off from 95 kDa to 44 kDa (Panda et al., 2015). Similar behavior is observed in the case of an alumina addition to the CAP mixed matrix membrane, where the cut-off decreases significantly from 110 kDa to 24 kDa (Chatterjee & De, 2014). As discussed above, this is due to blockage in the pores by the inorganic additive. There are some other instances where there is no particular trend observed in this case. For example, when nano-alumina is added to cellulose acetate phthalate polymer, the cut-off increases from 82 kDa to 122 kDa until a certain weight percentage (20 w/w%) (Mukherjee & De, 2014). However, upon a further increase in additive concentration to 25%, membrane cut-off decreases to 103 kDa. This is due to the agglomeration of the inorganic additives after a threshold concentration, resulting in a blockage of pores, thereby decreasing the MWCO (Mukherjee & De, 2014). But, when these nanoparticles are added to a PAN polymer, a steady increase in pore size from 79 Å to 100 Å occurs (Mukherjee & De, 2014). The arrangements of the polymer chain and its mobility might affect the pattern of the MWCO (Wara et al., 1995).

Porosity represents the number of pores in a unit cross sectional area and is measured as,

$$\varepsilon = \frac{w_0 - w_l}{\rho_w Al} \times 100$$ (4.3)

where, ε is the porosity of the membrane samples, w_o and w_l are the weights of the wet and dry membrane, respectively, ρ_w is the density of water, and A and l are the area and thickness, respectively. Permeability and porosity are interconnected parameters. For a membrane exhibiting high flux, the porosity is generally high. For example, an increase in porosity and permeability occurs simultaneously for the FMBO-embedded PES membrane (Gohari et al., 2013). In this case, the pure water flux increases from 39 L/m²h to 94 L/m²h for the MMM containing the highest additive percentage. Porosity also increases from 81% to 84% validating the rise in permeability. In the case of HFO-PSF MMM, the porosity increases from 49% to 89% (Abdullah et al., 2016). Porosity and permeability decrease with the additive concentration, at various instances. Such cases are reported for CAP-alumina, iron ore slime-PSF and laterite-PAN MMM (Chatterjee & De, 2014; Chatterjee & De, 2017; Chatterjee & De, 2015). For example, it is observed from Table 4.1, that the porosity decreases from 65% to 46% in the case of the PAN-laterite membrane. However, an anomaly is observed in such cases, also. In some cases, though the porosity increases, the permeability does not increase. For example, the porosity of the membrane increases when graphene oxide is added to polysulfone membrane, i.e., from 49% to 58%. But, the permeability is reduced from 3.3×10^{-11} m/s.Pa to 1.6×10^{-11} m/s.Pa, due to greater thickness of the top layer until a certain weight percentage (0.1 weight %), and then starts increasing with additive concentration at 0.5 weight%, to 4.2×10^{-11} m/s.Pa (Mukherjee et al., 2016). This might happen due to the rapid exchange rate of a nonsolvent, i.e., water from the gelation bath with the organic solvent in the membrane matrix.

Hydrophilicity is an important property of the membrane that determines its water uptake capacity. Water flux is directly a function of hydrophilicity of the membrane. It is measured by the contact angle of the membrane. A higher contact angle indicates a hydrophobic character (75°–80°), while a lower contact angle represents a hydrophilic character (45°–60°). A decrease in contact angle represents an increase in the hydrophilic character of the membrane. Hydrophilicity of the membrane is improved upon the impregnation of these inorganic particles. For example, contact angle decreases from 65° to 44° with the addition of titanium dioxide nanoparticles in the PES matrix (Vatanpour et al., 2012). An increase in hydrophilicity occurs from a higher exchange rate during phase inversion.

4.3.4 Improvement in Specific Rejection Capability of Different Ions

The addition of inorganic particles significantly increases the uptake efficiency of different ionic contaminants such as fluoride, arsenic, toxic metals, like, lead, cadmium, copper and chromium, dyes and organic solutes, which otherwise pollute the groundwater stream. The surface charge of the membrane and operating pH play a crucial role in determining its utility in the removal of cationic or anionic contaminant. An important parameter to judge in this case is isoelectric pH or pH at the point of zero charge (pH_{ZPC}). If the operating pH is higher than the isoelectric pH, the surface becomes negatively charged and it can easily adsorb positive charge bound contaminants like heavy metals and some cationic dyes. For example, the removal of lead by an HFO embedded PES membrane is found to be maximum at pH 8, which is higher than the isoelectric point of the membrane (Abdullah et al., 2016). However, the surface becomes positively charged as the operating pH becomes lower than the isoelectric pH (Chatterjee & De, 2014). This phenomenon is highly helpful in the removal of anions like fluoride, arsenic and some anionic dyes. For example, the isoelectric pH of the activated alumina-embedded cellulose acetate phthalate mixed matrix membrane is 8.6 (Chatterjee & De, 2014). Fluoride is an anionic contaminant that causes permanent deformation of skeletal and dental muscles, is removed at pH 7 by this membrane when the pH is lesser than the isoelectric point. Arsenic is also a hazardous carcinogenic contaminant and is found to be removed to the maximum extent at a neutral pH, which is lesser than the isoelectric point ($pH_{ZPC} = 7.9$) for laterite-PAN membrane (Chatterjee & De, 2015). However, the speciation of the contaminants also plays an important role in determining the removal percentage at different pHs. For example, most of the metallic hydroxides precipitate at basic pH. Therefore, the removal of heavy metals at basic pH is not always due to the attraction assisted adsorption phenomenon. However, most of the industrial effluents containing heavy metals fall in the acidic pH range. Mechanisms of solute separation by MMMs are: (i) specific adsorption of small sized ionic species like arsenic, fluoride, heavy metals, etc., facilitated by electrostatic attraction, (ii) charge-charge repulsion by the Donnan exclusion principle. For example, the removal of arsenate by the GO-PSF membrane takes place by charge-charge repulsion (Rezaee et al., 2015), and (iii) size exclusion of larger sized contaminants. This is particularly observed in case of filtration of suspended solids by MF membranes (Bodzek et al., 2011).

Along with the improvement in impurity rejection, the antifouling ability of these membranes is increased several folds, along with improvement in photocatalytic and antimicrobial activity. For example, when boehmite nano-fillers are added to PES membrane, flux recovery ratio increases from 58% to 96% up to 1% the weight of the fillers (Vatanpour et al., 2012). Similar behavior is noticed when GO is added to the PSF membrane (Ganesh et al., 2013). It is important to note in this study that the reversible fouling, which can be reduced by backwashing, is enhanced when irreversible fouling decreases upon the addition of these nanoparticles. Membranes embedded with inorganics showing photocatalytic activity, respond to ultraviolet radiation. For example, the flux is 21 kg/m²h after a time of 240 min for 4 wt% TiO_2 embedded PES membrane. Post-UV irradiation, the flux increases to 23 kg/m²h for the membrane for the same time interval (Rahimpour et al., 2008). UV radiation assisted cleaning process is also observed in the TiO_2-embedded PVDF membrane (Damodar et al., 2009). Membrane biofouling is another phenomenon that destroys membrane performance. Removal of pathogens and bacteria from water is an important step in the purification of drinking water. It is mostly inferred that upon the addition of silver nanoparticles, the antimicrobial activity shoots up several times, thereby helping in removal of pathogens and bacteria from drinking water. For example, when TiO_2 is added to PVDF membrane, the complete removal of pathogens (100%) occurred within 1 minute, whereas the removal of pathogens by the pure PVDF membrane was only 42%. The bacterial functioning of the membrane is due to reactive oxygen species generated during photo-catalysis of the additives in the presence of UV radiation (Damodar et al., 2009). Similar behaviour is observed when silver nanoparticles are embedded in a cellulose acetate (CA) membrane for antibacterial properties (Chou et al., 2005). However, silver nanoparticles are expensive and toxic in nature. Therefore, measures should be taken to arrest their leaching from the membrane surface to the permeate stream.

4.4 Cost Analysis of Mixed Matrix Membrane Processes

MMMs are UF membranes with comparatively open pores with respect to NF or RO membranes. Removal of ionic contaminants occurs through electrostatic interaction assisted adsorption on the surface of the inorganics. The interaction time between pollutant and inorganics on the membrane surface should be low in order to maximize removal performance, since adsorption is an equilibrium governed process. This explains maximum removal at a low pressure and flow rate. Therefore, the energy requirement for MMM is lower than any conventional membrane techniques and is given as (Bansal, 2005),

$$E = \frac{Q \times \Delta P}{\eta \times J_w} \tag{4.4}$$

where, Q are the cross-flow rates (m^3/s), ΔP is the transmembrane pressure drop (Pa), η is pump efficiency, and J_w is the permeate flux (m/s). A typical example shows that the organic-inorganic membrane made of hydrous zirconia nanoparticles embedded in the PVDF matrix membrane (Zheng et al., 2011) produced for filtration of arsenic and other water related contaminants consumes 15 Watt hours per unit area of the membrane (m^2) and per unit volume of water (m^3) produced. However, the energy required for a reverse osmosis plant for desalination is 4 Kwh/m^3 at a flow rate of 10 L/m^2h. Therefore, the energy required in this case is 1.4×10^6 Kwh per unit area of the membrane (m^2) and per unit volume of water (m^3) produced (Busch et al., 2009). Hence, the magnitude of energy requirement is high for a typical reverse osmosis plant. The cost of preparing such MMMs is quite low compared to conventional NF or RO membranes. For example, the cost of preparation of a laterite embedded PAN UF membrane is US\$ 20.00 per unit area (m^2). However, the cost of a typical RO membrane is US\$ 40.00 per unit area (m^2) (Zhu et al., 2009). This reveals the cost effectiveness of a prepared organic-inorganic membrane for water filtration.

4.5 Conclusions

Organic-inorganic membranes are a special class of membranes that have an organic polymer in which an inorganic particle is doped to modify its various characteristics and improve its productivity and selectivity. This type of membrane can be cast as a flat sheet module or hollow fiber module, depending on the type of application. Various properties like water flux, hydrophilicity, porosity is highly improved upon impregnation of such inorganic additives. Morphological analysis shows that the macrovoids gradually decrease in their thickness and the surface roughness of the membrane increases with their concentration in the polymeric solution. This typically helps in the surface adsorption of certain ions which otherwise pollute the groundwater stream. The pore size and molecular weight cut-off of the membrane decrease with concentration due to the restriction of free motion of polymeric chains by these particles. In some cases, the pore size and molecular weight cut-off may also increase, owing to increase in the mobility of polymer segments due to the addition of some specific additives in the membrane matrix. Removal efficiency of these membranes varies significantly with solution pH, as it controls the nature of surface charge. A wide variety of ionic contaminants such as arsenic, fluoride, nitrate, heavy metals and various charged dyes are removed by these special classes of membranes. Along with the removal of various ions, improvement in fouling, antimicrobial and photocatalytic activities are also observed. They are also economical and operate at a much lower energy consumption compared to conventional reverse osmosis membrane desalination.

Nomenclature

A	membrane area (m^2)
C_0	feed concentration of gelling solutes (mg/L)
C_p	permeate concentration of gelling solute (mg/L)
J_w	permeate flux (m/s)

L	membrane sample length (m)
L_p	membrane permeability (m/s.Pa)
Q	cross flow rate (m³/s)
R	rejection percentage (dimensionless)
w_0	weight of the wet membrane (g)
w_L	weight of dry membrane (g)

Greek Symbols

ΔP	transmembrane pressure drop (Pa)
$\Delta \pi$	osmotic pressure drop (Pa)
ε	porosity
ρ_w	density of water (kg/m³)
η	pump efficiency (dimensionless)

References

Abdullah, N., R.J. Gohari, N. Yusof, A.F. Ismail, J. Juhana, W.J. Lau, and T. Matsuura. 2016. Polysulfone/hydrous ferric oxide ultrafiltration mixed matrix membrane: Preparation, characterization and its adsorptive removal of lead (II) from aqueous solution. *Chemical Engineering Journal* 289: 28–37.

Aroon, M.A., A.F. Ismail, T. Matsuura, and M.M. Montazer-Rahmati. 2010. Performance studies of mixed matrix membranes for gas separation: A review. *Separation and Purification Technology* 75(3): 229–242.

Bansal, R.K. 2005. *A textbook of fluid mechanics*. Firewall Media.

Bhandari, D.A., P.J. McCloskey, P.E. Howson, K.J. Narang, and W. Koros. Hollow fiber membranes and methods for forming same. U.S. Patent 9,289,730, issued March 22, 2016.

Bodzek, M. and K. Konieczny. 2011. Application of membrane techniques in the removal of inorganic impurities from water environment-state of art. *Inżynieria Ekologiczna*: 18–36.

Bodzek, M., K. Konieczny, and A. Kwiecińska. 2011. Application of membrane processes in drinking water treatment–state of art. Desalination and Water Treatment 35(1–3): 164–184.

Bryjak, M., J. Wolska, and N. Kabay. 2008. Removal of boron from seawater by adsorption–membrane hybrid process: Implementation and challenges. *Desalination* 223(1–3): 57–62.

Busch, M., R. Chu, U. Kolbe, Q. Meng, and S. Li. 2009. Ultrafiltration pretreatment to reverse osmosis for seawater desalination—Three years field experience in the Wangtan Datang power plant. *Desalination and Water Treatment* 10(1–3): 1–20.

Celik, E., L. Liu, and H. Choi. 2011. Protein fouling behavior of carbon nanotube/polyethersulfone composite membranes during water filtration. *Water Research* 45(16): 5287–5294.

Chatterjee, S. and S. De. 2014. Adsorptive removal of fluoride by activated alumina doped cellulose acetate phthalate (CAP) mixed matrix membrane. *Separation and Purification Technology* 125: 223–238.

Chatterjee, S. and S. De. 2015. Adsorptive removal of arsenic from groundwater using a novel high flux polyacrylonitrile (PAN)–laterite mixed matrix ultrafiltration membrane. *Environmental Science: Water Research & Technology* 1(2): 227–243.

Chatterjee, S. and S. De. 2017. Adsorptive removal of arsenic from groundwater using chemically treated iron ore slime incorporated mixed matrix hollow fiber membrane. *Separation and Purification Technology* 179: 357–368.

Chou, W.-L., D.-G. Yu, and M.-C. Yang. 2005. The preparation and characterization of silver–loading cellulose acetate hollow fiber membrane for water treatment. *Polymers for Advanced Technologies* 16(8): 600–607.

Damodar, R.A., S.-J. You, and H.-H. Chou. 2009. Study the self cleaning, antibacterial and photo-catalytic properties of TiO2 entrapped PVDF membranes. *Journal of Hazardous Materials* 172(2): 1321–1328.

Daraei, P., S.S. Madaeni, E. Salehi, N. Ghaemi, H.S. Ghari, M.A. Khadivi, and E. Rostami. 2013. Novel thin film composite membrane fabricated by mixed matrix nanoclay/chitosan on PVDF micro-filtration support: Preparation, characterization and performance in dye removal. *Journal of Membrane Science* 436: 97–108.

De, S. 2010. Novel Separation Processes. Web Book. The National Programme on Technology Enhanced Learning, Ministry of Human Resource & Development, Government of India.

Ganesh, B.M., A.M. Isloor, and A.F. Ismail. 2013. Enhanced hydrophilicity and salt rejection study of graphene oxide-polysulfone mixed matrix membrane. *Desalination* 313: 199–207.

Gohari, R.J., W.J. Lau, T. Matsuura, and A.F. Ismail. 2013. Fabrication and characterization of novel PES/Fe–Mn binary oxide UF mixed matrix membrane for adsorptive removal of as (III) from contaminated water solution. *Separation and Purification Technology* 118: 64–72.

Han, M.-J. and S.-T. Nam. 2002. Thermodynamic and rheological variation in polysulfone solution by PVP and its effect in the preparation of phase inversion membrane. *Journal of Membrane Science* 202(1): 55–61.

Heberer, T. and D. Feldmann. 2008. Removal of pharmaceutical residues from contaminated raw water sources by membrane filtration. *Pharmaceuticals in the Environment* 427–453.

Li, N.N., A.G. Fane, WS Winston Ho, and T. Matsuura, eds. 2011. *Advanced Membrane Technology and Applications*. John Wiley & Sons.

Lohokare, H.R., M.R. Muthu, G.P. Agarwal, and U.K. Kharul. 2008. Effective arsenic removal using polyacrylonitrile-based ultrafiltration (UF) membrane. *Journal of Membrane Science* 320(1): 159–166.

Mukherjee, R. and S. De. 2014. Adsorptive removal of nitrate from aqueous solution by polyacrylonitrile–alumina nanoparticle mixed matrix hollow-fiber membrane. *Journal of Membrane Science* 466: 281–292.

Mukherjee, R. and S. De. 2014. Adsorptive removal of phenolic compounds using cellulose acetate phthalate–alumina nanoparticle mixed matrix membrane. *Journal of Hazardous Materials* 265: 8–19.

Mukherjee, R., P. Bhunia, and S. De. 2016. Impact of graphene oxide on removal of heavy metals using mixed matrix membrane. *Chemical Engineering Journal* 292: 284–297.

Mulder, J. 2012. *Basic Principles of Membrane Technology*. Springer Science & Business Media.

Nath, K. 2017. *Membrane Separation Processes*. PHI Learning Pvt. Ltd.

Nouzaki, K., M. Nagata, J. Arai, Y. Idemoto, N. Koura, H. Yanagishita, H. Negishi, D. Kitamoto, T. Ikegami, and K. Haraya. 2002. Preparation of polyacrylonitrile ultrafiltration membranes for wastewater treatment. *Desalination* 144(1): 53–59.

Panda, S.R., M. Mukherjee, and S. De. 2015. Preparation, characterization and humic acid removal capacity of chitosan coated iron-oxide-polyacrylonitrile mixed matrix membrane. *Journal of Water Process Engineering* 6: 93–104.

Pedersen, A.J., L.M. Ottosen, and A. Villumsen. 2003. Electrodialytic removal of heavy metals from different fly ashes: Influence of heavy metal speciation in the ashes. *Journal of Hazardous Materials* 100(1): 65–78.

Rahimpour, A., S.S. Madaeni, A.H. Taheri, and Y. Mansourpanah. 2008. Coupling TiO2 nanopar-ticles with UV irradiation for modification of polyethersulfone ultrafiltration membranes. *Journal of Membrane Science* 313(1): 158–169.

Rezaee, R., S. Nasseri, A.H. Mahvi, R. Nabizadeh, S. Abbas Mousavi, A. Rashidi, A. Jafari, and S. Nazmara. 2015. Fabrication and characterization of a polysulfone-graphene oxide nanocom-posite membrane for arsenate rejection from water. *Journal of Environmental Health Science and Engineering* 13(1): 61.

Safarpour, M., A. Khataee, and V. Vatanpour. 2014. Preparation of a novel polyvinylidene fluoride (PVDF) ultrafiltration membrane modified with reduced graphene oxide/titanium dioxide (TiO2) nanocomposite with enhanced hydrophilicity and antifouling properties. *Industrial & Engineering Chemistry Research* 53(34): 13370–13382.

Shimokawa, A. 2009. "Desalination plant with Unique Methods in FUKUOKA." Japan-US Governmental Conference on Drinking Water Quality Management and Wastewater Control, Las Vegas.

Singh, R. and N. Hankins, eds. 2016. *Emerging Membrane Technology for Sustainable Water Treatment.* Elsevier.

Taylor, J.S. and M. Wiesner. 2000. "Membranes." *Membrane Processes in Water Quality and Treatment.* (R.D. Letterman, ed.) New York: McGraw-Hill.

Ulbricht, M. 2006. Advanced functional polymer membranes. *Polymer* 47(7): 2217–2262.

Vatanpour, V., S.S. Madaeni, L. Rajabi, S. Zinadini, and A.A. Derakhshan. 2012. Boehmite nanoparticles as a new nanofiller for preparation of antifouling mixed matrix membranes. *Journal of Membrane Science* 401: 132–143.

Vatanpour, V., S.S. Madaeni, A.R. Khataee, E. Salehi, S. Zinadini, and H.A. Monfared. 2012. TiO2 embedded mixed matrix PES nanocomposite membranes: Influence of different sizes and types of nanoparticles on antifouling and performance. *Desalination* 292: 19–29.

Wang, Z.-G., L.-S. Wan, and Z.-K. Xu. 2007. Surface engineerings of polyacrylonitrile-based asymmetric membranes towards biomedical applications: An overview. *Journal of Membrane Science* 304(1): 8–23.

Wang, P., J. Ma, Z. Wang, F. Shi, and Q. Liu. 2012. Enhanced separation performance of PVDF/PVP-g-MMT nanocomposite ultrafiltration membrane based on the NVP-grafted polymerization modification of montmorillonite (MMT). *Langmuir* 28(10): 4776–4786.

Wara, N.M., L.F. Francis, and B.V. Velamakanni. 1995. Addition of alumina to cellulose acetate membranes. *Journal of Membrane Science* 104(1–2): 43–49.

Yin, J. and B. Deng. 2015. Polymer-matrix nanocomposite membranes for water treatment. *Journal of Membrane Science* 479: 256–275.

Yoon, Y., P. Westerhoff, S.A. Snyder, and E.C. Wert. 2006. Nanofiltration and ultrafiltration of endocrine disrupting compounds, pharmaceuticals and personal care products. *Journal of Membrane Science* 270(1): 88–100.

Zheng, Q.-Z., P. Wang, and Y.-N. Yang. 2006. Rheological and thermodynamic variation in polysulfone solution by PEG introduction and its effect on kinetics of membrane formation via phase-inversion process. *Journal of Membrane Science* 279(1): 230–237.

Zheng, Y.-M., S.-W. Zou, K.G.N. Nanayakkara, T. Matsuura, and J.P. Chen. 2011. Adsorptive removal of arsenic from aqueous solution by a PVDF/zirconia blend flat sheet membrane. *Journal of Membrane Science* 374(1): 1–11.

Zhu, A., P.D. Christofides, and Y. Cohen. 2009. On RO membrane and energy costs and associated incentives for future enhancements of membrane permeability. *Journal of Membrane Science* 344(1): 1–5.

Zhu, J., Y. Zhang, M. Tian, and J. Liu. 2015. Fabrication of a mixed matrix membrane with in situ synthesized quaternized polyethylenimine nanoparticles for dye purification and reuse. *ACS Sustainable Chemistry & Engineering* 3(4): 690–701.

Zinadini, S., A.A. Zinatizadeh, M. Rahimi, V. Vatanpour, and H. Zangeneh. 2014. Preparation of a novel antifouling mixed matrix PES membrane by embedding graphene oxide nanoplates. *Journal of Membrane Science* 453: 292–301.

5

Design of Highly Compact and Cost-Effective Water Purification Systems for Promoting Rural and Urban Welfare

B. Govardhan, Y.V.L. Ravikumar, Sankaracharya M. Sutar, and Sundergopal Sridhar

CONTENTS

5.1 Introduction

Safe drinking water is the second most important and essential requirement for sustenance of human life after clean air (oxygen) for breathing. Though access to safe drinking water has improved over the last decade in almost every part of the world, still one third of the world's population lacks access to safe water. Currently, about a billion people around the world habitually drink unhealthy water. In India, thousands of people including children die each year of waterborne diseases. Hence, accessibility to good quality drinking water to all Indians remains a major concern (Nikhil, 2015). Surface or ground water resources are extensively used for the supply of drinking water to urban and rural areas in India (Parmeshwar et al., 2016). However, the raw water available from surface water sources is not suitable for drinking purpose in most cases. This is because surface water intended to be used for drinking can vary manifestly in their organic and inorganic contents. High levels of variation can occur in parameters such as turbidity, total suspended solids (TSS), colloidal silica, total dissolved solids (TDS), color, odor, dissolved oxygen (DO), natural organic matter (NOM), microbial contamination and sometimes COD coming from illegal dumping of industrial and municipal wastewater. On the other hand, ground water bodies, particularly in arid or semi-arid regions, are known to contain harmful ions like fluoride and nitrate or heavy metals such as iron, arsenic, mercury and cadmium, etc., which is attributed to both hydro-geological conditions and human activity. Figure 5.1 illustrates the water scarcity problem in Latur, Maharashtra. Ground water is a key source

FIGURE 5.1
Picture depicting water scarcity and contamination in rural areas of India.

of water for domestic, industrial and agricultural purposes. It is estimated that more than 90% of the rural population uses ground water for domestic purposes (Govardhan et al., 2017). Fluoride concentration in some regions is as high as 15 ppm, predominantly in the Deccan Plateau and Rajasthan (Naaz et al., 2015). Bureau of Indian Standards (BIS) and the World Health Organization (WHO) have set the maximum permissible limit of fluoride in drinking water at 1.5 and 0.5 ppm, respectively (Manoj et al., 2012). The ingestion of excessive fluoride into the human body affects cell membranes and decreases the production of collagen, which leads to bone disorders, Genu-Valgum (knock-knee), dental fluorosis, gastroenteritis, heart problems, paralysis, and even cancer (Barbier et al., 2010). Northeast regions of India are the most affected by iron contamination in ground water, especially Assam (19 ppm) along with Bihar in the north and Chhattisgarh in central India (Roy et al., 2013). Iron contamination of water might lead to a condition known as iron overload which possibly will lead to hemochromatosis (Nadeem & Ronald, 1980), a severe disease that can damage internal organs. The permissible limit of iron in drinking water as per BIS standards is 0.3 ppm (Ramakrishnaiah et al., 2009). Arsenic contamination of water has been another health hazard that affects eight districts of West Bengal and 13 districts of Bihar (Badal et al., 2002). Those drinking arsenic contaminated water are affected by hyper pigmentation and keratosis, physical weakness, anemia, burning sensation of eyes, solid swelling of legs, liver fibrosis, chronic lung disease, gangrene of toes, neuropathy and skin cancer (Prosun et al., 2010). According to WHO, the maximum concentrations of arsenic in safe drinking water is 0.01 ppm (Joinal et al., 2002). In Bihar, groundwater in 13 districts has been found to be contaminated with arsenic of a concentration exceeding 0.05 ppm. Different methods are used for the purification of water containing arsenic, namely, coagulation/filtration, adsorption on iron oxide, ion exchange, reverse osmosis (RO) and electrodialysis (Choong et al., 2007). Orissa is severely afflicted by excess nitrate (> 45 ppm) that causes blue baby syndrome and thyroid cancer (Sunitha et al., 2013). Another source of contamination is the seepage of seawater into ground water resources in coastal areas, which produces brackish water with high salinity (≥ 2,000 ppm) that could cause renal failure by overloading the kidneys, which continuously filter out excess salts of from the body (Subba rao et al., 2005). Srikakulam is one such district in Andhra Pradesh that has exhibited a wide prevalence of chronic kidney disease (CKD) arising from regular consumption of brackish water.

CSIR-IICT has designed highly compact, low-cost water purification systems for producing safe drinking water, especially in remote fluoride affected regions of India. Installation of such compact plants in these areas helps in removing turbidity, TSS and colloidal silica as well as dissolved constituents such as toxic ions, heavy metals, TDS inclusive of excess hardness, etc. and enables disinfection by simply filtering out bacteria, viruses, amoeba, protozoa, and fungi, making the water safe for consumption. Figure 5.2 (a) provides an overview of drinking water purification plants installed by CSIR-IICT in various parts of India. Figure 5.2 (b) and Figure 5.2 (c) show pictures of senior citizens and school-children drinking pure water from the membrane plants installed by CSIR-IICT in villages and schools. Plant capacities have been tailored to suit small, moderate and large populations. Different water purification technologies have been devised for the removal of contaminants from surface, ground and other polluted water bodies (Ritu et al., 2010) with treatment methodology depending upon raw water composition and quality. One of the major processes that can effectively treat fluoride-contaminated water in Nalgonda is RO (Meenakshi et al., 2006). Iron and magnesium can be removed by precipitation followed by separation. Precipitation is achieved by coagulation or flocculation followed by separation using filtration or sedimentation (Bratby, 2006). Among several technology leads available for the removal of fluoride, iron, arsenic and nitrate from ground water, membrane-based

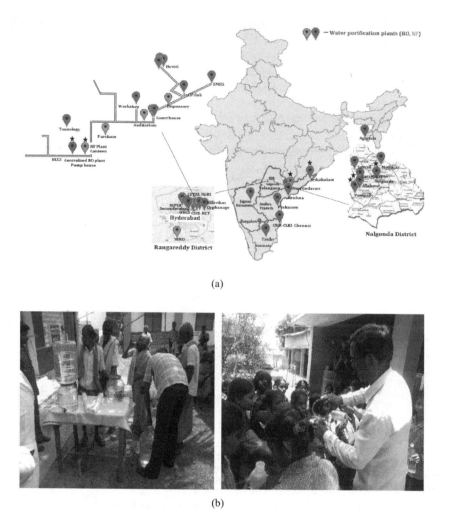

FIGURE 5.2
An overview of drinking water purification plants installed by CSIR-IICT, Hyderabad in various parts of India.
(a) An overview of drinking water purification plants installed by CSIR-IICT, Hyderabad in various parts of India;
(b) Photographs of senior citizens and school children drinking pure water from plants installed by CSIR-IICT.

processes are gaining importance due to their numerous advantages, such as removal of all the impurities in a single step, ease of operation and maintenance, small footprint, low capital investment for both domestic and community based systems and low energy consumption.

5.2 Overview of Water Purification Processes

5.2.1 Chemical Coagulation and Flocculation

Coagulation is a chemical process wherein a specific chemical or natural product is added to the water to destabilize the solubility of the contaminants and form a colloidal suspension or insoluble precipitate (suspended solids). A coagulant (typically a metallic salt) with the opposite charge is added to the water to overcome the repulsive charge and destabilize the suspension. For example, if colloidal particles are negatively charged, alum is added as a coagulant to generate positively charged ions. As a result the repulsive charges are neutralized and the van der Waals forces will cause the particles to cling together and form micro-flocs. It has been reported that pre-hydrolyzed metallic salts are frequently found to be more effective than hydrolyzing metallic salts including aluminum sulphate (alum), ferric chloride and ferric sulphate, which are readily soluble in water. Prehydrolyzed coagulants like polyaluminum chloride (PAC), polyaluminum ferric chloride (PAFC), polyferrous sulphate (PFS) and polyferric chloride (PFC) seem to offer better color removal even at low temperatures and also produce a lesser quantity of sludge. Flocculation is a process wherein colloids settle either due to gravitational settling or from the addition of reagents. Flocculation and sedimentation are widely employed in the purification of drinking water as well as in treatment of sewage, storm-water and industrial wastewater streams.

5.2.2 Adsorption

Adsorption refers to the selective collection and concentration of a particular type of molecule contained in a fluid phase onto a solid surface. Some of the common industrial adsorbents are activated carbon and charcoal, silica gel and alumina, because they present enormous surface area per unit weight (Dinesh et al., 2007). Adhesion occurs between the molecules of liquids, gases and dissolved substances on the surface of the adsorbent material. Some solids have the power to adsorb huge quantities of gases. For example, charcoal adsorbs large volumes of gases because of its high porosity. The adsorption efficiency is based on nature of adsorbent used, water composition and operating parameters. During water filtration through activated carbon, contaminants and small solids adhere to the surface of these carbon granules or become trapped in the small pores of the activated carbon. Adsorption is usually applied for the removal of micro pollutants such as color and odor both in drinking water purification and in waste water treatment prior to disposal.

   For fluoride separation, several adsorbents such as bentonite, activated alumina, amorphous alumina, activated carbon, zeolites, charcoal, fly ash, etc. have been tested in the past at different feed fluoride concentrations. Effective and low-cost materials such as calcite, fly ash brick powder, fishbone charcoal, sunflower plant dry powder, bagasse ash, burnt bone powder, etc. have been investigated for fluoride removal. However, due to lower efficiency or non-applicability on a mass scale, these techniques are not in much use.

5.2.3 Ion Exchange Resins

Ion exchange resins are widely used in separation, purification and decantation processes, particularly in water purification and water softening. Most ion exchange resins are based on polystyrene cross-linked with divinyl benzene. The cross-linking reduces the ion exchange capacity but increases the robustness of the resin. Activated charcoal filter mixed with the resin is commonly used to remove chlorine or even organic contaminants from water (Gary et al., 2003). Broadly speaking, there are four types of ion exchange resins: strong acid containing sulfonic acid groups, strong base (quaternary amino groups), weak acid (carboxylic acid groups) and weak base (primary, secondary or tertiary amino group).

The water is softened by using resins containing sodium ions. As the water passes through the resin it captures, calcium and magnesium ions and releases sodium ions making the water soft.

5.2.4 Disinfection by Chlorine, Ultraviolet Light and Ozonation

Chlorination is the most widely used method for the disinfection of public water distribution systems. Adoption of this method can be attributed to its low cost, convenience of dosing and highly satisfactory performance as a disinfectant. However, there is a limitation in chlorination as Cl_2 can react with naturally occurring organic matter to form disinfectant by-products such as trihalo methanes (THM), which are carcinogens. However, WHO has stated that health risks caused by these by-products are extremely small in comparison to the risks associated with inadequate disinfection (Hung et al., 2012). Excess chlorination can also render a strong odor and poor taste to the potable water.

Owing to safety and taste/odor issues associated with chlorination and advancements in ultraviolet light (UV) technology, the latter has experienced an increase in acceptance in both domestic and community-based systems. The type of UV system depends on the source and quality of water and is generally set at a wavelength of 254 nm (Silvio et al., 2008). Transmitted UV light distribution is affected by water clarity.

Ozone is produced from air using electrical discharge and is a more effective disinfectant than chlorine, due to its rapid action (500 times faster than Cl_2) and high oxidation potential arising from its ability to release nascent oxygen, which immediately deactivates the enzyme of bacteria and other microbes (Rakness, 2011). Ozone is widely used in drinking water purification and can be be added at several points throughout the treatment system, such as pre-oxidation, intermediate oxidation or final disinfection. However, ozone quickly decomposes in water as its life-span in aqueous solutions is quite short (less than one hour).

5.2.5 Membrane Processes

A membrane is a selective barrier that permits the separation of certain species in a fluid by a combination of sorption, sieving and diffusion mechanisms. Membranes can separate particles, molecules, over a wide particle size range, polar/non-polar nature and molecular weights. Membrane processes are classified on the basis of various driving forces such as pressure gradient (microfiltration (MF), ultrafiltration (UF), RO and piezo-dialysis) concentration difference (gas separation, pervaporation, liquid membrane and dialysis), thermal energy (membrane distillation, thermo-osmosis) and electric potential difference (Electrodialysis (ED)). Pressure driven membrane processes such as RO and nanofiltration (NF) are useful for the treatment of brackish, ground and surface water. All the impurities can be removed in a single step including excess total dissolved solids (TDS), salinity, hardness, turbidity, heavy metals and germs. The choice of membrane process is based on raw water quality, as shown in Table 5.1. Brackish water and ground water with high TDS ($\geq$ 700 ppm) can be treated by RO, whereas partial reduction of TDS from surface water (150–350 ppm) and moderately salty ground water (500–700 ppm) can be carried out by NF. Surface water requires only clarification and disinfection for which UF technique proves to be the most viable process due to its having the lowest operating pressure (2–3 bar), highest water recovery (95%) and lowest running cost amongst these three operations. However, NF of surface water can provide a better taste as it reduces

TABLE 5.1

Feed Limiting Conditions for Membrane Systems Used in Drinking Water Purification

Parameters	Reverse Osmosis	Nanofiltration	Ultrafiltration
Feed TDS (ppm)	700–2000	300–700	< 300
Feed pH	5.0–9.0	5.5–8.5	4.5–9.5
Feed Turbidity (NTU)	20	20	20
Silica (SiO_2) in Feed (ppm)	< 10	< 10	< 10
Iron (Fe^{+3}) in Feed (ppm)	< 0.01	< 0.01	< 0.01
Free Chlorine	Nil	Nil	Nil
Feed Flow Rate (LPH)	2000	2000	2000
Permeate Flow Rate (LPH)	1000–1200	1400	1800
Reject Flow Rate (LPH)	800–1000	600	200
Recovery (%)	50–60	70	90
Permeate TDS (ppm)	50–120	100–200	No significant change
Permeate Turbidity (NTU)	0	0	0
Permeate pH	6.5–8.5	6.5–8.5	6.5–8.5
Max Operating Pressure (kg/cm²)	21	18	9
Operating Temperature (°C)	25–30	25–30	25–30
Max Operating Temperature (°C)	40	40	45

the TDS to the 120–200 ppm range. It is worth mentioning that both NF and UF processes would not require remineralisation of the product water, while RO would. MF, on the other hand, can remove bacteria but not viruses, and separation of colloidal silica and turbidity would be incomplete. A process flow diagram of a NF plant for drinking water purification is shown in Figure 5.3 (a). A comparison of membrane filtration process with other water purification methodologies is provided in Table 5.2. A conventional RO system of 1000 L/h capacity was installed at the CSIR-IICT campus with additional pretreatment steps constituting sand filtration for removal of colloidal impurities, activated carbon filter for removal of odor, color and natural organic matter, antiscalant dosing for controlling fouling especially from $CaSO_4$ and micron filter cartridge for separation of fine suspended solids. The membrane assembly is horizontal and parallel to the ground as seen in Figure 5.3 (b) with five 4040 modules (4" dia × 40" long) providing an area of 7.5 m² each. Post-treatment involves UV light and ozonation to ensure long term storage apart from granular activated carbon to enhance taste. This particular design allows product water capacities of up to 20 m³/h and above.

5.3 Nanofiltration

NF is a technique in which membranes are intended to fractionate the components in nanodimensions such as sugars, small organic molecules and multivalent salts from monovalent ions (Cheryan et al., 1995). NF Membranes selectively split multivalent ions (Ca^{+2}, Mg^{+2}, etc.) and permeate monovalent ions (Na^+), thus softening the water for drinking purposes. However, some parts of the essential Ca and Mg salts are still available in permeate for consumption.

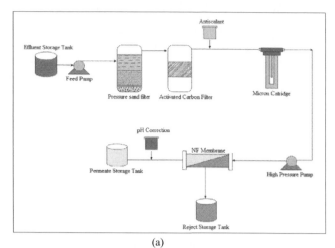

(a)

(b)

FIGURE 5.3
Process flow diagram of Nanofiltration plant. (a) Process flow diagram of Nanofiltration; (b) Photograph of Nanofiltration plant.

TABLE 5.2

Comparison of Membrane Filtration Processes with Other Water Purification Methods

Water Purification Processes	Suspended Solids	Heavy Metals	Oil/ Greases	Sulphates, Phosphates	Nitrates, Fluoride, As (III)	Organics COD,BOD	Micro Organisms	Ions
Coagulation, flocculation and sedimentation,	√	√	√	√				
& Distillation						√	√	√
Adsorption	√					√		
Oxidation					√	√		
Floatation			√					
Biology				√		√		
Ion exchanger		√		√	√	√		√
Membrane Filtration	√	√	√	√	√	√	√	√

Note: √ Applicable, √ Recommended and √ Highly recommended.

5.3.1 Principle and Applications

NF works on the principle of rejection of multivalent ions such as sulphate, phosphate, magnesium and calcium, according to the size and shape of the molecule, as shown in Figure 5.4 (a). The molecular weight cut-off (MWCO) of NF membranes is in the range of 100–1000 Daltons. The pore size of NF is larger than RO and smaller than UF. Applications of NF include removal of total organic carbon (TOC) and reduction of hardness, salinity, TDS and multivalent ions from surface water, ground water and wastewater. NF processes could play a vital role in isolation of valuable solvents and products during the manufacture of pharmaceuticals, fine chemicals, flavors and fragrances.

5.3.2 Nanofiltration of Ground/Surface Water Purification

Ground water contains considerable hardness, nitrates, and natural organic matter (NOM). Different processes are used for the treatment of ground water like coagulation, flocculation, adsorption and ion exchange processes which are less effective and cannot remove pathogens. NF can separate all these contaminants along with micro pollutants such as pesticides, VOCs, viruses, bacteria, salinity, nitrates, and arsenic, etc. These membranes also exhibit high retention of charged particles, especially bivalent ions making this technology suitable to remove hardness. NF is an innovative process having many applications for drinking water purification and wastewater treatment (Devin et al., 2012) and is advantageous over conventional methods such as clarification, coagulation, flocculation, sedimentation and adsorption, etc., as it can remove all impurities in a single step. Charged NF membranes exhibit high retention of bivalent ions of like charges making them suitable for the removal of hardness. NF membranes have very high and precise membrane selectivity, and the integrated process of NF and ED has emerged as the best technology for removal of hardness from ground water. Surface water usually contains impurities such as suspended solids, turbidity, organic matter, bacteria, virus and algae, etc., which cause diseases such as gastroenteritis, cholera, typhoid and jaundice, besides poor water taste and bad odor (Singh et al., 2010). However, the TDS in surface water is usually in the range 150–350 ppm, which is within the standards of potable water. NF is an important low-pressure membrane process which is ideal for treatment of surface water as well as low to moderate TDS ground water due to advantages of process simplicity, lower pressure (less energy consumption), higher recovery with partial TDS removal that provides the water with a good taste and sufficient mineral content. The separation of the components takes place by size exclusion and molecular sieving mechanisms. The permeability of the NF membrane is higher than that of RO, while the extent of removal of germs and heavy metals is similar due to their large molecular sizes. However, rejection of salts of Na, Ca and Mg in NF is marginally lower in NF, which facilitates adequate mineral content in the permeate, which is important for human consumption.

CSIR-IICT has designed and fabricated a highly-compact, low-cost skid mounted, vertical modular NF system to provide 1200 L/h of purified water from ground (moderate TDS) or surface water. This pilot plant removes all impurities, including excess TDS, salinity, hardness, turbidity, heavy metals and microbial content, in a single step. A schematic of the NF plant and the original photograph of the developed NF plant of 1000 L/h installed in a free drinking water camp at Science & Industrial exhibitions and trade fairs in India is shown in Figure 5.4 (b) and (c).

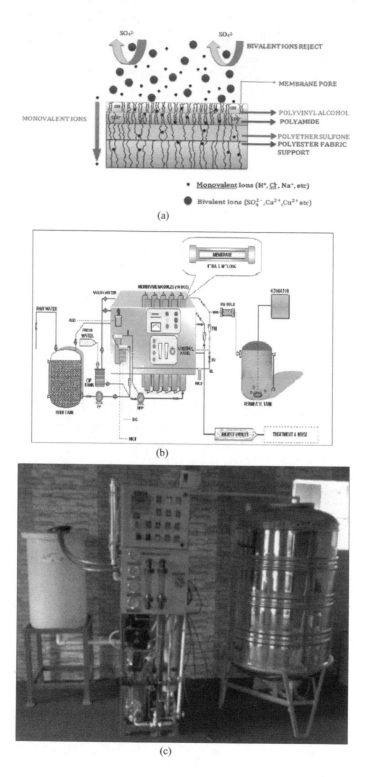

(a)

(b)

(c)

FIGURE 5.4
(a) Nanofiltration principle; (b) Schematic of NF plant of 1000 L/h capacity; (c) Nanofiltration plant of 1000 L/h installed at industrial exhibitions for providing free potable water.

NF provides sufficient minerals required for human consumption (60–150 ppm TDS) including sodium, magnesium, potassium, calcium, phosphorous, whereas RO removes all of the minerals from water and could be harmful in the long run if the water is not remineralised. Highly compact nature of the NF design enables its portability to remote villages where higher water recovery (70–80%) ensures an adequate supply to the entire village population. The reject water is reused for rest rooms, washing, laundry, and non-edible plantations. The design features of this system include: (i) Pretreatment using micron cartridge filter and activated carbon column for removing colloids, suspended particles, color and odor, (ii) NF membrane in vertical modular assembly, (iii) post carbon filter for enhancing the taste of purified water and (iv) UV system for removing any residual bacteria and virus, and (v) ozone treatment, for long-term storage or safe bottling of the water. Figure 5.5 (a) depicts a 1200 L/h highly compact NF system installed at the IICT

(a)

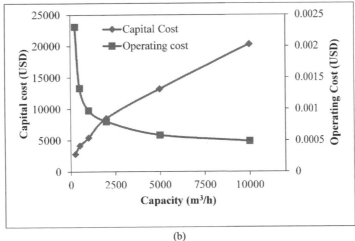

(b)

FIGURE 5.5
(a) Compact Nanofiltration system of 1000 L/h capacity installed at IICT canteen premises to provide purified drinking water for drinking and cooking purposes. (b) Capital and operating cost of 250–10,000 L/h capacities of Nanofiltration systems for drinking water purification systems.

TABLE 5.3

Operating Cost for Nanofiltration Pilot Plant of 5 m³/h Capacity

S.No	Type of Cost	Details	Cost/day (USD)	Cost/yr (USD)
1	Power (6 Rs./unit)	Feed pump – 5 Hp Dosing systems – 0.5 Hp High pressure pump – 10 Hp	21.49	7845.21
2	Membrane replacement	Life: 3 years	2.48	904.81
3	Chemicals	For cleaning, Dosing, Storage, UV	5.59	2041.16
4	Prefilter cartridge Replacement	Activated carbon, Micron filter	0.10	37.74
5	Utilities	Raw water	17.06	6227.70
6	Labour	For operation & maintenance	4.65	1698.46
7	Depreciation cost & interest on capital	20% (10% each)	7.22	2636.89

Note: Duration of operation per day: 20 h; Total purified water generated per year = 36,500,000 L. Cost of purified water generated: 5.86×10^{-4} USD/L; (1 USD = 64.47 INR). If sold at 0.003 USD/L, annual profit = 88108.03 USD; Payback period: 0.24 yr.

canteen premises to provide purified drinking water for drinking and cooking purposes. Detailed operating cost estimation of a 5 m³/h capacity NF pilot plant has been provided in Table 5.3, which includes replacement costs for membranes and prefilter cartridges, electric power consumption, chemicals for cleaning and storage of the membranes, labour, depreciation, interest on capital, etc. The total capital investment for processing 5 m³/h of NF plant is approximately US\$ 13,185 (INR 8.5 Lacs). The duration of operation per day was assumed to be 20 h and membrane life, 3 years. Depreciation and interest on capital reduced 10% each of total capital investment. The cost per liter of permeate produced was found to be Rs. 0.038/- (US\$ 5.86×10^{-4}) with a payback period of 0.24 years (3 months). Capital and operating costs of 250–10000 L/h capacities of NF systems for drinking water purification systems are shown in Figure 5.5 (b). It was found that with increasing capacity of the water purification system, the operating cost decreased, whereas capital cost increases gradually, but not linearly.

5.4 Reverse Osmosis

RO is a physical process in which the dissolved salts are separated by applying pressure on the feed water side that is greater than the natural osmotic pressure to force pure water through a semi-permeable membrane. RO works at a higher transmembrane pressure with more prominent rejection of dissolved solids. The membrane rejects the ions taking into account the size and polar nature. This process is famous for the desalination of sea and brackish water. RO is the long standing technique used in obtaining fresh water required in medical, industrial and domestic applications (Keizer et al., 1995).

5.4.1 Principle and Mechanism of Mass Transfer

The phenomenon of movement of pure water through a semi-permeable membrane from a region of high concentration of solvent (water) to the region of low solvent concentration is defined as "osmosis." The membrane rejects almost all ions and dissolved solids but is permeable to water and some ions. This process (movement of water) occurs until the osmotic equilibrium is reached, i.e., chemical potential is equal on both sides of the membrane. A difference of height is observed between both compartments when the chemical potential is equalized. The difference in height expresses the osmotic pressure difference between the the two solutions. RO is a process which occurs when pressure, greater than the osmotic pressure, is applied to the concentrated solution. Water is forced to flow from the concentrated side to the diluted side, and solutes are retained by the membrane as shown in Figure 5.6 (a).

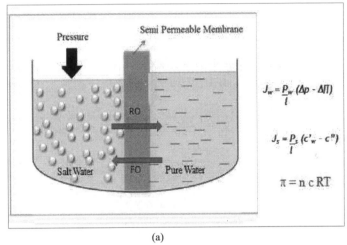

$$J_w = \frac{P_w}{l}(\Delta p - \Delta \Pi)$$

$$J_s = \frac{P_s}{l}(c'_w - c'')$$

$$\pi = n\, c\, RT$$

(a)

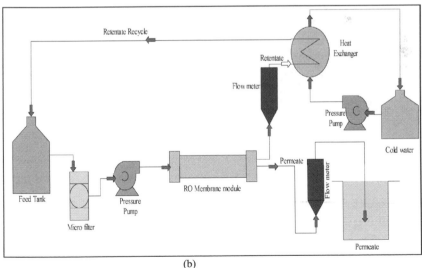

(b)

FIGURE 5.6

(a) Reverse osmosis principle; (b) Flow diagram of pilot RO system.   *(Continued)*

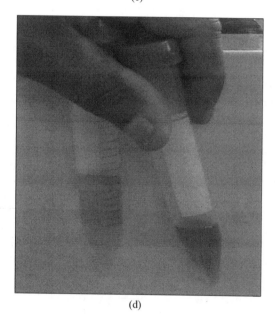

(c)

(d)

FIGURE 5.6 (CONTINUED)
(c) Chemical reaction of fluoride with zirconium complex, (d) Qualitative analysis of fluoride (Test tube to the left shows unacceptable level of fluoride, sample on the right is found acceptable for drinking).

5.4.2 Fluoride Contamination of Ground Water

Intake of a low level of fluoride (less than 0.5 ppm) is good for the body to prevent dental caries but long-term ingestion of fluoride is dangerous, which is responsible for fluorosis and is characterized by staining and pitting of teeth. High-level exposure to fluoride may lead to skeletal fluorosis. Fluoride levels in ground water have been reported to be high in almost 19 states in India, which is alarming. Fluoride contamination of ground water is mainly caused by hydrogeological conditions when ground water comes into contact with rocks rich in fluorite (CaF_2) mineral, leading to dissolution of fluoride up to its solubility limit of 20 ppm. Water used for drinking should not have fluoride in excess of 1.0 mg/L, as per BIS standards, and 0.5 ppm according to WHO. Fluoride-contaminated regions are mostly characterized by the presence of crystalline basement rocks/volcanic bedrocks with the dissolution of F^- promoted by arid/semi-arid climatic conditions, Ca deficient ground water (containing more of $NaHCO_3$), longer ground water residence time and distance from a recharge area (Raj et al., 2017).

5.4.3 Laboratory Experiments on Separation of Fluoride from Drinking Water

A skid mounted pilot RO system of 250 L capacity was connected to a feed tank of maximum 100 L capacity supply of raw water (feed) to the system. Figure 5.6 (b) provides a schematic representation of the pilot scale RO system. An inexpensive micron rope cartridge of polypropylene (PP) with 5 μm pore size is used at the upstream side of the spiral wound membrane module as a prefilter to prevent entry of suspended solid particles capable of damaging the membrane. The cartridge can be replaced once it gets fouled and recycled a couple of times by treating with dilute acid. A low-pressure centrifugal pump of discharge pressure 3.5 bar is provided for pumping the feed to prefilters, whereas a high-pressure pump (Hironisha, Japan) capable of maintaining a pressure of upto 50 bar is installed for pumping the feed coming from prefiltration assembly to the RO module. Both pumps would run in tandem and their flow rates are matched so that there is no starvation of fluid for the second pump. In fact, flooding in the inlet line of the high-pressure pump is desirable. Owing to a pressure gradient developed on the membrane, the contaminants, like excess salts and hardness, heavy metals, harmful ions, bacteria, viruses and pyrogensare, are rejected while pure water diffuses through the membrane as permeate. The retentate is recycled back to the feed tank and a needle valve is placed in the retentate line to read the pressure on the membrane. The feed tank has a provision for recycle of the reject which passes through a glass coil type heat exchanger for maintaining constant temperature (28–30°C). Ice-cold water is circulated through the shell side of the exchanger using a water circulation pump whereas the reject flows through the tube side. Reject coming out of the exchanger is sent to the feed tank as a concentrated stream with total recycle. Permeate and concentrate flow rates are measured using glass rotameters containing metal floats which enable the determination of percentage of recovery by helping in continuous recording of flow rate permeate with respect to that of the feed. Permeate is collected in marked buckets with the help of a polyethylene tube.

5.4.4 Experimental Procedure for Reverse Osmosis System

The thin film composite (TFC) polyamide RO membrane is cleaned and kept moist with deionized water before starting the experiments. The system is initially run to remove 2.2 L of the deionized water present as dead volume in the system. Initially, the feed tank is filled with 100 L of feed water and passed through the micron prefilter cartridge using a low-pressure pump to remove coarse solids. The pretreated water is then pressurized through the spiral wound membrane module using the high-pressure pump and the system pressure is maintained constant by means of the needle valve. The retentate flow rate is maintained at constant value of 10 L/min throughout the experiment to ensure similar hydrodynamic conditions inside the membrane module. The feed pressure varies from 5 to 25 bar. The initial permeate is recycled until a steady permeate flux is obtained. On the other hand, the reject is circulated back to the feed tank through the heat exchanger. Permeate flow is recorded and samples of reject and permeate are collected for the analysis of conductivity, TDS and flux values of feed and permeate. The flux values are also recorded as a function of time until the desired water recovery is achieved.

5.4.5 Equations for Calculation of Operating Parameters

5.4.5.1 Permeate Flux

The permeate volume is measured as a function of time during the separation process. The permeate flux is calculated by dividing the permeate volume by the effective membrane area and sampling time as seen in Eq. (1):

$$Flux\ (J) = \frac{Volume\ of\ permeate\ collected}{Membrane\ area \times Time} \tag{1}$$

5.4.5.2 Rejection Efficiency

This is one of the important parameters by which the separation performance of the membrane can be rated. The performance of the membrane is denoted in terms of percentage rejection of TDS. The percentage of rejection is calculated by Eq. (2):

$$\%\ Rejection = \left(1 - \frac{C_p}{C_f}\right) \times 100 \tag{2}$$

where, C_p = concentration of the solute in permeate, C_f = concentration of the solute in feed.

5.4.5.3 Water Recovery (%)

The percent (%) of water recovery for a single element of membrane system is the ratio of permeate flow rate to feed flow rate as shown in Eq. (3):

$$\%\ Recovery = \frac{Q_p}{Q_f} \times 100 \tag{3}$$

where, Q_p and Q_f are the flow rates of permeate and feed, respectively.

5.5 Analytical Procedures

5.5.1 Fluoride Analysis

The fluoride content in water is analyzed using SPADNS reagent, which is (4,5-Dihydroxy-3-(p-sulfophenylazo)-2,7-naphthalene disulfonic acid, trisodium salt), in a colorimetric method which is based on the reaction between fluoride ion and a zirconium-dye lake. Fluoride in water reacts with the dye lake, dissociating a portion of it into colorless complex anion (ZrF_6^{2-}) and the dye. As the amount of fluoride increases, the sample color progressively decreases. The reaction rate between fluoride and zirconium ions takes place in an acidic condition. The selection of dye for the rapid fluoride method is governed by the resulting tolerance to these ions. SPADNS reagent is a dark red, zirconium-dye complex. The fluoride combines with the zirconium complex as shown in Figure 5.6 (c).

This method is applicable to potable, surface and saline water as well as domestic and industrial wastewater. After the experiment is complete, the sample is treated with SPADNS reagent where the loss of color resulting from the reaction of fluoride with the zirconyl-SPADNS dye is a function of the fluoride concentration. The SPADNS reagent is more tolerant to interfering materials than other accepted fluoride reagents. The addition of the highly colored SPADNS reagent must be done with the utmost accuracy because the fluoride concentration is measured as a difference of absorbance in the blank and the sample. Apart from SPADNS method, qualitative analysis can be carried out using an acidic reagent developed by Nuclear Fuel Complex, Hyderabad. One part the reagent is added to four parts of the water sample, wherein brown coloration indicates the presence of excess fluoride, whereas pink color shows an acceptable concentration for consumption, as shown in Figure 5.6 (d).

5.5.2 Potable Quality Analysis by H$_2$S Vial Method

The H$_2$S vial test is very useful technique to check the presence of bacteria in water. First, the sample water is filled up to a marked line and swirled to dissolve the powder completely followed by incubation at 32–35°C for 24–28 h in a BOD incubator. If the medium shows a brown color, the water is considered fit for drinking. If the medium shows turbidity with bluish-green/purple or black color, the water is not suitable for drinking. Black color with turbidity indicates the presence of *Salmonella* or *Citrobacter* species, bluish-green color of medium turbidity indicates *Escherichia coli*, bluish-purple color with turbidity indicates *Vibrio* species and dark purple color with turbidity indicate the presence of *Klebsiella* species. Figure 5.7 (a) shows a sample which is bluish-green in color indicating presence of *E. coli*, which makes the water unsuitable for drinking purpose.

FIGURE 5.7
(a) H$_2$S Vial Method; (b) E-Coli analysis in feed, permeate and reject of RO process.

5.5.3 Analysis for E. Coli and Total Coliform Bacteria

Total Coliform bacteria (TC) are a group of bacteria that are regularly present in environmental (surface) waters. Fecal coliform (FC) and E. coli are a sub-group of TC that are more associated with feces of humans and warm-blooded animals. The presence of FC or E. coli can indicate contamination of water supplies resulting in an increased risk of waterborne pathogens. Bacterial indicators for TC and E. coli are valuable in assessing the performance of drinking water treatment processes and the integrity of distribution systems. In E. coli analysis, samples which are to be tested are collected in clean vials without any contamination. Required amount of fresh EMB Agar solution is prepared using distilled water and the solution is mixed thoroughly. The solution is kept in autoclave for 60 min for sterilization. The hot agar solution is poured in autoclaved Petri plates and is cooled. After the agar becomes solidified, different samples are taken and streaking is done using inoculation loops. Para film is wrapped around the Petri dish. These samples are kept in an incubator for 24 h for the growth of E. Coli, if present. If E. Coli have grown, then the water sample is not safe for drinking. In Figure 5.7 (b), the plates can be observed for the presence or absence of fluorescence indicative of TC and blue color indicating the presence of E. coli, or actual counts can be made to monitor distribution lines or the treatment method's effectiveness. Furthermore, MI agar is capable of recovering E. coli from water samples containing high particulate concentrations and is less expensive than the liquid media containing chromogens and/or fluorogens.

5.6 Effect of Operating Parameters

5.6.1 Effect of Feed Pressure on Pure Water Flux

The effect of feed pressure on pure water flux for RO processes is shown in Figure 5.8 (a). Deionized water is used as the feed at ambient temperature to study the permeation characterics of the TFC RO polyamide membrane. As expected, a rise in feed pressure results in an increase of flux. With an increase in feed pressure from 5 to 20 bar, increment in flux is observed to be in the range of 18–115 L/m^2h for RO. Since the driving force of the process increases, it results in enhancement of water flux due to increased affinity between H_2O molecules and the polar groups of polyamide RO membrane. The flux is zero at applied pressures less than 5 bar for the effluent due to high osmotic pressure arising from substantial concentration of dissolved solids in the effluent feed. However, significant flux is obtained with pure water feed wherein the trans-membrane pressure gradient is quite high.

5.6.2 Flux and Rejection for Synthetic Fluoride Feed

Experiments were conducted using TFC polyamide spiral wound RO membranes to determine flux and fluoride rejection in permeate. Feed solutions were prepared using CaF_2 and transported at a pressure of 10 bar across the membrane. The TDS of feed sample was 1,250 ppm. The effect of pressure on flux and permeate fluoride concentration is shown in Figure 5.8 (b) for a feed fluoride concentration of 10.6 mg/L prepared by addition of CaF_2 salt in tap water having 1,250 ppm TDS. An increase in flux from 4 to 45 L/m^2h is observed when the pressure is rised from 5 to 20 bar with simultaneous decrease in fluoride ion concentration in permeate from 3.5 to 0.28 mg/L. The observations can be attributed to

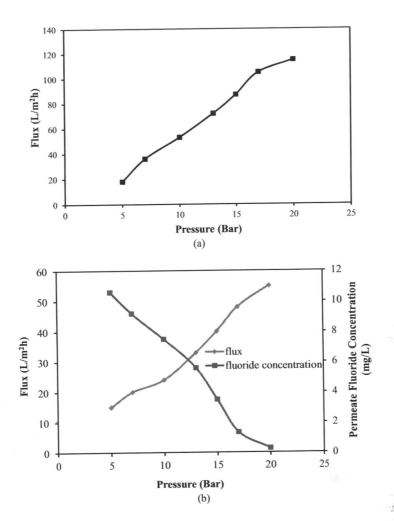

FIGURE 5.8
(a) Effect of feed pressure on pure water flux; (b) Effect of pressure on flux and permeate fluoride concentration.

solution-diffusion mechanism of mass transfer wherein sorption of H_2O molecules rises due to their affinity for the hydrophilic membrane resulting in greater flux. Water molecules are also much smaller than fluoride and other ions present in the feed and are expected to have greater diffusivity resulting in higher water flux and improved fluoride and TDS rejections at higher pressures.

5.7 Defluoridation in Rural Areas

Presence of excess fluoride in ground water is a major problem in most parts of rural and urban areas of India. As per the WHO standards, the recommended fluoride content in drinking water should be less than 1.5 mg/L. The rapid growth in industrialization and urbanization, as well as natural sources like geo-chemical dissolution of fluoride containing minerals, has increased the fluoride concentration in the environment even including surface water. Various districts of Telangana and Andhra Pradesh, especially the villages of Nalgonda, Warangal, Mahboobnagar and Prakasham districts, etc., are severely affected due to their high ground water fluoride concentration, which has reached a maximum of even 15 ppm at certain locations. CSIR-IICT, Hyderabad has put signficant effort in research and the application of RO process for defluoridation of ground water in affected areas of Telangana, A.P. and other states of India. IICT's Membrane group has installed twelve RO-based defluoridation pilot plants of 1000 L/h capacity and twenty-five smaller units of 100 to 250 L/h productivity in and around various remote villages of the four southern states of Telangana, Andhra Pradesh, Karnataka and Tamil Nadu to promote rural welfare by the prevention of fluorosis.

RO technology has major advantages over conventional processes, as it removes all impurities in a single step at a low operating cost and high-production capacity. Details of the plants installed in various places of Telangana and Andhra Pradesh are listed in Table 5.4. CSIR-IICT has implemented these ventures by conducting a survey of villages, detailed water analysis, plant design, procurement of various accessories/equipment, assembling of the system and commissioning. This process is followed by providing hands-on operation of plant training to the villagers, periodic maintenance and troubleshooting, with comprehensive reject treatment and disposal. The first model RO pilot plant of 600 L/h capacity was designed and commissioned at Mylaram Village, Nalgonda District, Telangana in July 2005 and proved to be a model for replication by other organizations, and widely appreciated by press and masses due to the meticulous efforts of IICT in long-term sustenance. Following this, IICT successfully installed, commissioned and demonstrated five defluoridation plants of 1000 L/h capacity each during 2009 to 2010, which were funded by TDT Division of Department of Science and Technology (DST), New Delhi. Three of these plants were installed in remote villages of the Nalgonda District, namely, Rachakonda tribal habitat, Allahpur and Marlapadu, in addition to one plant each in Nerimetta Ashram, Warangal District and Gaggalapally Village, Mahboobnagar District.

TABLE 5.4

Water Analysis before and after Membrane Treatment for Pilot Plants Installed in Telangana and Andhra Pradesh States of India

S. No.	Name of Village	Feed Characteristics				Permeate Parameters			
		Fluoride (mg/L)	TDS (ppm)	Turbidity (NTU)	pH	Fluoride (mg/L)	TDS (ppm)	Turbidity (NTU)	pH
1.	Allahpur	2.46	615	1	8.2	0.23	53	0	7.1
2.	Marlapadu	3.20	512	2	7.9	0.31	46	0	7.3
3.	Gaggalapally	1.80	825	5	8.3	0.15	33	0	6.8
4.	Nerimetta	2.03	622	1	8.4	0.27	45	0	7.1
5.	Rachakonda	3.60	637	6	8.9	0.22	38	0	7.2
6.	Mylaram	5.00	820	3	8.5	0.50	42	0	7.1
7.	Karamchedu	3.20	650	5	8.2	0.20	35	0	7.3
8.	Sethurapatti	2.1	830	3	8.1	0.28	36	0	7.2
9.	Chandrapadu	–	500	0	7.2	–	160	0	7
10.	Srikakulam	–	170	0	8	–	33	0	7.1
11.	Vatipally	4.72	1061	1	6.8	0.12	68	0	6.2
12.	Khammam	0.78	784	0	8.1	0	75	0	7.8
13.	IICT Campus	0	240	0	7.8	0	57	0	7.2

CSIR-IICT oversaw the design and installation of a 4000 L/h RO plant at Karamchedu Village, Prakasam District in 2007 and provided consultancy for the installation of a 1000 L/h RO-based defluoridation system in Chandrapadu Village, Cheemakurti Mandal of Prakasam District, which has been running successfully since July 2011. Subsequently, the first model compact defluoridation plant in Tamil Nadu State of 1000 L/h capacity was installed for the fluoride-affected Sethurapatti Village in Srirangam Constituency with a comprehensive scheme for proper treatment and reuse of the reject water in 2012. Original photographs of some of the installed plants are provided in Figure 5.9 (a, b, c, d). In the year 2016, a 1000 L/h RO plant funded by DBT, India under Inno-Indigo European collaboration, was installed in the worst affected area, Vattipally Village of Marriguda Mandal (5–15 ppm fluoride) in Nalgonda District, along with a smaller plant of 250 L/h at Jigani in Karnataka, funded by the Kewaunee Scientific Corporation. In 2017, two RO plants were installed in Madupalli and Desinenipalem Villages of Khammam Districts, elucidated by the report given in Table 5.5. For the Srikakulam District, affected by chronic kidney disease (CKD), a UF pretreated RO plant was installed for ground water purification at Tippanaputtuga Village for the treatment of brackish water. The cost of the equipment for a 1000 L/h plant is around Rs. 4.0 Lakh (US$ 6,154) at the time of installation. All these plants have been successfully running from past five years. Some of the villages that surround the RO plants also have been provided access to the purified water. The water is given free of cost to the poor and underprivileged and is being distributed equitably to populations that can afford a cost varying from Rs. 2/- to 5/- (US$ 0.054) per 20 L can, to enable self-substainabilty.

(a)   (b)

(c)   (d)

FIGURE 5.9
Reverse osmosis pilot plant installed in (a, b) Khammam District, (c) Vattipally village and (d) Mylaram Village (Editor is seen in two pictures).

TABLE 5.5

Water Analysis of Pilot Plant Installed in Khammam District

S. No.	Parameter	Madupalli Village, Khammam District		Desinenipalem School, Khammam District			Permissible Limits (BIS)
		Feed	Permeate	Feed	Permeate	Reject Water	
1	TDS (mg/L)	784	75	1920	97	2857	50–500
2	pH	8.14	7.85	7.50	7.8	7.70	6.5–8.5
3	Turbidity (NTU)	0	0	1	0	0	< 5
4	Conductivity (mS/cm)	1.75	0.25	4.26	0.20	6.38	0.781
5	Fluoride (mg/L)	0.78	0	2.1	0	3.1	< 1
6	Hardness (mg/L)	300	50	250	45	225	< 250

5.7.1 Methods for Treatment of Reject Stream for Water Recycle for Safe Disposal

The reuse of untreated RO reject water has raised serious environmental concerns. Direct disposal of this water into a water-body may invigorate the growth of aquatic masses causing environmental problems. Disposing on land has the potential of concentrating salts and nutrient compounds in soil and contamination of ground water with excess fluoride. Reject water produced from these processes requires further treatment for removing concentrated fluoride and other salts before its safe disposal. A study was carried out to investigate a process that has potential of removing the salts from the reject wastewater.

Figure 5.10 exhibits a comprehensive method for treatment and reuse of the reject stream coming from a defluoridation plant. The reject can be pumped through an activated

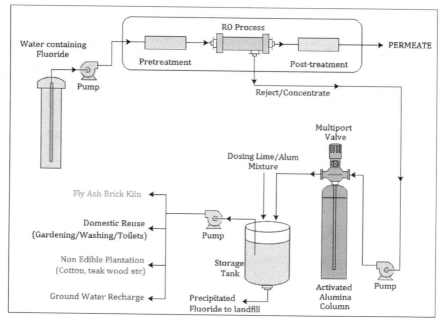

FIGURE 5.10

Method for treatment and reuse of RO reject stream.

alumina column to remove particularly fluoride present in the water. Alternately, the water is stored in a tank/sump where fluoride and other salts present are precipitated by adding lime + alum mixture. The clear water is decanted at regular intervals and solid precipitate is safely disposed to a landfill. The treated water may be reused for domestic purposes like gardening, washing, fly ash brick kilns or for non-edible plantations (like cotton, teak plantation, etc.) and occasionally in ground water recharge.

5.8 Urban Deployments

CSIR-IICT started a free drinking water camp in April 2016, near the NGRI Metro Railway Station on Uppal Road which leads to the Hyderabad airport. This venture was aimed at promoting the welfare of the urban population in peak summer (April–May), when temperatures soar up to 45°C. Three compact RO units were designed and installed at this camp to generate healthy drinking water from a bore water source of 829 ppm TDS content. After RO process, the water contains a TDS of 70–120 ppm, which is ideal for consumption as it provides sufficient minerals. Post-treatment includes UV and Ozonation for disinfection with its taste being enhanced by activated carbon columns. This water is being consumed by pedestrian, bus drivers and passengers, cars, two-wheelers, as well as senior citizens and children at the rate of 3000 L/day. So far, 20 Lakh liters of free drinking water have been served to the community, which corresponds to a consumptive population of 30 Lakh people. A shed with a water cooler has been installed at the outlet of this free drinking water camp, as shown in Figure 5.11, to enable the elderly to sit in the shade and have water in the unbearably hot summer. Based on requests from pedestrians, drivers and commuters of public and private transport, this project has continued beyond summers and has been running successfully for 20 months, and the camp has become a virtual bus stop. Such camps are expected to continue further due to their popularity amongst the urban community amid growing public demand, especially with the recent inauguration of the metro railway service.

FIGURE 5.11
Free drinking water camp of CSIR-IICT under operation since April 2016 at NGRI Metro Railway Station on Uppal Road, Hyderabad to promote welfare of urban population.

5.9 Ultrafiltration for Purification of Surface Water

Ultrafiltration (UF) is a pressure driven membrane separation process, where suspended solids and macro solutes of higher molecular weight will be rejected while low molecular weight micro solutes and water will pass through the membrane of MWCO in the range from 1000 Da to 1100 kDa. The basic separation principle involved in UF is size exclusion. The rejection is determined mainly by the size and shape of solutes relative to the pore size of the membrane and transport of the solvent is proportional to the applied pressure. UF principle is illustrated in Figure 5.12 (a). Schematic of integrated UF + RO process for the plant installed at Tippanaputtuga Village of the Srikakulam District for drinking water purification is depicted in Figure 5.12 (b), whereas the original photograph of the installed plant is shown in Figure 5.12 (c). Though the TDS is low at 170 ppm, some other ingredients such as colloidal silica and heavy metals present in this raw water have wreaked havoc on the local population. The process provides purified drinking water of 53 ppm after remineralization using a solution of rock salt. UF is used as a pretreatment stage where it helps to remove turbidity and other suspended solids from the feed water and its permeate is sent as feed to the RO assembly for the removal of heavy metals and other harmful constituents of purified drinking water. Integration of UF with RO helps to reduce the load of colloidal silica, considered one of the factors responsible for chronic kidney disease (CKD). The recovery is 90% in UF step and subsequently 70% in RO. Both reject streams are combined together for treatment before disposal for irrigation of non-edible crops.

5.9.1 Hollow Fiber Membranes for Surface Water Treatment

The shortage of fresh water resources is a direct consequence of rapid industrial expansion and exponential population growth (Bjorn et al., 2002). Rivers and lakes, which are the principle drinking water sources across the globe, are being severely polluted by the discharge of domestic and industrial effluents. A massive population explosion and rising environmental pollution become two key factors leading to severe water scarcity. The quality of water is influenced by seasonal changes and surface water contains natural organic matter such as humic and fulvic substances, inorganic salts of monovalent and multivalent metal ions besides microbes like bacteria, viruses, protozoa, etc., that can be removed by means of UF. Hollow fiber ultrafiltration membranes are highly practical and cost-effective alternatives to conventional separation processes for treatment of surface water, which has TDS below the permissible limits of 500 ppm. Hollow fiber membranes can clarify and disinfect surface water which proves sufficient. They occupy less space owing to high surface area per unit volume ratio, involve low operating cost due to low operating pressure (< 5 bar) and generate minimum quantity of reject (< 10%) compared to other membrane processes. Feed may flow from outside-in with feed flowing in the shell side or inside out with feed flowing through the lumen of hollow fibers and permeate passing through the walls of the fiber into the housing of the hollow fiber membrane module. Hollow fiber spinning machine, shown in Figure 5.13 (a), creates polymeric fibers using a technique called diffusion induced phase separation method (DIPS). A homogeneous polymer solution (dope) and bore fluid (or inner coagulant) are simultaneously fed into a spinneret, as shown in Figure 5.13 (b), by a precision gear pump and a liquid pump, respectively. The co-extruded solution is immediately immersed in a coagulation bath to induce phase separation and the formation of a fiber. The nascent fiber is pulled out, collected on a roller and washed

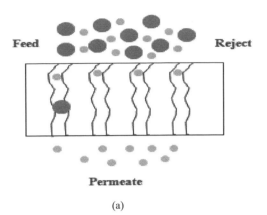

(a)

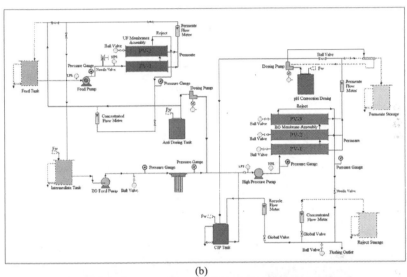

(b)

(c)

FIGURE 5.12
(a) Ultrafiltration principle; (b) Schematic of drinking water purification plant installed at Tippanaputtuga village of Srikakulam district; (c) Original photograph of drinking water purification plant installed in Srikakulam district.

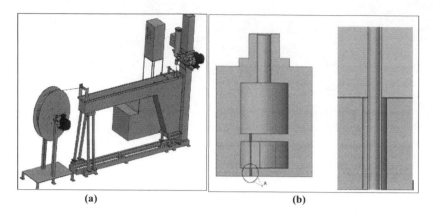

FIGURE 5.13
Schematic of hollow fiber spinning machine. (a) Schematic of manual hollow fiber spinning machine; (b) Design drawing of spinneret.

for complete removal of residual solvents. Additionally, a custom-designed potting station is available for making modules to test the performance of the fibers. Hollow fiber module is designed such that it is highly flexible and easily handles large volumes for circulation, dead-end, and single pass operations. Based on the structural integrity and construction, hollow fiber membranes can withstand permeate back pressure, thus allowing back wash for maintenance and flexibility in system design and operation.

Hollow fiber membrane module consists of thin, fine capillaries collected together and made into a bundle. This bundle is perfectly potted on both the ends to arrest any leakage between feed and permeate chambers and then placed inhousing as shown in Figure 5.14. Depending on the feed and permeate flow patterns, modules can be operated in cocurrent or countercurrent mode. The fibers generally have an outer diameter of 1 mm to 1.5 mm and are arranged parallel to one other to pass through tube-sheets at either one or both ends of the device. Seals are provided between the tube-sheet exterior and the pressure vessel to isolate the high-pressure feed from low-pressure permeates. The bore of the fibers are generally 1 mm in diameter. The important advantage of hollow fiber membrane is the high packing density of 10,000–30,000 m^2/m^3.

5.9.2 Hand Pump Operated Ultrafiltration Membrane Systems

Surface water is a major source of drinking water but is contaminated in several parts of India due to presence of pathogens and turbidity. Electricity is still not available in several regions, especially remote villages. Many states of India are prone to floods. Hence, there is a requirement of an efficient filtration system that can clarify and disinfect the water in the absence of electric power. UF is sufficient in such cases as the water has low TDS but high turbidity and microbial content. On the other hand, RO is not required as it removes most of the essential minerals from the low TDS water. Besides, RO requires electric power to generate high hydrostatic pressure essential for forcing pure water through the membrane. The hand pump-operated purification system is low-cost and operates without electricity. Portable hand pump-operated UF membrane systems are designed for conversion of surface water to potable water. Both submerged and external membrane configurations can be arranged depending upon the requirement. Since UF is a low pressure process, a manually applied differential of 0.5 bar is sufficient in such situations.

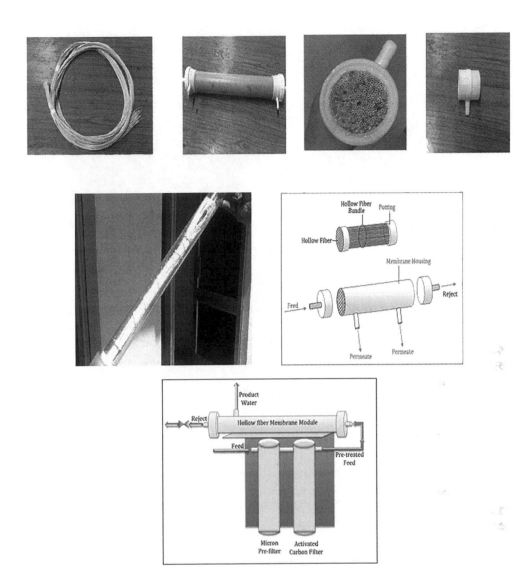

FIGURE 5.14
Representation of various stages of fabrication of hollow fiber membrane system used in surface water treatment.

5.9.2.1 Submerged Membrane Module with Suction Mode of Operation

A schematic view of a hand pump with submerged membrane configuration module is shown in Figure 5.15 (a). Both spiral wound and hollow fiber membrane modules can be arranged in submerged mode to be exposed to feed.

5.9.2.2 External Membrane Module for Positive Hydrostatic Feed Pressure

A configuration with external membrane module is also available for surface water treatment. At the time of operation due to positive hydrostatic feed pressure, feed water enters into the housing (shell side) of the external UF membrane module and final permeate can be collected as purified water, as shown in Figure 5.15 (c). Computational Fluid Dynamics

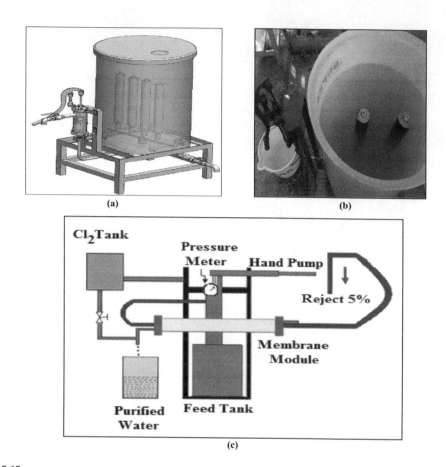

FIGURE 5.15

Hand-operated ultrafiltration membrane system; (a) Schematic of submerged ultrafiltration module arrangement; (b) Photograph of the submerged system operated using suction on permeate side; (c) System with external ultrafiltration module operated under positive feed pressure.

(CFD) simulation is performed using solid works flow simulation which uses the EFD solver from Flow Herm Mentor Graphics Inc., Wilsonville, Oregon, United States. The computational domain used for CFD study is depicted in Figure 5.16 (a), whereas the meshed computational domain is shown in Figure 5.16 (b). To calculate the force and effort required to operate the hand-operated membrane system, porous media with 0.9 porosity is applied to the candle and the pre-filtration membrane of the candle is assumed to have a pore size of 0.01 mm. Ambient environmental pressure and ambient temperature of the water is considered for the analysis. Fluid is filled inside the tank and water is sent through the porous media and eventually passing through four pipes which are connected to a single pipe leading to the exit of the pump. Water enters into the candle through the porous media within the candle and eventually fluid enters into the six openings of the pipe and finally leaves axially through the candle. The hand pump is assumed to provide a flow rate of 0.1 L/sec at the outlet. Figure 5.16 (c) shows the velocity distribution inside the tank, connecting pipe and piston face. The velocity is almost zero at the top of the tank. As the fluid gushes from the candles due to a developed pressure difference, the water is forced out of the pipe.

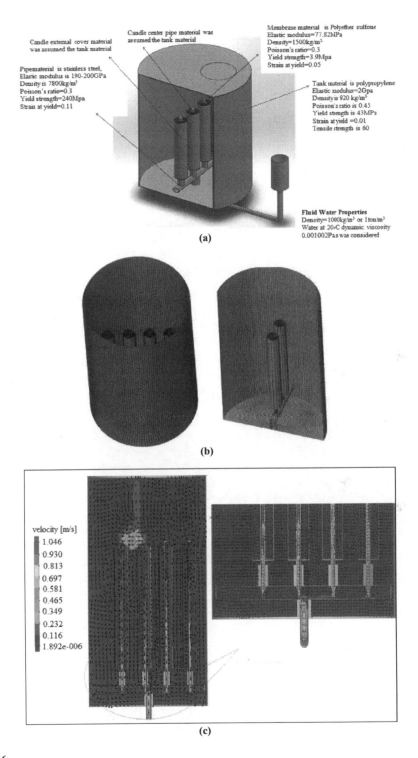

FIGURE 5.16
CFD study of hand pump–operated ultrafiltration membrane; (a) Computational domain; (b) Meshed domain; (c) Velocity distribution profile inside the tank and connecting pipe from a frontal viewpoint.

5.10 Concentration Polarization and Fouling

Concentration polarization is defined as an accumulation of solute deposit on a membrane surface. Consequences of concentration polarization are increase in osmotic pressure, resistance to solvent flow, increase in solution viscosity and subsequently fouling or blocking of membrane pores, thereby reducing the flux, rejection and membrane life. Concentration polarization cannot be avoided in the membrane process, but it can be minimized by various methods. Chemical treatment methods can minimize fouling, and in situations involving gel or cake formation on the membrane surface, hydrodynamics has to be changed in the feed channel to improve mass transfer. This can be done either by steady state technique using high cross-flow velocity or unsteady state by introducing turbulence promoters such as spacers, inserts, etc., in the feed flow path. Fouling is a term that indirectly describes the performance loss of a membrane, which becomes physically or chemically changed by the process fluids leading to reduced and poor quality output. Fouling can also be called loss in flux, which cannot be reversed while the process is in operation.

5.11 Membrane Cleaning and Storage

To overcome fouling, the membrane should be washed with acid or alkali. Membrane cleaning can be done every alternate day, once a week or every fortnight, depending upon feed water TDS and silt density index (SDI), as per the following protocols which prolong membrane life: Acid wash is to be given with 1% (w/v) citric acid or 1% HCl in tap water for 15 min. in recirculation mode to remove mineral scales and metal precipitates. This is followed by a tap water wash for 5 min. Alkaline wash can be performed by an aqueous solution containing 1% (w/v) of NaOH + 0.5% (w/v) of tetra sodium salt of EDTA + 0.1% (w/v) of the surfactant, sodium lauryl sulphate (SLS), for 15 min. to remove organic foulants. Trisodium phosphate (TSP) can be used in place of NaOH in the alkaline wash step. It may be noted that the efficiency of these washing steps would improve if the acidic and alkaline solutions are prepared in deionized water or RO permeate water, instead of tap water. To prevent biological fouling by fungal growth or blue-green algae, which brings about permanent degradation, the membrane can be preserved safely in 0.5% (w/v) aqueous sodium metabisulphite solution (SMBS) for months together, especially during shutdown. Flushing with 0.5% SMBS once a week when the plant is under continuous operation also helps.

5.12 Conclusions

Highly compact vertical modular membrane systems are portable and convenient for installation and maintenance in interior villages. Reverse Osmosis (RO) process needs remineralization either through blending or addition of mixture of salts since the RO permeate is devoid of important minerals and low TDS levels need to be boosted to sufficiently healthy concentrations (60–300 ppm) for the human heart and body.

Sometimes bicarbonate dosing or aeration is required to neutralize permeate water pH, which falls into acidic range of 4.5–6.5 pH due to dissociation of carbonates and bicarbonates into CO_2 while salts penetrate the membrane. The reject from a RO process needs to be effectively treated either through chemical precipitation or adsorption such that the water can be recycled for domestic washing, non-edible plantation, fly ash brick kiln or ground water recharging, to conserve water. Without an effective reject treatment scheme, the RO- or NF-based processes would be incomplete solutions since the concentrated contaminants present in the waste stream would go back into the ground water. Hand pump-operated ultrafiltration module will be helpful to produce safe drinking water in remote villages and flood affected regions. Computational fluid dynamics approach can be used to scale-up the membrane processes by optimizing flow and pressure profiles. With proper awareness schemes, waterborne diseases can be gradually reduced in the developing world and finally eradicated using the technology leads projected in this work.

Acknowledgments

The authors would like to acknowledge the funding provided by Department of Biotechnology (DBT), New Delhi for the installation of three pilot plants in villages and hospitals in POMACEA project (GAP-542), which comes under Inno Indigo's Indo-European collaborative scheme.

References

Abedi, M.J., H.J. Cotter, and M.M. Andy. 2002. Arsenic uptake and accumulation in rice (*Oryzasativa* L.) irrigated with contaminated water. *Plant and Soil* 240: 311–319.

Ayoob, S. and A.K. Gupta. 2007. Fluoride in Drinking Water: A Review on the Status and Stress Effect. *Critical Reviews in Environmental Science Technology* 12: 433–487.

Badal, K.M. and K.T. Suzuki. 2002. Arsenic round the world: A review. *Talanta* 58: 201–235.

Barbier, O., L. Arreola-Mendoza, and L.M. Del Razo. 2010. Molecular mechanisms of fluoride toxicity. *Chemico-Biological Interactions* 188(2): 319–333.

Bratby, J. 2006. *Coagulation and Flocculation in Water and wastewater Treatment*. London, Seattle: IWA Publishing.

Cherya, M. 1998. *Ultrafiltration and Microfiltration Handbook*. Lancaster, PA: Techno Economic Publishing Company.

Cheryan, M. and F.R. Alvarez. 1995. Food and Beverage Industry Applications. In *Membrane Separation Technology: Principles and Applications*. Noble, R.D. and S.A. Stern, eds. 415–460. Amsterdam: Elsevier Science.

Choong, T.S.Y., T.G. Chuah, Y. Robiah, F.L. Gregory Koay, and I. Azni. 2007. Arsenic toxicity, health hazards and removal techniques from water: An overview. *Desalination* 217: 139–166.

Devin, L., S. Ngai, Y.Y. Gilron, and J.E. Menachem. 2012. Seawater desalination for agriculture by integrated forward and reverse osmosis: Improved product water quality for potentially less energy. *Journal of Membrane Science* (415–416): 1–8.

Dinesh, M., U. Charles, and J. Pittman. 2007. Arsenic removal from water/wastewater using adsorbents: A critical review. *Journal of Hazardous Materials* 142: 1–53.

Gary, W.B. 2003. The use of organo-clays in water treatment. *Applied Clay Science* 24: 11–20.

Govardhan, B., S.S. Chandrasekhar, and S. Sridhar. 2017. Purification of surface water using novel hollow fiber membranes prepared from polyetherimide/polyethersulfone blends. *Journal of Environmental Chemical Engineering* 5: 1068–1078.

Heikens, A., S. Sumarti, M. Bergen, B. Widianarko, L. Fokkert, K. Van Leeuwen, and W. Seinen. 2005. The impact of the hyperacid Ijen Crater Lake: Risks of excess fluoride to human health. *Science of The Total Environment* 346: 56–69.

Hung, Y.T., L.K. Wang, and N.K. Shammas. 2012. *Handbook of Environment and Waste Management: Air and Water Pollution Control*. USA: World Scientific.

Keizer, K.U., R.J.R. Uhlhorn, T.J. Burggraaf, R.D. Nobel, and S.A. Stern. 1995. *Gas Separation Using Inorganic Membranes, in Membrane Separation Technology, Principles and Applications*. Amsterdam: Elsevier. 553–584.

Khan, A.H., S.B. Rasul, K. Munir, M. Habibuddowla, M. Alauddin, and M.S.S. Newaz. 2000. Appraisal of a simple arsenic removal method for ground water of bangladesh. *Journal of Environmental Science and Health* 35: 1021–1041.

Malmqvist, B. and S. Rundle. 2002. Threats to the running water ecosystems of the world. *Environmental Conservation* 29(2): 134–153.

Manoj, K. and P. Avinash. 2012. A review of permissible limits of drinking water. *Indian Journal of Occupational & Environmental Medicine* 16(1): 40–44.

Meenakshi, and R.C. Maheshwari. 2006. Flouride in drinking water and its removal. *Journal of Hazardous Materials* 137(1): 456–463.

Naaz, A., B. Kumar, C. Narayan, and K. Shukla. 2015. Assessment of fluoride pollution in ground waters of arid and semi-arid regions of tonalite – Trondjhemite series in central India. *Water Quality, Exposure and Health* 7(4):545–556.

Nadeem N.M.D. and C.B.M.D. Ronald. 1980. Disorders of Iron Metabolism. *Medical Clinics of North America* 64: 631–645.

Nikhil, J.Z., P.S. Vimal, T. Sachin, and K. Vijith. 2015. Design and fabrication of hand pump operated water purification system using reverse osmosis. *International Journal for Innovative Research in Science & Technology* 1: 48–54.

Parmeshwar, U., C. Yutaka, N. Takashi, S. Ning, I. Hiroshi, and F. Futaba. 2016. Rural drinking water issues in India's drought-prone area: A case of Maharashtra State. *Environmental Research Letters* 11: 1–13.

Porter, M.C. 1990. *Handbook of Industrial Membrane Technology*. Park Kidge, New Jersey: Elsevier.

Prosun, B., C. Debashis, and J. Gunnar. 2010. Occurrence of arsenic-contaminated ground water in Alluvial Aquifers from Delta Plains, Eastern India: Options for safe drinking water supply. *International Journal of Water Resources Development* 13: 79–92.

Raj, D. and E. Shaji. 2017. Fluoride contamination in groundwater resources of Alleppey, Southern India. *Geoscience Frontiers* (117–124): 1–8.

Rakness, K.L. 2011. *Ozone in Drinking Water Treatment: Process Design, Operation, and Optimization*. USA: American Water Works Association.

Ramakrishnaiah, C.R., C. Sadashivaiah, and G. Ranganna. 2009. Assessment of water quality index for the ground water in Tumkur Taluk. *E-Journal of Chemistry* 6(2): 523–530.

Ritchie, S.M.C., and D. Bhattacharyya. 2002. Membrane-based hybrid processes for high water recovery and selective inorganic pollutant separation. *Journal of Hazardous Materials* 92: 21–32.

Ritu, D. and S.M. Ambashta. 2010. Water purification using magnetic asistance: A review. *Journal of Hazardous Materials* 180: 38–49.

Roy, D., T.K. Das, and S. Vaswani. 2013. Arsenic: It's extent of pollution and toxicosis: An animal perspective. *Vet World* 53–58.

Silvio, C., L. Meunier, and U. Gunten. 2008. Phototransformation of selected pharmaceuticals during UV treatment of drinking water. *Water Research* 42: 121–128.

Singh, J. 2010. Health hazards of water pollution and its legal control with special reference to State of Punjab. *Journal of Environmental & Healthcare Law* 9: 19–34.

Subba Rao, N., I. Saroja Nirmala, and K. Suryanarayana. 2005. Groundwater quality in a coastal area: A case study from Andhra Pradesh, India. *Environmental Geology* 49(5): 543–550.

Sunitha, V. 2013. Nitrates in groundwater: Health hazards and remedial measures review article. *Indian Journal of Advances in Chemical Science* 1(3): 164–170.

6

Ceramic Membrane Based Community Model Plants for Arsenic Decontamination from Ground Water and Quality Drinking Water Supply

Sibdas Bandyopadhyay and Mainak Majumder

CONTENTS

6.1 Introduction

Arsenic occurs in mineral ore deposits, soil and water. The presence of inorganic arsenic compounds has an adverse effect on living systems, including human beings, flora and fauna and animal life. On the other hand, organic forms of arsenic are primarily used in making insecticides and weed killers. Increasing health risks related to arsenic presence in ground water has led to the development of innovative treatment techniques for the production of potable water. A long-term option is possible through construction of deep tube wells with depth of 200 meters for drawing water from aquifers. The tapped aquifer is underneath a thick clay barrier where there is less possibility of arsenic contamination. However, unconfined aquifers are not found to be advantageous, despite construction of proper tube wells. Therefore, treatment of arsenic contaminated ground water seems to be an essential need in absence of any alternate source of drinking water.

For community-based water purification systems, doubts prevail that membrane-based water purification technologies may not be viable in rural areas due to their high operational and recurring costs. This chapter deals with the development and demonstration of a sustainable technology based on low-cost ceramic membrane for point of use (POU) application and its acceptance by those who use it.

6.1.1 Global Issue

Excess amounts of arsenic found in ground water have posed serious health problems in various parts of the globe. Recent studies have revealed that over 20 countries worldwide have reported high levels of arsenic contamination in ground water (Bordoloi, 2012). Elevated levels of arsenic in ground water have been well documented in Chile, Mexico, China, Argentina, the United States, and Hungary, as well as in the Indian State of West Bengal, Bangladesh, and Vietnam (Shankar et al., 2014). The major arsenicosis affected areas have been reported in large deltas and/or along major river basins across the world, such as the Paraiba do Sul Delta, Brazil, the Bengal Delta, Mekong Delta, Cambodia, the Danube River Basin, Hungary, the Hetao River Basin, Mongolia, Duero Cenozoic Basin, Spain, Zenne River Basin, Belgium and Tulare Lake, U.S. About 150 million people around the world are estimated to be affected globally, and more than 45 million individuals, especially in developing countries from Asia, are susceptible to exposure to arsenic levels of > 50 parts per billion (ppb). Arsenic contamination in ground water in the Ganga-Brahmaputra fluvial plains in India and the Padma-Meghna fluvial plains in Bangladesh have been reported to be one of the world's biggest natural ground water calamities.

6.1.2 Permissible Limits of Arsenic Content in Drinking Water

Considering its toxicity, occurrence and frequency of exposure to humans, vis-à-vis the accessibility and cost-effectiveness of water treatment technologies, administrative organizations have proposed the greatest contaminant level for arsenic in drinking water to be in the range of 10–50 ppb. Even though the toxicity of arsenic is largely dependent on its chemical form, the guideline values target only the total arsenic concentration. USEPA normally considers a cancer risk of no more than 1 per 10,000 individuals as acceptable. New studies suggest the cancer risk from exposure to levels as low as 3 ppb of arsenic is closer to 4–10 fatalities per 10,000 individuals. In October 2001, the United States Environmental Protection Agency (USEPA) issued an additional ruling on the allowable concentration of arsenic in drinking water, affirming its lowering from 50 ppb to 10 ppb. This step taken to safeguard the public health has important implications for water treatment professionals and their customers. The 10-ppb level was chosen because decreasing the acceptable limit to 5 ppb or less would stretch the existing arsenic removal technologies and possibly cause economic hardships for water suppliers who would be out of compliance.

6.2 Nature of the Problem on Arsenic Contamination in Ground Water

Widespread arsenic contamination in ground water has been reported in the eastern part of India and adjacent districts of Bangladesh. In India, the states located near the flood plains of the Ganga, Brahamaputra and Imphal Rivers are reported to be severely affected by arsenic presence in ground water (Chakraborti et al., 2017; Bordoloi, 2012). The arsenic belt lies in the new-delta region of the Ganga and Bramhaputra rivers bounded by the Rajmahal hills to the northwest, Darjeeling Hill to the north and Shillong Plateau to the northeast. One third of the delta lies in West Bengal, India while the rest constitutes the area of Bangladesh. In the mid-1990s, arsenic contamination was reported in parts of the Bengal Delta Plain covering the state of West Bengal in India and Bangladesh. A 400-km-long and 60-km-wide belt of the Bengal basin was reported to be contaminated with arsenic and in its drinking water (International Conference 1995 and 1998; International Workshop, 2000; Dhar et al., 1997).

Recently, arsenic poisoning in drinking water was also reported in the northeastern state of Bihar, Bhojpur District and Rajnandgaon Village in the Chhattisgarh State of India. New findings also suggest that arsenic may be found as far away as Chandigarh, India and Nepal. There are several reports on arsenic contamination in Ganga-Meghna-Brahmaputra (GMB) Plain, like arsenic contamination in Middle Ganga Plain, Bihar (Chakraborti et al., 2003), arsenic affected areas in Uttar Pradesh and Jharkhand States in the Gangetic Plain and the State of Assam in the Brahmaputra Plain of India. A publication of North Eastern Land and Water Management reported that arsenic contamination occurs in five states of North Eastern Region (Singh, 2007). Chakraborti et al. reported some arsenic contamination in even deep (depth range 100–415 m) tube wells in Bangladesh (Chakraborti et al., 1999), as did the British Geological Survey (2001). Several tube wells that were more secure a couple of years ago in West Bengal region have now reported high arsenic levels of > 50 ppb (Rahman et al., 2001).

The above overview on arsenic contamination of ground water demonstrates that a significant number of states and countries within the GMB plain of 500,000 km^2 area and a population assessed to be more than 450 million might be under high risk of arsenic exposure (Chakraborti et al., 2004). In West Bengal alone, about 5.36 million people have been identified as being exposed to the risk of arsenic contamination, and about 7.87% population of the state is under the threat of contamination. Thus, an approach towards sustainable solution to ensure a safe drinking water supply is a serious public health issue, primarily in the rural sector of developing countries, as ground water offers drinking water.

6.2.1 Arsenic Contamination in Ground Water—Arsenic Speciation

Arsenic species are an essential viewpoint to contemplate while applying treatment methods for their removal. The main source of arsenic distribution into an environment is by natural and anthropogenic activities. The toxic and mobile nature of arsenic is associated with its speciation (Vecorena & Rominna, 2016). Arsenic is mainly classified into two forms, (a) organic and (b) inorganic. Among the two, inorganic arsenic was found to be more carcinogenic to humans and the ecosystem. Studies on temporal variability of arsenic in ground water revealed the presence of both trivalent (As(III)), and pentavalent (As(V)) inorganic arsenicals. Among these two inorganic species, As(III) arsenite form is considered to be more soluble, toxic and mobile than the As(V) arsenate. In natural waters, arsenic usually exists either in As(III) or As(V) oxidation states. As(V) in the forms of $H_2AsO_4^{-1}$, $HAsO_4^{-2}$ and AsO_4^{-3} is dominant in oxygenated water. Under anoxic conditions, As(III) exists in anionic ($H_2AsO_3^-$) form at a pH > 9.22 and nonionic form (H_3AsO_3) at a pH < 9.22. The total average arsenic level in feed water from the Spiro Tunnel Bulkhead (Park City, Utah) was found to be 60 ppb. The results showed 70% of arsenic presence in dissolved form in the feed water sample, whereas the arsenic speciation revealed 76% of As(V) in dissolved state in the source water (Pawlak et al., 2002). As(III) species under atmospheric conditions was found to be unstable resulting in the formation of As(V).

6.2.2 Co-Contaminants in Ground Water

Dissolution and desorption of arsenic from arsenic-rich iron hydroxides in reducing soil environment at low redox potential was reported to be the possible cause of arsenic contamination in ground water. The co-existence of iron (Fe) and arsenic in ground water supports this hypothesis. Consequently, high Fe concentration is also associated with arsenic in ground water in Bangladesh and India. The arsenic removal efficiency is highly dependent on the initial concentrations of Fe and As in ground water.

A simple ceramic filter for Fe removal was developed for use in rural households of Bangladesh. The performance of the filter for removal of Fe and As was investigated by introduction of additional source of Fe (iron net) and iron oxidizing bacteria to the filter. The results showed effective removal of Fe through oxidation and co-precipitation (Shafiquzzaman et al., 2011).

6.3 Technologies for Arsenic Removal in Drinking Water

In general, techniques such as chemical treatment, adsorption or ion exchange are used for treatment of arsenic polluted ground water. Chemical treatment is based on coagulation-flocculation-clarification-sand filtration, while activated alumina or granular ferric hydroxide columns are used for arsenic adsorption. These techniques have been tested under field conditions using ground water with 0.1–0.2 ppm arsenic content in most cases. High-pressure, membrane-based technologies like nanofiltration (NF) and reverse osmosis (RO) have also been used for arsenic removal.

6.3.1 Conventional Technologies for Arsenic Removal

Arsenic separation from drinking water by Fe removal plants is well-known. The treatment process consists of aeration, chlorination, sedimentation and filtration (USEPA, 2000). However, for the complete removal of Fe, a two- or three-stage filtration is often necessary when the iron concentration in the raw water is high. Simultaneous Fe and As(III) removal using oxidation, coagulation, settling, filtration, and membrane filtration, was studied in tap water spiked with an As(III) concentration of 0.2 mg/L and Fe(II) concentration of 1 mg/L. The order of effectiveness of oxidants for the simultaneous removal of As(III) and Fe was found to be: $KMnO_4 > O_3 > NaOCl > O_2$. The most effective inorganic coagulants for arsenic removal were found to be $CuSO_4$, $CuCl_2$, $ZnCl_2$, $FeSO_4$, and $Fe_2(SO_4)_3$. Removal of As(III) was reported to be in the range 75–88% after oxidation with $KMnO_4$, NaOCl, or O_3 followed by coagulation, settling and membrane filtration. Pilot scale studies were conducted using raw water from a contaminated well containing 0.8–0.9 ppm of arsenic. Addition of about 20 mg/L of chlorine during the aeration process, followed by $FeCl_3$ (about 60 mg/L) coagulation, settling and filtration, resulted in an effluent with arsenic below the detectable range. Based on the outcome of the experimental results, a full-scale plant with a capacity of 150 m^3/day to serve 1,500 people was constructed, the cost of which was estimated at US$ 0.055 per m^3 of treated water (Chen, 1973). Pilot plant studies were conducted at Naval Air Station (NAS) Fallon, Nevada, to study the removal of arsenic (0.08–0.116 mg/L) from NAS's drinking water supply using granular activated alumna (AA) with and without pH adjustment and estimate the operation and maintenance costs for the activated alumna plants.

6.3.2 Shortcomings of Conventional Technologies

The co-precipitation technique suffers from various constraints like controlling the dosage of chemicals added, the low rate of arsenic removal at higher doses due to ineffective separation of fine impurities using a sand filter, etc. Similarly, the lower separation efficiency of the adsorption technique may be attributed to the limitation towards iron content in ground water leading to the formation of iron-coated particles over the adsorbent and minimal duration of exposure for adsorption during the flow of raw water through the column of the granular adsorbent media. The granular ferric hydroxide media leads to generation of fines, which comes out with the treated water. As a result, the efficiency of arsenic removal is lower and although the arsenic content in the treated water meets the requirements for Indian Standard Specification (0.05 ppm), the levels are still higher than the recommended limit set by WHO.

6.4 Membrane-Based Processes for Decontamination of Arsenic

Membrane-based technologies for separation of arsenic from ground water may be categorized into two types:

  i. Oxidation and Membrane Filtration
  ii. Treatment with RO/NF Membrane

Typical membrane materials include polymeric, ceramic oxide or silicon carbide (SiC) membranes.

6.4.1 High-Pressure Membrane Processes

Membrane-based processes like RO and NF can also separate arsenic (Fox & Sorg, 1987; Sato et al., 2002; Huxstep, 1981). Huxstep reported the removal of inorganic contaminants including arsenic from water using high-pressure (400 psi) and low-pressure (200 psi) RO systems, having a rated capacity of 1.82 L/s of product water. He studied the removal of As(III) and As(V) in spiked ground water in test runs lasting from 1 to 5 days. The high-pressure system, though requiring almost twice the energy as the low-pressure system, was found to be more effective in the removal of arsenic. High-pressure RO systems achieved 91–98% removal of As(V), whereas removal of low-pressure systems ranged from 77 to 87%. The high-pressure systems indicated 63–70% removal of As(III) with separation in the low-pressure systems ranging between 12 to 35%. In summary, the results showed that both the high- and low-pressure RO systems were effective in removal of As(V), but not satisfactory for removal of the more toxic As(III). Thus, prior to using RO for drinking water treatment, the speciation chemistry of arsenic in the water should be studied.

Laboratory studies have been conducted for the removal of inorganic arsenic using an RO system with a 5 μm prefilter to remove suspended material and an activated carbon prefilter to remove chlorine. The RO membrane used was a spiral would polyamide film. Tap water spiked with As(III) to an initial concentration of 0.101 mg/L was pumped at a pressure of 42 ± 2 psi and arsenic removal of 73.3% was achieved using the system. A 2005 field study was also conducted in San Ysidro, New Mexico for evaluating RO as a point-of-use (POU) treatment system.

Clifford and Lin (1991) reported removal of As(III) and As(V) using various types of RO membranes and recommended the oxidation of As(III) to As(V) prior to the use of the process. Removal of As(III) by the various types of membranes varied widely with rejection values ranging from 46 to 75% for initial As(III) concentrations in the range from 0.04 to 1.3 mg/L. High rejections (98–99%) were observed using the same membranes for As(V) at initial concentrations ranging from 0.11 to 1.9 mg/L. Nanofiltration membranes were found to be effective for arsenic removal at low (15–30%) recovery levels. They suggested the use of low-pressure RO membrane for high arsenic removal with high recovery. However, these high-pressure methods lead to an increase in capital and production costs. Also, pretreatments like microfiltration and ultrafiltration prior to RO and NF are necessary for efficient operation of the process.

6.4.2 Low-Pressure Point-of-Use (POU) Systems

Chang et al. (1994) found microfiltration to be more effective than conventional filtration for arsenic removal. Microfiltration (MF), a low-pressure membrane process effectively separates large colloidal and suspended particles of sizes ranging from 0.1–10 μm. Flow through porous membranes follows Darcy's law and can be expressed as:

$$J_w = \Delta P / \mu . R_m \qquad (6.1)$$

Where, J_w is the permeate flux of pure water (ms^{-1}), ΔP is the trans-membrane pressure (Pa), and μ is the viscosity of water (Pa.s).

Flux decline during membrane filtration has been attributed to the formation of a cake layer on top of the primary membrane. Such a cake layer, termed "dynamic membrane," has been often been used to augment separation processes (Nomura et al., 1980). The formation of a filter cake can be limited by the shear flow of the fluid parallel to the membrane surface such as in cross-flow filtration conditions.

The water flux using the dynamic membrane layer in place at a certain cross-flow velocity is measured as follows:

$$J/w = \Delta P / \mu.(Rm + Rc) \tag{6.2}$$

Where, R_m is membrane resistance, and R_c is cake resistance. The adsorption characteristics of particulate ferric hydroxide adsorbent have been determined in cross-flow microfiltration condition to study the scope of application of this combined adsorption–membrane filtration technique for enhancing the separation efficiency of arsenic from ground water at a lower operating pressure.

6.5 Ceramic Membrane-Based Process for Decontamination of Arsenic

Ceramic microfiltration membranes have a niche market in water treatment applications due to their chemical and mechanical ruggedness, hydrophilicity and ease of flux regeneration by back pulsing or chemical washing. The sorption of arsenate/arsenite on ferric hydroxide/aluminum hydroxide particles has a potential for its removal when used in combination with a suitable solid/liquid separation technique.

The iron oxide nanoparticles derived from carboxylate-FeOOH, termed ferroxane-AA, were deposited on a tubular porous alumina support to create an adsorbent ceramic membrane for the removal of arsenic from drinking water. The experiments performed using alumina based ceramic MF tube supported with iron oxide resulted in adsorption capacities of 0.128 mg As/m^2 Fe$_2$O$_3$ [10 ppm As(V), 0.5 g of iron oxide] and 0.008 mg As/m^2 Fe$_2$O$_3$ [86 ppb As(V), 0.3 g of iron oxide] (Sabbatini et al., 2016).

A hybrid process combining adsorption and ultrafiltration has been proposed for the removal of As(V) ions. Fe$_2$O$_3$ nanoparticles (Zaspalis et al., 2007) with an average pore diameter of 20 to 30 nm have been used as adsorbent while γ-Al$_2$O$_3$ based membrane with an average pore diameter between 3 and 4 nm was used for ultrafiltration. The experimental results showed effective removal of As(V) ions from an initial feed concentration of 1 ppm to a final concentration of 10 ppb, for Fe$_2$O$_3$ adsorbent loading of 0.2 wt% (w/v). The subsequent ultrafiltration step enabled 100% rejection of sorbent nanoparticles with water flux of 156 kg h^{-1} m^{-2}, under a transmembrane pressure difference of 3×10^5 N m^{-2} (3 bar) (Pagana et al., 2008). U.S. patent 5,556545 reported the use of fine particles of alumina (below 200 micron) in a slurry form, contacting the contaminated water with the slurry and separation of the alumina particles from water using microfiltration membranes to achieve arsenic removal of 50 ppb or less. However, the drawbacks associated with the use of activated alumina remains, when natural ground water containing high amount of iron is used, due to coating/preferential adsorption of iron hydroxides on the alumina particles. Separation achieved by arsenic adsorption on amorphous aluminum hydroxide showed that a plateau of 0.3 ppm arsenic can be attained, but further reduction of arsenic content is difficult (Anderson et al., 1976). Recently, flocculation process coupled with an ultra-/microfiltration (UF/MF) step demonstrated a process for arsenic removal from feed water. Polymeric, ceramic oxide and silicon carbide based MF membranes have been investigated as potential candidates for the MF treatment step. Saint-Gobain manufactures recrystallized silicon carbide (R-SiC) dead-end filters with R-SiC membranes of a controlled pore size for the MF range. SiC membranes were found to be efficient in removal of arsenic from flocculated water (Vincent et al., 2016). In comparison to other MF membrane materials, R-SiC membrane technology in dead-end mode promises to be a more stable process with appreciable backwash efficiency, minimal drop in flow rate over the filtration time, higher flux and favorable filtration efficiency.

6.6 Hybrid Process for Decontamination of Arsenic Using Low-Cost Ceramic Membrane

MF was found to be the most promising technique for separation of colloids and suspended particles from liquid streams. CSIR-CGCRI developed a clay-alumina based, low-cost ceramic membrane for decontamination of arsenic utilizing the concept of a hybrid process combining adsorption and microfiltration (Bandyopadhyay et al., 2007, 2002, 2006). The hybrid process was based on the principle of adsorption followed by cross-flow microfiltration using a low-cost ceramic membrane made from clay-alumina mix formulation. The contaminated water is pumped to a slurry reactor tank containing nanosized adsorbent media and re-circulated through tubular ceramic membrane module under pressure. Arsenic-free water permeates through the walls of the ceramic membrane, while the arsenic-laden adsorbent is retained and re-circulated to the reactor tank. When the adsorbent becomes saturated with arsenic, the spent adsorbent is separated by passing through a ceramic candle filter and a fresh batch of adsorbent is charged in the reactor tank.

6.6.1 Adsorbent Preparation and Selection

The method of preparation of the particulate adsorbents is based on self-hydrolysis of iron and aluminum salts without using any base for precipitation of the hydroxides. The efficiency of two different types of particulate adsorbent, viz., one containing only iron and the other containing iron and aluminum using stirred tank reactor in spiked distilled water, is shown in Figure 6.1. The arsenic adsorption efficiency of the mixed adsorbent at 1 ppm arsenic concentration is low, while the single component based particulate adsorbent is consistent in the 0.5–5 ppm arsenic spiking range. Hence, the adsorbent based on iron having a higher adsorption capacity has been selected and utilized for subsequent experiments.

6.6.2 Bench Scale Studies Under Cross-Flow Microfiltration Conditions Using Ceramic Membrane

Adsorption studies under dynamic conditions were conducted in an experimental set-up (Figure 6.2) using a colloidal suspension of iron based adsorbent in 70 L tap water (Table 6.1) at pH 7.4. The concentration of the adsorbent was measured gravimetrically by drying at 110°C, after collecting the sample through the by-pass outlet while conducting the experiment under cross-flow mode. The experiments were carried out using about 1,450 ppm of the colloids. Figure 6.3 shows a bimodal particle size distribution of the adsorbent. The mean particle size is ~ 6 microns, but there are some fractions less than 0.4 microns (lower than the mean pore size of the membrane), which may pass through prior to the formation of the dynamic layer.

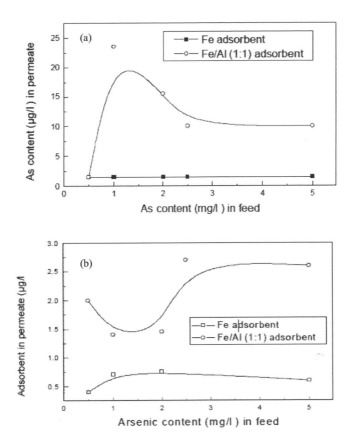

FIGURE 6.1
Efficiency of two different types of particulate adsorbent. (a) Arsenic content in supernatant after treatment with particulate adsorbent and (b) concentration of adsorbent ions in supernatant after treatment with particulate adsorbent and overnight settling of the suspension.

A measured amount of 1,000 ppm sodium arsenate solution was added gradually in the feed stream to vary the arsenic concentration in the range of 3–80 ppm. Retentate and permeate samples were collected 2 min after each addition. The amount of arsenic adsorbed was calculated from the difference between the feed and permeate concentration. Using these experimental values, the adsorption isotherm was plotted and parameters extracted from the Freundlich Isotherm model. The temperature of the water was in the range 25–27°C. Permeate was recycled during the experiment to maintain the concentration of the adsorbent at a fixed level.

6.6.3 Adsorption Studies Under Dynamic Condition Using Arsenic-Spiked Tap Water

The effect of arsenic-spiked tap water on removal efficiency is shown in Figure 6.4. The figure indicates that about 97.8% separation efficiency is obtained using 80 ppm arsenic-spiked tap water. Figure 6.4 also reveals that treated water with lower arsenic content (< 10 ppb), as per the WHO recommendation, could be achieved from spiked tap water containing up to 40 ppm As.

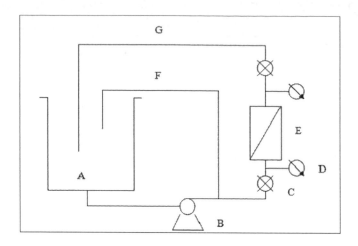

FIGURE 6.2

Set-up for conducting the adsorption and microfiltration experiments: (A) Feed tank (70 L) containing the As(V) spiked water or natural ground water and the particulate adsorbent; (B) Centripetal pump; (C) valve; (D) pressure gauge; (E) Ceramic membrane module containing a single tubular membrane with 20 mm O.D. and 7 no. of channels of 4.2 mm diameter and filtration area 150 cm^2; (F) By-pass pipe; (G) Retentate stream.

TABLE 6.1

Characteristics of Tap Water Used for Spiking Arsenic

Parameters	Tap Water Characteristics
pH	7.4
Turbidity (NTU)	15.0
Iron (ppm)	0.6
Total hardness (ppm as CaCO$_3$)	350
Chloride (ppm)	200
Methyl Orange Alkalinity (ppm as CaCO$_3$)	490
Sulphate (ppm)	6.7
Silica (ppm)	13.5

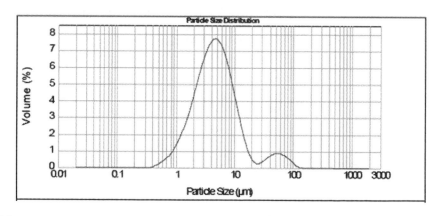

FIGURE 6.3

Bimodal particle size distribution of the adsorbent.

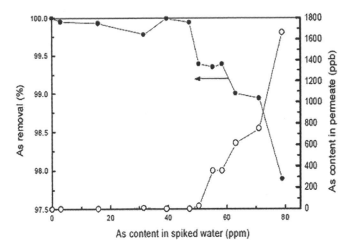

FIGURE 6.4
Removal efficiency of As(V) using a combination of particulate adsorbent and cross-flow microfiltration.

6.6.4 Adsorption Capacity Under Cross-Flow Condition

The adsorption capacity of the colloids was found to be 53 mg As(V) per gm of the colloidal adsorbent using 80 ppm arsenic in spiked water. Figure 6.5 shows that the non-equilibrium adsorption behavior of the SPA can be represented by a Freundlich type equation:

$$X = K_F C^{1/n} \tag{6.3}$$

Where, X is mg of adsorbate/gm of adsorbent and C is the concentration in ppm of the adsorbate, n is the exponential factor and K_F is the proportionality constant. The values of K_F and $1/n$ were found to be 0.68 and 1.0013, respectively. A larger value of K_F indicates a larger adsorption capacity and a smaller value of $1/n$ indicates a stronger adsorption bond.

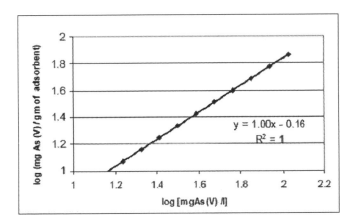

FIGURE 6.5
Freundlich adsorption isotherm of the colloidal adsorbent determined using cross-flow microfiltration process.

TABLE 6.2

Comparison of the Freundlich Isotherm Parameters of Granular Adsorbent Media and Ferric Hydroxide Based Particulate Adsorbent at Comparable pH

pH	Activated Alumina		Granular Ferric Hydroxide		Ferric Hydroxide Based Adsorbent (Present Study)	
	K_f	$1/n$	K_f	$1/n$	K_f	$1/n$
7.0	0.148	0.88	0.09	0.60	–	–
7.4	–	–	–	–	0.68	1.00
7.7	0.1679	0.452	–	–	–	–
8.0	0.221	0.52	0.071	0.72	–	–

The Freundlich Isotherm parameters of activated alumina, granular ferric hydroxide and the fine particle amorphous ferric hydroxide adsorbent are shown in Table 6.2. The high K_F values indicate that the amount of adsorption in fine particle form is indeed large. The adsorption bond between arsenic and a metal oxide adsorbent is likely to be a two-stage process. In the first stage, arsenate ions are electrostatically attracted to the positively charged surface of ferric hydroxide. The adsorption of arsenic on ferric-hydroxide is found to increase with time. The high values of $1/n$ in our experiments may be due to the smaller contact time (~ 2 min. after each spiking) between the adsorbate and adsorbent. A survey of the literature revealed that As(V) adsorption values determined using batch tests were lower: 30 mg/g at pH 7 and 18.75 mg/g at pH 8 (Bose & Sharma, 2002).

6.6.5 Effect of Dynamic Membrane Formation during Cross-Flow Microfiltration Experiments Using Natural Ground Water

The multichannel ceramic membrane (O/D of 20 mm, 7 nos. of channels of diameter 4.2 mm each) with filtration area of 0.015 m^2 and pore size 0.4 μm (measured on Capillary Flow Porometer, PMI, Ithaca, United States) was used to study the effect of dynamic membrane formation during cross-flow microfiltration experiments. The permeability of the ceramic membrane was determined using D.I. water. The clean water permeability of these membranes is about 800 LMH. Short-term (2 h) tests were conducted to find the dynamic membrane formation and initial flux decline characteristics using D.I. water, aerated ground water and ground water with particulate adsorbent. New membranes were used during each study. Long-term (30 h) filtration tests were also conducted to find the pseudo-steady state values of filtration rate. The short- and long-term filtration experiments were carried out at 1 kg/cm^2 and a constant cross-flow velocity of 1.3 m/s.

6.6.6 Role of Dynamic Membrane Formation on Removal of Arsenic

During microfiltration of particulate suspensions, a pronounced decline in flux due to the formation of a cake layer of the deposited particles on the membrane surface is observed. To evaluate the effect of the formation of the dynamic membrane layer on arsenic removal from natural ground water, filtration experiments were carried out with ~ 4,500 ppm of the particulate adsorbent. Arsenic in the feed is adsorbed by the particulate media and some of these particles are smaller than the pore size (Figure 6.4) of the membrane. The color of the initial permeate is brownish, indicating the presence of colloidal particles in the permeate. However, within 30 min., the dynamic membrane layer is formed which reduces the flux from about 680 LMH to 350 LMH, but helps in retaining colloidal particles. Formation of a cake layer over the membrane surface is a phenomenon which has been investigated especially for its role in controlling the permeate flux (Belfort et al., 1994). The hierarchy of particle deposit during cake formation is of interest to filtration technology. During experimentation, the deposition of larger particles occurs on a filter surface at rates faster than the finer particles, and, as a result of which, the pore size of the cake layer formed initially is large and cannot retain the smaller colloidal particles (Václavíková et al., 2009). Once the cake layer starts building and the membrane surface is further modified, the colloidal particles are completely retained, which brings down the arsenic and iron levels in permeate to BDL. Thus, the deposited cake behaves as a dynamic membrane for filtration of fine particulate adsorbent. After this dynamic membrane layer is formed, both the permeation rate and the separation remain steady both in short-term and long-term filtration (Figure 6.6). The dynamic membrane layer formed from FeOOH particles thus increases particle retention characteristics and helps to stabilize the flux. After the 30-hour filtration experiment, the cake resistance was calculated to be 6.95×10^{11} m^{-1}. The membrane resistance was calculated from the D.I. water permeability and found to be 2.97×10^{11} m^{-1}. An increase in membrane resistance of 2.3 times has been found to occur due to the deposition of the cake layer over the membrane surface when the flux reaches its pseudo-steady state value.

6.6.7 Findings of Bench Scale Trial

The specific observations of the investigation conducted on the bench scale are as follows:

i. The particulate adsorbent consisting of amorphous ferric hydroxide has very high adsorption capability compared to GFH, which means that a smaller amount of sludge is generated compared to co-precipitation and fixed bed adsorption process.

ii. Higher adsorption efficiency compared to granular adsorbents has been observed which may be attributed to higher available surface area and better mass transport dynamics in the cross-flow filtration conditions.

iii. Ceramic microfiltration membrane with a dynamic layer of FeOOH can be used for complete separation of As and Fe required for obtaining drinking water as per the new MCL (< 10 μg/L As).

6.7 Pilot Plant Trials for Treatment of Arsenic Contaminated Natural Ground Water

A pilot trial was also conducted under field conditions to demonstrate that the hybrid process consisting of adsorption by ferric hydroxide particulates and cross-flow microfiltration using low cost ceramic membrane is an efficient hybrid process for separation of arsenic (< 10 ppb As) for production of water as per WHO MCL (Bandyopadhyay et al., 2002, 2004, 2006).

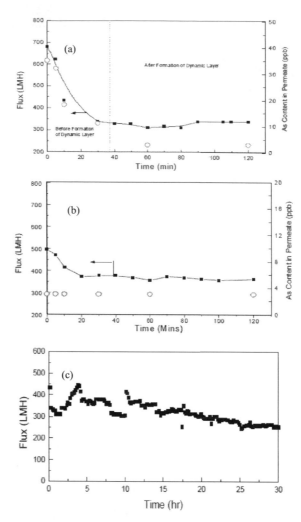

FIGURE 6.6

Water Flux and Arsenic content in permeate (a) during formation of the dynamic membrane and (b) after formation of the dynamic membrane using a feed containing 4500 ppm of the particulate adsorbent in 70 litres of natural ground water containing 1.2 ppm As and 10 ppm Fe (c) Long term flux decline studies under similar conditions.

6.7.1 Performance Evaluation for Production of Quality Drinking Water

The main objectives of the pilot plant are to study the feasibility for production of potable water using highly contaminated natural ground water and social acceptability of the treated water for drinking purposes through the adoption of the pay-and-use system leading to sustainability of the technology. The pilot plant was attached to a hand pump tube well at Akrampur Talikhola, Ward No. 26, Barasat Municipality, North 24 Parganas, West Bengal, India (Figure 6.7). The pilot plant consisted of a reservoir tank made of food grade PVC into which raw water from a nearby hand pump tube well, uncontaminated with arsenic, was pumped for the preparation of a suspension of amorphous ferric hydroxide particulates by self-hydrolysis of a ferric salt. Contaminated ground water (0.9–1.2 ppm

As, 10–12 ppm Fe) was pumped from the selected tube well identified by PHED, Barasat Division, North 24 Parganas district, Govt. of West Bengal, and the raw water tank was filled. The raw water was kept in contact with the adsorbent media for complete precipitation of the soluble iron present in the ground water. The adsorbent media and raw ground water were mixed for about 30 min. (mixing mode) followed by filtration through the membrane module (filtration mode). The transmembrane pressure was kept below 1 kg/cm² as higher pressure leads to greater compression of the filter cake and decreased permeability. The raw water and the filtered water were collected daily for estimation of arsenic and iron and the results are shown in Figure 6.8.

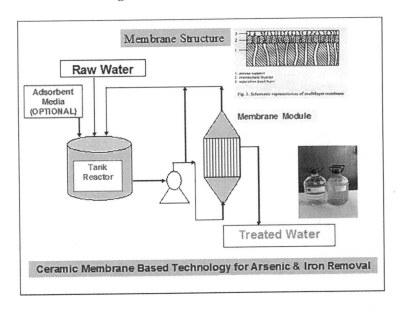

FIGURE 6.7
Pilot plant attached to a hand pump tube well installed at Akrampur Tali Khola, Barasat, North 24 Parganas District, West Bengal, India.

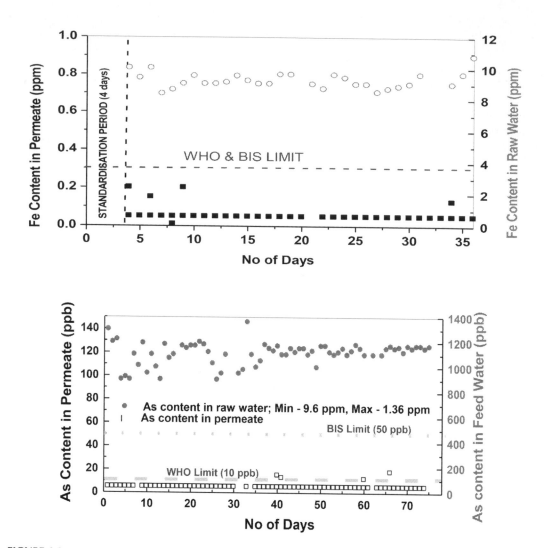

FIGURE 6.8

(a) Iron content with time in raw water and permeate after first media charging during pilot plant trial (b) Arsenic removal.

6.7.2 Sustainability of the Technology

The pilot plant installed in an arsenic affected area (Talikhola Village, Barasat, North 24 Parganas District, West Bengal, India), under a project sponsored by the former Dept. of Drinking Water Supply, Ministry of Rural Development, Govt. of India, has been working for the past 15 years. It has demonstrated the sustainability of the technology for the simultaneous removal of higher concentrations of arsenic (up to 1.5 ppm) and iron (up to 15 ppm) from contaminated ground water, achieving the level of purification as per WHO recommended limits for arsenic (< 0.01 ppm) with low iron content (< 0.1 ppm) in the treated water (Roy et al., 2005; Government of India, 2007; Majumdar et al., 2016; Bandyopadhyay, 2012). Social acceptance of the drinking water produced by CGCRI Technology has been quite remarkable due to its quality, comparable to bottled water. The operation and maintenance of the plant are sustained by collection of a nominal fee from consumers.

6.8 Conclusions

This study clearly demonstrates the effectiveness of the combined adsorption and microfiltration technique for simultaneous separation of arsenic and iron from highly contaminated natural ground water for the production of potable water as per WHO specifications (< 10 ppb As and < 0.3 ppm Fe). The following are the specific conclusions of the pilot plant study:

i. Unlike adsorption techniques based on activated alumina or ferric hydroxides, the iron present in natural ground water can also be utilized for the removal of arsenic from ground water.

ii. The hybrid system can remove iron and arsenic completely from contaminated ground water and no other step is required for treating the water further to make it potable.

iii. Downsizing of ceramic membrane-based system for community application is a user-friendly, techno-economically viable and sustainable process with peoples' participation particularly due to ease of operation and maintenance and longer life of low cost ceramic membrane.

iv. The hybrid system is a low-pressure membrane-based process for separation of arsenic and iron from ground water compared to other membrane-based processes like nanofiltration or reverse osmosis.

The backwashing operation is used for cleaning and regeneration of the ceramic membrane. The pilot plant (500 LPD capacity) is under operation for more than 15 years in an arsenic affected area of West Bengal. The quality of the filtered water being equivalent to natural mineral water proves its marketability, thus leading to the generation of employment and sustainability of the technology.

Acknowledgments

The authors are thankful to the Director, CSIR-CGCRI, Kolkata for extending necessary facilities and to the scientists and staff members of Ceramic Membrane Division and Analytical Chemistry Section of CSIR-CGCRI, for extending necessary assistance while undertaking the laboratory investigation and field work at different stages. We also thank M/s Entech Metals Pvt. Ltd. for designing the Pilot Plant, Ministry of Drinking Water, Govt. of India for sponsoring the projects and Public Health Eng. Directorate, Barasat Division, Govt. of West Bengal for offering cooperation to carry out the pilot scale studies under field conditions.

References

Anderson, M.A., J.F. Ferguson, and J. Gavis. 1976. Arsenate adsorption on amorphous aluminum hydroxide. *Journal of Colloids and Interface Science* 54: 391–399.

Bandyopadhyay, S. and H.S. Maiti. 2004. "Arsenic Removal Plants based on Ceramic Membrane Technology for Production of High Quality Drinking Water from Contaminated Ground Water." Presented at 7th International Symposium on Water, Arsenic and Environmental Crisis in Bengal Basin organized by International Institute of Bengal Basin (IIBB). Dhaka, Bangladesh.

Bandyopadhyay, S. 2006. "A Global Issue: Arsenic Contamination in Water and Case Studies for Remediation." Presented at Workshop on Arsenic contamination of ground water and its mitigation. University of Leeds, U.K.

Bandyopadhyay, S., D. Kundu, S.N. Roy, B.P. Ghosh, and H.S. Maiti. 2006. *Process for preparing water having an arsenic level of less than 10 PPB.* U.S. Patent 7,014,771.

Bandyopadhyay, S., D. Kundu, S.N. Roy, B.P. Ghosh, and H.S. Maiti. 2007. *Apparatus for the preparation of arsenic free water.* U.S. Patent 7,309,425.

Bandyopadhyay, S., D. Kundu, S.N. Roy, B.P. Ghosh, and H.S. Maiti. *A process for preparing arsenic free (<10ppb) water from arsenic contaminated ground water and an equipment therefor.* Indian Patent 23,177,768.

Bandyopadhyay, S., M. Majumder, and H.S. Maiti. 2002. "Potabilisation of Arsenic Contaminated Ground Water using Ceramic Membrane." Presented in Seventh International Conference on Inorganic Membranes (ICIM-7) held at Dalian Institute of Chemical Physics. Dalian, China.

Bandyopadhyay, S., M. Majumder, U.B. Adhikary, H.S. Maiti and G. Roychowdhury. 2002. "Field trial studies on performance evaluation of pilot plant unit for arsenic removal using ceramic filter module: Process of Workshop on Arsenic Hazards in Ground Water of West Bengal – Steps for Ultimate Solution." Presented to Central Ground Water Board. Kolkata.

Bandyopadhyay, S. 2012. "Ceramic Membrane Technology for Arsenic Mitigation: A sustainable community model working for about 10 years." Presented at International Conference on Water Quality with special reference to Arsenic. Kolkata.

Belfort, G., R.H. Davis, and A.L. Zydney. 1994. The behaviour of suspensions and macromolecular solutions in cross flow microfiltration. *Journal of Membrane Science* 96: 1–58.

Government of India. 2007. Report of the task force on formulating action plan for removal of arsenic contamination in West Bengal. Government of India, Planning Commission. New Delhi, India.

Chakraborti, D. et al. 2017. Groundwater arsenic contamination and its health effects in India. *Hydrogeology* 25: 1165–1181.

Bordoloi. 2012. Arsenic. India Mart. http://www.indiawaterportal.org/topics/arsenic.

Bose, P. and A. Sharma. 2002. Role of iron in controlling speciation and mobilization of arsenic in subsurface environment. *Water Research* 36: 4916.

Chakraborti, D., B.K. Biswas, G.K. Basu et al. 1999. Possible arsenic contamination free ground water source in Bangladesh, *Journal of Surface Science and Technology* 15: 180–188.

Chakraborti, D., M.K. Sengupta, M.M. Rahman et al. 2004. Ground water arsenic contamination and its health effects in the Ganga-Meghna-Brahmaputra plain. *Journal of Environmental Monitoring* 6: 75N–83N.

Chakraborti, D., S.C. Mukherjee, S. Pati et al. 2003. Arsenic Ground water Contamination in Middle Ganga Plain, Bihar, India: A Future Danger. *Environmental Health Perspectives* 111: 1194–1201.

Chang, S.D., W.D. Bellamy, and H. Ruiz. 1994. "Removal of arsenic by enhanced coagulation and membrane technology." Presented at AWWA 1994 Annual Conference. American Water Works Association. New York.

Chen, Y.S. 1973. Study of arsenic removal from drinking water. *Journal of American Water Works Association* 65: 543–548.

Clifford, D. and C.C. Lin. 1991. Arsenic III and arsenic V removal from drinking water in San Ysidro, New Mexico. *National Service Center for Environmental Publications*. United States.

Dhar, R.K., B.K. Biswas, G. Samanta et al. 1997. Ground water arsenic calamity in Bangladesh. *Current Science* 73: 48–59.

Fox K.R. and T.J. Sorg. 1987. Controlling arsenic, fluoride and uranium by point-of-use treatment. *Journal of American Water Works Association* 79: 53.

Harvey et al. 2002. Arsenic mobility and groundwater extraction in Bangladesh. *Science* 298: 1602–1606.

Huxstep, M.R. 1981. Inorganic contaminants removal from drinking water by reverse osmosis. EPA-600/2-81-115, USEPA Municipal Environmental Research Lab.

Václavíková, M., K. Vitale, G.P. Gallios, and L. Ivanicová. 2009. *Water Treatment Technologies for the Removal of High-Toxity Pollutants*. The Netherlands: Springer Science & Business Media.

Kinniburgh, D.G. and P.L. Smedley. 1995. Arsenic contamination of ground water in Bangladesh. *British Geological Survey* 1: 630.

Majumdar, S., S. Bandyopadhyay, S. Sensharma, S. Sarkar, S.K. Sarkar, T. Prasad Sahoo and K. Das. 2014. Ceramic membrane based technology for purification of water—Potential for rural development for community supply of quality drinking water. *Technologies for Sustainable Rural Development*. Jai Parkash Shukla, ed. 388–396. New Delhi: Allied Publishers.

Mustaque, A. and R. Chowdhuri. 2004. Arsenic crisis in Bangladesh. *Scientific American* 291.

Nomura, T. and S. Mimura. 1980. Properties of dynamically formed membranes. *Desalination* 32: 57–63.

Pagana, A.E., S.D. Sklari, E.S. Kikkinides, and V.T. Zaspalis. 2008. Microporous ceramic membrane technology for the removal of arsenic and chromium ions from contaminated water. *Microporous and Mesoporous Materials* 110: 150–156.

Pagana, A., S. Sklari, E.S. Kikkinides, and V. Zaspalis. 2007. Microporous ceramic membrane technology for the removal of arsenic and chromium ions from contaminated water. *Desalination* 217: 167.

Pawlak, Z., L. Scanlan, and P. Cartwright. 2002. "Speciation of arsenic in ground water and technologies for removal of arsenic in drinking water in the Spiro Tunnel Bulkhead, Park City, Utah." Presented at XVII-th ARS Separatoria – Borówno, Poland.

Rahman M.M., U.K. Chowdhury, S.C. Mukherjee et al. 2001. Chronic arsenic toxicity in Bangladesh and West Bengal—A review and commentary. *Journal of Toxicology: Clinical Toxicology* 39: 683–700.

Rahman, M., 2002. Arsenic and contamination of drinking-water in Bangladesh: A public-health perspective. *Journal of Health, Population and Nutrition* 20: 193–197.

Roy, S.N., T. Dey, M. Majumder, S. Majumdar, B.P. Ghosh, S. Bandyopadhyay and H.S. Maiti. 2005. Ceramic membrane and water purification. *Science and Culture* 71: 5–6.

Sabbatini, P., F. Yrazu, F. Rossi, G. Thern, A. Marajofsky, and M.M. Fidalgo de Cortalezzi. 2010. Fabrication and characterization of iron oxide ceramic membranes for arsenic removal. *Water Research* 44: 5702–5712.

Sato, Y., M. Kang, T. Kamei, and Y. Magara. 2002. Performance of nanofiltration for arsenic removal. *Water Research* 36: 3371–3377.

Shafiquzzaman, Md., Md. Shafiul Azam, J. Nakajima, and Q. Hamidul Bari. 2011. Investigation of arsenic removal performance by a simple iron removal ceramic filter in rural households of Bangladesh. *Desalination* 265: 60–66.

Shankar, S., U. Shanker, and Shikha. 2014. Arsenic contamination of ground water: A review of sources, prevalence, health risks, and strategies for mitigation. *The Scientific World Journal* 2014: 1–18.

Singh, A.K. 2007. Approaches for removal of arsenic from ground water of northeastern India, *Current Science* 92: 11.

Smith, A.H., 1998. "Arsenic pollution of ground water in Bangladesh: Causes, Effects and Remedies." Presented at International Conference at Dhaka Community Hospital Trust, Dhaka, Bangladesh and SOES.

Tomar, N.S. 2000. Report submitted at International Workshop on Arsenic contamination in Ground Water. PHED, Govt. of West Bengal, Kolkata.

Vecorena, V. and E. Rominna. 2016. Arsenic Analysis: Comparative Arsenic Ground water Concentration in Relation to Soil and Vegetation. California State University, San Bernardino. 279.

Vincent, A., N. Elkhiati, M. Moeller, D. Ragazzon, and M. Santalucia. 2016. "Arsenic Removal Process for Drinking Water Production: Benefits of R-SiC Microfiltration Membranes." Presented at Membrane Technology Conference and Exposition, MTC16 – San Antonio, Texas.

Volchek, K., S. Mortazavi, and H. Whittaker. 1996. *Removal of arsenic from aqueous liquids with selected alumina*. U.S. Patent 5,556,545 A.

7

Forward Osmosis: An Efficient and Economical Alternative for Water Reclamation and Concentration of Food Products & Beverages

Ravindra Revanur

CONTENTS

7.1 Introduction and Background

Forward osmosis (FO) is a membrane technique best suited for water reuse, waste minimization, product concentration, and, in some cases, water purification and desalination. While it is often labeled as a membrane technology to replace reverse osmosis (RO) membranes for desalination, a more accurate description of FO is as a unique tool to provide solutions not possible with conventional competing processes, including RO. In some cases, this means that FO can compete directly with other processes (e.g., RO or evaporators), but in other cases, FO complements these processes by expanding potential applications and making them more cost-effective.

In other words, FO is not simply a new type of process, but a tool that, at its best, can remove water from a product or waste stream where other membrane processes typically fail. FO can also result in significant savings in the form of capital expenditure, operating cost, energy use, or often and more importantly, the cost of transporting and storing this 'additional water' in the waste or product stream. Based on this introduction, a common question is "when would a user consider FO over competing processes?"

7.1.1 Example Benefits

An FO process, when designed and used appropriately, can provide new solutions for food and beverage concentration or "difficult to treat" waste and waste minimization projects. For example, FO can provide the following benefits when concentrating a natural beverage product, such as fruit juice:

1. A higher sugar concentration than a RO process, which is impeded by a hydrostatic pressure limitation. This is because FO is driven by osmotic pressure instead of hydrostatic pressure as in the case of RO.
2. A higher quality product than competing evaporators that utilize heat that degrades flavors, colors and nutrients. FO does not use heat, thereby preserving the quality of the product.

As for "difficult to treat" wastewater, FO offers the following benefits:

1. Reduced fouling, cleaning frequency, system downtime, and process complexity for wastes that have high oil and grease contents (not suitable for other types of polymeric membranes), fibrous materials (e.g., algae, pulp and paper wastes, etc.), sugars, or other organics that cause rapid fouling of other types of polymeric membranes (e.g., MF, UF, NF, RO, MD, etc.).
2. FO can be used as a dilution step (i.e., osmotic dilution) to transport clean water from a difficult waste stream (e.g., tertiary treated municipal wastewater) into seawater to reduce salinity by dilution, which reduces energy use and the cost of seawater desalination.

FO can also provide the following benefits when treating high-salinity wastes that exceed the limitations of RO membranes:

1. FO can concentrate aqueous solutions to higher salinity than RO.
2. FO can also utilize a membrane (non-traditional forms of NF or RO, membrane distillation, etc.) or a thermal draw recovery process that can operate more efficiently on a clean draw stream than a high-fouling, high-salinity feed in a suitable way to reduce capital, operating, and/or energy costs.
3. FO can occasionally use an onsite draw stream or brine, when available, without draw recovery, thus significantly reducing capital and energy costs.

7.1.2 Commercial Adoption

Although the theory of FO has been circulated since the 1960's, more recent advances in FO membranes and modules have been necessary for its commercial applications. Over the last few years, commercial FO-based systems have begun operating around the world, including in several waste minimization systems in China (FTSH2O, Oasys) and food and beverage processing in North America (Porifera). Additionally, pilot scale systems are being evaluated for industrial and municipal water reuse (Porifera, Modern Water, and Trevi Systems) and seawater desalination (Hyorim, Modern Water, and Trevi Systems).

7.1.3 Technical Background

FO is an osmotic process that uses a semi-permeable membrane to separate water from a feed solution. Unlike pressure-driven membrane separation processes, such as reverse osmosis (RO), nanofiltration (NF), etc., FO requires two solutions of a different nature on each side of the membrane to generate the driving force for filtration. The driving force for this separation is an osmotic pressure gradient between a solution of high concentration, referred to as the "draw solution," or simply the "draw," and a solution of lower concentration, referred to as the "feed." The osmotic pressure gradient is used to induce a net flow of water through the membrane into the draw, thus concentrating the feed, as shown in Figure 7.1. The draw solution can consist of a single or multiple solutes that may be simple salts (e.g., sodium chloride, magnesium chloride, magnesium sulfate, etc.) or can be a substance specifically tailored to FO applications, often referred to as a "designer draw" (e.g., ammonium bicarbonate, magnetic nanoparticles, switchable polarity solvents, etc.). The feed solution can be a dilute product stream (e.g., fruit or vegetable juice, dairy produce, etc.), waste stream, seawater, or high-salinity brine.

FO has been shown to reduce fouling rates and prevent irreversible fouling, when compared to pressure-driven membrane processes. Since osmosis is a spontaneous, chemically-driven process, FO requires zero or low hydraulic pressure. While it is often theorized that low hydraulic pressure reduces membrane fouling, recent studies suggest that a different mechanism(s) may be the reason for reduced fouling observations (Tow et al., 2017). Whatever the cause of reduced fouling, FO has shown advantages in treating challenging feeds where competing membranes are limited by either foulants or hydrostatic pressures.

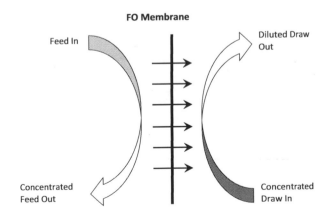

Water flows from the feed to the draw
solution by osmotic power

FIGURE 7.1
Schematic of forward osmosis separation process.

FO processes have gained attention in numerous applications, including water reuse and desalination, food and beverage (Jiao et al., 2004), mining, oil and gas (Bryan et al., 2015; Kerri et al., 2013), power (Bruce et al., 2012) and pharmaceutical (Xue et al., 2012; Fan-xin et al., 2015) industries. Most FO applications fall into two broad categories: product concentration and water reuse. Product concentration is more commonly used in the food and beverage, fragrance, dye industrial chemical, and mining industries. Water reuse encompasses both high-salinity and near-zero liquid discharge (ZLD) applications as well as low-salinity industrial and potable purposes.

In some cases, a "FO only" process can be used to concentrate the feed if there is an existing draw solution available for dilution. In such cases, the diluted draw may be disposed of or used at a diluted concentration. Such cases typically require that a site has multiple process streams and/or an infrastructure suitable for this arrangement, which may be rare. However, most FO applications require that the draw be recovered for reuse. This additional step processes the draw into a draw concentrate to power the FO process and generates a clean water permeate. This usually involves a combination of two different separation processes such as FO+RO, FO+MD (membrane distillation), or FO+Thermal Draw Recovery combinations, with the second process acting as the draw recovery step. In some cases, FO can concentrate a waste (e.g., high BOD waste) turning it into a product (e.g., biofuel feedstock), while simultaneously producing clean water for reuse.

The relationship among forward osmosis, pressure retarded osmosis (PRO) and conventional reverse osmosis—three closely related separation processes—is shown in Figure 7.2 (Lee et al., 1981). As explained earlier, in the FO process, a semipermeable membrane separates two solutions of different concentrations, such that water flow from the feed side into the draw solution occurs due to the difference in osmotic pressure between the two

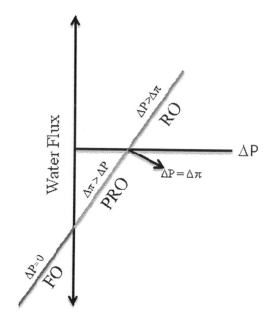

FIGURE 7.2
The illustration showing the relationship between FO, PRO and RO for a membrane. Modified image adopted from (Lee et al., 1981).

solutions. However, when a gradually increasing pressure is applied to the salt solution (in the direction from draw to feed), the rate of the water flow (i.e., flux) begins to decrease and will eventually become zero once the applied pressure equals the osmotic pressure difference across the membrane. The process in which water flows by osmosis across a membrane into a pressurized salt solution is known as PRO. Once the applied hydrostatic pressure into the salt solution exceeds the osmotic pressure difference across the membrane, the direction of water flow reverses and is referred to as RO. In conventional RO, hydrostatic pressure exceeding the osmotic pressure is applied to the feed to produce a clean water permeate.

7.2 Key Membranes and Desired Properties for FO

Commercial FO membranes are available in both asymmetric and thin film composite (TFC) forms. Asymmetric membranes are generally made in a single-step phase inversion process from cellulose triacetate (CTA) polymer solution. CTA is an ideal material for FO membranes because of its hydrophilic nature, a higher chlorine resistance compared to TFC membranes, and antifouling properties due to neutral surface charges. On other hand, these membranes have relatively low water permeability and higher reverse salt flux values than TFC, which impacts commercial system footprint and cost. Additionally, CTA membranes degrade when exposed to certain draw solutions (Baker, 2004) and undergo hydrolysis in harsh environments, including alkaline conditions. Alternatively, commercial TFC-FO membranes exhibit higher salt rejection and water flux values with better chemical stability and mechanical strength. TFC-FO membranes are also reportedly more stable than CTA membranes over a broad pH range of 2–12. TFC membranes are relatively hydrophobic and may have more surface charges. TFC membranes will have different fouling properties depending on the composition of the feed and draw solutions. Additionally, TFCs are sensitive to free chlorine. There are several companies that manufacture FO membranes (Bryan et al., 2014), including Porifera, Fluid Technology Solutions (formerly known as HTI), Oasys Water, and Aquaporin, most of which sell commercial membranes, modules, and/or systems.

As previously mentioned, the properties of the draw solution (high concentration) and feed solution (low concentration) are very important in terms of characterizing membrane performance. Deionized water is commonly used as the feed to compare performance with different draw solutions. Draw solution compounds, according to their physio-chemical properties, can be classified into inorganic salts, organic salts, volatile compounds, non-volatile compounds, polymers and oligomers. The compounds in these classifications may overlap with each other.

The other property often referred to in terms of membrane performance is the FO membrane orientation. The FO process can be operated with the membrane in two different membrane orientations: FO mode (forward osmosis mode) and PRO mode (pressure retarded osmosis mode). In FO mode, the membrane active (or skin layer) will be in direct contact with the feed solution and the draw solution flows against the porous support layer. When the draw solution is in direct contact with the membrane active layer and the feed flows against the membrane's porous support, the process is referred to as PRO mode. In general, permeate flux values during PRO mode are higher than in FO mode; however, the membrane is typically much more prone to fouling in PRO mode. The flux difference between FO mode and PRO mode operations is mainly due to external concentration polarization (ECP) and internal concentration polarization(ICP). A detailed description of concentration polarization is included in the following section.

7.3 Concentration Polarization

In this section, we discuss the reasons for lower permeate water flux through a membrane than theoretical calculations based on the osmotic pressure difference (i.e., driving force) across the membrane. The general equation to quantify the water flux (J_W) through a semi-permeable membrane is:

$$J_W = A\sigma(\Delta\pi - \Delta P) \tag{7.1}$$

where A is the membrane water permeability constant, σ is the reflection coefficient, $\Delta\pi$ is the osmotic pressure difference across the membrane, and ΔP is the applied pressure. For FO, ΔP is typically assumed to be zero, such that Equation 7.1 simplifies to:

$$J_W = A\sigma\Delta\pi \tag{7.2}$$

Assuming that $\sigma = 1$ would mean that the membrane completely rejects the solutes, which is usually assumed in FO studies, due to the high rejection property of the membrane (Baker, 2004). With this assumption, Equation 7.2 simplifies to:

$$J_W = A\Delta\pi = A(\pi_{D,b} - \pi_{F,b}) \tag{7.3}$$

From the above equation it is clear that there is generally a linear relationship between the osmotic pressure difference ($\Delta\pi$) of the bulk draw ($\pi_{D,b}$) and the bulk feed ($\pi_{F,b}$). However, actual water flux values are lower than theoretical water flux values calculated from osmotic pressure difference (Mehta and Loeb, 1978; Loeb et al., 1997). This is mainly because the effective osmotic pressure difference across the membrane ($\pi_{D,m} - \pi_{F,m}$) is lower than the bulk fluid osmotic pressure difference ($\pi_{D,b} - \pi_{F,b}$) referenced in Equation 7.3. This phenomenon is called concentration polarization (CP). CP may take several forms depending on the type of membrane and the operating conditions used during separation. Figures 7.3 and 7.4 illustrate the basic concepts that explain CP for dense and asymmetric membranes respectively.

7.3.1 External Concentration Polarization (ECP)

Consider a simple example of water and salt solutions used in a FO process where both water (from feed to draw) and salt (from draw to feed) pass through the membrane in opposite directions. As shown in Figure 7.3, movement of water from the feed to the draw effectively dilutes the draw side salt solution. Assuming a dense membrane, the osmotic power of the bulk draw ($\pi_{D,b}$) is higher than the draw concentration adjacent to the membrane surface ($\pi_{D,m}$), i.e., $\pi_{D,b} > \pi_{D,m}$. Similarly, the diffusion of salts from the draw to the feed

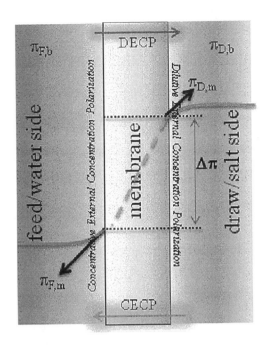

FIGURE 7.3
Pictorial representation of external concentration polarization with a dense membrane.

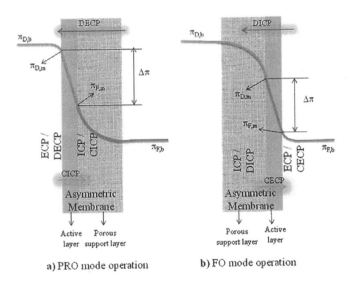

a) PRO mode operation   b) FO mode operation

FIGURE 7.4
Pictorial representation of external and internal concentration polarization with an asymmetric membrane during (a) PRO mode, (b) FO mode operations.

causes the accumulation of salts on the feed side, hence $\pi_{F,m} > \pi_{F,b}$ where $\pi_{F,b}$ and $\pi_{F,m}$ are feed osmotic power at bulk and adjacent to membrane surface, respectively. The changes in osmotic power occurring on the outer membrane surfaces are termed external concentration polarization (ECP). Because of this, ECP can be alleviated by mixing, which is generally created by increasing flow rates to create turbulence.

The process of the draw being diluted on the membrane surface by feed water transport through a membrane is called dilutive external concentration polarization (DECP). Similarly, solutes diffuse through the membrane from the draw to the feed surface of the membrane, which is referred to as concentrative external concentration polarization (CECP). The effects of DECP and CECP are limited to the active membrane surface where solute rejection is occurring.

7.3.2 ECP Model

In FO, ECP occurs at the surface of the membrane active skin layer, hence the models mentioned below are applicable to dense membranes and double skin membranes only. Understanding CECP and DECP is very important for accurately estimating the effective osmotic driving force to precisely determine FO membrane flux performance. However, it is not easy to estimate the effective feed concentration at the membrane surface. McCutcheon and Elimelech (2006; 2007) modeled ECP in FO using boundary layer film theory to calculate CP in pressure-driven membrane processes (Mulder, 1996) which was expressed as:

$$\frac{C_{F,m}}{C_{F,b}} = \exp\left(\frac{J_W}{k}\right) \tag{7.4}$$

where J_W is the water flux, k is the mass transfer coefficient, and $C_{F,m}$ and $C_{F,b}$ are the feed solution concentrations at the membrane surface and in the bulk phase, respectively.

The mass transfer coefficient (k) mentioned in Equation 7.4 is related to the Sherwood number (Sh) such that:

$$k = \frac{ShD}{d_h} \tag{7.5}$$

where D is the solute diffusion coefficient and d_h is the hydraulic diameter.

In Equation 7.4, when the FO feed solution is relatively dilute, a reasonable assumption can be made to replace $C_{F,m}$ and $C_{F,b}$ with corresponding solute osmotic pressures as demonstrated below.

$$\frac{\pi_{F,m}}{\pi_{F,b}} = \exp\left(\frac{J_W}{k}\right) \tag{7.6}$$

Equation 7.6 explains CECP in which $\pi_{F,m}$ and $\pi_{F,b}$ are the osmotic pressure of feed solution at membrane surface and in bulk phase, respectively. A positive exponent in the above equation is due to $\pi_{F,m} > \pi_{F,b}$. DECP is observed in FO during operation in PRO mode. In DECP, the draw concentration at the membrane surface is much lower than that of the bulk fluid i.e., $\pi_{D,b} > \pi_{D,m}$, such that Equation 7.6 for DECP can be rewritten with a negative exponent, as shown below.

$$\frac{\pi_{D,m}}{\pi_{D,b}} = \exp\left(-\frac{J_W}{k}\right) \tag{7.7}$$

Substituting Equations 7.6 and 7.7 in Equation 7.3 yields:

$$J_W = \left[\pi_{D,b} \exp\left(-\frac{J_W}{k} \right) - \pi_{F,b} \exp\left(\frac{J_W}{k} \right) \right] \tag{7.8}$$

Both concentrative ECP and dilutive ECP are considered in the above equation. However, for most of the osmotic processes, asymmetric membranes are commonly used today and thus, the usefulness of this flux model is limited. We therefore need to understand flux models for asymmetric membranes in which significant ICP effects are observed.

7.3.3 Internal Concentration Polarization

Most commercial FO membranes are less than 100 μm thick. Thinner TFC membranes are more permeable and consist of an ultra-thin active skin layer on a relatively thick porous support known as a thin film composite (TFC). Considering the two different membrane orientations for FO operation (FO mode and PRO mode), the CP concept becomes much more complex to understand, for FO operations using TFC membranes. In this case, the existence of two distinct boundary layers creates ICP on both sides of the membrane and reduces the effective flux. The second boundary layer occurs within the porous support layer of the membrane and is generally referred to as internal concentration polarization (ICP). Figure 7.4 illustrates both ECP and ICP effects in FO operations conducted in PRO mode (Figure 7.4 (a)) and FO mode (Figure 7.4 (b)).

Contrary to ECP, ICP is inversely proportional to membrane thickness, as shown in Figure 7.4. Flow turbulence has little effect in minimizing ICP because it is hard to remove or mix the solute molecules trapped inside the pores. ICP will contribute to significant water flux decline in FO (Mehta & Loeb, 1978; McCutcheon & Elimelech, 2006). Previous FO studies reported that ICP could reduce water flux by over 80%, however, understanding the details is necessary (Mehta & Loeb, 1978) as operating conditions matter significantly.

Similar to ECP, ICP can also be subdivided into two types depending on membrane orientation. In PRO mode (see Figure 7.4 (a)), where the draw solution is in contact with the active skin layer of a TFC membrane, water enters the porous support layer and permeates across the active layer into the draw solution, which can be explained as DECP. At the same time, the salt in the feed freely enters the open structure of the membrane porous support and can become trapped. Salt cannot easily penetrate the pores in the active layer basically designed to reject salts—thus, the concentrations in the pores will increase. This process is generally referred to as concentrative internal concentration polarization (CICP).

7.3.4 Model for Concentrative Internal Concentration Polarization

Lee et al., (1981) and Loeb et al., (1987) describe the following expression for CICP and how it relates to water flux and other membrane constants:

$$K = \left(\frac{1}{J_W} \right) \ln \frac{(B + A\pi_{D,m} - J_W)}{B + A\pi_{F,b}} \tag{7.9}$$

where, B is the salt permeability coefficient of the active skin layer of the membrane and K is the diffusion resistivity of solute within pores, defined by:

$$K = \left(\frac{t\tau}{D\varepsilon} \right) = \frac{S}{D} \tag{7.10}$$

Here, D is the diffusion coefficient of the solute, whereas t, τ, and ε, respectively, represent the thickness, tortuosity, and porosity of the porous support layer of the membrane; and S is the structural parameter of membrane. The importance of the S value on membrane performance will be discussed in following sections.

The assumptions are: that (1) the membrane is operated at very high flux, and (2) the solute transport from the feed to the draw side is negligible indicating a high percentage of solute (salt) rejection by the membrane, which means B is negligible compared to the other terms in Equation 7.9. With these assumptions and upon rearranging the parameters in Equation 7.9, the membrane flux (J_W) can be derived as:

$$J_W = A \left[\pi_{D,m} - \pi_{F,b} \exp(J_W K) \right] \tag{7.11}$$

Therefore, the product of the water permeability coefficient and the effective osmotic driving force gives the water flux through the membrane. The exponential term in Equation 7.11 refers the CICP is defined as:

$$\frac{\pi_{F,i}}{\pi_{F,b}} = \exp(J_W K) \tag{7.12}$$

where $\pi_{F,i}$ is the osmotic pressure of the feed underneath the active layer within the porous support. As shown in Figure 7.4 (a), the effect is concentrative when $\pi_{F,i} > \pi_{F,b}$.

By substituting $\pi_{F,m}$ values from Equation 7.7 into Equation 7.11, we can obtain a model for the effect of ICP and ECP on permeate flux as:

$$J_W = A \left[\pi_{D,b} \exp\left(-\frac{J_W}{k} \right) - \pi_{F,b} \exp(J_W K) \right] \tag{7.13}$$

One can solve Equation 7.13 by incorporating the inputs obtained from experiments or calculations to predict water flux through an asymmetric membrane in PRO mode. Expression 7.13 is coupled with concentrative ICP and dilutive ECP contributions. In PRO mode operation, dilutive ICP has little effect and hence it has not been included in Equation 7.13.

7.3.5 Model for Dilutive Internal Concentration Polarization (DICP)

In the case of FO mode operation as shown in Figure 7.4 (b), where the feed solution is in contact with the active skin layer of the membrane and the draw solution is in contact with the porous support side of the membrane, dilutive ICP typically dominates water flux. In this case, water permeates through the membrane, dilutes the salt in the pores in boundary layers, drastically reducing the net driving force between the feed and the draw. In

FO mode, dilutive ICP plays a major role and concentrative ECP can be ignored ($\pi_{F,b} = \pi_{F,a}$). Loeb et al. considered the effect of DICP on water flux in the FO mode operation as:

$$K = \left(\frac{1}{J_W}\right) \ln \frac{B + A\pi_{D,b}}{B + J_W + A\pi_{F,m}} \tag{7.14}$$

The permeate water flux expressed in Equation 7.15 was obtained by rearranging Equation 7.14 and assuming that salt permeability is negligible (i.e., $B = 0$, $\sigma = 1$).

$$J_W = A\left[\pi_{D,b} \exp(-J_W K) - \pi_{F,m}\right] \tag{7.15}$$

in which $\pi_{D,b}$ is given by

$$\frac{\pi_{D,m}}{\pi_{D,b}} = \exp(-J_W K) \tag{7.16}$$

Here, $\pi_{D,m}$ is the concentration of draw on the inside of the active layer within the porous support. The negative exponent indicates that $\pi_{D,b} > \pi_{D,m}$.

By substituting Equation 7.6 into Equation 7.15, we get:

$$J_W = A\left[\pi_{D,b} \exp(-J_W K) - \pi_{F,b} \exp\frac{J_W}{k}\right] \tag{7.17}$$

Equation 7.17 models water flux for asymmetric membranes during FO mode operation. The equation includes both DICP and CECP terms. The expressions clearly indicate that both ECP and ICP contribute negatively to the overall osmotic driving force. The severity of negative impact increases with an overall increase in driving force, i.e., the higher the membrane flux, the higher the CP effect. Estimation of various parameters from the expressions subsequently helps to minimize the CP contributions by changing operating conditions, membrane characteristics and boundary conditions.

7.4 Forward Osmosis Membrane Properties

FO processes have inherent advantages in treating challenging feeds with high amounts of suspended solids or high osmotic pressures. However, until recently FO membranes and systems had not seen broad adoption. For example, early TFC-FO membranes, typically the RO membranes, were too thick for efficient FO such that internal concentration polarization would significantly impact performance. An efficient FO membrane should possess: (1) high chemical resistance to potential feed and draw solutions for stability, and (2) high water permeability and solute rejection for performance. To maximize performance the ideal membrane structure is built on a thin, microporous substrate to minimize ICP with hydrophilic material with a neutral charge to enhance water flux and improve antifouling properties. Researchers have considered a wide variety of materials and structures for FO membranes in their pursuit of the optimal FO membrane. Early commercial deployments of FO utilized flat sheet cellulose triacetate asymmetric (CTA) FO membranes. As with the case of the first RO membranes, these CTA membranes suffered from poor water permeability and were susceptible to degradation due to their low chemical stability. Polyamide TFC-FO membranes of 50–100 μm thickness, specifically designed for FO, have been introduced commercially as an alternative to the existing asymmetric CTA membranes as they exhibit higher water flux and salt rejection.

7.4.1 TFC-FO Membrane Structure and Modification

TFC membrane preparation involves two steps: first, making a porous support by phase inversion and second, incorporating an ultra-thin dense active layer on the porous support by interfacial polymerization (IP). The properties of the support layer dictates ICP, and the IP layer dictates water permeability and salt rejection. The ideal FO membrane would be manufactured in such a way to minimize ICP and to maximize water permeability and salt rejection. The best parameter used to describe these characteristics is to target a low membrane structural parameter (S).

A recent effort to evaluate FO membrane performance was primarily focused on reducing S. The solute's ability to diffuse into or out of the membrane support layer and membrane structural parameter are derived from Equation 7.10:

$$S = \frac{t\tau}{\varepsilon} \tag{7.18}$$

where t, τ and ε are the thickness, tortuosity, and porosity of the membrane, respectively. The structural parameter, S, is an important intrinsic parameter of a membrane and it determines ICP in the membrane support layer. The S value of a membrane can be obtained according to Equation 7.18 and by correlating FO test results (Tiraferri et al., 2013; Phillip et al., 2010). Equation 7.18 indicates that a smaller S value can be obtained by reducing tortuosity, increasing porosity and/or reducing the thickness of the support layer structure. The phase inversion step, or procedures involved during phase inversion process, plays an important role in controlling S values of the final membrane.

In addition to reducing the S value, alternative methods have been made to reduce the effects of ICP. These include modifying the polymeric substrate by incorporating nanomaterials using electrospun nanofiber substrates, casting porous supports onto vertically aligned nanotube substrates, and preparing double skin FO membranes.

In the FO process, reverse solute diffusion through the membrane from the draw to the feed solution is inevitable. This reverse solute diffusion not only reduces water flux through the membrane but may also enhance membrane fouling when the chemistry is not compatible (Lee at al., 2010). It can also alter the characteristics of the concentrated product especially when used in food and beverage applications. The specific reverse solute flux, which is defined as the ratio of the reverse solute flux to the forward water flux, is considered a primary measure of FO membrane performance (Hancock & Cath, 2009). Many membranes achieve high flux by reducing the membrane selectivity, resulting in a high reverse solute flux. However, in most cases, the high reverse solute flux results in either increased operational costs or reduced product quality. The ideal FO membrane should have both a high water flux and a low reverse solute flux.

7.4.2 Hydrophilic TFC-FO Membranes for Enhanced Antifouling Properties

Many studies have focused on post-treatment methods for improving membrane properties. One way to reduce membrane fouling is to increase the hydrophilicity, or reduce the hydrophobic nature of the membrane. Hydrophilization is a membrane post-treatment process that makes the membrane more hydrophilic, increasing the membrane permeability and chlorine resistance. There are multiple methods (e.g., synthesis, solvent treatment, coatings, etc.) that can be used; however, some methods perform better than others.

Synthesizing modified membranes during the IP step has had little success because the additives used to enhance the hydrophilicity of the IP layer typically affect the quality of IP adversely, which reduces membrane performance by increasing the reverse solute flux. Other post-treatment procedures, such as treating the membranes with various water-soluble solvents (e.g., acids such as HF and HCl and alcohols such as ethanol and isopropanol) have been shown to increase membrane flux, but only temporarily. Hydrophilization has also been achieved by coating the membrane with more hydrophilic compounds through chemical reactions. In such cases, changes in membrane performance typically last longer than solvent treatment.

Recent developments of high-flux TFC-FO membranes have led to the increased use of FO in various applications. However, the impact of operating at higher permeation fluxes can potentially increase the fouling and/or scaling rates on the membrane surface (Xie et al., 2014; Gimun & Seungkwan, 2017). This creates the need for adapting antifouling strategies including modifying the membrane surface to minimize fouling. Membrane surface modification procedures reported in the literature for RO and NF membranes can similarly be employed on FO membranes to enhance antifouling properties. However, there is no single modification or method that can address all fouling problems associated with a wide variety of different foulants.

It is commonly reported that increasing hydrophilicity (Revanur et al., 2007; Li et al., 2017), makes the membrane surface charge neutral, reduces membrane surface roughness, optimizes flow and flux rates to consequently reduce membrane fouling. In some reports, it was demonstrated that double skin layer FO membranes have a lower fouling propensity than conventional single skin membranes (Duong et al., 2014). Membrane surface modification using polyethylene glycol or incorporating nanoparticles (Nguyen et al., 2014) also showed a substantial reduction in fouling.

A quick and effective coating method using an aqueous dopamine solution was proposed by Lee et al. (2007) and Kim et al. (2013). A non-selective coating of thin polydopamine on any surface, including polymer membranes, is also gaining attention. Dopamine coatings unlike other antifouling coating methods will not drastically reduce membrane flux because they enhance the wetting property of polysulfone-polyamide hydrophobic TFC membranes. A major disadvantage of dopamine coating is that it cannot withstand harsh chemicals and may delaminate under high cross-flow conditions (McCloskey et al., 2010).

7.5 Forward Osmosis Membrane Module Configurations: Advantages and Disadvantages

There are several different types of membrane modules on the market, and recent innovations have contributed to significantly lower cost and higher system efficiency compared to conventional module types.

The ideal membrane elements and modules maximize the membrane surface to volume ratio to enhance cross-flow velocity across the membrane surface for both feed and draw channels. This high cross-flow minimizes ECP on both sides of the membrane, resulting in higher specific flux. To achieve high flow rates, the element should exhibit low head-loss, allowing multiple elements to be connected within a module in series without a high pressure drop across the module. To achieve an efficient FO process, it is very important for the module to be able to operate in both co-current and counter-current configurations. A co-current operation occurs when both the feed and draw solutions enter and exit on the same sides of a module. A counter-current operation occurs when the feed and draw enter and exit on opposite sides of a module. A co-current configuration is suitable for low osmotic pressure feeds and low water recovery applications, while counter-current has advantages for high osmotic pressure feeds and high water recovery applications. The counter-current operation results in constant flux across the membrane module as the feed gets concentrated and the draw is diluted. This is ideal for FO processing because it reduces the required average draw concentration to achieve a target feed concentration. A lower average draw concentration and consistent osmotic driving force reduces fouling and RSF and improves solute rejection compared to co-current operation. This is because the flux in the first few elements is very high and very low in the final few elements in series is in co-current operation. The higher flux exacerbates fouling, the lower flux can allow higher solute passage, and the higher average draw concentration increases RSF.

Early FO modules were based on conventional RO designs. Spiral wound are most commonly used due to their popularity in seawater desalination industry as well as the wide availability of pressure vessels modules (Bamaga et al., 2011; Gu et al., 2011). To convert the RO spiral wound module with three ports (feed, draw, permeate) into an FO module with four ports (feed in, feed out, draw in, draw out), an extra glue line is needed to create the required flow path. As a result, FO spiral modules contain a lower packing density than RO spiral modules. The added glue lines also can create low flow zones within the module that reduce mixing and contribute to higher head loss than in similar RO systems. While spiral wound modules are relatively easy to operate in co-current mode, it is difficult to operate them in counter-current mode, especially at high recovery due to high head loss and difficulties in designing and operating multi-stage units.

The other type of module commonly found in RO is the hollow fiber membrane (Liu et al., 2013; Setiawan et al., 2011). Hollow fiber (HF) membranes have the potential to offer the highest packing density. However, there is a trade-off between packing density and fouling potential. This is because small diameter fibers allow a high packing density, but also intrinsically have a small flow path for solids and foulants to pass through. Small flow paths increase head-loss, which increases energy use and restricts mixing at the membrane surface. Reduced mixing enhances fouling, scaling and concentration polarization. Hollow fibers have also traditionally been limited to using asymmetric membranes with lower flux and rejection than TFCs. Thin film composite HF membranes are an active area of research showing some success (Panu & Tai-Shung, 2012), yet thus far, TFC hollow fibers have only been made in the laboratory. Furthermore, these lab-scale HFs have only been possible with large fibers, which nullifies the advantage of high packing density compared to flat-sheet membrane modules. One benefit is that small HFs can often withstand relatively high hydraulic pressures without deformation, which can be useful for certain applications, such as PRO.

New FO elements and modules have recently been developed specifically for FO applications (Gu et al., 2011). These modules are similar to plate and frame designs, but have been optimized for FO and have a higher packing density and smaller footprint than traditional plate and frame types. The advantage of this new design is that it provides large flow paths without dead zones on both the feed and draw sides of the membrane. Additionally, these modules can be designed to operate at high cross-flow velocities with significantly lower rates of head loss (pressure drop) than other types of modules. This improves mixing, which is the key to reducing CP and maximizing membrane performance, as previously described. Another benefit of these types of modules is that they can be designed to have a uniform surface velocity across an entire array of membrane modules in series and in parallel, which minimizes CP and is virtually impossible to achieve with other module types due to flow path, housing, and system design considerations. These modules are relatively easy to operate in both co-current and counter-current modes.

7.6 Effect of Draw

Selection of a suitable draw solution is a key factor in selecting FO operational conditions. In general, a good draw solute should be non-toxic, inexpensive, readily available, relatively easy to regenerate and highly soluble in various conditions to avoid precipitation. Amongst many varieties of available draw solutes, sodium chloride is used most often due to its high diffusivity (high specific flux) and because it is relatively easy to re-concentrate using conventional desalination processes such as RO. Apart from inorganic salts and sugars, recent publications indicate that polymers, poly-electrolytes, nanoparticles, dendrimers, volatile solutes, switchable polarity solvents/polymers, and organic ionic salts are also being used to make draw solutions (Lutchmiah et al., 2014).

   Zhao and Zou (2011) reported that FO water flux through membranes was different with various draw solutes even under the same net osmotic driving force. Using sucrose as the draw solution caused a severe dilutive ICP effect, while a NaCl draw provided higher flux values due to lesser ICP effects. These differences, in reference to the dilutive ICP and effects of different draw solutions, may be reasonably explained on the basis of diffusivity and viscosity of the draw solution (Hancock & Cath, 2009). A solution with higher viscosity would be more difficult to diffuse into and out of the membrane porous layer, leading to more severe ICP. By considering variations in solute molecular size, solution viscosity and diffusion of solutes, Equation 7.10 can be modified by introducing the concept of constrictivity.

$$K = \frac{t\tau}{\delta \varepsilon_{eff} D} \qquad (7.19)$$

   Here, a new parameter (δ) is the constrictivity factor, and ε_{eff} is the effective transport through porosity. The value of ε_{eff} is always less than the overall porosity of the membrane. In other words, the ratio of the solute diameter to the pore diameter, i.e., ϕ influences δ. The empirical relationship between ϕ and δ can be expressed as (Beck & Schultz, 1970):

$$\delta = (1 - \phi)^4 \qquad (7.20)$$

$$\phi = \frac{molecular\ diameter}{pore\ diameter} < 1 \qquad (7.21)$$

From Equations 7.19 and 7.20, it is inferred that larger solute size will result in larger ϕ, but smaller δ. It also reduces the effective porosity of the membrane due to size exclusion and restricted diffusion. As a result, the solute resistivity, K, will become higher based on Equation 7.19.

These relationships, along with constrictivity (as described earlier), indicated that a draw solution with lower viscosity and higher diffusivity allows the draw solute to diffuse into and out of the membrane support layer more easily, thus reducing ICP. Therefore, when selecting an ideal draw solution, it is necessary to consider: solute molecule/ion size, diffusivity and viscosity. Otherwise, permeate flux will be limited significantly by ICP while operating in FO mode. Additional factors that also need to be considered when selecting and comparing draw solutions for commercial applications include but are not limited to the following:

- Membrane vs. thermal draw recovery methods and related efficiencies.
- Chemistry (toxicity, corrosiveness, scaling potential, ion exchange between feed and draw, etc.).
- Point of entry (e.g., co-current vs. counter-current operation) and pressure drop.
- Integrity monitoring (e.g., dye addition or other monitoring method).
- Draw conditioning (e.g., blowdown; oxidant, antifoulant and/or antiscalant addition).
- Availability and ease of use for adding draw solute that is lost from the system.

7.7 Conclusions

In this chapter, the FO process has been described as a new tool for product concentration and water reuse applications. FO membranes, modules, draw solutions and FO processes were also presented and discussed. Furthermore, the theory behind the effects of internal and external concentration polarization were presented and their impact on permeate flux was discussed in detail.

The use of hybrid systems, i.e., FO-RO, FO-MD, etc., are common and often necessary to implement FO for real-world applications. Advances in FO technology, including membranes, modules, systems and processes, are enabling many different applications with its most attractive implementations occurring wherever conventional membrane technology fails due to fouling or high osmotic pressures.

References

Baker, R.W. 2004. *Membrane Technology and Applications: 2nd Edition.* New York: Wiley.

Bamaga, O.A., A. Yokochi, B. Zabara, and A.S. Babaqi. 2011. Hybrid FO/RO desalination system: Preliminary assessment of osmotic energy recovery and designs of new FO membrane module configurations. *Desalination* 268: 163–169.

Beck, R.E., and J.S. Schultz. 1970. Hindered diffusion in microporous membranes with known pore geometry. *Science* 170: 1302–1305.

Bruce, E.L. and M. Elimelech. 2012. Membrane-based processes for sustainable power generation using water. *Nature* 488: 313–319.

Bryan, D.C., N. Almaraz, and Y.C. Tzahi. 2015. Forward osmosis desalination of oil and gas wastewater: Impacts of membrane selection and operating conditions on process performance. *Journal of Membrane Science* 488: 40–55.

Bryan, D.C., X. Pei, G.B. Edward, H. Jack, L. Keith., T.H. Nathan, and Y.C. Tzahi. 2014. The sweet spot of forward osmosis: Treatment of produced water, drilling wastewater, and other complex and difficult liquid streams. *Desalination* 333: 23–35.

Duong, P.H.H., T.S. Chung, S. Wei, and L. Irish. 2014. Highly permeable double-skinned forward osmosis membranes for anti-fouling in the eulsified oil-water separation process. *Environmental Science and Technology* 48: 4537–4545.

Fan-xin, K., Y. Hong-wei, W. Yu-qiao, W. Xiao-mao, and F.X. Yuefeng. 2015. Rejection of pharmaceuticals during forward osmosis and prediction by using the solution–diffusion model. *Journal of Membrane Science* 476: 410–420.

Gimun, G. and H. Seungkwan. 2017. New approach for scaling control in forward osmosis by using in antiscalant-blended draw solution. *Journal of Membrane Science* 530: 95–103.

Gu, B., D.Y. Kim, J.H. Kim, and D.R. Yang. 2011. Mathematical model of flat sheet membrane modules for FO process: Plate-and frame module and spiral-wound module. *Journal of Membrane Science* 379: 403–415.

Hancock, N.T. and T.Y. Cath. 2009. Solute coupled diffusion in osmotically driven membrane processes. *Environmental Science and Technology* 43: 6769–6775.

Jiao, B., A. Cassano, and E. Drioli. 2004. Recent advances on membrane processes for the concentration of fruit juices: A review. *Journal of Food Engineering* 63: 303–324.

Kerri, L.H., T.H. Nathan, R.H. Nathan, W.A. Eric, G.B. Edward, X. Pei, and Y.C. Tzahi. 2013. Forward osmosis treatment of drilling mud and fracturing wastewater from oil and gas operations. *Desalination* 312: 60–66.

Kim, H.W., B.D. McCloskey, T.H. Choi, C. Lee, M.J. Kim, B.D. Freeman, and H.B. Park. 2013. Oxygen concentration control of dopamine-induced high uniformity surface coating chemistry. *ACS Applied Material Interfaces* 5: 233–238.

Lee, H., S.M. Dellatore, W.M. Miller, and P.B. Messersmith. 2007. Mussel-inspired surface chemistry for multifunctional coatings. *Science* 318: 426–430.

Lee, K.L., R.W. Baker, and H.K. Lonsdale. 1981. Membranes for power generation by pressure retarded osmosis. *Journal of Membrane Science* 8: 141–171.

Lee, S., C. Boo, M. Elimelech, and S. Hong. 2010. Comparison of fouling behavior in forward osmosis (FO) and reverse osmosis (RO). *Journal of Membrane Science* 365: 34–39.

Li, X., X. Hu, and T. Cai. 2017. Construction of hierarchical fouling resistance surfaces onto poly(vinylidene fluoride) membranes for combating membrane biofouling. *Langmuir* 33: 4477–4489.

Liu, C., W. Fang, S. Chou, L. Shi, A.G. Fane, and R. Wang. 2013. Fabrication of layer-by-layer assembled FO hollow fiber membranes and their performances using low concentration draw solutions. *Desalination* 308: 147–153.

Loeb, S., L. Titelman, E. Korngold, and J. Freiman. 1997. Effect of porous support fabric on osmosis through a Loeb–Sourirajan type asymmetric membrane. *Journal of Membrane Science* 129: 243–249.

Lutchmiah, K., A.R.D. Verliefde, K. Roest, and L.C. Rietveld. 2014. Forward osmosis for application in wastewater treatment: A review. *Water Research* 58: 179–197.

McCloskey, B.D., H.B. Park, H. Ju, B.W. Rowe, D.J. Miller, B.J. Chun, K. Kin, and B.D. Freeman. 2010. Influence of polydopamine deposition conditions on pure water flux and foulant adhesion resistance of reverse osmosis, ultrafiltration, and microfiltration membranes. *Polymer* 51: 3472–3485.

McCutcheon, J.R. and M. Elimelech. 2006. Influence of concentrative and dilutive internal concentration polarization on flux behavior in forward osmosis. *Journal of Membrane Science* 284: 237–247.

McCutcheon, J.R., and M. Elimelech. 2007. Modelling water flux in forward osmosis: Implications for improved membrane design. *American Institute of Chemical Engineers* 53: 1736–1744.

Mehta, G.D., and S. Loeb. 1978. Internal polarization in the porous substructure of a semi-permeable membrane under pressure-retarded osmosis. *Journal of Membrane Science* 4: 261–265.

Mehta, G.D. and S. Loeb. 1978. Performance of permasep B-9 and B-10 membranes in various osmotic regions and at high osmotic pressures. *Journal of Membrane Science* 4: 335–349.

Mulder, M. 1996. *Basic Principles of Membrane Technology.* Dordrecht: Kluwer Academic.

Nguyen, A., L. Zou, and C. Priest. 2014. Evaluating the anti-fouling effects of silver nanoparticles regenerated by TiO2 on forward osmosis membrane. *Journal of Membrane Science* 454: 264–271.

Panu, S. and C. Tai-Shung. 2012. High performance thin-film composite forward osmosis hollow fiber membranes with macrovoid-free and highly porous structure for sustainable water production. *Environmental Science and Technology* 46: 7358–7365.

Phillip, W.A., J.S. Yong, and M. Elimelech. 2010. Reverse draw solute permeation in forward osmosis: Modeling and experiments. *Environmental Science and Technology* 44: 5170–5176.

Revanur, R., B. McCloskey, K. Breitenkamp, B.D. Freeman, and T. Emrick. 2007. Reactive amphiphilic graft copolymer coatings applied to poly(vinylidene fluoride) ultrafiltration membranes. *Macromolecules* 40: 3624–3630.

Setiawan, L., R. Wang, K. Li, and A.G. Fane. 2011. Fabrication of novel poly(amide-imide) layer. *Journal of Membrane Science* 369: 196–205.

Tiraferri, A., N.Y. Yip, A.P. Straub, S. Romero-Vargas Castrillon, and M. Elimelech. 2013. A method for the simultaneous determination of transport and structural parameters of forward osmosis membranes. *Journal of Membrane Science* 444: 523–538.

Tow, E.W., V. Lienhard, and H. John. 2017. Unpacking compaction: Effect of hydraulic pressure on alginate fouling. *Journal of Membrane Science* 544: 221–233.

Xie, M., L.D. Nghiem, W.E. Price, and M. Elimelech. 2014. Impact of organic and colloidal fouling on trace organic contaminant rejection by forward osmosis: role of initial permeate flux. *Desalination* 336: 146–152.

Xue, J., S. Junhong, W. Can, W. Jing, and Y.T. Chuyang. 2012. Rejection of pharmaceuticals by forward osmosis membranes. *Journal of Hazardous Materials* 227–228: 55–61.

Zhao, S. and L. Zou. 2011. Relating solution physicochemical properties to internal concentration polarization in forward osmosis. *Journal of Membrane Science* 379: 459–467.

Section III

Health

8

Low-Cost Production of Anti-Diabetic and Anti-Obesity Sweetener from Stevia Leaves by Diafiltration Membrane Process

Shaik Nazia, Bukke Vani, Suresh K. Bhargava, and Sundergopal Sridhar

CONTENTS

8.1 Introduction

Globally, 3% of the total population is affected by diabetes mellitus, a chronic disorder characterized by high blood sugar (glucose) levels. The outset of diabetes has already affected more than 40.9 million people, thus urging the need for a natural sweetener (Midmore et al., 2002). Stevia plants originated from Paraguay and belong to the Asteraceae family, which bears medicinal and commercial importance (Pourvi et al., 2009).

Due to its natural sweetness, stevia can replace the consumption of sugar—one of the main causes of obesity (Alhady, 2011). The major role of stevioside in reducing blood sugar level is significant to diabetes affected people. Even the regular consumption of stevioside helps in regulating glucose and cholesterol levels, increases cell growth, and strengthens blood vessels (Pourvi et al., 2009). Attracted by its non-toxic, zero-calorie and stabilizing characteristics, numerous research studies are being conducted on this "sweet herb" for bringing out stevioside as a natural, alternative sweetener with a zero glycemic index (Das et al., 2011; Stephen et al., 2010). For decades, stevia leaves had been used as sweeteners by natives of Brazil and Paraguay for medicines and tea (Kinghorn et al., 1985; Kennelly, 2002; Geuns, 2003; Ahmad et al., 2014), but research reports have also revealed the structure of stevia glycosides (Soejarto et al., 1983; Brandle et al., 1998). Some of the food industries are also including stevia as a natural sweetening agent (Shi et al., 2000). The reason for its typical regulatory properties is attributed to its release mechanisms, wherein the steviol glycosides are usually broken down into steviol and glucose by the bacteria colonizing the large intestine, rather than digested in the alimentary tract. Hence, glucose is prevented from absorption to the bloodstream, but metabolized by the intestinal flora; however, consumption not exceeding 5 mg/kg per day is reported to be safe (Atteh et al., 2011). Its hypertensive effect in adult women, especially in the age group of 28–75 years, has been studied experimentally to understand its blood-regulating properties (Chen et al., 2005). Research further reinforces that the concentrate of stevia leaves could help to improve not only the nutritional value of food products, but their functional properties, as well (Ulbricht, 2010). The efficiency of steviosides in treating patients suffering from carbohydrate metabolic diseases, such as diabetes, obesity, high blood pressure and stimulated cell regeneration, is also observed to be promising (Jain et al., 2012), provided the concentration of stevioside is higher (Mogra et al., 2009).

8.1.1 History of Stevia Glycosides

The inimitable flavor of stevia leaves was first identified by the people of Paraguay and Brazil and later, the Japanese used the leaves to sweeten their ancient herbal tea, "yerba mate." In the 1800s, the use of stevia was commonly enjoyed not only in Brazil, but also by adjacent South American communities. In 1887, Moises Santiago Bertonihad discovered stevia from Indian guides while exploring Paraguay's eastern forests (Lewis, 1992). The use of stevia as a sweetener was first described in a botanical journal in 1900. In 1931, two French chemists reported the isolation of stevioside, initiating the eventual production of stevioside (Stuart et al., 1987). This remarkable discovery in the early 1970s has enabled large-scale cultivation of stevia for its use as an artificial sweetener. In 1971, Japan became the first nation to commercially produce a stevia sweetener, also using it in a variety of food products (Parashar et al., 2013). From then onward, stevioside has been regularly used as a natural sweetener in Japan and has established its role as a natural sugar substitute (Barathi et al., 2003).

8.1.2 Botanical Description of the Plant

Stevia rebaudiana, a new member species of Asteraceae family can grow up to 1 m height and bears 2–3 cm long, elliptical leaves in an alternate arrangement (Shizhen, 1995). Leaves are generally sessile, opposite lancoelate in shape and are serrated above the middle (Yadav et al., 2011). The plant has brittle stem with an extensive root system, and its flowers are of a smaller size (7–8 mm), white in color with a pale purple throat (Sivaram et al., 2003). The amount of steviol glycoside in different parts of the plant is found to decline in the following order: leaves > flower > stems > seeds > roots. The highest content of steviol glycosides is found in the upper young, actively growing shoot sections (Soejarto et al., 1983). Stevia leaves are thus a rich source of sweet steviol glycosides (SG), which are low-calorie, non-toxic and non-mutagenic in nature.

8.2 Applications of Stevia

Nowadays, many soft drink manufacturers are introducing various health drinks and food supplementary beverages with natural sweeteners for diabetics by emphasizing the need for fiber and protein content (Manjusha et al., 2010). Stevia, being a rich source of proteins, calcium, phosphorus, other antioxidants, flavonoids and trace elements, is finding attractive applications in energy drinks and functional food preparation (Prakash et al., 2008). Some of the stevia based beverages, which are already available in market, are described in Table 8.1.

8.2.1 Milk and Food Products

Various food preparations, including milk, milk shakes, curd, lemon water, lapsi (sweet dalia), custard, halwa, kheer, carrot halwa, tea, and coffee, are being prepared using stevia extract and are also found to possess characteristics superior to table sugar (Deshmukh and Ade, 2012; Agarwal et al., 2010). Japan has a long history of using the extract for seafood, fish and meat products and vegetables.

TABLE 8.1

List of Some Stevia Based Beverages Available in Market

S. No.	Stevia Sweetened Beverages	Stevia Brand	Company
1	Sprite Green, Sprite Select	Truvia	Coke
2	Zevia (Zero Calorie Diet Soda)	Truvia	Zevia LLc
3	Fruit Juices	Truvia	Del Monte (Uk)
4	Sobe Life Water With Coconut Water, Virgil Diet Soda, Diet Root Beer Purevia	Purevia	Pepsi
5	Blue Sky Free Soda	Truvia	Blue Sky Beverage
6	Thomas Kemper Natural Diet Soda	Stevia Sweetener	Beverage Company
7	Organic Tea, Stevia Bankable Blends, Juices	Stevia Based Sweetener	Pyure Brand
8	Glaceau Vitamin Water	Stevia Sweetener	Coca Cola

8.2.2 Essential Oil and Fatty Acids

The presence of essential oils in stevia plant extract is also useful in preparation of perfumes, drugs as well as food products. Essential fatty acids, important building sources of cellular structures, are rich in stevia, enabling its use as proprietary food additives (Singh et al., 2005). Palmitic, palmitoleic, stearic, oleic, linoleic and linolenic acids are some of the fatty acids that have been identified in Stevia leaves.

8.2.3 Health Benefits of Stevia

The beneficial feature of stevia usually is achieved by declining the glucose absorption, which eventually helps in reducing the sugar levels. Also, the chlorogenic acid in stevia reduces the conversion of glycogen to glucose leading to the control of obesity, high blood pressure and digestive problems with no side effects (Amzad-Hossain et al., 2010; Prakash et al., 2008). Stevia can also help to regulate blood pressure at slightly higher doses than normal (Singh et al., 2005).

8.3 Separation Processes for Stevia Isolation

In order to get a purified product with higher commercial acceptability, membrane-based separation processes have emerged as potential alternatives for clarification and purification of aqueous stevia extract (Zhang et al., 2000; Chhaya et al., 2012). Stevia extraction using a green solvent like water is depicted as an eco-friendly and energy efficient method that yields no waste (Chemat et al., 2015). Extraction was also successfully performed using polar solvents such as methanol and water for complete withdrawal of the polar stevioside molecule from stevia leaves, whereas nonpolar or medium polar solvents showed a lesser affinity for taking out the sweetener. Misra et al. (2011) investigated the hypoglycemic effect of medium polar solvent mixture comprising of benzene and acetone for the extraction of stevioside from dried leaves of *Stevia rebaudiana*. Metallic ion extraction of stevia, followed by deionization by electrolysis also holds well, owing to the removal of toxic residues; however, the cost remained a limiting factor (Zhang et al., 2000). Silva et al. (2007) studied the application of adsorption and membrane filtration process for purification of crude stevia extract. The combined process of adsorption in CaX-modified zeolites and ultrafiltration enabled 91.75% stevioside recovery (Silva et al., 2007). Mantovaneli et al. (2004) clarified stevia extract solution using fixed bed column of calcium zeolites. The results showed that the mass transfer coefficient increased with a rise in solution flow rate and length of the bed (Mantovaneli et al., 2004). Water mixed with 70–90% alcohol was used as an efficient solvent for extraction. The resultant extract was filtered three times at intervals of 2 h each, but the fourth and final step was run for 12 h. Finally, stevia was quantified using high-performance liquid chromatography (Martinsa et al., 2016). Enzymatic extraction (Puri et al., 2012), a novel method for stevia extraction has been reported wherein many green solvents like methanol (Kienle et al., 2010), ethanol (Martins et al., 2016) and bleaching agents (Martins et al., 2017) were employed. Other significant methods combining the extraction step and membrane process are summarized in Table 8.2.

TABLE 8.2

Literature Survey on Extraction Methods of Stevia

S. No.	Method	Results	Reference
1	Water Extraction → UF → DF → RO → Ion Exchange	Stevia Purity 46% Efficiency 90%	Fuh and Chiang (1990)
2	Water Extraction → MF → UF	Recovery 78%	Kutowy et al. (1999)
3	Water extraction → MF → UF→ NF	Efficiency 89%	Zhang et al. (2000)
4	Water extract → Adsorption in Zeolite column → Membranes	Recovery 90 to 95%	Silva et al. (2007)
5	Water extract → MF → UF → NF	Stevia Purity 37% Recovery 30%	Vanneste et al. (2011)
6	Water extract → Centrifugation → UF	Recovery 45%	Chhaya et al. (2012)
7	Water extract → UF	Recovery 58 to 70%	Chhaya et al. (2012)
8	Hexane extraction → Cloth filtration → UF → NF → Solvent extraction → Crystallization	Stevia Purity 97%	Rao et al. (2012)
9	Water extract → Blend membrane filtration	Recovery 55% Stevia Purity 30%	Roy et al. (2015)

8.4 Role of Membranes in Stevia Glycosides Isolation

Membrane-based processes have been devised as significant cost-effective methods for clarification and concentration of stevia leaves extract owing to reduced use of chemicals, energy and unit operations, when compared to conventional processes. Diafiltration (DF) process, involving intermittent dilution of membrane retentate in a stage-wise manner, can help in removing contaminants with a simultaneous increase in the sweetening effect (Siddhartha et al., 2015; Rao et al., 2012). In several reports, ceramic MF membranes have been employed to remove suspended solids and other impurities (Chhaya et al., 2013). Nevertheless, the UF membrane module made of high flux polyethersulfone (PES) exhibited greater throughput without compromising the quality of the permeate (Sridhar et al., 2014). By using specialized UF clarification, a stevioside rich permeate was recovered. While using NF, concentration of reject was primarily focused to recover higher sweetener content with permeate usually free of stevia (Figure 8.1). NF reject was also subjected to final processing steps by rota-evaporator and crystallizer to obtain the final product (Rafika et al., 2015).

8.4.1 Important Equations

The flux of permeating solution through the membrane is calculated using Equation 8.1:

$$J = \frac{W}{t \times A} \tag{8.1}$$

where, W is the amount of permeate collected, t is time taken and A is cross-sectional area of the membrane.

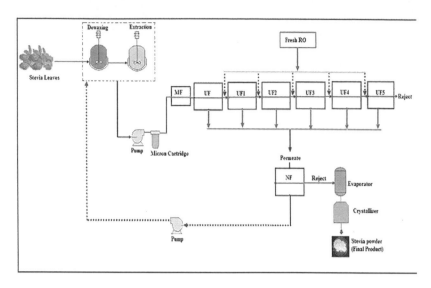

FIGURE 8.1
Process flow diagram for the production of stevia sugar using multistage membrane process.

The recovery percentage of steviosides can be determined using the following equation:

$$\% \; Recovery \; of \; Steviosides = \frac{C_p}{C_f} \times 100 \qquad (8.2)$$

where, C_p and C_f are conductivities of permeate and feed streams, respectively.

8.5 Hexane Extraction

The leaves of stevia rebaudiana (5 kg) were initially dried and crushed into fine powdered form of size ranging from 180–850 microns. The stevia leaf powder was then charged into a stainless-steel batch reactor of 50 L capacity containing hexane as the solvent (Figure 8.2 (a)). The extraction process was carried out at 60°C for a period of 2–3 h with total reflux for removal of color, plant pigments and waxy material, etc., from the stevia leaf. The solution was cooled to room temperature and subjected to rota-evaporator for recovery of hexane (recycled for the next batch) from leaf extract and further air-dried to remove trace amounts of hexane.

8.6 Water Extraction

The dried stevia powder obtained from hexane extraction step was fed into a stainless-steel batch reactor of 100 L capacity containing distilled water (Figure 8.2 (b)). The mixture

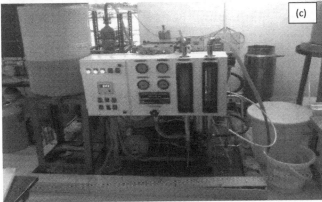

FIGURE 8.2
Pilot scale batch reactors used for (a) hexane and (b) water extraction and (c) experimental set-up of multistage membrane system.

was continuously stirred at 120°C for 1 h to extract steviol glycosides into water by removing color-imparting impurities and plant pigments. The extract solution was cooled to room temperature and filtered using cotton cloth for separating large particles such as solid wastes/biomass.

8.7 Case Study 1: Bench Scale Experimental Trials

8.7.1 Performance of Ceramic Tubular Microfiltration Module

The stevia extract solution was allowed to pass through a polypropylene micron cartridge rope pre-filter followed by ceramic tubular MF module of 0.4 m² area and a 0.2 µm pore size, for removal of higher molecular weight colloidal particles, suspended solids and bacteria (Figure 8.2 (c)). Figure 8.3 (a) shows the variation of permeate flux with time for a ceramic tubular membrane module at a constant operating pressure of 2 kg/cm² and temperature of 26–29°C. Experiment was carried out until a permeate recovery of 92–95% was achieved. With increment in time, a gradual reduction in flux

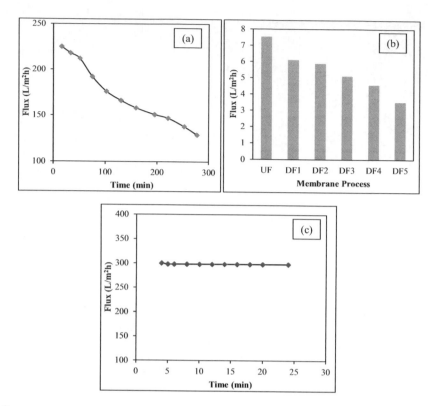

FIGURE 8.3
Variation of flux with time for (a) MF (b) UF–DF and (c) NF processes for bench scale system.

from 225 L/m²h to 129.96 L/m²h was observed. This is attributed to membrane pore clogging, accumulation and deposition of higher molecular weight particles over the surface of the membrane.

8.7.2 Performance of Ultrafiltration Based Diafiltration Process

The pre-filtered stevia solution from MF system was allowed to pass through spiral wound membrane module based on high flux PES ultrafiltration membrane (HF-UF) of 10 kDa MWCO with 1.2 m² effective membrane cross-sectional area at a constant operating pressure of 10 kg/cm² and ambient room temperature. The UF membrane allowed most of the steviol glycoside molecules to pass through, while the complete recovery of steviosides was achieved by concentrating the UF retentate stream by a series of five diafiltration (DF) units. The recovery of water was maintained constant at 70% at all operating pressures. Figure 8.3 (b) shows the flux variation in UF-DF integrated process. A gradual decline in the permeate flux was observed due to concentration polarization of higher molecular weight solute molecules over the membrane surface, followed by fouling of the pores. However, intermittent addition of deionized water after each UF based diafiltration step enabled further processing to recover the remainder of stevia left over in the retentate, such that losses of the sweetener in final retentate was at a bare minimum (0–1%). In this process, 80% clarity in color and impurity removal from the extracted medium was observed.

8.7.3 Performance of Hydrophilized Polyamide Nanofiltration Membrane

Permeate obtained from the UF-DF membrane process were combined as feed and passed through a hydrophilized polyamide NF spiral wound membrane (150 Da MWCO) of 1 m^2 membrane area, operating at a feed pressure of 21 kg/cm^2. Variation of flux with time is shown in Figure 8.3 (c). Membrane exhibited constant water flux of 300 L/m^{2}h with higher rejection of steviosides (~ 90%), without any loss of steviosides in the permeate. Prior to crystallization, the stevioside-rich NF retentate solution was extracted into *n*-butanol organic solvent to isolate maximum amount of steviosides. High purity steviosides were obtained, with a recovery yield of 70–85%, of which Stevioside (80–85%) and Rebaudioside-A (8–10%) were present, as confirmed by HPLC analysis. However, diafiltration-aided NF using hydrophilized polyamide membrane could enhance product purity further, by permeating smaller impurities like salts during nanofiltration. The actual photograph of feed, permeate, reject samples and final stevia product obtained during overall MF-UF/DF-NF integrated process is shown in Figure 8.4. The material balance for all membrane processes operated using bench scale system was done and results are provided in Figure 8.5.

FIGURE 8.4
Actual photograph of feed, permeate and reject samples obtained from (a) MF (b) UF-DF (c) NF processes and (d) Final stevia product.

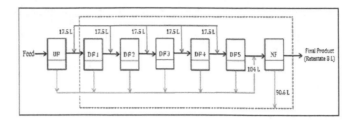

FIGURE 8.5
Overall material balances for all membrane processes operated using bench scale system.

8.8 Case Study 2: Pilot Scale Experimental Trials

125 kg of crude *Stevia rebaudiana Bertoni* leaf powder was initially treated with 800–1000 liters of hexane for a period of 2–3 hours at 50–60°C with total reflux to remove plant pigments and waxes from the leaf powder. After hexane was separated, the dry leaves were subjected to pressurized hot water for extraction of a crude solution rich in steviosides. The obtained aqueous extract solution of 800–1000 liters was initially pre-filtered using a micron cartridge (polypropylene rope) to remove higher molecular weight colloids and suspended solids. The filtrate was then passed through a ceramic microfiltration tubular module of 0.1–0.5 m² area to remove suspended particles. The experiment was carried out in batch mode at an operating pressure of 0–2 kg/cm² and temperature of 25–30°C, until a permeate recovery of 95–98% containing steviosides 4–5% was registered.

To improve the purity levels, a tighter UF membrane of 5 kDa MWCO was chosen for improved separation of high molecular weight impurities, while a slightly looser NF membrane (400 Da MWCO) was selected to permeate low molecular weight impurities including mono- and bivalent salts, while simultaneously concentrating the stevia product. Further purity can be attained by diafiltration-aided NF involving a stage-wise addition of water to extract smaller impurities and obtain a concentrate enriched with stevia, not including even trace impurities.

8.8.1 Extraction of Stevia by Ultrafiltration

Pilot scale experimental studies for clarification of stevia extract solution were done on a larger scale using spiral wound HF-UF membrane module of 5 kDa MWCO and dimensions of 4″ dia. x 40″ length housing a membrane area of 7.5 m². The experiment was carried until a permeate recovery of 80–90% was achieved with a steviosides concentration of 6200–6650 gm. There was 1–2% loss of steviosides in reject during ultrafiltration. Variation in flux with time at a constant operating pressure of 3.5 kg/cm² is depicted in Figure 8.6 (a). A gradual reduction in flux from 16.67 to 13.10 L/m²h was observed within 70 min. The results also showed that the HF-UF 5 kDa membrane exhibited better separation when compared to the 10 kDa UF membrane used in bench scale studies. The retentate from UF system was fed to a series of 4 DF units for a near-complete recovery of stevia glycosides into permeate. The recovery of water was maintained in the range 60–87% at all operating pressures.

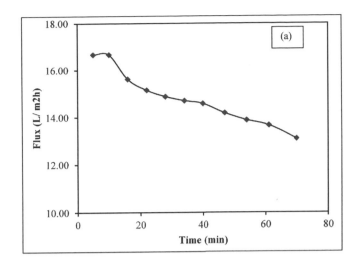

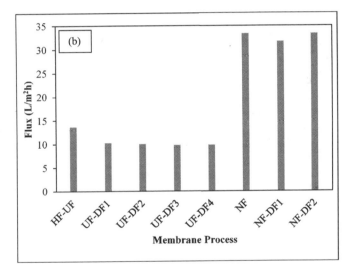

FIGURE 8.6
Variation of flux for (a) HF-UF and (b) Multistage pilot membrane systems.

8.8.2 Concentration of Stevia by Nanofiltration

The permeate from UF-DF integrated system was collected and fed to a hydrophilized polyamide NF membrane of 400 Da MWCO at an operating pressure of 3.5 kg/cm^2. The concentrated retentate stream from NF system was subjected to a series of two DF units by recovering 50% of water in each step, to facilitate permeation of small-sized impurities. The stevioside product was concentrated in the retentate by removing 80–90% of water as permeate. The obtained water solution in the permeate was free from sweet glycosides. It was observed that NF membrane showed high retention of steviosides from 80–95%. The results of flux for HFUF, HFUF-DF and NF-DF for integrated process are provided in Figure 8.6 (b). The final retentate solution from the second stage of DF system undergoes

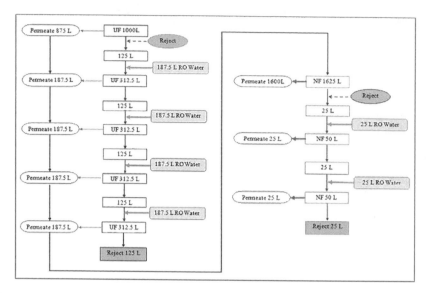

FIGURE 8.7
Overall material balances for all membrane processes operated using pilot scale system.

solvent extraction, followed by crystallization to obtain steviosides of a high purity of 95–98%, with recovery of sweet steviol glycosides from the stevia leaves in the range of 90–95%. All the experiments were carried out in batch mode with the reject being recycled into the feed tank. The process, developed by combining MF, UF and NF processes, exhibits a great potential for steviosides recovery. The pictorial representation of material balance for UF, NF and DF systems is shown in Figure 8.7.

8.9 Operation and Maintenance Costs

Operation and maintenance costs for processing crude stevia extract solution using MF-UF and NF systems on a pilot scale including costs for power consumption, pre-filter cartridges replacement, membrane replacement besides chemicals for cleaning and storage of the membranes are provided in Table 8.3. The feed capacity and recovery percentage of permeate were assumed to be 1 m³/h and 87.5% for UF system and 1.6 m³/h and 95.5% for NF system, respectively. The operating time was assumed to be 8 h per day and the duration of membrane module replacement, 1 year. The total operating cost for MF/UF and NF hybrid system was found to be Rs. 294.64/– per h, which is approximately US$ 4.54 for producing 1 kg of 95–98% pure stevia.

TABLE 8.3

Operation and Maintenance Costs for Processing Crude Stevia Extract Solution Using UF-DF and NF Systems

	UF-DF	NF
Feed Capacity (m^3/h)	1	1.625
Permeate Capacity (m^3/h)	0.87	1.6
% Recovery	87.5	98
Power Consumption Cost		
Feed pump (kW)	0.55	3.7
UF pump (kW)	5.5	–
Total power consumption (kW)	6.05	3.7
Price per unit (Rs.)	6	6
Hourly cost (Rs.)	36.3	22.2
Cartridge Replacement Cost		
Nos. of cartridges	1	–
Price per cartridge (Rs.)	25000	–
Total cartridge replacement cost (Rs.)	25000	–
Duration of replacement (Year)	1	–
No. of working hrs/day	8	–
Hourly cost (Rs.)	8.56	–
Module Replacement Cost		
No. of modules	12	4
Price per module	35000	30000
Total module replacement cost (Rs.)	420000	120000
Duration of replacement (Year)	1	1
No. of working hrs/day	8	8
Hourly cost (Rs.)	143.84	41.10
CIP Chemicals		
Caustic soda (kg) (Rs. 40 per kg)	2	1
Sodium EDTA (kg) (Rs. 295 per kg)	0.5	0.25
Frequency of washing (Days)	1	1
Total cost of chemicals per hour (Rs.)	28.44	14.22
Operating cost per hour (Rs.)	217.13	77.51
*Operating cost per hour in USD (per kg stevia)**	3.34	1.19

*Membrane process operating cost: US $ 4.53 per kg of 95–98% stevia product.

8.10 Conclusions and Future Scope

The present study established an eco-friendly and inexpensive process for the isolation of stevia glycosides from crude stevia leaf extract solution using inexpensive microfiltration, ultrafiltration and nanofiltration membranes. Micro- and ultrafiltration were carried out to clarify the stevia solution by removing color, pigments and higher molecular weight impurities like chlorophyll, whereas nanofiltration enabled the concentration of the natural sweetener with maximum retention of steviosides, confirming recovery of most of the stevia glycosides from crude stevia leaf extract. Both ultra- and nanofiltration were carried out in diafiltration mode to maximize removal of small- and large-sized impurities. Nanofiltration enabled reduction in the evaporation process load. The final purification step involved evaporation followed by crystallization using an organic solvent to obtain steviosides of a high purity of 95–98%, with a yield of 90–95%. The study revealed that the use of multistage membrane processes for clarification and concentration of stevia glycosides was found to be a commercially viable, cost-effective and eco-friendly process, when compared to conventional technologies.

Nomenclature

SrB	Stevia Rebaudiana Bertoni
MF	Microfiltration
UF	Ultrafiltration
DF	Diafiltration
NF	Nanofiltration
RO	Reverse Osmosis
PV	Pervaporation
ED	Electrodialysis
SWM	Spiral Wound Membrane
SMBS	Sodium metabisulphate
EDTA	Ethylenediamine tetraacetic acid
NaOH	Sodium Hydroxide
SLS	Sodium lauryl sulphate
HCl	Hydrochloric Acid
PES	Polyether sulfone
PVA	Polyvinyl alcohol
DMF	Dimethyl formamide
TFC	Thin Film Composite
GA	Glutaraldehyde
HPA	Hydrophilized Polyamide
MWCO	Molecular Weight Cut-Off

References

Agarwal, V., A. Kochhar, and R. Sachdeva. 2010. Sensory and Nutritional Evaluation of Sweet milk products prepared using Stevia Powder for Diabetics 4: 9–13.

Ahmad, S., A.K. Farooq, H. Abdul, and S.N. Muhammad. 2014. A review on potential toxicity of artificial sweetners vs safety of Stevia: A natural bio-sweetner. *Journal of Biology, Agriculture and Healthcare* 4: 137–147.

Alhady, M. 2011. Micro propagation of Stevia Rebaudaina Bertoni. A new sweetening crop in Egypt. *Global Journal of Biotechnology & Biochemistry* 6: 178–192.

Amzad, H., M.A. Siddique, S.M. Rahman, and M.A. Hussain. 2010. Application of stevia and research and development data. *Asian Journal of Traditional Medicines* 5: 56–61.

Atteh, J., O. Onagbesan, K. Tona, J. Buyse, E. Decuypere, and J. Geuns. 2011. Potential use of Stevia Rebaudiana in animal feeds. *Arch Zootec* 60: 133–136.

Barathi, N. 2003. Stevia-the calorie free natural sweetener. *Natural Product Radiance* 2(3): 120–122.

Brandle, E., A.N. Starratt, and M. Gijzen. 1998. Stevia rebaudiana: Its agricultural, biological and chemical properties. *Journal of Plant Science* 78: 527–536.

Chemat, F., F. Tixier, A.S. Vian, M.A. Allaf, and E. Vorobiev. 2015. Solvent free extraction of food and natural products. *Trends in Analytical Chemistry* 71: 157–168.

Chen, T.H., S.C. Chen, P. Chan, Y.L. Chu, H.Y. Yang, and J.T. Cheng. 2005. Mechanism of the hypoglycemic effect of stevioside, a glycoside of Stevia rebaudiana. *Planta Medica* 71: 108–113.

Chhaya, C.S., G.C. Majumdar, and D. Sirshendu. 2013. Primary clarification of Stevia extract: A comparison between centrifugation and microfiltration. *Separation Science and Technology* 48: 113–121.

Das, A., S. Gantait, and N. Mandal. 2011. Micropropagation of an elite medicinal plant: Stevia rebaudiana Bert. *International Journal of Agricultural Research* 6: 40–48.

Deshmukh, S. and R. Ade. 2012. In vitro rapid multiplication of Stevia rebaudiana: An important natural sweetener herb. *Bioscience* 4: 105–108.

Fuh, W.S., and B.H. Chiang. 1990. Purification of steviosides by membrane and ion exchange process. *Journal of Food Science* 55: 1454–1457.

Jain, P., S. Kachhwaha, and S.L. Kothari. 2012. Optimization of micronutrients for the improvement of in vitro plant regeneration of Stevia Rebaudaina (bert.) Bertoni. *Indian Journal of Biotechnology* 11: 486–490.

Kienle, U. 2007. Stevia rebaudiana. Natural sweetness in the bureaucratic jungle. *Journal of Clinical Investigation* 5: 241–250.

Kutowy, O., S.Q. Zhang, and A. Kumar. 1999. Extraction of sweet compounds from Stevia Rebaudiana Bertoni. U.S. Patent 9,116,925.

Kinghorn, A.D., D.D. Soejarto, H. Hikino, and N.R. Famsworth. 1985. Current status of stevioside as a sweetening agent for human use. *Economic and Medicinal Plant Research* 1: 1–52.

Lewis, W.H. 1992. Early uses of Stevia redaudiana (Asteraceae) leaves as a sweetner in Paraguay. *Economic Botany* 46: 336–337.

Manjusha, A.V.M. and B.N. Sathyanarayana. 2010. Acclimatization studies in Stevia (Stevia rebuadiana Bert. *Acta Horticulturae* 865: 129–133.

Mantovaneli, I.C.C., E.C. Ferretti, M.R. Simoes, and C. Ferreira da Silva. 2004. The effect of temperature and flow rate on the clarification of the aqueous stevia-extract in a fixed-bed column with zeolites. *Brazilian Journal of Chemical Engineering* 21: 449–458.

Martins P.M., B.N. Thorat, A.D. Lanchote, and Luis A.P.F. 2016. Green extraction of glycosides from Stevia rebaudiana (Bert.) with low solvent consumption: A desirability approach. *Resource-Efficient Technologies* 2: 247–253.

Misra, H., M. Soni, N. Silawat, D. Mehta, B.K. Mehta, and D.C. Jain. 2011. Antidiabetic activity of medium-polar extract from the leaves of Stevia rebaudiana Bertoni on alloxan-induced diabetic rats. *Journal of Pharmacy and Bioallied Sciences* 3(2): 242–248.

Mogra, R. and V. Dashora. 2009. Exploring the use of Stevia rebaudiana as a sweetener in comparison with other sweeteners. *Journal of Human Ecology* 25: 117–120.

Parashar, B., V. Yadav, K. Amrita, L. Sharma, and B. Thomas. 2013. Stevia (Meethi patti): Prospects as an emerging natural sweetener. *International Journal of Pharmaceutical and Chemical Sciences* 2(1): 214–225.

Pourvi, J., S. Kachhwaha, and S.L. Kothari. 2009. Improved micro propagation protocol and enhancement in biomass and chlorophyll content in stevia rebaudaina (bert.) bertoni by using high copper levels in the culture medium. *Scientia Horticulturae* 119: 315–319.

Prakash, I., G. Dubois, J. Clos, K. Wilkens, and L. Fosdick. 2008. Development of rebaudiana, a natural, non-caloric sweetener. *Food and Chemical Toxicology* 46: S75–S82.

Puri, M., D. Sharma, C.J. Barrow, and A.K. Tiwary. 2012. Optimisation of novel method for the extraction of steviosides from Stevia rebaudiana leaves. *Food chemistry* 132: 1113–1120.

Rafika, M., H. Qablia, S. Belhamidia, F. Elhannounia, A. Elkhedmaouib, and A. Elmidaoui. 2015. Membrane separation in the sugar industry. *Journal of Chemical and Pharmaceutical Research* 7(9): 653–658.

Rao, B., A.E. Prasad, G. Roopa, S. Sridhar, and Y.V.L. Ravikumar. 2012. Simple extraction and membrane purification process in isolation of steviosides with improved organoleptic activity. *Advances in Bioscience and Biotechnology* 3: 327–335.

Roy, A., S. Moulik, S. Sridhar, and S. De. 2015. Potential of extraction of Steviol glycosides using cellulose acetate phthalate (CAP)–polyacrylonitrile (PAN) blend hollow fiber membranes. *Journal of Food Science and Technology* 52: 7081–7091.

Shi, Q.Z., K. Ashwani, and O. Kutowy. 2000. Membrane-based separation scheme for processing sweeteners from stevia leaves. *Food Research International* 33: 617–620.

Shizhen, S. 1995. A study on good variety selection in Stevia rebaudiana. *Journal of the Science of Food and Agriculture* 28: 37–41.

Siddhartha, M., P. Vadthya, K. Yamuna Rani, C. Sumana, and S. Sridhar. 2015. Production of fructose sugar from aqueous solutions: Nanofiltration performance and hydrodynamic analysis. *Journal of Cleaner Production* 2: 44–53.

Silva, F.V., R. Bergamasco, C.M.G. Andrade, N. Pinheiro, N.R.C.F. Machado, M. Reis, A.A. Araujo, and S.L. Rezende. 2007. Purification process of Stevioside using zeolites and membranes. *International Journal of Chemical Reactor Engineering* 5(40): 1–6.

Singh, S., and G. Rao. 2005. Stevia: The herbal sugar of 21st Century. *An International Journal of Sugar Crops and Related Industries* 71: 17–24.

Sivaram, L. and U. Mukundan. 2003. In vitro culture studies on Stevia rebaudiana. *In Vitro Cellular and Developmental Biology* 39(101): 520–523.

Soejarto, D., D.C.M. Compadre, P.J. Medon, S.K. Kamath, and A.D. Kinghorn. 1983. Potential sweetening agents of plant origin II. Field search for sweet-tasting Stevia species. *Economic Botany* 37: 71–79.

Stuart, R.T. and D. Horton. 1987. *Advances in Carbohydrate Chemistry and Biochemistry*. United States: Elsevier.

Ulbricht, C. et al. 2010. An evidence-based systematic review of Stevia by the natural standard research collaboration. *Cardiovascular & Hematological Agents in Medicinal Chemistry* 8: 113–127.

Vanneste, J., A. Sotto, C.M. Courtin, V. Craeyveld, V. Bernaerts, K. Van Impe, J. Vandeur, J. Taes, and S.V. Bruggen. 2011. Application of tailor-made membranes in a multi-stage process for the purification of sweeteners from Stevia rebaudiana. *Journal of Food Engineering* 103: 285–293.

Yadav, A.K., S.C. Singh, D. Dhyani, and P.S. Ahuja. 2011. A review on the improvement of stevia Stevia rebaudiana (Bertoni). *Canadian Journal of Plant Science* 91: 1–27.

Zhang, S.Q., A. Kumar, and O. Kutowy. 2000. Membrane-based separation scheme for processing sweeteners from stevia leaves. *Food Research International* 33: 617–620.

9

Microfiltration Membranes: Fabrication and Application

Barun Kumar Nandi, Mehabub Rahaman, Randeep Singh, and Mihir Kumar Purkait

CONTENTS

9.1 Introduction

Microfiltration (MF) is an important and widely used separation process among other membrane processes such as ultrafiltration (UF), nanofiltration (NF), and reverse osmosis (RO). MF has average pore size or size of retained particles in the range of 0.1–10 μm. In terms of pore size, MF lies in-between process cloth filtration and UF. MF comes under the category of low-pressure membrane processes due to the use of an applied pressure below 2.5 bar. Most of the suspended solids, micro emulsions, droplets, and bacteria can be successfully separated by employing MF, since the size of these entities fall in the size range (0.1–10 μm) of MF (Mulder, 1991). A few applications of MF membranes include the removal of suspended solids from various water sources, primary treatment of industrial wastewaters; fruit juice clarification, the removal of fine particle from flue gases, removal of bacteria, fat and protein from food, and pharmaceutical and biotechnological industry feeds.

The microfiltration and ultrafiltration membranes have grown and developed concurrently. However, initially the MF membranes are at the center of focus. During 1920–1930, work was done on the development of nitrocellulose MF membranes. GmbH was the only company founded in the year 1926 for the production of nitrocellulose MF membranes on a large scale. Later, in the 1940s, companies like Sartorius and Schuell came into existence, which also prepared nitrocellulose based MF membranes (Baker, 2004). Initially, MF membranes were used for the assessment of drinking water quality by culturing micro-organisms on them. The MF membranes were used to filter the drinking water and then the membrane was placed on a nutrient-rich gel pad for almost 24 h. The micro-organisms, if present in the water, started to proliferate on the membrane surface, and thus the water quality was monitored. This technique was widely used during World War II for the assessment of drinking water quality. Later, Millipore took over most of the membrane business and became the largest company to manufacture MF membranes. In the last two decades, MF membranes have been widely used for the production of safe and clean drinking water and treatment of municipal wastewaters. Companies like Koch, Norit, Hydranautics, and US filter manufacture MF membranes for the said purposes.

Initially, MF or other membrane processes were mainly used on a laboratory scale but with the continuous growth and development in materials and membrane technology, membranes are widely used on an industrial scale, as well. Now membranes are extensively used in food, pharmaceutical and biotechnological industries for the production of clean water, separation and purification of heat-labile products and treatment of the wastewater. Mainly two configurations are used viz., dead end and cross-flow filtration. In the case of dead end filtration, the feed is permeated through the MF membrane under pressure. The pressure needs to maintain the prerequisite flow, which decreases over time due to the accumulation of feed components over the membrane surface. Therefore, the MF membrane is to be cleaned at regular intervals and to be replaced in cases where the prerequisite flow can't be attained (even after cleaning). On the other hand, the cross-flow MF process involves feed flow across the membrane surface, which gets divided into two streams viz., permeate and retentate. In this case, the efficiency and effectiveness of the overall membrane process is high due to low fouling tendency. Depending upon the needs of the application, any of the two available membrane modes can be used but mostly cross-flow membrane filtration is preferred for industrial scale applications.

Initially, MF membranes were prepared from cellulose and its derivatives, but today, various other polymers such as polysulfone, polyvinylidenefluoride, polyamides, polytetrafluoroethylene, polyolefins, etc., and ceramics such as kaolin, quartz, etc., are used. These improvements increased the reach and depth of MF membrane processes in various fields. Therefore, MF membranes are now an integral part of many industrial processes.

This chapter is written in a simple and systematic manner so as to provide in-depth knowledge about the vast area of MF membranes. The subsequent sections of this chapter discuss important aspects of MF membranes with descriptions of various types of materials, preparation and characterization techniques along with some important applications of MF. Lastly, the chapter ends with a brief summary elucidating the basics of MF membranes and their usefulness in various fields including future perspectives.

9.2 Membrane Materials and Trade-Offs

In the case of industrial scale membrane applications, polymeric and ceramic materials are the most widely used functional materials to obtain symmetric MF membranes. The asymmetric membranes are usually prepared from symmetric polymeric or ceramic membranes, or both. In an asymmetric membrane, usually the porous support layer provides desired mechanical strength whereas the thin skin layer comprising of either polymeric or ceramic material caters toward the desired separation characteristics. However, these materials have various pros and cons and hence trade-offs, as elaborated in subsequent sections.

9.2.1 Ceramic Membranes

Generally, ceramic membranes comprise of various inorganic materials such as α-alumina, γ-alumina, zirconia, silica, titania, kaolin, and fly ash (Nandi et al., 2008). Compared to polymeric membranes, ceramic membranes possess superior chemical, thermal, and mechanical stabilities. The usual thickness of the ceramic membrane is in the range of 2–5 mm and sometimes higher depending on the specific application. Asymmetric ceramic membranes constitute of a thin film (10–100 μm) of ceramic coating over a thick porous symmetric support, which enhances the selectivity of the membrane. Some common advantages for ceramic membranes are:

a. High corrosion resistance. Except for very few chemicals like hydrofluoric acid and phosphoric acid, ceramic membranes show a very high degree of tolerance to strong doses of acids, bases, solvents and other chemicals like chlorine (up to 2000 mg/L in certain cases).

b. Applicability to wider pH ranges (0.5–14).

c. Higher mechanical strength.

d. Resistance to vast temperature ranges (up to 500°C). Therefore, they can be utilized for industrial scale separations without the need of any feed pre-conditioning steps.

e. Longer life span (5–10 years). There are many examples available where a ceramic membrane system is operational even after 10–14 years of installation.

f. Low fouling propensities.

Industrial applications foul the membranes aggressively and therefore stringent cleaning agents and regimes are commonly used. The said properties of the ceramic membranes render them resistant to these highly aggressive cleaning procedures and methods. Therefore, they can be subjected to these prevalent industrial cleaning agents and methods in chemical processing units.

However, there exist a few drawbacks of ceramic membranes—they are given below:

a. Generally, ceramic membranes are not applicable for the separation schemes related to NF and RO, since most of the ceramic membranes are available with pore diameters within the MF and UF range (0.10–10 μm).

b. They are comparatively costly due to the requirement of higher sintering temperatures, materials and conditions with high purity, higher amount of inorganic precursor compared to very less amount of polymeric raw materials required in the fabrication of polymeric membranes and involvement of tedious membrane fabrication procedures.

c. They are very brittle or fragile in nature. If dropped or subjected to undue vibrations, they may get damaged.

9.2.2 Polymeric Membranes

Polymeric membranes are thin films of 10–100 μm thickness. Different types of polymers, such as polysulphone (PSU), cellulose acetate (CA), polyamide (PA), polyethersulphone (PES), polyvinylidenefluoride (PVDF), polyacrylonitrile (PAN), polytetrafluoroethylene (PTFE), polyetherimide (PEI), and polypropylene (PP), are widely used to fabricate polymeric membranes. Some of the common advantages of polymeric membranes are given below:

a. Availability in a wide range of pore sizes varying from those of MF to those of RO.

b. Availability of both hydrophobic as well as hydrophilic polymeric membranes so as to minimize fouling.

c. Easy to fabricate and use.

d. Feasibility of scale-up.

e. Inexpensive compared to ceramic membranes.

However, there are some basic disadvantages associated with polymeric membranes, such as:

a. Low resistance toward organic solvents.

b. Narrower pH range of applicability.

c. Low resistance to high temperatures.

d. Fouling.

e. Low life span (12–18 months).

Recently, modified polymeric membranes are developed to perform under wider pH and solvent containing media. However, their resistance against corrosion and organic solvents are yet to be solved to the extent of providing confidence for their successful use in industrial applications.

9.2.3 Polymeric vs. Ceramic Membranes

Considering the advantages and disadvantages of both ceramic and polymeric membranes, it can be observed that polymeric membranes are more suitable for laboratory use wherein the particular separation performance is the main objective, as compared to life span and cost. However, for industrial scale applications, cost and life span are significant matters in addition to separation efficiency. Most of the polymeric membrane applications have shown an average life span of 12–18 months, extendable up to 36 months by adopting optimal cleaning schemes, as compared to the 10-year life-span for ceramic membranes. Therefore, the ability of ceramic membranes to provide higher flux and applicability to wide range of temperatures and chemical processing conditions make them favorable in comparison to polymeric membranes. Although, ceramic membranes possess separation characteristics similar to polymeric membranes, their higher initial costs restrict their widespread application on a commercial scale.

9.2.4 Polymer-Ceramic Composite Membranes

Polymer-ceramic asymmetric composite membranes partly overcome some of the disadvantages associated with ceramic and polymer symmetric membranes. The asymmetric membranes possess wider pore size ranges achieved through polymer film deposition on top of a ceramic support, which provide greater chemical, mechanical, and thermal stability. Additionally, asymmetric membranes employ an optimal combination of both polymeric and ceramic membrane layers. Therefore, polymeric-ceramic composite membranes can perform better than symmetric polymeric membranes along with a marginally longer life span. These features, accompanied by a marginally higher cost of ceramic support are beneficial for applications where polymeric membranes are favored over ceramic membranes. The fine tuning of membrane pore size is also possible in these types of composite membranes, furthering their appeal. The possibility of fabrication of tailor-made MF membranes makes it possible to use them as either a NF/RO membrane or selective solute passing membranes, in the case of pervaporation or selective separation of CO_2 or H_2 from gas mixtures.

9.3 General Methods of Preparation of Microfiltration Membranes

9.3.1 Ceramic Membrane Preparation Techniques

Symmetric ceramic membranes can be fabricated either by using paste or the uni-axial method (Mulder, 1991). Both of these methods involve the preparation of an inorganic mixture using suitable organic and inorganic pore forming materials along with binders. The membrane preparation process is initiated by thorough mixing of dry inorganic raw materials. Typical raw materials and their compositions for the fabrication of ceramic membranes are shown in Table 9.1. In the case of the paste method, a paste is prepared from the raw materials using distilled water. The paste is then cast over gypsum in the shape of a circular compact disk or tubular configuration using a suitable casing. Subsequently, the membrane casing is carefully removed and the paste is kept under distributed pressure for 24–48 h to prevent the propagation of deformation and ensure homogeneity in the inorganic matrix. The paste is then subjected to different sequential heat treatment steps.

TABLE 9.1

Typical Raw Materials and Compositions for Ceramic
Membrane Preparation

Raw Material	Typical wt%
Kaolin, Flyash	40–60%
Quartz	10–15%
Calcium Carbonate, Sodium Carbonate, etc.	20–25%
Boric Acid, Feldspar, Pyrophyllite, etc.	10–15%
Sodium meta-silicate	5–10%

The first step involves drying at an ambient temperature for 24 h. During the second heat treatment step, the membrane is dried at 100°C for 12 h in a hot air oven. The third step consists of drying at 250°C for 24 h. During the transition from 100–250°C, a low heating rate is maintained in order to eliminate the induction of thermal stresses generated due to loss of moisture. The final heat treatment step involves heating of the membrane from 250°C to desired sintering temperature at a heating rate of 2–3°C per minute. Eventually, the membrane is kept for 3–5 h for sintering at the final sintering temperature. The membranes are sintered at different temperatures and times to impart desired morphological properties. The selection of minimum sintering temperature is based on thermogravimetric analysis (TGA) and x-ray diffraction (XRD) analysis. Subsequent cooling of the membrane is carried out by using either atmospheric cooling or controlled cooling procedure to avoid any thermal stress generation. On the completion of sintering process, a membrane with a hard, rigid and porous texture is obtained. Finally, the prepared membrane is polished with silicon carbide abrasive paper (C-220) to obtain a smooth membrane.

In the case of the uni-axial method, the casting of an inorganic mixture is carried out in a suitable disk or tubular shape under very high pressure (30–50 MPa). Subsequently, the disk or tubular type mold is sintered in a similar way as in the case of the paste method. The properties of the ceramic membranes are largely influenced by the composition of raw materials, sintering temperature, and procedures involved for heating, sintering, and cooling.

9.3.2 Materials for Ceramic Membranes

Researchers throughout the world are working on ceramic membranes to improve their capability and efficiency. There are various reports related to the preparation of ceramic membranes using alumina (Lee et al., 2014), zirconia, kaolin, titania, and silica (Kumar et al., 2015). Also, vast literature is available regarding the fabrication of ceramic membranes using cheaper raw materials such as apatite powder (Masmoudia et al., 2007), cordierite (Saffaj et al., 2004), fly ash (Zhu et al., 2015), natural raw clay (Saffaj et al., 2006), dolomite (Bouzerara et al., 2006), and kaolin (Nandi et al., 2008, 2010). Table 9.2 summarizes the details of raw materials, average pore size and sintering temperatures reported by various researchers.

9.3.3 Important Parameters Influencing Ceramic Membrane Structure

The formation of stable and porous ceramic membrane is the result of solid phase chemical reactions among raw materials. During the sintering of raw materials at high temperature, different precursors play different roles to yield the desired ceramic membrane

TABLE 9.2

Summary of Literature on the Preparation of Symmetric MF Membranes

Materials	Sintering Temperature	Average Pore Diameter	Reference
Alumina	1200–1600°C	<0.20 μm	Lee et al. (2014)
Bauxite	1100–1500°C	0.27–2.64 μm	Lü et al. (2014)
Cordierite	1275°C	7.0μm	Saffaj et al. (2004)
Dolomite	1150–1300°C	1.65–48.53 μm	Bouzerara et al. (2006)
Titania	1100–1500°C	1.6–2.8μm	David et al. (2014)
Zirconia	1400°C	0.66 μm	Kumar et al. (2015)
Fly ash	1100–1400°C	0.32–0.37 μm	Zhu et al. (2015)
Kaolin	900°C	2.16 μm	Emani et al. (2014)
Apatite powder	1150–1200°C	5.0–8.0μm	Masmoudia et al. (2007)
Natural clay	1100–1250°C	9.3–10.75 μm	Saffaj et al. (2006)

morphology. The role of precursors is generally categorized as structural material, pore forming material, binder material along with another important operating condition, i.e., sintering temperature. These are explained in detail in the following sub-sections.

9.3.4 Structural Material

Structural materials are those which form the basic structure of ceramic membranes. Characteristically, these materials are alumina, bauxite, clays, quartz, kaolin and flyash. The contributions of these materials is to provide maximum raw material (50–60%) for preparation of ceramic membranes. Kaolin and alumina provide low plasticity and high refractory properties to the membrane. Quartz and fly ash contribute mechanical and thermal stability. These materials post-sintering, create a solid and rigid structure. They are also responsible for providing resistivity to the membranes from heat, chemicals, solvents and mechanical shock. Primary cost of the ceramic membranes mainly depends on the cost of these materials. Therefore, for the preparation of low-cost ceramic membranes, inexpensive structural materials are proposed to be used.

9.3.5 Pore Forming Material

The creation of porous textures in the ceramic membrane is realized by using different types of pore forming materials categorized broadly as inorganic carbonate compounds that include calcium carbonate, magnesium carbonate, sodium carbonate, and organic compounds, such as poly-ethyl glycol, starch, etc. During sintering, inorganic carbonates dissociate into oxides and CO_2 at the corresponding decomposition temperature of the carbonate. The generated CO_2 gas from the system is released to the atmosphere. Similarly, for organic compounds, combustion occurs and CO gas is released. The path taken by the released CO_2 and CO gas thereby creates a porous texture in the ceramic membrane and contributes to membrane porosity during the sintering process.

9.3.6 Binder Material

Mechanical strength or structural stability of a ceramic membrane is controlled by a large number of factors, such as particle size and their distribution, chemical structure and composition of raw materials, sintering temperature, and preparation methods. Among all the different parameters, the composition of raw material plays a significant role in deciding the strength of the membrane. Therefore, to obtain a stable structure from various raw materials, such as alumina, kaolin, fly ash, etc., the addition of binding materials is necessary. The major role of binding materials is to improve plasticity during mixing and casting of membrane. It also helps in the uniform mixing of different types of raw materials and thus, brings homogeneity in the cast membrane.

9.3.7 Sintering Temperature

During membrane fabrication, the cast ceramic material is heated at a slow rate of 2–5°C/ min up to the desired sintering temperature of about 800–1600°C and kept at this sintering temperature for 2–5 h. Selection of heating rate, final sintering temperature and time of sintering depend on the chosen raw materials and desirable mechanical strength, pore size and porosity of the membrane. Generally, it is observed that sintering at a high temperature provides compact ceramic structure with lower porosity and higher mechanical strength. Final sintering temperature is mainly designed on the basis of raw and binding materials used. Normally, alumina, zirconia, and titania have fusion temperatures above 1700°C, therefore, membranes can be sintered at temperatures across 1200–1600°C to acquire excellent mechanical strength. On the other hand, raw materials like kaolin and fly ash have fusion temperatures in the range of 1200–1400°C. Therefore, these membranes need to be sintered below 1100°C.

9.4 Characterization Techniques

The membranes are applicable in any separation process only if they have the desired properties for the particular application. Therefore, it is important to characterize the prepared membranes to confirm the presence of desired properties. One important factor that makes membrane characterization crucial is the fact that various changes take place during membrane preparation, which may or may not be suitable for an application. Thus, it is mandatory to characterize the prepared membranes extensively. There are various morphological and permeation based techniques available for the precise characterization of membranes. These techniques are explained in detail in this section under various appropriate sub-sections.

9.4.1 Characterization Techniques for Membranes

The morphology of a membrane is the most important factor related to its performance. Therefore, it is very important to characterize a membrane for the analysis of its morphological as well as permeation parameters for which various techniques are available. For example, electron microscopy is used to analyze the morphology of a membrane; permeation based techniques such as pure water flux, liquid-liquid displacement porosimetry are used for the analysis of the permeation properties of the membranes; techniques, such as FTIR, XRD, and XRF are used for the functional analysis of the membrane; and techniques such as TGA and the three point bending flexural test are employed so as to analyze the temperature as well as mechanical stability of the membranes. In this section, some of the important membrane characterization techniques are discussed briefly. Also, their basics, importance and inference regarding the membranes is discussed with fitting examples.

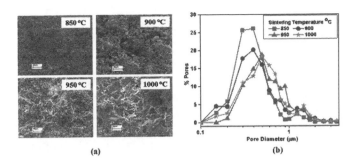

FIGURE 9.1

(a) Top surface SEM images of prepared membranes at various sintering temperatures (Nandi et al., 2008) and (b) pore size distributions (based on SEM micrographs) of prepared ceramic membranes at various sintering temperatures.

9.4.2 Surface Morphology

Membrane morphological studies are carried out using scanning electron microscopy (SEM) to analyze the presence of possible defects and estimate pore size. Figure 9.1 (a) shows SEM analysis of membranes sintered at four different temperatures. The surface morphology of the membranes (Figure 9.1 (a)) shows that all the membranes have a rough surface. The ceramic substrates sintered at lower temperatures (850°C and 900°C) possess highly porous structures compared to membranes sintered at higher temperatures (950°C and 1000°C). Membranes sintered at 950°C and 1000°C are more consolidated due to the fact that sintering temperatures above 900°C enables greater agglomeration of particles, which yields a denser structure. Due to this, the porosity of the membrane reduced with a rising sintering temperature.

9.4.3 Average Pore Size and Pore Size Distribution

The SEM micrographs of membranes are generally used for the analysis of membrane pore size and pore size distribution by using ImageJ software. The estimation of average membrane pore size (d_s) and pore size distribution is carried out by measuring the sizes of different pores visible in the SEM micrographs through ImageJ software. The obtained results are presented in the form of average membrane pore size and pore size distribution (Singh et al., 2016). Since average pore size and pore size distribution values are critically dependent upon the sampling procedure, usually more than one SEM micrograph is evaluated using the software.

Assuming the membrane pores to be of cylindrical nature, the average pore size (d_s) is calculated by using the data obtained from SEM micrographs by using the following equation:

$$
d_s = \left[\frac{\sum\limits_{i=1}^{n} n_i d_i^2}{\sum\limits_{i=1}^{n} n_i} \right]^{0.5}
\tag{9.1}
$$

where, n is the number of pores and d_i is the pore diameter (μm) of the ith pore.

Figure 9.1 (b) shows pore size distribution calculated for membranes sintered at different temperatures using SEM micrographs and ImageJ software.

9.4.4 Porosity and Structural Density

The functional attributes of a membrane depend mainly on its porosity, which is therefore a crucial parameter. There are a number of techniques and methods available for the accurate measurement of membrane porosity. Generally, the total porosity of a membrane is estimated by using Archimedes principle. The experimental procedure involves the measurement of the volume of the wetting liquid that displaces air in a dry membrane after equilibrating the membrane with water for 12 h. The dry weight, as well as the wet weight, of the membranes are measured to find the total porosity (ε_m) and structural density (ρ_{mem}) of the membrane by using the following equations (Mulder, 1991):

$$\varepsilon_m(\%) = \left[\frac{w_1 - w_2}{\rho_{water}} \right] \times \frac{100}{v_{mem}} \quad (9.2)$$

$$\rho_{mem} = \frac{w_2}{v_{mem}} \quad (9.3)$$

where, w_1 and w_2 represent the wet and dry weight of the membrane, respectively, whereas ρ_{water} represents the water density and v_{mem}, the membrane volume. The variation of membrane porosity and structural density with varying sintering temperatures is shown in Figure 9.2.

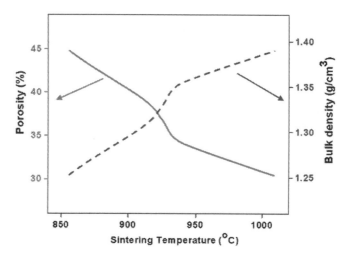

FIGURE 9.2
Variation of membrane porosity and structural density with sintering temperatures.

9.4.5 Liquid Permeation Characterization

The assessment of the efficiency and effectiveness of a membrane's purification capability is done on the basis of permeation based characterization, which is carried out for the determination of the hydraulic permeability (P_m) and hydraulic pore diameter (d_l) of a membrane by using deionized water. The permeation tests involve the measurement of permeate liquid volume as a function of time at a specific trans-membrane pressure differential (ΔP). The liquid flux is measured at regular intervals to verify the variation of flux with time. Prior to the liquid permeation experiments, the membranes are compacted at pressures higher than the operating pressures to get a membrane with uniform pores. The hydraulic permeability (P_m) and hydraulic pore radius (d_l) of the membranes are evaluated by assuming the presence of cylindrical pores in the membrane matrix using the following expressions (Mulder, 1991):

$$J_w = \frac{Q}{S \cdot \Delta t} = \frac{\Delta P}{\mu_w} \frac{\varepsilon_m d_l^2}{32\, l_m} = P_m \times \Delta P \tag{9.4}$$

$$P_m = \frac{\varepsilon_m d_l^2}{32\, l_m \mu_w} \tag{9.5}$$

$$d_l = \left[\frac{32\, l_m \mu_w P_m}{\varepsilon_m} \right]^{0.5} \tag{9.6}$$

In Equation 9.4, $\varepsilon_m d_l^2$ corresponds to the effective permeable area factor that determines the actual permeable area available during filtration. The other terms, ΔP, μ_w, and l_m represent the transmembrane pressure differential, feed viscosity and membrane thickness, respectively.

9.4.6 Gas Transport Characteristics

Ceramic microfiltration membranes have diverse applications, one of them being air purification application in case of flue gas cleaning. Therefore, it is important to characterize the ceramic MF membranes for gas purification capabilities by conducting membrane permeation experiments with gases to quantify the membrane morphological parameters, such as average pore size (d_g) and effective porosity (ε/q^2), that contribute to the gas transport. Therefore, gas permeation experimentation is adopted to observe the distribution of percentage of pores in the macroporous (pore dia. >50 nm) and mesoporous (pore dia. <50 nm) range within the membrane matrix. The contribution of macropores and mesopores to the overall membrane flux is determined by the result of gas permeation trials. The average pore diameter and effective porosity of the membrane for gaseous phase are determined from the plot of effective permeability factor (K) of the membrane and average pressure on the membrane (P) using the following expressions (Marchese et al., 1991):

$$K = 2.133 \times \frac{r \times v_g}{l_m} \times \frac{\varepsilon}{q^2} + 1.6 \times \frac{r^2}{l_m \times \eta} \times \frac{\varepsilon}{q^2} \times P \qquad (9.7)$$

$$d_g = 2 \times r = 2.666 \times \frac{B}{A} \times v_g \times \eta \qquad (9.8)$$

where,

$$A = 2.133 \times \frac{r \times v_g}{l_m} \times \frac{\varepsilon}{q^2} \qquad (9.9)$$

$$B = 1.6 \times \frac{r^2}{l_m \times \eta} \times \frac{\varepsilon}{q^2} \qquad (9.10)$$

and

$$K = \frac{Q \times P_2}{S \times \Delta P} \qquad (9.11)$$

The average pore diameter (d_g) can be calculated by using Equation 9.8 for known values of η (permeant viscosity), v_g (molecular speed of the permeant), A, and B. A and B are obtained from the slope (B) and intercept (A) of the linear plot of K vs. P. The effective porosity (ε/q^2) of the membrane can be calculated by using Equation 9.7 for known values of A after calculating the values of d_g. In Equation 9.7, the first term (intercept) corresponds to Knudsen permeance and the second term (slope) corresponds to the viscous permeance. Henceforth, the values of the slope and intercept obtained from the graph can be used to evaluate the percentage contribution of pores (and pore sizes) toward viscous and Knudsen flow transport mechanisms. In other words, gaseous flux characterization of the ceramic membrane can yield qualitative information with respect to the pore sizes contributing to Knudsen or viscous flow regimes.

9.5 Applications of Microfiltration Membranes

Presently, both polymeric as well as ceramic membranes are found to be suitable for various pressure driven membrane process (MF, UF, NF and RO), such as desalination, food processing, effluent treatment, drinking water purification and treatment of industrial wastewater. Among these, major applications of MF membranes are treatment of industrial oily wastewaters, fruit juice processing, separation of coagulated water and turbid water purification. These applications are important for membrane technology because of their highly diversified nature, which conventional processes are unable to handle appropriately. The tasks ahead for MF are different for each application viz., the treatment of oily and turbid wastewater streams requires the production of permeate streams with a lower concentration of waste product, while juice processing involves careful permeation of all desired components through the membrane filters to achieve a high quality permeate juice that is bereft of pectins and other colloidal suspensions. Therefore, each application is considered exclusively in subsequent sections.

9.5.1 Treatment of Oily Wastewater Using Ceramic Membrane

Process industries such as petroleum refineries, petrochemical, metallurgical, transportation and food processing enterprises produce large volumes of oily wastewater. Typical concentration ranges of produced oil-in-water (o/w) emulsions vary between 50–1000 mg/L of total oil and grease besides 50–350 mg/L of total suspended solids (Nandi et al., 2009[a]). Existing tolerance limits of total oil and grease concentrations in wastewater streams is about 10–15 mg/L. The desired discharge limits can be achieved by using conventional processes, such as thermal de-emulsification, biological methods, and chemical treatment methods. These processes are effective for the treatment of oily wastewater streams with high feed concentrations (500–5000 mg/L). On the other hand, due to the existence of smaller droplet sizes (<1 μm) of the emulsions in low oil concentrations (50–500 mg/L), these methods are ineffective. Amongst the various alternative, conceivable technologies, membrane technology is promising. The advantages of membrane technology, such as lower capital cost, higher separation factors, compact design, and the elimination of other chemical and mechanical treatment units like mechanical separation, filtration and chemical de-emulsification, render membranes a better choice for this application.

Generally, on laboratory scale either dead end or cross-flow mode of MF is used to verify the efficiency of a particular membrane. Figure 9.3 (a) shows the experimental set up for a dead-end MF process for the treatment of an oily wastewater (Nandi et al., 2010). The oily wastewater with different concentrations is introduced as feed to the membrane system at different trans-membrane pressure differentials (ΔP), for example, Nandi et al. (2010) has used given oil concentrations of 50, 75, 100 and 150 mg/L at different trans-membrane pressures so as to observe the effect of oil concentration as well as ΔP on permeate flux and oil rejection efficiency of the membrane. A particle size analyzer is used to measure

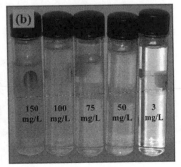

FIGURE 9.3
(a) Experimental set up for MF experiments with oil in water emulsions. (b) Variation of oil color with oil concentration.

the droplet sizes and their distribution in the prepared o/w emulsions. This analysis helps in the selection of a suitable MF membrane with appropriate pore size with reference to the droplet size distribution of the o/w emulsions. The permeate collected in a dead end MF system at regular interval is analyzed for its oil content. The permeate flux (*J*) and the percentage oil rejection (*R*) is calculated using the following equations:

$$J = \frac{V}{S \cdot \Delta t} \tag{9.12}$$

$$R(\%) = \left(1 - \frac{C_P}{C_F}\right) \times 100 \tag{9.13}$$

where, V represents the volume of the permeant (L/m²h), S the effective membrane area (m²), t the permeation time (h), and C_P and C_F, the oil concentrations in the permeant and feed, respectively. Oil concentrations in permeate and feed are usually determined by using a UV-Vis spectrophotometer or a total organic carbon (TOC) analyzer. Membranes are cleaned after each experimental run for their proper assessment in terms of fouling. This is done by measuring pure water flux (PWF) of each membrane before and after the permeation of oily wastewater through the cleaned membranes. Figure 9.3 (b) shows the ceramic MF treated oily wastewater for different feed compositions.

9.5.1.1 Effect of Trans-Membrane Pressure on Flux

Theoretically, membrane flux is directly proportional to trans-membrane pressure and increases linearly with a rise in trans-membrane pressure. However, during the actual run, an exactly linear relationship may not be observed due to membrane intervention, feed properties, concentration polarization and membrane fouling.

9.5.1.2 Effect of Trans-Membrane Pressure on Oil Separation

In this section, the effect of trans-membrane pressure on the separation of oil from a feed is discussed. The rejection efficiency of the membrane slightly increases with time due to the fact that the oil droplets get adsorbed into the membrane pores, causing a reduction in overall membrane pore size. Also, the rejection efficiency is observed to decrease with an increase in trans-membrane pressures since higher pressures facilitate enhanced wetting and coalescence of oil droplets, thereby imposing some oil droplets to pass through the membrane pores and reach the permeate stream (Nandi et al., 2010), thus reducing overall oil rejection at increased trans-membrane pressures.

9.5.1.3 Identification of Flux Decline Mechanism

Membrane flux decline is important to understand, for developing better and improved membranes in the future. There are various membrane flux decline mechanisms reported by various researchers. Here, the most famous membrane flux decline mechanisms, given by Hermia (1982), are discussed in detail. Hermia proposed four empirical models to present membrane fouling mechanisms in dead-end filtration based on constant pressure, namely these models are: complete pore blocking, standard pore blocking, intermediate pore blocking, and cake filtration. Parameters associated with these models subsequently have a physical relevance. The models were developed using the constant pressure filtration law:

$$\frac{d^2t}{dV^2} = K_p \left(\frac{dt}{dV} \right)^n \tag{9.14}$$

where the selection of values for parameter $n = 2, 1.5, 1$ and 0, corresponds to complete pore blocking, standard pore blocking, intermediate pore blocking, and cake filtration, respectively.

Figure 9.4 shows the schematics of four different membrane fouling mechanisms proposed by Hermia as explained below:

- *Complete pore blocking model:* In the case of complete pore blocking, the pore is fouled by the feed components when their molecular sizes are greater than the size of the membrane pores. As a result, pore blocking occurs over the membrane surface and not inside the membrane pores as shown in Figure 9.4 (a).

- *Standard pore blocking model:* This model hypothesizes that the feed components enter the membrane pores and deposit over the pore walls due to the irregularity of pore passages, thereby reducing pore volume. This type of fouling is caused by particles smaller in size than the membrane pores. According to this model, pore blocking occurs inside the membrane pores as shown in Figure 9.4 (b). Thereby, the volume of membrane pores decreases proportionally with the filtered permeate volume.

- *Intermediate pore blocking model:* Intermediate blocking occurs when the feed component size is similar to the membrane pore size. In this model, it is assumed that a membrane pore is not necessarily blocked by the feed component and some

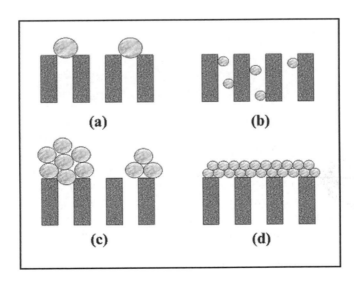

FIGURE 9.4
Schematic representation of various pore blocking mechanisms (a) complete pore blocking, (b) standard pore blocking, (c) intermediate pore blocking and (d) cake filtration.

feed components may settle over others. Therefore, the non-blocked membrane surface area diminished with time and membrane pore entrance is expected to be obstructed by feed components without blocking the membrane pore completely, as shown in Figure 9.4 (c).

- *Cake filtration model:* The cake filtration model corresponds to a scenario where feed components larger than the average membrane pore size accumulate over the membrane surface forming a "cake," as shown in Figure 9.4 (d). Thereby, the cake grows with time and provides an additional porous barrier and subsequently hydraulic resistance to the permeating liquid.

Substituting $n = 2$, 1.5, 1 and 0 in Equation 9.14, various pore blocking models can be represented in terms of time dependent membrane fluxes using the following linearized expressions (Nandi et al., 2010):

i. $n = 2.0$: *Complete pore blocking*:

$$ln(J^{-1}) = ln\left(J_0^{-1}\right) + k_b t \tag{9.15}$$

ii. $n = 1.5$: *Standard pore blocking*:

$$J^{-0.5} = J_o^{-0.5} + k_s t \tag{9.16}$$

iii. $n = 1.0$: *Intermediate pore blocking*:

$$J^{-1} = J_o^{-1} + k_i t \tag{9.17}$$

iv. $n = 0.0$: *Cake filtration*:

$$J^{-2} = J_o^{-2} + k_c t \tag{9.18}$$

9.5.2 Clarification of Sweet Lemon Juice by Microfiltration

Fruit and vegetable juices are beverages of high nutritional value. These beverages constitute of several key components beneficial for human health such as minerals, vitamins, and antioxidants. Traditional methods for juice processing involve filtration using gelatin or diatomaceous earth to remove suspended and colloidal particles and low-pressure evaporation. Unfortunately, the conventional juice processing methods involves thermal and chemical treatment steps during which a major portion of the compounds that contribute toward the quality of the beverage (such as sugar content, acidity, flavor, and aroma) get deteriorated. Hence, the application of membrane technology is found to be beneficial when compared to conventional methods due to absence of thermal and chemical processing steps, low energy requirements, lower processing times and ease of scale-up without any significant change in juice quality. Citrus fruits primarily constitute of both lower molecular weight compounds such as sugar, acid, salt, flavor, and aroma as well as higher molecular weight polysaccharides, such as pectic material, cellulose, and hemicellulose in addition to haze producing proteins and microorganisms. The presence of pectic material and protein in fruit juice is responsible for the cloudiness, post-bottling haziness as well as their fermentation during long storage. The objective of fruit juice clarification by MF is to eliminate the high molecular weight pectic material and their derivatives but retain low molecular weight solutes, with palatable and nutritional values, such as sucrose, acid, salt, aroma, and flavor compounds.

9.5.2.1 *Juice Preparation and Pretreatment*

Juice preparation and pretreatment are important steps in juice clarification considering membrane fouling tendency. Firstly, the pulp of the freshly purchased sweet lemon fruits (*Citrus sinensis (L.) Osbeck*) of proper maturity and ripeness is used for the extraction of the fruit juice (FJ). A manually operated screw type juice extractor is used for this purpose. Secondly, the juice is centrifuged with or without enzyme treatment at 4000 rpm for 20 min. to prepare centrifuged juice (CJ) or enzyme treated centrifuged juice (ETCJ), respectively. Generally, enzyme treatment is used so as to degrade the pectic material. As discussed previously, the presence of pectic materials is not beneficial for the either nutritional value or shelf life. Also, the centrifugation process is not efficient in the effective removal of pectic materials from the juice. Therefore, enzymes like pectinases are used for pretreatment of sweet lemon juice. It is important to use an enzyme with good activity values, for example, around 3.5 units/mg. The centrifugation speed and duration is set based upon data obtained from the trial and error method. The main purpose of centrifugation is the removal of a maximum number of suspended solids from the juice and achieve juice with the highest clarity and minimum color. In addition to this, the enzymatic pretreatment of the juice is carried out by heating the juice at 42°C for 100 min. with an enzyme concentration of 0.0004 w/v%. Subsequently, the suspension is heated to 90°C for 5 min. in a water bath to inactivate the remaining enzyme in the juice. Finally, the juice is cooled to ambient temperature (25°C) and centrifuged.

9.5.2.2 *Microfiltration Studies*

MF for fruit juice clarification is commonly carried out by following standard procedures to evaluate membrane performance and quality of the clarified juice:

The overall permeate flux in different types of juices such as FJ, CJ, ETCJ as well as permeate of FJ, CJ, and ETCJ in case of sweet lemon juice is measured along with the particle size distributions of the juices using a laser particle size analyzer. The common parameters for which the fruit juices should be analyzed are color, clarity, total soluble solids (TSS), pH, acidity, viscosity, density, and alcohol insoluble solids (AIS). Color and clarity of the sweet lemon juice is evaluated by a UV-Vis spectrophotometer for measuring the absorbance at 420 nm and transmittance at 660 nm. The TSS of the juice is measured (in °Brix) using a digital refractometer. The fruit juice acidity is measured by titrating 10 mL of the sample with 0.1 N NaOH until the solution pH reaches 8.2 and expressed as wt % anhydrous citric acid equivalent. The viscosity of the fruit juice samples can be measured by using a viscometer and density by a 25-mL pycnometer. AIS of the fruit juice denotes the total amount of pectic material present in the fruit juice. It is measured by using the Hart and Fisher method, which states that 20 g of fruit juice should be mixed with 300 mL of 80% methanol, which is later simmered for 30 min. (Hart & Fisher, 1971). Subsequently, the final solution is filtered and the obtained residue is washed with an 80% alcohol solution, followed by its drying at 100°C for 2 h. The final residue is weighed and expressed as AIS

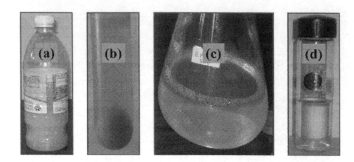

FIGURE 9.5
(a) Fresh juice, (b) centrifuged suspended particles, (c) centrifuged juice and (d) permeate of enzyme treated centrifuged juice.

in wt%. Figure 9.5 shows the images of the obtained FJ (Figure 9.5 (a)), centrifuged suspended particles (Figure 9.5 (b)), CJ (Figure 9.5 (c)), and permeate of ETCJ (Figure 9.5 (d)).

9.5.2.3 Effect of Operating Pressure on Permeate Flux

Nandi et al. (2009[b]) performed MF studies with both CJ and ETCJ sweet lemon juices by using ceramic membranes. They evaluated the effect of pore size on the permeation characteristics of the membrane at different trans-membrane pressures, as shown in Figures 9.6 (a) and (b). Permeate flux declined sharply within 5–10 min of membrane operation and thereafter became consistent. This decline in membrane flux is due to the formation of a gel layer, made up of pectic material present in the fruit juice over the membrane surface, which results in the blocking of the membrane pores. The study shows that the flux decreased from 12.67×10^{-6} to 5.65×10^{-6} m^3/m^{2}s for CJ and from 58.49×10^{-6} to 21.45×10^{-6} m^3/m^2 s for ETCJ at the end of 45 min. of the experimental run at a trans-membrane pressure of 137.9 kPa. The figures also illustrate that flux increased with enhancement in the trans-membrane pressure. The initial permeate flux increased from 12.67×10^{-6} to 33.15×10^{-6} m^3/m^2 s for CJ and 58.49×10^{-6} to 117×10^{-6} m^3/m^2 s for ETCJ when the trans-membrane pressure was increased from 137.9 to 344.74 kPa due to greater driving force. Also, the permeate flux is higher for ETCJ compared to CJ. This is because of the presence of higher amounts of pectic materials in the CJ that cause higher amount of gel layer formation as well as higher viscosity of the CJ. The observed flux data using the ceramic membrane have been found to be better than those obtained using polymeric membranes. The ceramic membrane provided a flux decline of 138.6×10^{-6} to 20×10^{-6} m^3/m^{2}s within 45 min. of experimental run

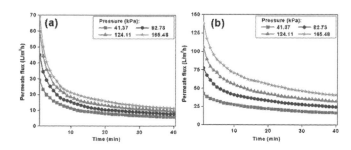

FIGURE 9.6
Variation of permeate flux with time at various trans-membrane pressure drops for (a) CJ and (b) ETCJ.

over a trans-membrane pressure differential of 137.9–44.7 kPa, which is comparable to the flux reported (Rai et al., 2006) using polymeric membranes that lowered from 250×10^{-6} to 1.94×10^{-6} m^3/m^{2}s at 138 kPa within 27 min. of experimental run. Further, the polymeric membranes exhibited 98% reduction in the overall membrane flux due to fouling phenomena which is quite high in comparison to the ceramic membrane (84%). In other words, the ceramic membranes performed better than the polymeric membranes.

9.5.3 Separation of Bio-Molecules, Proteins, and Bacteria by Microfiltration

Nowadays, MF is increasingly used for the separation or purification of bio-products, such as microbial cells, proteins, bacteria, and extracellular polymer substances in biochemical processes. The most prominent advantage of bio separation by using MF membrane process is the availability of efficient and effective separation without the necessity of heating. Since most of the bio-molecules or compounds are heat sensitive, it therefore becomes a boon for bio-separations. In other separation methods, such as distillation and solvent extraction materials are either exposed to extreme conditions like high temperature, pH or harsh solvents, where biomaterials or molecules may get denatured. However, membrane separation processes perform bio-separations without exposing the biomolecules to such extreme conditions. As a result, use of membrane for isolation of bio-molecules is increasing by the day.

9.5.4 Pretreatment of Drinking Water

Clean drinking water is a basic need of every species on earth. Therefore, mankind is always in search of clean water sources or processes that make the available raw water clean and safe for consumption. Membrane processes have become very popular due to their effective capability to remove a variety of impurities from raw water. MF and UF membranes have capabilities to remove suspended particles. On the other hand, nanofiltration (NF) membranes are effective for the removal of dissolved organic contaminants with molecular weights more than 200 Da. The mechanisms involved during removal of contaminants from raw water are electrostatic repulsion, size exclusion and a combination of other factors including preferential affinity. MF is a favorable choice for pretreatment due to its ability to remove contaminants and without addition of chemicals. Also, the availability of ceramic microfiltration membranes makes them more popular due to their low susceptibility to fouling and durability. Other pretreatment processes such as coagulation–flocculation fail to remove the broad spectrum of contaminants present in raw water. The commonly present contaminants in the raw water are macromolecular and dissolved or sparingly soluble organic substances, soluble inorganic compounds, colloidal and particulate matter like silica and microorganisms. The time-tested MF membrane provides a platform for the perfect pretreatment process. Thereafter, NF or RO could be applied to get clean drinking water making the whole process economical, swift and smooth.

9.6 Conclusions

This chapter presented a discussion on the basic fundamentals, membrane materials, different terminologies and applications involved in the field of microfiltration. The different preparation and characterization methods for ceramic, polymer, and polymer-ceramic composite membranes are also discussed along with important applications. The importance and role of different membrane materials, process parameters and membrane types on the microfiltration membrane process are also discussed. This includes the effects of membrane morphology, sintering temperatures, trans-membrane pressures, pore blocking models, and fouling on microfiltration membranes. The considerations given in the chapter definitely provide better insights on the assessment, use, and importance of the said parameters for the development and growth of microfiltration membranes. The role of different membrane parameters based on morphological and permeation properties are crucial for efficient and effective performance of a membrane for a particular application. Therefore, it is important to assess these parameters perfectly. Ceramic microfiltration membranes reveal greater competence in terms of applicability, performance, and efficiency as compared to polymeric membranes for different microfiltration applications. However, the cost of the ceramic membrane is a challenge for its widespread use on an industrial scale.

References

Baker, R.W. 2004. *Membrane Technology and Applications*. West Sussex, England: John Wiley & Sons Ltd.

Bouzerara, F., A. Harabi, S. Achour, and A. Larbot. 2006. Porous ceramic supports for membranes prepared from kaolin and doloma mixtures. *Journal of the European Ceramic Society* 26: 1663–1671.

David, O., Y. Gendel, and M. Wessling. 2014. Tubular macro-porous titanium membrane. *Membrane*: 139–145.

Emani, S., R. Uppaluri, and M.K. Purkait. 2014. Microfiltration of oil–water emulsions using low cost ceramic membranes prepared with the uniaxial dry compaction method. *Ceramics International* 40: 1155–1164.

Hart, F.L. and H.J. Fisher. 1971. *Modern Food Analysis*. Berlin: Springer.

Hermia, J. 1982. Constant pressure blocking filtration laws-application to power-law non-newtonian fluids. *Transactions of the American Institute of Chemical Engineers* 60: 183–187.

Lee, M., Z. Wu, R. Wang, and K. Li. 2014. Micro-structured alumina hollow fibre membranes–Potential applications in wastewater treatment. *Journal of Membrane Science* 461: 39–48.

Lü, Q., X. Dong, Z. Zhu, and Y. Dong. 2014. Environment-oriented low-cost porous mullite ceramic membrane supports fabricated from coal gangue and bauxite. *Journal of Hazardous Materials* 273: 136–145.

Marchese, J. and C.L. Pagliero. 1991. Characterization of asymmetric polysulphone membranes for gas separation. *Gas Separation & Purification* 5: 215–221.

Masmoudia, S., A. Larbot, H. El Feki, and R.B. Amara. 2007. Elaboration and characterisation of apatite based mineral supports for microfiltration and ultrafiltration membranes. *Ceramics International* 33: 337–344.

Mulder, M. 1991. *Basic Principles of Membrane Technology*. Dordrecht: Kluwer Academic Publishers.

Nandi, B.K., B. Das, R. Uppaluri, and M.K. Purkait. 2009[b]. Microfiltration of mosambi juice using low cost ceramic membrane. *Journal of Food Engineering* 95: 597–605.

Nandi, B.K., R. Uppaluri, and M.K. Purkait. 2010. Microfiltration of stable oil-in-water emulsions using kaolin based ceramic membrane and evaluation of fouling mechanism. *Desalination and Water Treatment* 22: 133–145.

Nandi, B.K., R. Uppaluri, and M.K. Purkait. 2008. Preparation and characterization of low cost ceramic membranes for microfiltration applications. *Applied Clay Science* 42: 102–110.

Nandi, B.K., R. Uppaluri, and M.K. Purkait. 2009[a]. Treatment of oily waste water using low-cost ceramic membrane: Flux decline mechanism and economic feasibility. *Separation Science and Technology* 44: 2840–2869.

Rai, P., G.C. Majumdar, G. Sharma, S. Das Gupta, and S. De. 2006. Effect of various cutoff membranes on permeate flux and quality during filtration of mosambi (citrus sinensis (l.) Osbeck) juice. *Food and Bioproducts Processing* 84: 213–219.

Saffaj, N., M. Persin, S.A. Younsi, A. Albizane, M. Cretin, and A. Larbot. 2006. Elaboration and characterization of micro-filtration and ultra-filtration membranes deposited on raw support prepared from natural Moroccan clay: Application to filtration of solution containing dyes and salts. *Applied Clay Science* 31: 110–119.

Saffaj, N., S. Alami Younssi, A. Albizane, A. Messouadi, M. Bouhria, M. Persin, M. Cretin, and A. Larbot. 2004. Elaboration and properties of TiO2–ZnAl$_2$O$_4$ ultrafiltration membranes deposited on cordierite support. *Separation and Purification Technology* 36: 107–114.

Singh R. and M.K. Purkait. 2016. Evaluation of mPEG effect on the hydrophilicity and antifouling nature of the PVDF-co-HFP flat sheet polymeric membranes for humic acid removal. *Journal of Water Process Engineering* 14: 9–18.

Vinoth Kumar, R., A.K. Ghoshal, and G. Pugazhenthi. 2015. Fabrication of zirconia composite membrane by in-situ hydrothermal technique and its application in separation of methyl orange. *Ecotoxicology and Environmental Safety* 121: 73–79.

Zhu, Z., J. Xiao, W. He, T. Wang, Z. Wei, and Y. Dong. 2015. A phase inversion casting process for preparation of tubular porous alumina ceramic membranes. *Journal of the European Ceramic Society* 35: 3187–3194.

10

Hemodialysis Membranes for Treatment of Chronic Kidney Disease: State-of-the-Art and Future Prospects

N.L. Gayatri, N. Shiva Prasad, and Sundergopal Sridhar

CONTENTS

10.1 Introduction

Chronic Kidney Disease (CKD) is a worldwide crisis and its treatment costs as high or more than any other disease for an individual as per the government medical policy in a given country. CKD, at an early stage, is not easily detectable as there are no visible symptoms of the illness in patients. Due to this, diagnosis of CKD by location, disease stage, age and gender are not recorded properly. Renal therapy and kidney transplantation are used in CKD treatment. Since treatment is expensive, most patients in developing countries cannot afford it and subsequently succumb (World Kidney Day, 2017). More than 2 million individuals overall at present get treatment with dialysis or a kidney transplant to stay alive, yet this number may just speak to 10% of individuals who really require treatment to live (Couser et al., 2011). CKD treatment is received by older individuals, mostly in developed countries. In developing countries like India and China, the estimated kidney failure is high and disproportionate. The cost of CKD treatment is a huge economic burden to any country's health care budget.

In people aged between 65 and 74 worldwide, it is estimated that one in five men, and one in four women, have CKD (World Kidney Day, 2017). For example, in 2005, 35 million ascribed to CKD, as per World Health Organization (Levey et al., 2007). Chronic kidney disease can be treated with early identification and treatment making it possible to slow down or stop the advancement of the illness.

10.1.1 Kidney Function

The kidney performs one of the critical tasks in human body, i.e., removal of uremic wastes, preventing their accumulation. It also helps in maintaining the whole-body homeostasis (Hall, n.d.), i.e., a relatively constant internal environment in terms of temperature and acid-base balance (Sinnakirouchenan & Holley, 2011). Thus, it is evident that a dysfunctional kidney can disrupt the normal functioning of the body as a whole, disturbing the delicate balance, causing the accumulation of toxic wastes in the body and consequently, death.

10.1.2 Kidney Dysfunction

Kidney dysfunction, termed in the medical field as "Renal Failure," includes two types: acute kidney injury (AKI) and chronic kidney disease (CKD). AKI grossly refers to the rapid loss of kidney function, whereas CKD refers to the gradual loss of kidney function. The reasons for such kidney failure can be multiple. The biochemical indication of such problem is an elevated urea and creatinine concentration in blood.

Functional ability of the kidney is quantified in terms of Glomerular Filtration Rate (GFR), which is defined as the volume of fluid filtered from the renal glomerular capillaries into the Bowman's capsules per unit time. Mathematically it is defined by Equation 10.1:

$$GFR(ml/min) = \frac{\left(\dfrac{urea}{creatinine\ concentration\ in\ urine}\right) \times (urine\ flow\ rate)}{concentration\ of\ urea/creatinine\ in\ plasma} \qquad (10.1)$$

Based on GFR, CKD can be further divided into five stages of kidney dysfunction.

Patients suffering from end-stage renal disease (ESRD) undergo dialysis, where an artificial setup is used to filter uremic wastes, salts and excess fluids from the body, restoring it to its normal healthy balance. Dialysis can be done in two ways: one is hemodialysis (HD) and other peritoneal dialysis (PD). Removal of uremic toxins carried out using a hollow fiber membrane module is called hemodialysis. In PD, dialysate fluid is filled in the abdomen where the peritoneum (a thin walled cavity made of tissues) acts as a filtrating media and permeates uremic toxins and excess fluids into the dialysate solution, thus purifying the blood within the body.

10.1.3 Hemodialysis

Dialysis is a process of the diffusion of solutes brought about by creating a concentration gradient across the membrane. Hemodialysis is a process of diffusion of uremic toxins, such as creatinine and urea from the blood, into the dialysate flowing across the membrane. When kidney function is impaired or fails in a patient, it leads to the accumulation of high levels of uremic toxins in the blood. Hemodialysis is a process where these uremic toxins in blood are removed using a hemodialyzer. The other two available treatments are peritoneal dialysis (PD) and kidney transplantation. Hemodialysis provides a life support system for patients suffering from end-stage renal syndrome (ESRD). The reasons for HD's being a more common choice for ESRD treatment compared to PD are:

i. More information and clinical data are available on HD than on PD.

ii. HD is versatile and suitable for all patients, whereas PD is not adaptable to all stages of CKD.

iii. HD has acquired more social awareness and is less difficult to perform.

Thus, since the past few decades, individuals are practicing HD process in their pursuit to treat renal failure. A nephrologist (a therapeutic kidney expert) chooses when hemodialysis is required as well as the different parameters for treatment. These parameters incorporate recurrence (number of sessions every week), length of every treatment, blood and dialysis solution stream flow rates, and, in addition, the particular size of the dialyzer (surface area). The concentrations of the dialysis solution are also varied at the time of dialysis in terms of its sodium, bicarbonate and potassium levels. The larger the body size of an individual, the more dialysis he/she will require. Four sessions for each week are frequently endorsed for larger patients.

A simplistic HD process is described in Figure 10.1 (a). The hemodialysis machine pumps the patient's blood and the dialysate through the cartridge/dialyzer. The blood is drawn from the patient, circulated into the tube side of the hollow fibers and sent back from the

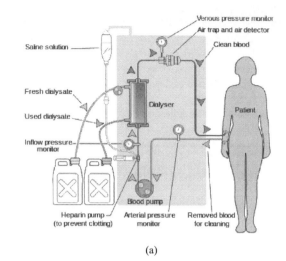

(a)

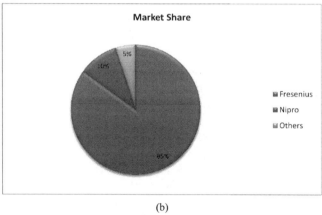

(b)

FIGURE 10.1
(a) Schematic of a hemodialysis circuit (Yassine Mrabet, 2008) [Reproduced with permission from Wikipedia], (b) Market share of dialyzer manufacturers, worldwide.

exit stream. The dialysate solution containing sodium and potassium salts is pumped in the shell side of the dialyzer. A concentration gradient is created between the blood and dialysate solution resulting the diffusion of uremic toxins from the blood into dialysate. The latest machines are fully automated with monitoring of flow rates of blood and dialysate, temperature, air presence, pH and blood leakage. Manufacturers of dialysis machines include companies such as Nipro, Fresenius, Gambro, Baxter, B. Braun, NxStage and Bellco (Wikipedia contributors, "Hemodialysis").

10.1.3.1 Dialyzer

Dialyzer plays a vital role in HD process which actually removes uremic toxins from the blood. All dialyzers, which are available commercially in the market are of a hollow fiber variety and works on the principle of membrane contactors. Design configuration of dialyzer consists of semi-permeable membranes mounted in a transparent polymeric cylinder. The very fine cylindrical membranes synthesized from biocompatible synthetic polymers using the wet-spinning method. Bundles of these fibers anchored at each end with biocompatible resin and this assembly is then put into a clear plastic cylindrical shell with four openings. One opening or blood port at each end of the cylinder communicates with each end of the bundle of hollow fibers. This forms the "blood compartment" of the dialyzer. Two other ports are cut into the side of the cylinder. These communicate with the space around the hollow fibers, the "dialysate compartment." Blood is pumped via the blood ports through this bundle of very thin capillary-like tubes, and the dialysate is pumped through the space surrounding the fibers to move fluid from the blood to the dialysate compartment.

10.2 Hemodialysis Membranes

Dialyzers are classified into two types based on membrane pore size and porosity. Devices with a smaller pore size and low molecular weight cut off (MWCO) are called "low-flux dialyzers". Those with large pore diameter and high MWCO are termed "high-flux dialyzers". Low-flux dialyzers effectively remove uremic toxins of size below 5 kDa such as urea, but large molecules like beta-2-microgobulinare are not removed at all. The efficiency of dialysis process depends on the type of dialyzer used. At first, cellulose acetate was used as the base polymeric material for making dialysis fibers, but later polyacrylonitrile, polysulfone and other polymers were used. All the above fibers can be made in either low-or high-flux mode, yet most are high-flux. Nanotechnology is being utilized as a part of the development to synthesize high-flux fibers of uniform pore size. The objective of high-flux membranes is to conveniently pass larger particles such as beta-2-microglobulin (MW 11,600 Da), but not to pass albumin (MW ~ 66,400 Da). The key players who manufacture dialyzers worldwide are presented in Figure 10.1 (b). The acceptability of a particular membrane material for dialysis is not based entirely on the issue of biocompatibility. The uremic toxin flux is an equally important factor when recommending a particular material for use as a dialysis membrane. An ideal dialysis material yields a membrane with high flux of uremic toxins with minimum biological reactions.

The polymeric membrane behaviour is very important in terms of compatibility when it comes into contact with blood. The second aspect is the hydrophilicity of the membrane. More hydrophilic membranes exhibit less fouling and hence unhindered flow. The third aspect is the flux through the membrane. It was already reported that shorter dialysis periods (enhanced uremic toxin flux) helps in improving the quality of life of a patient (Kolff, 1956a). Lastly, the membrane material must be robust to withstand sterilization by steam, gamma radiation or ethylene oxide, as available. All these aspects together decide the suitability of a dialysis membrane. Adding PVP also improves biocompatibility. Membrane material selection is a part of the whole gamut of dialysis membrane spinning and module manufacturing.

Dialysis is classically defined as a diffusion-driven process. In conventional membrane separation processes, a driving force is applied across a semi-permeable membrane, separating two phases. The selective passage of solutes is carried out by a simple-size exclusion principle governed by the membrane pore size. The solutes smaller than the pores permeate and the larger ones are retained. In the case of hemodialysis, the chemical potential gradient (μ) between the two phases (μ_A and μ_B), transports the solute through the membrane. No external pressure gradient is applied. Hemodialysis utilizes counter current flow where the dialysate is flowing in the opposite direction to blood flow in the extracorporeal

TABLE 10.1

Global Manufacturers and Suppliers of Hemodialysis Modules

S. No	Company	Surface Area Range (m²)	Priming Volume (ml)	Wall Thickness (μm)	Inner Diameter (μm)
1	Fresenius	0.6–2.5	32–132	35	185–210
2	Nipro	0.9–2.5	62–149	40	200
3	Baxter	0.5–2.4	35–165	15–50	190–215
4	B Braun	1–2.3	58–121	38–40	195–200
5	Allmed	1–2	59–109	40	200
6	Asahi Kasei Medical	1–2.5	66–139	25–45	175–200
7	Bain Medical	Not Specified	Not Specified	Not Specified	Not Specified
8	Bellco	1.1–2.2	73–132	30–35	200
9	Medica	1.1–2.7	66–156	30	200
10	Browndove	1.3–1.4	Not Specified	Not Specified	Not Specified

circuit. Counter-current flow maintains the concentration gradient across the membrane at a maximum and increases the efficiency of the dialysis.

10.2.1 Dialyzer Size and Efficiency

Dialyzers come in a wide range of sizes. A dialyzer with a larger surface area (A) will remove more uremic toxins than a smaller dialyzer, especially at high blood flow rates. Different capacity dialyzers having different membrane surface areas are available in market as seen in Table 10.1.

10.2.2 Reuse of Dialyzers

The dialyzer may either be disposed of after every treatment or reused. Reuse requires a broad technique of abnormal state sanitization. Reused dialyzers are not shared between patients. Today, doctors recommend single use of dialyzers.

10.3 History of Dialysis Process

Thomas Graham of Scotland coined the term dialysis and described the process in 1861 (Graham, 1861). John Abel of the United States developed the first artificial kidney in 1913 (Abel et al., 1914). George Haas of Germany performed a successful human dialysis in 1924 (Haas, 1925; Benedum, 1986). Kolff and Berk developed a practical human dialysis machine (Kolff et al., 2017; Kolff, 1956b). In 1946, Nils Alwall was able to produce a dialyzer with controllable ultrafiltration (UF) rates (Alwall, 1947; Alwall, 1986). The field of HD really experienced a paradigm shift with the invention of Quinton and Scribner arterio venous (AV) shunt in 1960 (Quinton et al., 1960) along with Cimino and Brescia's native AV fistula for chronic vascular access (Brescia et al., 1966). However, from mid-1960s, hollow fiber membranes started to replace the earlier dialyzer designs (Bailey, 1972; Gotch et al., 1969) including celloidin tubes (Haas, 1925; Benedum 1986), twin-coil dialyzer (Kolff, 2017; Kolff 1956b) and rotating drum models (Alwall, 1947; Alwall, 1986).

10.4 Hemodialysis Module Design

The evolution has been in the direction of membrane materials and from flat sheet to hollow fiber configuration. Flat sheet membranes are cast on a support, usually a non-woven fabric. The macro voids are clearly visible below the selective skin layer. The skin layer is responsible for the filtration performance of the membranes. Contrastingly, in hollow fiber membranes, filtration occurs radially outwards. The blood stream flows through the hollow section of the fibers (tube side), the skin layer filters the stream and the filtrate flows radially outwards and is ultimately collected outside of the lateral surface. A general rule of thumb for hollow fiber membranes is that the fiber diameter should be at least ten times lower than the diameter of the largest feed particle.

Dialysis membranes are hollow fiber and self-supporting membranes allowing convenient cleaning mechanisms (e.g., back flushing). Hollow fibers used for dialysis have an inner diameter in range of 180–220 microns with thickness of 35–40 microns. The hollow fiber module design gives very high packing densities in m^2/m^3. Around 10,000–15,000 of such fibers are packed in a single cartridge, yielding a minimum area of around 1 m^2 per filter. The "one-square meter" hypothesis came into existence as proposed by the Seattle group, who later presented the hypothesis of the removal of the middle molecule as well, leading to modification in dialysis protocols (von Hartitzsch et al., 1973; Shaldon et al., 1976). Figure 10.2 (a) exhibits a bundle of hemodialysis fibers stored in a water bath prior to module fabrication, while Figure 10.2 (b) provides a comparative display of dialysis fibers of smaller dimensions in comparison with larger hollow fibers used in water purification. Typical SEM images of dialysis fibers are shown in Figure 10.2 (c).

The modern-day dialysis procedure involves the following paraphernalia:

1. A HD cartridge.
2. A dialysis machine, including heater to warm the dialysate to body temperature; dialysis pump with the provision to regulate flow rate and transmembrane pressure; sensors; alarms to detect leakages; etc.
3. Dialysate fluid.

10.4.1 Design of Dialysis Cartridge

The cartridges comprise of housing, fibers, spacers, nozzles for flow and potting material. The housing design is as important as the membrane. A good housing should be small in design to minimize transport, storage costs and blood hold-up volume. Compact designs have been possible in recent decades due to the development of thinner fibers with reduced wall thickness. The material used for housing has also changed from polycarbonate to polypropylene, since polypropylene can be disposed in an environment-friendly manner (Mandolfo et al., 2003). The fibers are potted in the housing using resins like polyurethane (PU). The resin has to be inert to blood contact as well as β- and γ-radiation, which are used during sterilization. It is clear that, within a bundle of hollow fiber membranes, there are spacers to improve dialysate flow distribution, thereby increasing the mass transfer coefficient and enhancing the transport rate of uremic toxins (Unger et al., 2006). The trick is to maintain the dialysate flow in a cross-flow pattern rather than a parallel flow pattern along the fibers.

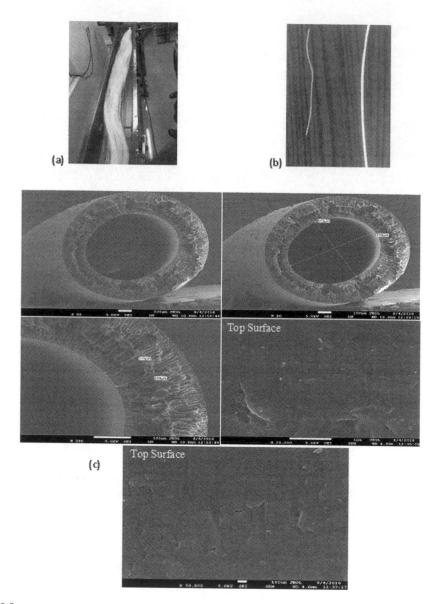

FIGURE 10.2
(a) Bundle of hemodialysis fibers, (b) A comparative view of dialysis fiber of smaller dimensions as against hollow fiber used in water purification and (c) SEM images of cross-section and surface of hemodialysis membrane.

10.4.2 Mechanism of Dialysis

In the dialyzer, blood flows through tube side of hollow fibers and uremic toxins diffuse through the membranes into the dialysate compartment. The dialysate is heated to the same temperature as the body and pumped into the inlet port. This washes the uremic toxins permeating through the membranes, replenishing the body of its lost salts and thereby maintaining the blood homeostasis. The dialysate fluid composition is a very important aspect in the dialysis procedure, and the concentration of each ion is carefully adjusted to minimize the losses of plasma and body fluid but maximize the toxins' transport rates.

For example, high concentration of Na helps minimize hypo-osmotality, and lactic or ace-tic acid concentration in dialysate helps prevent precipitation of calcium and magnesium. These concentrations are reported by Locatelli et al. (2004). The clinical procedure involves the use of catheters or fistula to draw blood to the extracorporeal circuit and return it back. For younger children and patients undergoing less frequent HD sessions, catheters are used. They have two lumens—one draws blood from the body to the extracorporeal circuit and the other returns the blood into the body. Fistula is an arrangement wherein a vein and an artery are joined together facilitating a larger flow of blood and mechanical strength to withstand multiple arduous dialysis sessions. This is used for older children and patients undergoing frequent dialysis.

10.4.3 Parameters Influencing Hemodialysis

The efficiency of dialysis depends on clearance, UF coefficient (K_{UF}), volumetric mass transfer coefficient (k_0A), dialysis adequacy (Kt/V) and transport mechanisms.

10.4.3.1 Clearance

Clearance (C_L in ml/min) is defined as the removal rate of uremic toxins in a single pass.

10.4.3.2 UF Coefficient (K_{UF})

K_{UF} of a dialyzer is calculated in vitro, whereby bovine blood is ultrafiltered at various trans-membrane pressures (TMPs).

10.4.3.3 Mass Transfer Coefficient (k_0A)

The volumetric mass transfer coefficient of a dialyzer is indicated by the product of linear mass transfer coefficient k_0, and area of membrane A, and expressed in ml/min.

10.4.3.4 Dialysis Adequacy (Kt/V)

Kt/V is a term indicating the adequacy of the dialysis session, where K is the clearance in ml/min, t is the time of dialysis in minutes and V is the volume of water in the patient's body, in liters.

10.4.3.5 Transport Mechanisms

Transport mechanisms involved in dialysis can be diffusion, adsorption or convection, or all of them occurring simultaneously. The field of HD membranes continues to be vibrant, posing challenges for membrane engineers and material scientists.

10.4.4 Features of Hemodialysis Fibers

Hollow fiber membranes find the greatest application in the field of HD. These UF hollow fiber membranes present in hemodialyzer mimic natural filtration for biomedical applica-tions. Functional stability under a wide range of biological conditions and minimal foul-ing are some of the properties of these membranes. Along with biocompatibility issues, the flux of HD fibers has also proven to be as important an aspect as the nature of polymers.

Dialysis fibers pass small biomolecules present in blood through their pores into the dialysate solution. This way, all small uremic toxins are removed from blood. Most of the known uremic retention solutes in human blood are reported as 90 in number, with their concentration ranging from 2.3 g/L (urea) to 0.32 ng/L (methi-onine-enkephalin). Uremic toxins are categorized on the basis of their molecular weights since the membrane functions by molecular sieving mechanism. The size of the uremic toxins influences removal pattern potentiality during dialysis. We segregate, compounds of low molecular weight (<500 Da) as low molecules, and compounds with moderate molecular weight, such as small protein-bounds (>500 Da) as middle molecules. Low-flux membranes, which have small pore size, do not remove middle molecules, hence high-flux membranes, which have large pore size (60 kDa), are found to be more efficient.

10.4.5 Cost Estimation

The cost of a dialyzer depends completely on the scale of manufacture. The approximate cost estimation for one dialyzer with a surface area of 1.2 m^2 is shown in Table 10.2 on a production scale of 1 million dialyzers per annum in India. Bulk production could ensure a reasonable cost of just US$ 2.52 per module, allowing for the affordability of dialysis in developing and under developed nations.

10.5 Membrane Formation

The technology to spin dialysis-grade fibers is a challenging problem. The basic physics of the hollow fiber membrane formation involve the polymer solution coming into contact with the bore fluid, and then undergoing phase inversion to form the hollow core. The inner and outer diameters of the spinneret determine the diameter and thickness of the resultant hollow fiber membranes. The HD grade hollow fiber spinning technology poses a twofold challenge. The first set of issues are the composition of the dope solution and the spinning conditions like viscosity, thermodynamics of the solution, along with operating conditions like temperature, humidity, take-up speed and post-treatment methodologies. The second challenge is the design of the extruder. These two aspects in tandem helped to successfully design the spinning process for dialysis grade hollow fibers.

10.5.1 Efficient Casting Machine for Spinning Hollow Fibers

A laboratory hollow fiber spinning machine is shown in Figure 10.3 (a). This casting machine comprises of a polymer reservoir, bore fluid container, spinneret mounted on a movable jack, coagulation bath, a pulley/or a fiber collector. Polymer reservoir supplies the polymer solution during fiber preparation. Bore fluid container, which supplies bore fluid can be similar to or different from the coagulating solution in the bath. The bore fluid can be incorporated with additive to tailor the desired properties or by changing its temperature. The spinneret is mounted on a movable jack, which varies the distance between the extrusion points of polymer to the coagulation bath surface. Coagulation or non-solvent bath provides the required residence time for the polymer to complete phase inversion. By varying the temperature and using additives in the bath, we can attain the desired pore size. Pulley is used in fiber casing machine to stretch the fibers and changing their dimensions.

TABLE 10.2

Cost of Hemodialysis Modules When Manufactured in Bulk

	Input		
Quantity Served	10,00,000 Dialysis Modules		
Time Frame	12 Months		

FIXED COSTS

Item	Total Cost		Allocated Unit Cost	
Plant Establishment	$	38,500.00	$	0.04
Plant cost	$	3,07,700.00	$	0.31
Skilled labour	$	60,000.00	$	0.06
Semi skilled labour	$	44,300.00	$	0.04
Utilities	$	42,000.00	$	0.04
Administration, Overheads marketing	$	78,000.00	$	0.08
Plant maintains	$	18,500.00	$	0.02
Certification and R&D	$	36,000.00	$	0.04
Waste management	$	55,000.00	$	0.06
Packing & Storage	$	1,54,000.00	$	0.15
Average Fixed Cost per Unit			$	0.83
TOTAL FIXED COSTS	$	8,34,000.00		

VARIABLE COSTS

Item	Total Cost		Cost per Unit	
Raw material	$	16,90,000.00	$	1.69
a. Polymer				
b. Solvent				
c. Additives				
d. Resin				
e. Housing				
f. Sterilization				
Average Variable Cost per Unit			$	1.69
TOTAL VARIABLE COSTS	$	16,90,000.00		
AVERAGE COST PER UNIT			$	2.52
TOTAL COSTS	$	25,24,000.00		

By varying the polymer concentration in the dope and spinning conditions, such as distance between spinneret tip and coagulation bath, composition of coagulation bath or residence time inside the bath, the fibers pore size distribution and diameter can be controlled/altered.

10.5.2 Spinneret for Dialysis Fibers

The spinneret designed for making synthetic polymeric hollow fibers comprises of a reservoir nozzle and a cylindrical pin assembled coaxially are shown in Figure 10.3 (b) along

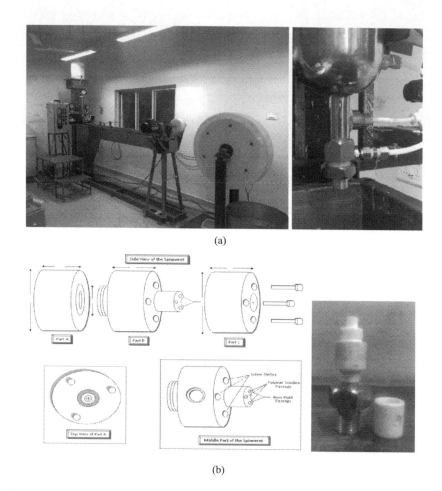

FIGURE 10.3
(a) Hollow fiber spinning machine and (b) Schematic design drawing of the dialysis hollow fiber spinneret with photograph.

with a picture of the spinneret. The dimensions of both the reservoir nozzle and the cylindrical pin determine the size (inner and outer diameter) of the fibers. By varying the size of the reservoir nozzle and cylindrical pin of the above said spinneret, we can achieve hemodialysis fiber dimensions. The other parameters influencing dialysis fibers are volumetric flow rates of the polymer and bore fluids besides pulling rate of the winding roller.

10.6 Drawbacks of Hemodialysis

Hemodialysis often involves fluid removal (through UF) as patients pass little urine. Removal of fluid during or post dialysis causes low blood pressure, fatigue, chest pains, leg-cramps, nausea and headaches.

Hemodialysis is performed outside of the body, which could expose a patient's circulatory system to microbes. making him prone to serious infections if sterilization standards during dialysis process are not maintained. Heparin is used as an anticoagulant during e dialysis, and if the patient is allergic to heparin, it can cause low platelet count. Patients on long term hemodialysis treatment may be susceptible to amyloidosis, neuropathy and heart disease.

10.7 Emerging Trends in Dialysis Research

Future research poses challenging and intriguing avenues towards developing better and more efficient methodologies in dialysis therapy to mimic the human kidney. The future scope of dialysis consists of three different avenues (Humes et al., 2006).

10.7.1 Heparinization of Fibers

Hemodialysis requires anticoagulation agents to avoid clotting of blood on the internal surface of fibers. Actuated blood thickening phenomena are utilized to gauge the concentration of anticoagulant required amid dialysis. The span of clotting time acquired for a given level of anticoagulant agents relies upon the test utilized, and every dialysis unit must set up its own particular target range of dosage. The dosage of anticoagulant agent depends on bio-compatibility of fiber material, physical condition of the patient and administration methods. Disadvantages of these anticoagulant agents include lack of consistency in their action, and the expanded volume stack which must be expelled during dialysis. Usage of heparin may cause **osteoporosis**, especially in case of prolonged exposure. Anticoagulant-free hemodialysis can be achieved by increasing the biocompatibility of the dialyzer through specific surface modification or integration. Certain biomolecules, including linoleic acid (Kung & Yang, 2006), dextran (Yu et al., 2007), chitosan (Lin et al., 2004), apart from heparin (Li et al., 2013; Lin et al., 2005), have been employed as anticoagulants to improve the membrane's hemo-compatibility.

Heparin has been particularly prominent as an exceptionally sulfated polysaccharide, which is adversely charged in view of the high level of SO^{-2}_4/SO^{-2}_3 in the molecule. This negative charge can disturb the blood coagulation process and curtail thrombosis (Lin et al., 2005). However, excess heparin in the blood stream could cause severe problems such as thrombocytopenia (Arepally & Ortel, 2006; Cuker & Cines, 2012) and abnormal haemorrhage. To solve such problems, there is a need for antithrombogenic continuous hemofilters, which function without any systemic anticoagulation for at least one week, achieved by the modification of inner surfaces of the hollow fibers with heparin and other antithrombogenic substances. Researchers have endeavored to immobilize heparin onto the dialysis fibers rather than directly infusing it into the blood (Lin et al., 2004; Li et al., 2013; Kang et al., 2001). The goal is to make a non-thrombogenic hemodialysis system, eliminate the risk of severe complications and enhance hemo-compatibility by decreased platelet adhesion and decreased blood cell loss in patients with a high risk of bleeding (Ren et al., 2013). Heparin immobilization on to membranes can be done using varies methods (Ren et al., 2013).

10.7.1.1 Progress on Antithrombogenic Continuous Hemofilter

Ishihara et al. (1994) developed methacryloyloxyethyl phosphorylcholine (MPC) polymers, which mimic phospholipids as the component of cell membrane and heparin covalently bound to specific group of filter membrane surface (Ishihara et al., 1994; Tatsumi et al., 1999).

They may be the most appropriate substances to enhance surface hemo-compatibility of hollow fiber modules.

10.7.2 Biocompatible Membranes

Heparinization involves developing more biocompatible membranes with an enhanced ability to reduce oxidative stress development. Decrease in chances of blood coagulation is also an important aspect, as it reduces the use of heparin during dialysis, thereby reducing treatment costs. Achieving higher GFRs by engineering larger pore sizes with sharp cut-offs is the target of next-generation membranes.

10.7.3 Innovative Membranes

Two groups have proposed possible mechanisms that can eliminate the use of membranes. Edward and his team (Humes et al., 2006) proposed microfluidic principles to carry out dialysis between two streams, while Nissensson and team (Humes et al., 2006) developed a novel replacement device called the "human nephron filter." The former technology discusses two liquid streams coming into contact with each other at a low Reynolds number, thereby inhibiting mixing, and maintaining two distinct phases, with mass transfer occurring from one stream to the other. The latter discusses a device that functions similar to a glomerulus and tubule and has two filters, the G membrane (mimicking glomerulus) and T membrane (mimicking tubule). There is no use of dialysate and theoretically, 1/10th of 1 m^2 surface area is required. Nanotechnology has been utilized to achieve the pores of the two membranes with both membranes being one molecule thick. Silicone nonporous membranes are being developed with the help of the sacrificial layer technique (Humes et al., 2006). Nature-inspired slit-shaped pores are also being explored as an alternative to round pores as they have potential for better performance.

10.7.4 Living Membranes

Scientists, nephrologists and engineers are working together to approach the ideal kidney functioning with possible wearable devices, which would function continuously to reject uremic toxins. Ongoing research, specifically to address this problem, has paved the way for endothelial cell seeded micro-fabricated capillary network devices (Humes et al., 2006). Additionally, cells cultured from adult human kidneys are grown in the inner surface of hemofiltration membranes to help maintain blood homeostasis and re-absorption of selective solutes during filtration.

Hemodialysis treatment has grown impressively, particularly in the fields of hollow fiber configuration, bio-compatibility, and dialysate composition, which have been improved by addressing a few entanglements and unwanted reactions. Hemofiltration, hemodiafiltration, and day to day hemodialysis, have proven to boost renal replacement treatment (RRT). The up and coming of age of RRT ought to tackle the deficiency of antithrombogenic property of hemofilters and provide tubular metabolic capacity. The metabolic and endocrinological capacity of tubules have never been supplanted by hemodialysis, albeit glomerular capacity is mostly supplanted by these modalities. As a consequence, hemodialysis patients experience the ill effects of serious long-term issues such as atherosclerosis and cardiovascular ailments (Foley et al., 1998), secondary hyperparathyroidism (Madias Cohen and Harrington, 1995), β2-microglobulin amyloidosis (Gejyo et al., 1985) and so on, because it is very difficult to replace natural kidney function.

In 1987, Aebischer et al. introduced the primary idea of a bioartificial kidney, which comprised of hollow fiber membranes and tubular epithelial cells (Aebischer et al., 1987; Ip & Aebischer, 1989). Since 1996, there has been the expectation to build up a bioartificial kidney comprising of a persistent hemofilter and a bioartificial tubule gadget comprising of a hollow fiber module and tubular epithelial cells to avoid serious dialysis difficulties in long-term hemodialysis patients, in spite of the fact that a bioartificial kidney for long-term treatment is a more troublesome choice than acute treatment (Saito et al., 1998; Kushida et al., 2000; Terashima et al., 2001; Asano et al., 2002; Saito, 2004; Fujita et al., 2004).

10.7.5 A Bioartificial Kidney System

Bioartificial kidneys for chronic renal failure patients require a better hemofilter and longer function of a tubule device. Thus, the first step towards the development of bioartificial kidneys is to develop a system with continuous hemofiltration for at least one week. The second step is preparation of an artificial membrane with tubular epithelial cell layers, which maintain tubular function for at least one month. Blood is first sent through the continuous hemofilter and then perfused outside of the hollow fibers in the tubule device (Saito, 2003).

10.7.5.1 Artificial Membrane for Bioartificial Tubule Devices

Polysulfone hollow fiber membrane modules have been used as scaffolds for artificial tubule devices (Humes et al., 1999; Humes et al., 2004; Ozgen et al., 2004). The attachment and the proliferation of HK-2 cells on a polyimide based membrane were found to be superior to those on polysulfone membrane.

10.7.5.2 Formation of Tubular Epithelial Cells-Monolayer

Tubular epithelial cells are polarized by which substances in the glomerular filtrate are actively transported from the apical side to the basal side of a tubule. A tubule must form a confluent monolayer in order to transport biological substances from the apical to the basal side, or from the basal side to the apical side.

The primary necessity for development of a bioartificial kidney for dialysis patients, would be a membrane which does continuous hemofiltration yielding 10 L/day of filtrate. The bioartificial tubule devices with specific gene transfection is required for keeping contact inhibition of tubular epithelial cells to maintain confluent mono layers for a long duration. Transfection of functional protein genes into tubular epithelial cells could facilitate the active transportation of useful biological substances, leading to a *compact artificial tubule device.*

10.8 Conclusions

Renal dysfunction is a fast growing disease which is partially treated by artificial dialyzers based on hollow fiber membranes. Hemodialyzers in the current market are polymeric membranes which effectively remove uremic toxins along with some vital molecules present in blood. This leads to various side effects in dialysis patients which prevents them from leading a normal life. These membranes can closely mimic natural kidneys if new properties are incorporated in hollow fiber polymeric membranes such as heparinization of fibers or bioartificial deposition of living membranes on to the hollow fiber walls. This could bring about a revolution in combating chronic kidney disease and improve the quality of life of renal patients.

References

Abel, J.J., L.G. Rowntree, and B.B. Turner. 1914. On the removal of diffusible substances from the circulating blood of living animals by dialysis ii. some constituents of the blood. *Journal of Pharmacology and Experimental Therapeutics* 5(6): 611–623.

Aebischer, P., T.K. Ip, G. Panol, and P.M. Galletti. 1987. The bioartificial kidney: Progress towards an ultrafiltration device with renal epithelial cells processing. *Life Support Systems* 5(2): 159–168.

Aebischer, P., T.K. Ip, L. Miracoli, and P.M. Galletti. 1987. Renal epithelial cells grown on semipermeable hollow fibers as a potential ultrafiltrate processor. *ASAIO Journal*.

Alwall, N. 1949. On the artificial kidney xi. *Acta Medica Scandinavica* 133(S229): 20–21.

Asano, M., Y. Fujita, Y. Ueda, D. Suzuki, T. Miyata, and H. Sakai. 2002. Renal proximal tubular metabolism of protein-linked pentosidine, an advanced glycation end product. *Nephron*.

Ayus, J.C., M. Reza Mizani, S.G. Achinger, R. Thadhani, A.S. Go, and S. Lee. 2005. Effects of short daily versus conventional hemodialysis on left ventricular hypertrophy and inflammatory markers: A prospective, controlled study. *Journal of the American Society of Nephrology* 16(9): 2778–2788.

Bailey, G.L. 1972. *Hemodialysis: Principles and Practice*.

Barrington, J.T. 1995. The pathogenesis of parathyroid gland hyperplasia in chronic renal failure principal discussant. *Kidney International* 48: 259–72.

Benedum, J. 1986. Georg Haas (1886–1971), pioneer in hemodialysis. *Schweizerische Rundschau Fur Medizin Praxis* 75(14): 390–394.

Birdee, G.S., R.S. Phillips, and R.S. Brown. 2013. Use of complementary and alternative medicine among patients with end-stage renal disease. *Evidence-Based Complementary and Alternative Medicine*.

Bott, R. 2014. *Guyton and Hall Textbook of Medical Physiology, 13th ed.* Igarss.

Brescia, M.J., J.E. Cimino, K. Appell, and B.J. Hurwich. 1966. Chronic hemodialysis using venipuncture and a surgically created arteriovenous fistula. *The New England Journal of Medicine* 275(20): 1089–1092.

Colice, G.L. 1994. "Historical Perspective on the Development of Mechanical Ventilation."*Principles and Practice of Mechanical Ventilation,* Second edition.1–36. McGraw Hill Education.

Couser, W.G., G. Remuzzi, S. Mendis, and M. Tonelli. 2011. The contribution of chronic kidney disease to the global burden of major noncommunicable diseases. *Kidney International* 80(12): 1258–1270.

Cuker, A. and D.B. Cines. 2012. How I treat heparin-induced thrombocytopenia. *Blood*.

Foley, R.N., P.S. Parfrey, and M.J. Sarnak. 1998. Epidemiology of Cardiovascular Disease in Chronic Renal Disease. *Journal of the American Society Nephrology* 9(12 Suppl.): S16–23.

Fujita, Y., M. Terashima, T. Kakuta, J. Itoh, T. Tokimasa, D. Brown, and A. Saito. 2004. Transcellular Water Transport and Stability of Expression in Aquaporin 1-Transfected LLC-PK1 Cells in the Development of a Portable Bioartificial Renal Tubule Device. *Tissue Engineering* 10(5/6): 711–722.

Gejyo, F., T. Yamada, S. Odani, Y. Nakagawa, M. Arakawa, T. Kunitomo, H. Kataoka, M. Suzuki, Y. Hirasawa, and T. Shirahama. 1985. A new form of amyloid protein associated with chronic hemodialysis was identified as beta 2-Microglobulin. *Biochemical and Biophysical Research Communications* 129(3): 701–706.

Gotch, F., B. Lipps, and J. Weaver. 1969. Chronic Hemodialysis with the Hollow Fiber Artificial Kidney (HFAK). *Transactions– American Society for Artificial Internal Organs* 15: 87–96.

Graham, T. 1861. Liquid Diffusion Applied to Analysis. *Philosophical Transactions of the Royal Society of London* 151: 183–224.

Greinacher, A. 2015. "Heparin-Induced Thrombocytopenia." *New England Journal of Medicine*: 5–7.

Haas, G. 1925. Versuche der blutauswachung am lebenden mit hilfe de dialyse. *Wiener Klinische Wochenschrift* 4: 13–14.

Hernández, R., J. Margarita, C. García, N. Vega, M. Macía, D. Hernández, A. Rodríguez, B. Maceira, and V. Lorenzo. 2011. Diálisis Peritoneal Actual Comparada Con Hemodiálisis: Análisis de Supervivencia a Medio Plazo En Pacientes Incidentes En Diálisis En La Comunidad Canaria En Los Últimos Años. (Eng.: Current peritoneal dialysis compared with haemodialysis: Medium–term survival analysis of incident dialysis patients in the Canary Islands in recent years.) *Nefrologia* 31(2): 174–184.

Humes, H.D. A tissue engineering\nof a bioartificial renal tubule assist device: in vitro transport and\nmetabolic characteristics.future of hemodialysis membranes clinical results of the bioartificial kidney containing human celpatients with acute renal. Sf.M. MacKay, A.J. Funke, and D.A. Buffington. 1999. Tissue engineering\nof a bioartificial renal tubule assist device: In vitro transport and\nmetabolic characteristics. *Kidney International* 55: 2502–2514.

Humes, H.D., W.H. Fissell, and K. Tiranathanagul. 2006. The future of hemodialysis membranes. *Kidney International* 691G(7): 1115–1119.

Humes, H.D., W.F. Weitzel, R.H. Bartlett, F.C. Swankier, E.P. Paganini, J.R. Luderer, and J. Sobota. 2004. Initial clinical results of the bioartificial kidney containing human cells in ICU patients with acute renal failure. *Kidney International* 66(4): 1578–1588.

Ip, T.K. and P. Aebischer. 1989. Renal epithelial-cell-controlled solute transport across permeable membranes as the foundation for a bioartificial kidney. *Artificial Organs* 13(1): 58–65.

Ishihara, K., K. Fukumoto, H. Miyazaki, and N. Nakabayashi. 1994. Improvement of hemocompatibility on a cellulose dialysis membrane with a novel biomedical polymer having a phospholipid polar group. *Artificial Organs* 18(8): 559–564.

Ishihara, K., T. Tsuji, T. Kurosaki, and N. Nakabayashi. 1994. Hemocompatibility on graft copolymers composed of poly(2-methacryloyloxyethyl Phosphorylcholine) side chain and poly(n-butyl methacrylate) backbone. *Journal of Biomedical Materials Research Part A* 28(2). Wiley Online Library: 225–232.

Jha, V., G. Garcia-Garcia, K. Iseki, Z. Li, S. Naicker, B. Plattner, R. Saran, A.Y. Moon Wang, and C.W. Yang. 2013. Chronic kidney disease: Global dimension and perspectives. *The Lancet* 382(9888): 260–272.

Kang, I.K., E.J. Seo, M.W. Huh, and K.H. Kim. 2001. Interaction of blood components with heparin-immobilized polyurethanes prepared by plasma glow discharge. *Journal of Biomaterials Science, Polymer Edition* 12(10): 1091–1108.

Kolff, W.J. 1956. Further development of a coil kidney, disposable artificial kidney. *Journal of Laboratory and Clinical Medicine*.

Kolff, W.J., H.T.J. Berk, and L.J.W. Welle Mter, A.J.W. van der Ley , E.C. van Dijk, and J. van Nordwijk. 1944. The artificial kidney: A dialyzer with a great area. *Acta Medica Scandinavica*.

Kung, F.C. and M.C. Yang. 2006. Effect of conjugated linoleic acid grafting on the hemocompatibility of polyacrylonitrile membrane. *Polymers for Advanced Technologies* 17(6): 419–425.

Kushida, A., M. Yamato, C. Konno, A. Kikuchi, Y. Sakurai, and T. Okano. 2000. Temperature-responsive culture dishes allow nonenzymatic harvest of differentiated Madin-Darby Canine Kidney (MDCK) cell sheets. *Journal of Biomedical Materials Research* 51(2): 216–223.

Levey, A.S., R. Atkins, J. Coresh, E.P. Cohen, A.J. Collins, K.-U. Eckardt, M.E. Nahas et al. 2007. Chronic Kidney Disease as a global public health problem: Approaches and initiatives—A position statement from kidney disease improving global outcomes. *Kidney International* 72(3): 247–259.

Li, R., H. Wang, W. Wang, and Y. Ye. 2013. Immobilization of heparin on the surface of polypropylene non-woven fabric for improvement of the hydrophilicity and blood compatibility. *Journal of Biomaterials Science, Polymer Edition* 24(1): 15–30.

Lin, W.C., T.Y. Liu, and M.C. Yang. 2004. Hemocompatibility of polyacrylonitrile dialysis membrane immobilized with chitosan and heparin conjugate. *Biomaterials* 25(10): 1947–1957.

Lin, W.C., C.H. Tseng, and M.C. Yang. 2005. In-vitro hemocompatibility evaluation of a thermoplastic polyurethane membrane with surface-immobilized water-soluble chitosan and heparin. *Macromolecular Bioscience* 5(10): 1013–1021.

Locatelli, F., A. Covic, C. Chazot, K. Leunissen, J. Luño, and M. Yaqoob. 2004. Optimal composition of the dialysate, with emphasis on its influence on blood pressure. *Nephrology Dialysis Transplantation* 19(4): 785–796.

Madias, N.E., J.J. Cohen, and J.T. Harrington, 1995. The pathogenesis of parathyroid gland hyperplasia in chronic renal failure. *Kidney International*, 48: 259–272, Nephrology Forum.

Mandolfo, S., F. Malberti, E. Imbasciati, P. Cogliati, and A. Gauly. 2003. Impact of blood and dialysate flow and surface on performance of new polysulfone hemodialysis dialyzers. *International Journal of Artificial Organs* 26(2): 113–120.

Ozgen, N., M. Terashima, T. Aung, Y. Sato, C. Isoe, T. Kakuta, and A. Saito. 2004. Evaluation of long-term transport ability of a bioartificial renal tubule device using LLC-PK1 cells. *Nephrology Dialysis Transplantation* 19(9): 2198–2207.

Quinton, W., D. Dillard; H. B. Scribner. 1960. "Cannulation_of_blood_vessels_for_prolonged.19.pdf." *ASAIO Journal*.

Ren, X., L. Xu, J. Xu, P. Zhu, L. Zuo, and S. Wei. 2013. Immobilized heparin and its anti-coagulation effect on polysulfone membrane surface. *Journal of Biomaterials Science, Polymer Edition* 24(15): 1707–1720.

Roozbeh, J., M.H. Hashempur, and M. Heydari. 2013. Use of herbal remedies among patients undergoing hemodialysis. *Iranian Journal of Kidney Diseases* 7(6): 492–495.

Saito, A. 1998. Regeneration of peritoneal effluent by madin–darby canine kidney cells-lined hollow fibers. *Materials Science and Engineering* C6(4): 221–226.

Saito, A. 2003. Development of bioartificial kidneys. *Nephrology* 8 (Suppl.).

Saito, A. 2004. Research into the development of a wearable bioartificial kidney with a continuous hemofilter and a bioartificial tubule device using tubular epithelial cells. *Artificial Organs* 28(1): 58–63.

Sato, Y., M. Terashima, N. Kagiwada, T. Tun, M. Inagaki, T. Kakuta, and A. Saito. 2005. Evaluation of proliferation and functional differentiation of LLC-PK1 cells on porous polymer membranes for the development of a bioartificial renal tubule device.*Tissue Engineering* 11(9–10): 1506–1515.

Shaldon, S., P. Florence, P. Fontanier, C. Polito, and C. Mion. 1976. Comparison of two strategies for short dialysis using 1m² and 2m² surface area dialysers." *Proceedings of the European Dialysis and Transplant Association. European Dialysis and Transplant Association* 12: 596–605.

Sinnakirouchenan, R. and J.L. Holley. 2011. Peritoneal dialysis versus hemodialysis: Risks, benefits, and access issues. *Advances in Chronic Kidney Disease* 18(6): 428–432.

Tatsumi, E., H. Takano, Y. Taenaka, T. Nishimura, Y. Kakuta, M. Nakata, T. Tsukiya, and T. Nishinaka. 1999. Development of an ultracompact integrated heart-lung assist device. *Artificial Organs* 23(6): 518–523.

Terashima, M., Y. Fujita, K. Sugano, M. Asano, N. Kagiwada, Y. Sheng, S. Nakamura, A. Hasegawa, T. Kakuta, and A. Saito. 2001. Evaluation of water and electrolyte transport of tubular epithelial cells under osmotic and hydraulic pressure for development of bioartificial tubules. *Artificial Organs* 25(3): 209–212.

Unger, J.K., A.-J. Lemke, and C. Grosse-Siestrup. 2006. Thermography as potential real-time technique to assess changes in flow distribution in hemofiltration. *Kidney International* 69(3): 520–525.

von Hartitzsch, B., N.A. Hoenich, R.J. Peterson, T.J. Buselmeier, D.N. Kerr, and C.M. Kjellstrand. 1973. Middle molecule clearance in current dialysers. *Proceedings of the European Dialysis and Transplant Association. European Dialysis and Transplant Association* 10: 522–527.

Weinreich, T., T. De losRios, A. Gauly, and J. Passlick-Deetjen. 2006. Effects of an increase in time vs. frequency on cardiovascular parameters in chronic hemodialysis patients. *Clinical Nephrology* 66(6): 433–439.

Wikipedia contributors, "Hemodialysis," *Wikipedia, The Free Encyclopedia,* https://en.wikipedia.org/w/index.php?title=Hemodialysis&oldid=829498004 (accessed March 18, 2018).

World Kidney Day. 2017. "Chronic Kidney Disease–World Kidney Day." ISN–Global Operations Center. 2017. http://www.worldkidneyday.org/faqs/chronic-kidney-disease/.

Yu, D.G., C.H. Jou, W.C. Lin, and M.C. Yang. 2007. Surface modification of poly (tetramethylene adipate-co-terephthalate) membrane via layer-by-layer assembly of chitosan and dextran sulfate polyelectrolyte. *Colloids and Surfaces B: Biointerfaces.*

11

Design of Cost-Effective Membrane Devices for Production of Potable Alkaline Ionized Water

Pavani Vadthya, M. Praveen, C. Sumana, and Sundergopal Sridhar

CONTENTS

11.1 Introduction

Alkaline ionized water has pH in the range of 8.0–10.5. It has a negative oxidation reduction potential (ORP) of −200 to −600 mV, which means that the water is antioxidant rich due to the presence of essential minerals such as calcium, potassium, magnesium and sodium and bicarbonate ions such as ($-HCO_3^-$), which tastes lighter than purified RO and NF drinking water. The water clusters of alkaline water contain only 4–6 molecules, while that of normal ground water contains 12–16 molecules. All of these special properties of alkaline water make it a high-value product.

Normal drinking water is neutral with a pH nearly equal to 7. However, consumption of slightly alkaline water with pH ranging from 8 to 10 has many health benefits. It may be noted that both BIS and WHO drinking water standards exhibit restriction in acidic range (minimum pH 6.5) but flexibility in the alkaline range (up to 8.5). The pH of normal or neutral drinking water can be increased by the process of ionization and the water generated in this process is called electrically ionized water or alkaline ionized water (AIW). Ionized water has a wide range of applications and is proves useful for the average person to deal with dreadful health hazards such as diabetes, cancer, hyperacidity (Hidemitsu, 2012). For a healthy human being, AIW with pH 8.5 is advisable to maintain good health and prevent diseases, whereas AIW with pH 9.0–9.5 is advisable for diabetic patients, and a pH 10.0–10.5 for people suffering from cancer (Schwalfenberg, 2012). The present chapter gives an overview of the principle, design and operation of a cost-effective electrolyzer developed in-house and made affordable for the average person.

11.1.1 Background

For the past 200 years, research on water electrolysis has been progressing continuously (Zeng & Zhang, 2010) beginning with the invention of phenomenon of electrolytic splitting of water into oxygen and hydrogen to the more recent evolution of modern electrolyzers. By the evolution of electrochemistry, the constituent relation between the quantities of gases produced and the amount of electricity consumed were established through Faraday's laws of electrolysis (Remer & Manz, 1995); and by 1902, more than 400 water electrolyzers were installed and put into operation. The half-millennium between 1920 to 1970 was the golden age for development of water electrolysis technology and it was mainly used in production of hydrogen (Zeng & Zhang, 2011). Post-1970, there were several improved versions of AIW units with selected electrode material having high conductivity, good corrosion resistance and a low cost. Research efforts in this direction have led to the invention of solid polymer electrolysis, which can also be termed PEM water electrolysis and small-scale PEM electrolyzers, were used in military and space applications in the early 1970s (Sequeira & Santos, 2010).

11.1.2 Alkaline Water: Health Benefits

Our daily diet contains a high content of proteins in the form of eggs, meat, alcohol, soft drinks and sugar, which lead to acid load in the body and urine of low pH. The ingestion of a high dietary acid load can lead to chronic disorders. Therefore, there exists a need for neutralization of the acid load to balance the body's pH. The pH of human blood is alkaline, which is in the range of 7.35–7.45, and that of urine in the range of 4.5–8.0 (Remer & Manz, 1995). The pH of selected bodily tissues is given in Table 11.1. Consumption of alkaline water has more benefits than regular mineral drinking water as it acts as a powerful anti-oxidant by giving oxygen and electrons to cells and thereby prevents DNA damage (Shirahata et al., 1997). During food intake, the stomach secretes hydrochloric acid to aid digestion. However, a regular intake of acidic food results in excess acidity, which becomes taxing on the organs to maintain alkaline blood pH by releasing carbonates and bicarbonates of Ca, Mg and other minerals. Consumption of AIW reduces the free radicals produced in the human body by its antioxidant activity and brings about many health benefits in terms of elimination of constipation, prevention of body fat accumulation, reduction of ultraviolet radiation induced skin damage, etc. Alkaline water was proven to serve as an adjunct therapy for diabetes, chronic diseases and kidney disorders

TABLE 11.1

pH of Selected Body Tissues

Body Tissue	pH
Blood	7.35–7.45
Muscles	6.1
Liver	6.9
Saliva	6.35–6.85
Urine	4.5–8.0
Gastric juice	1.2–3.0
Pancreatic juice	7.8–8.0

and also helps in reducing the blood cell stickiness, blood clotting and aggregation (Rosa et al., 2012). Enagic's SD501 produces alkaline and acidic water of different pH, they suggest a high pH (< 11.5) water to drink, while lower pH water for cleaning, grooming regimens, etc. Further, many other uses of strong alkaline Kangen water were reported in the study made by David et al. (2009). Several studies are reported in literature demonstrating the usefulness of AIW. Regulation of alkalinity near the kidney tubules is found to result in active detoxification of the human body (Minich & Bland, 2007). A comparative study that was performed on diabetic rats reported that the AIW fed rats have shown lower blood sugar levels compared to tap water fed rats. Thus, consumption of AIW may decrease the risk of one getting diabetes mellitus (Jin & Kazuhito, 1997). In another study, the effect of AIW was observed on mice. The blood electrolyte studies on mice revealed less neutrophils. The histological studies conducted on stomach, small intestine, heart and liver exhibited no changes in the tissue. The observations proved that the electrolytically reduced AIW does not cause any side effects on blood nor bodily organs and is completely safe for drinking (Han et al., 2008). Rapid hydration through micro-clusters of water molecules (4–6 molecules per cluster) present in ionized water along with anti-oxidants are the added benefits of AIW. Elucidation of the water microstructure was carried out by nuclear magnetic resonance (NMR) (Hidemitsu, 2012). Long-term ingestion of AIW is found to decrease cecal fermentation in rats that were given highly fermentable commercial diet (Keijiroo, 2000).

Though consumption of AIW is proven to be advantageous to human health, excess intake of AIW may lead to metabolic alkalosis. Consuming AIW of pH $\geq$ 11 is found to cause irritation of eyes, skin and mucous membranes. However, these reports need proper validation. Moreover, excess intake of alkaline water may dehydrate the body dehydrate rather than hydrating it. Hence it may preferable to follow BIS and WHO standards of pH 8.5 during regular intake and below a pH 10 for infrequent users.

On the other hand, electrolyzers produce acidic water, i.e., electrolyzed oxidizing (EO) water as a by-product from anode chamber that has several applications. A study conducted on rats proved that the reactive oxygen present in anode chamber water quickens the wound healing process by aiding the migration and propagation of fibroblasts (Naoki et al., 2000). The water produced by an electrolyzer that uses water and NaCl produces HOCl that acts as a sanitizer. EO water is advantageous compared to the traditional cleaning agents as it is safe, eco-friendly and cost-effective in terms of facilitating effective disinfection, as well as its ease of operation. Though EO water is a strong acid, it is not corrosive to either skin, mucous membrane or any other organic material, unlike hydrochloric acid or sulfuric acid. EO water can be effectively used for domestic applications including floor cleaning, dish washing, cleaning vegetables and other food products, etc. EO water

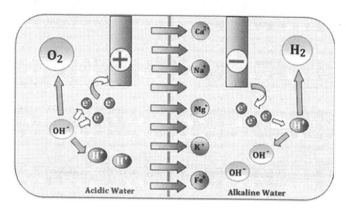

FIGURE 11.1

Mechanism of electrolysis and production of alkaline water in the electrolyzer.

has also been tested and used as a disinfectant in the food industry and other applications. Combinations of EO water with other cleaning and disinfecting measures are also possible. The article by Huang et al. (2008) gives an overview on EO water and its applications in effective cleaning of surfaces of raw foods and animal products in food processing plants.

11.1.3 Principle of Electrolytic Ionization of Water

Electrolysis of water is the process of decomposition of water into hydrogen and oxygen gases due to the passage of an electric current through water. Alkaline water electrolyzer consists of two electrodes immersed in a water chamber, which are separated by a barrier for extrication of the product gases and transportation of hydroxyl ions from one electrode to the other, as shown in Figure 11.1. The presence of minerals in feed water is a necessary condition for ionization, and therefore ground water, tap or nanofiltration water are the suitable feeds as they contain various minerals. In contrast, deionized water cannot be an ideal feed for AIW production units.

11.2 State-of-the-Art Water Electrolysis and Electrolyzers

Electrolysis is an electrochemical process in which electrical energy drives chemical reactions to decompose the substances by passing a current through them. Firstly, the phenomenon of electrolysis was observed in 1789, and Nicholson-Carlisle developed the technique in 1800. More than 400 types of industrial electrolyzers were established by the twentieth century. Later on, many advances took place in the development of cathode, anode and membrane technologies to aid effective electrolysis. Comparative studies were also carried out on AIW production by electrochemical and physicochemical methods (Derek & Xiaohong, 2011). Studies conducted in polypropylene tube electrolyzer revealed that the alkaline pH of AIW was intact until three months in an anaerobic environment, although ORP values varied widely both in aerobic and anaerobic conditions. Material of construction of electrolyzer was also found to affect the electrolysis process (Seung et al., 2004).

11.3 Design of Device for Alkaline Ionized Water Production

In this section, complete details of the design and operation of the indigenously developed AIW units, both in batch and continuous modes, is presented. The working principle of AIW unit is discussed, followed by the development of various membranes for different applications. Furthermore, the details of three models including table top, batch and continuous AIW units are also given.

11.3.1 Operating Mechanism of AIW Device

The AIW device works on the principle of electrolysis process as shown in Figure 11.1. Before starting the electrolyzer device, the anodic tank is filled with tap water while the cathode tank is filled with purified NF or RO water, since that of NF will have sufficient minerals. By supplying power to the electrolyzer device, electric potential difference develops across anode and cathode electrodes due to the occurrence of oxidation and reduction reactions in the respective electrode chambers. The water gets decomposed to positively charged hydrogen and negatively charged hydroxyl ions and the electric current forces hydrogen ions to migrate towards anode, where reduction reaction takes place to form hydrogen atoms. The atoms formed then combine to form gaseous hydrogen molecules (H_2). Similarly, oxygen is formed at the cathode due to the reduction reaction. As a result, the water present in the anode cell becomes acidic with increased concentrations of O_2, H^+, H_2O and anionic minerals resulting in formation of electrolyzed oxidized water. Similarly, the water present in the cathode chamber becomes alkaline with increasing concentration of H^+, OH^-, H_2, H_2O and cationic minerals resulting in alkaline ionized water (AIW). The reactions occurring at the anode and cathode cells are given below:

Reaction in anodic cell:

$$4OH^- -4e^- \rightarrow 2H_2O + O_2\uparrow$$
$$2H_2O -4e^- \rightarrow 2H^+ + 2OH^- -4e^-$$
$$2H^+ + 2OH^- -4e^- \rightarrow 2H^+ + 2H^+ + O_2\uparrow$$
$$2H^+ + 2H^+ + O_2\uparrow \rightarrow 4H^+ + O_2\uparrow$$

Reaction in cathode cell

$$2H^+ + 2e^- \rightarrow H_2\uparrow$$
$$2H^+ + 2e^- \rightarrow 2H \text{ (Active Hydrogen)}$$
$$2H_2O + 2e^- \rightarrow 2OH^- + 2H^+ + 2e^-$$
$$2OH^- + 2H^+ + 2e^- \rightarrow 2OH^- + H_2\uparrow$$

11.3.2 Membrane Synthesis

The most necessary component of the AIW unit is the membrane that separates the anode and cathode chambers and allows the passage of either ions or electrons, but not the oxygen or hydrogen atoms. The membrane also prevents the mixing of gases produced in the electrode chambers that may form an explosive mixture upon contact (Emmanuel et al., n.d.). The membrane plays a major role in designing of an electrolyzer. Polymer based cation exchange (CE) membrane and conventional ultrafiltration (UF)membranes, in both flat sheet and hollow fiber (HF) geometry, were synthesized in the lab for incorporation as barriers in the AIW units. RO and NF membranes were used for the production of feed water to the AIW unit. CE and flat sheet UF membranes were synthesized in-house for utilization in batch AIW units, whereas hollow fiber UF membranes of larger surface area per unit volume were prepared for application in continuous AIW units.

11.3.2.1 Synthesis of Nonporous Cation Exchange Membrane

Nonporous sulfonated polyethersulfone polymer (SPES) was prepared by solution casting and solvent evaporation method. 15% PES was dissolved in concentrated sulphuric acid, as shown in Figure 11.2 (a). After complete sulfonation, the polymer solution was poured into an ice-cold water bath to obtain a solid precipitate as illustrated by Figures 11.2 (b) and 11.2 (c). The precipitate was collected and washed with water until the pH of the solution became neutral. The SPES was then dried completely and dissolved in N-methyl-2-pyrrolidone (NMP) solvent to obtain a 15% w/v SPES solution, as shown in Figure 11.2 (d). To enhance the strength of SPES membrane and to control the degree of swelling, it was blended with polyvinyl chloride (PVC). PVC was prepared in similar method from a 12% w/v solution prepared in NMP. Both polymer solutions were blended in 1:1 volumetric ratio and the resultant homogeneous mixture was cast in a Petri dish and dried at 40–50°C to completely remove the solvent and acquire SPES-PVC ion exchange blend membrane.

11.3.2.2 Synthesis of Flat Sheet Ultrafiltration Membrane

15% w/v of polyethersulfone (PES) was mixed with 3% propionic acid in dimethyl formamide (DMF) solvent, and the resultant polymer solution was de-aerated to remove air bubbles before casting on a nonwoven polyester fabric support using a hand casting machine shown in Figure 11.3 (a) to achieve the desired thickness, followed by immersion in ice cold water bath. The PES substrate prepared by the phase inversion method was ultraporous in nature with a MWCO of 50 kDa.

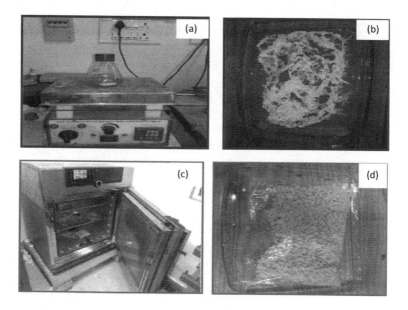

FIGURE 11.2
(a) Preparation of polymer solution, (b) Precipitation in ice-cold water bath, (c) Removal of residual solvent by heating in an oven, (d) Dried polymer.

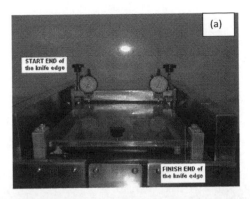

FIGURE 11.3
(a) Flat sheet membrane casting unit, (b) Hollow fiber spinning machine.

11.3.2.3 Synthesis of Hollow Fiber Ultrafiltration Membrane

Hollow fiber membranes have several advantages: they are cost effective, they have self-supporting structures that does not require any nonwoven backing, they can undergo easy cleaning protocols (backwashing) and possess high surface area per unit volume for a more rapid mass transfer. HF membrane provides permeate with minimal pressure drop. HF is prepared by three different methods such as wet spinning, melt spinning, and dry spinning. Most of the HF membranes are generally prepared by using polymers such as polyamide, polyimide, polyethersulfone (PES), polyurethane (PU), polyethylene (PE), polytetrafluoroethylene (PTFE) and polyvinylidene fluoride (PVDF), etc. The HF membrane used for continuous mode AIW unit was prepared from 15% PES with dimethyl formamide (DMF) solvent by using a spinning machine shown in Figure 11.3 (b). The viscosity of polymer is usually increased by raising the percentage of polymer in the dope. Before subjecting to spinning of the fiber, the required polymer solution should have sufficient internal strength to hold together during extrusion and coagulation.

11.3.3 Design and Operation of Electrolyzers

Three electrolyzer models designed for the production of alkaline ionized water are discussed in detail as follows:

11.3.3.1 Table Top Electrolyzer

Figure 11.4 shows the actual photograph of indigenous low-cost table top electrolyzer device of 200–500 mL capacity for the production of AIW from ground water/filtered water sources. The device comprises of a stainless-steel container (cathode), torsion test tube with perforations, stainless steel (SS316) rod used as anode, the flat sheet polyethersulfone based UF membrane and a small DC adapter (36 V) for power supply. The stainless-steel rod of surface area 12 cm^2 was initially affixed to the center of the torsion test tube with perforations, which was wrapped with an ultrafiltration membrane and connected to a positive power supply, whereas the stainless-steel container (surface area 302 cm^2) was connected to the negative terminal of DC adapter. Steel container and torsion tube were both filled with reverse osmosis (RO)/nanofiltration (NF) grade water generated by a laboratory water purification unit. When DC current was applied across the two electrodes, electrochemical reactions occur. The positively charged ions present in the torsion tube tend to migrate toward the negative cathode side (i.e., steel container side) through a porous membrane.

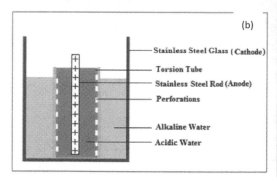

FIGURE 11.4
(a) Actual photograph of indigenous low-cost table top electrolyzer, and DC power supply unit (b) Schematic of table top AIW unit and (c) View from top of cathode container and anode rod separated by an ultrafiltration membrane supported on a perforated plastic bottle.

11.3.3.2 Batch Electrolyzer

A higher capacity device for batch-wise production of alkaline water from ground water or filtered water sources was designed and developed in-house as shown in Figure 11.5 (a). The device consists of two HDPE tanks that were interconnected with slots for incorporation of a flat sheet membrane and electrodes. The connection between the two tanks was tightened with a stainless-steel clamp to avoid leaks as displayed in Figure 11.5 (b). The electrolyzer was of flexible design, flexible for manual replacement of any flat sheet membrane. A DC adapter of 36 V was used to provide DC current across the electrodes that are immersed into the coupled HDPE tanks. The tanks were filled with tap water and filtered NF water produced from laboratory unit as shown in Figure 11.5 (c). SS washers

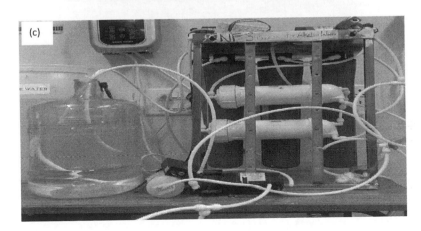

FIGURE 11.5
(a) Photograph of 20 L capacity batch electrolyzer, (b) Membrane assembly between the electrode chambers, (c) Cascaded nanofiltration unit.

were connected to a 36 V DC adapter to behave as anode and cathode. Anode was dipped in tap water, while cathode was inserted in NF water of 80 ppm TDS as shown in Figure 11.5 (c). The applied potential resulted in electrolysis reactions at electrodes causing water to split into H⁺ and OH⁻ while salts present in the tap water got ionized and split into cations and anions. The potential difference across the electrodes drove the cations to move toward the cathode, and anions to move toward the anode. Washers made of stainless steel containing extra molybdenum (corrosion resistant SS316L) with an effective surface area of 15.3 cm² were used as electrodes. PVC taps were provided at the bottom of the tanks for collection of AIW water.

11.3.3.3 Continuous Electrolyzer

The continuous electrolyzer device for production of alkaline water from ground water and filtered water sources is designed and developed in-house as shown in Figure 11.6 (a).

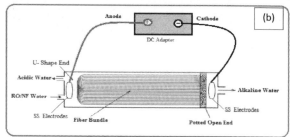

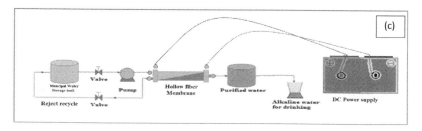

FIGURE 11.6
(a) Picture of electrolyzer for continuous production of alkaline water, (b) Blow up of hollow fiber module and (c) Process flow diagram of continuous electrolyzer.

TABLE 11.2

Features of Three Differently Designed AIW Units

Electrolyzer	Capacity	Membrane Type & Area	Electrode Material		Voltage/Power Consumption	Cost	
			Cathode	Anode		Capital	Operating
Table top device	200 mL	UF & 40 cm²	Stainless steel container	Stainless - 316 L rod	36 V/0.0036 kWh	410/-	0.963/L
Batch AIW unit	20 L	Cation exchange & 50.24 cm²	Stainless steel washers	Stainless steel washers	36 V/0.6048 kWh	1025/-	1.463/L
Continuous AIW electrode	3.3 L/min	PES-HF & 445 cm²	Stainless steel 316 L	Stainless steel 316 L	36 V/0.4284 kWh	2400/-	1.272/L

The design of high flux polyethersulfone based hollow fiber module with a MWCO of 30 kDa is illustrated in Figure 11.6 (b) and a flow sheet of the process is given in Figure 11.6 (c). Two slots each were made on either side of the membrane module for the incorporation of electrodes made of SS 316. The electrode sheets are rectangular in shape with an effective surface area 4.84 cm² each. Initially, NF/RO filtered water was continuously allowed to pass through the shell side of the hollow fiber module, whereas permeate was collected from lumen (tube side) of the fiber as indicated in Figure 11.6 (b). The potted open end separates the anode chamber containing hollow fibers from the cathode chamber. The membrane purifies the feed water from the anode chamber and the permeate water enters the cathode chamber. Application of direct current across the electrodes results in electrochemical reactions and enrichment of OH⁻ ion concentration in the cathode chamber. The generation of H_2 gas at the cathode side ensures that the bulk of OH⁻ ions remain at the negative electrode side for drinking purpose, along with essential ions that have been transferred. The electrolyzer enables continuous production of alkaline water having pH in the range 7.5–10 at a continuous flow rate of 3–4 L/h. The features of the above-mentioned three ionizers are given in detail in Table 11.2.

11.4 Performance of Electrolyzers

11.4.1 Table Top Electrolyzer

The performance of the table top electrolyzer was tested with PES flat sheet membranes and SS electrodes. Filtered water of volume 200 mL was taken as feed. The pH, total dissolved solids (TDS) and oxygen reduction potential (ORP) values of the solutions were recorded at time intervals of 10 min. each. The experiment was continued until a constant value of pH was attained, as revealed in Figure 11.7 (a), with the respective ORP values plotted in Figure 11.7 (b). It was observed that within 20 min. of operation, the pH of the cathode chamber reached nearly 10.06 from an initial value of 7.83. Similarly, the corresponding ORP values varied from +110 mV to −600 mV. The prepared alkaline water was kept for storage (under anaerobic conditions) for 17 days and analyzed for the effect of pH and ORP with time. From the results shown, it is noted that the pH of the AIW decreased from 10 to nearly 9 and corresponding ORP values changed from −666 mV to −200 mV

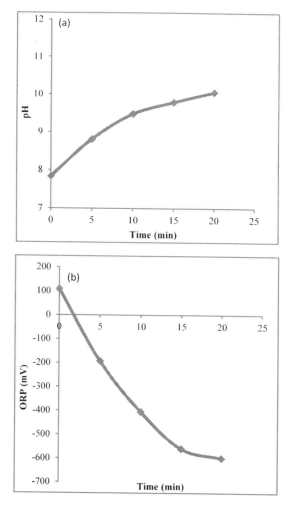

FIGURE 11.7
Variation of (a) pH and (b) ORP with operating time for production of AIW.

over a period of 1 week. The device shows capability to produce alkaline water having better ORP when compared to ionizers available on the market, at a much lower investment, using a simple apparatus. Further, the device allows the elimination of expensive corrosion-resistant electrodes since the membrane does not leak the anode chamber water into cathode chamber. Compared to the 20 L batch mode electrolyzer and continuous mode electrolyzer, the specific surface area of the cathode available was higher in this electrolyzer. Moreover, the volume of water used was less and enabled faster electrochemical reactions near the electrodes. This model of electrolyzer is highly portable.

11.4.2 Batch Electrolyzer

The behavior of indigenously synthesized SPES-PVC blend ion exchange membrane and high flux UF membrane were examined using stainless steel washers as electrodes, at 36 VDC potential. 20 L of NF filtered water and tap water were filled in the cathode

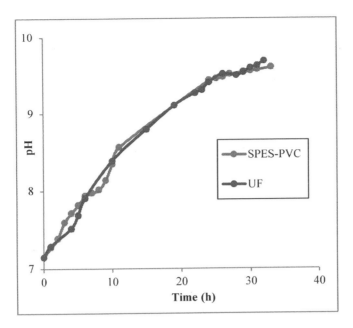

FIGURE 11.8
Comparison of membrane performance: Variation of pH with operating time.

chamber and anode chamber, respectively. The experiment was conducted for 33 h in a 20 L capacity batch electrolyzer. The observations reveal that the water in the cathode chamber increased from 7.14 pH to 9.6 pH AIW in 33 h and 30 h, using SPES-PVC and UF membranes as the respective barriers (Figure 11.8). Similarly, the pH on the anode side marginally reduced to 7.51 and 7.23, respectively. The highest flux obtained was 0.60 L/h with an ORP of –153 mV with SPES-PVC membrane, whereas UF membrane exhibited the highest flux of 0.62 L/h. Corrosion generally occurs on the anode, which is the sacrificial electrode, whereas the cathode has shown no signs of degradation. Hence, the possibility of contamination by harmful ions at the alkaline water side (cathode) of the electrolyzer is minimal.

11.4.3 Continuous Hollow Fiber Based Electrolyzer

The performance of hollow fiber based continuous mode electrolyzer is evaluated by observing the variation of pH at different applied voltages. The experiments were conducted on an electrolyzer built with SS316 electrodes in continuous recirculation mode of operation. Figure 11.9 (a) reveals the effect of voltage on operating time for attaining alkaline pH in the permeate water. To get 10 pH of AIW as permeate, the operating time required was 11 h, 9 h and 7.5 h for an applied electric potential of 50 V, 60 V and 70 V. As voltage was increased the operating time decreased as shown in Figure 11.9 (a). Higher flux can be obtained by regulating the applied voltage.

Hollow fiber membranes developed from PSf were also tested in the continuous mode electrolyzer. The performance of the PSf membrane based electrolyzer was compared with that of the PES membrane, as shown in Figure 11.9 (b). The rise in pH of permeate water was higher using the PES membrane compared to the PSf membrane. PES membranes offered high flux and less resistance between electrodes compared to PSf. Thus, the PES membrane was used for further studies.

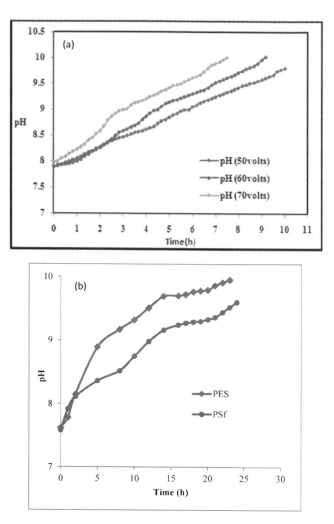

FIGURE 11.9
Effect of (a) voltage and (b) polymer type on pH of alkaline water for hollow fiber based continuous electrolyzer.

11.5 Economic Analysis

Very few ionizers are available in India and around the world, in general, with most of them being imported from selected suppliers. Though there are several advantages associated with the developed devices for the production of AIW, these devices are very cost-intensive and have become a luxury product that no average person can afford. The current study focuses more on a cost-effective construction of AIW production units.

The capital cost of the electrolyzers is estimated by considering the cost of material of construction, the membrane and the electrodes, given in Table 11.3. The cost of table top, batch and continuous electrolyzers is estimated to be Rs. 410/- (US$ 7.75), Rs. 1025/- (US$ 15.90) and Rs. 2400/- (US$ 37.22) respectively. Table 11.4 presents the approximate cost of some popular electrolyzer devices available globally. The price of these commercially available ionizers ranges from Rs. 40,000/- to Rs. 4,00,000/- (US$ 615.00–6,154).

TABLE 11.3

Capital and Operating Cost of the Alkaline Water Electrolyzer Devices

S. No	Items	Unit Price (INR)	Quantity	Amount (INR)
(a) Batch Electrolyzer				
Capital Cost				
1	20L Plastic water can	150	2	300
2	PVC tap	50	2	100
3	Stainless steel washers	10	2	20
4	DC adapter 36V/48V	500	1	500
5	Membrane	50	1	50
6	PVC coupling	45	1	345
7	Rubber washers	5	2	10
			Total	1025/-
Operating Cost for Production of 1 L of Alkaline Water				
1	Electricity cost	5 per KWH	1	0.36
2	RO/NF water cost	0.05	1	0.05
3	Membrane replacement cost	0.013	1	0.013
4	Electrode replacement cost	0.50	2	1.00
			Total	1.463/-
(b) Continuous Electrolyzer				
Capital Cost				
1	Hollow fiber membrane	900	1	900
2	Deng yuan pump	1	1	1000
3	Adapter	1	1	300
4	Stainless steel washers	10	2	20
5	Connectors	5	6	30
6	Piping	50	1	50
7	Stand	1	1	100
			Total	2400/-
Operating Cost for Production of 1 L of Alkaline Water				
1	Electricity cost	5 per KWH	1	0.012
2	RO/NF water cost	0.05	1	0.05
3	Membrane replacement cost	0.21	1	0.21
4	Electrode replacement cost	0.50	2	1.00
			Total	1.272/-
(c) Table Top Electrolyzer				
Capital Cost				
1	Stainless steel glass	20	1	50
2	Stainless steel rod	10	1	10
3	Adapter	300	1	300
4	Membrane	50	1	50
			Total	410/-

(Continued)

TABLE 11.3 (CONTINUED)

Capital and Operating Cost of the Alkaline Water Electrolyzer Devices

S. No	Items	Unit Price (INR)	Quantity	Amount (INR)
Operating Cost for Production of 1 L of Alkaline Water				
1	Electricity cost	5 per KWH	1	0.018
2	RO/NF water cost	0.05	1	0.05
3	Membrane replacement cost	0.065	1	0.065
4	Stainless steel rod	0.20	1	0.50
5	Stainless steel glass	0.33	1	0.33
			Total	0.963/-

TABLE 11.4

Market Cost Statistics

Cost of Electrolyzers	
Ionizer Brand	**Approximate Cost in Rs.**
Hydrojal	40,000
KYK HISHA Ionizer	80,123
Panasonic TK-7585E	86,000
Chanson Ionizer	2,00,000
Leveluk Kangen 8	4,00,000

Cost of AIW per Liter	
Bottled Water Brand	**Cost in Rs. per Litre of Bottled Alkaline Water**
Bottled Alkame® Water	2000
QURE Alkaline Water	3000
Neo Super Water – Alkaline	4529
Essential Drinking Water	1800
AQUA hydrate Electrolyte Enhanced Water	2010

The operating cost of AIW is estimated by considering the cost of energy, membrane/electrode replacement and the NF feed water. The operating cost per L of AIW production for table top, batch and continuous electrolyzers is estimated to be Rs. 0.963/-, Rs. 1.463/- and Rs. 1.272/- (< US$ 2.00) respectively, while the commercially available bottled alkaline water costs as high as Rs. 1,800/- to 4,500/- per L (US$ 23.00–46.00), as shown in Table 11.4.

11.6 Conclusions

In the present study, a cost-effective electrolyzer incorporating a membrane has been designed to produce alkaline water wherein acidic water is also produced as a by-product. The continuous enrichment of hydroxyl ions (OH$^-$)in the cathode chamber water provides negative oxidation-reduction potential and anti-oxidants to fight diseases and improve immunity through supply of extra oxygen to the cells, as well as immediate neutralization of toxins including reactive oxygen species. Three different designs, namely a batch electrolyzer, table top device and continuous electrolyzer, were designed for the production of alkaline water, which can be used for drinking, whereas the acidic water generated at the anode side can be utilized for washing purposes. Alkaline ionized water of pH up to 8.5 is permissible for drinking as per WHO and BIS standards. However, further studies are necessary to verify the safety of consuming alkaline water with a pH in the range of 9–10.5. The electrolyzers exhibited promising results with both flat sheet and hollow fiber membranes. The capital cost of all the devices, especially the table top AIW, is easily affordable to the average person, and the operating cost is also quite low. Nevertheless, a detailed analysis of the water that contains any undesirable ions, arising from electrode reactions, needs to be carried out before the electrolyzer can proliferate in the market.

References

Bowen, C.T. and H.J. Davis. 1984. Developments in advanced alkaline water electrolysis. *International Journal of Hydrogen Energy* 9: 59–66.

Carpenter, D., P. Parker, and L. Tauscher. 2009. Water therapies: Change your water, change your life. United States.

Derek, P. and L. Xiaohong. 2011. Prospects for alkaline zero gap water electrolysers for hydrogen production. *International Journal of Hydrogen Energy* 36: 15089–15104.

Emmanuel Z., E. Varkaraki, N. Lymberopoulos, C.N. Christodoulou and G.N. Karagiorgis. n.d. A review on water electrolysis. *Electrochemistry* 1–18.

Han, S.J., H.K. Dong, S.Y. Yang, C.T. Yung, S.C. Byung, and K.J. Lee. 2008. The effects of electrolyzed reduced water on blood and organ tissues of mice. Department of Environmental Medical Biology. *Journal of Microscopy Korean* 38: 321–328.

Hidemitsu, H. 2012. Understanding alkaline antioxidant water-ionised water. Japan. https://www.healthline.com.

Huang, Y.R., Y.C. Hung, S.Y. Hsu, Y.W. Huang, and D.F. Hwang. 2008. Review: Application of electrolyzed water in the food industry. *Food Control* 19: 329–345.

Jin, M.K. and Y. Kazuhito. 1997. Effects of alkaline ionised water on spontaneously diabetic GK-rats fed sucrose. *Korea Journal of Laboratory Animal Research* 13: 187–190.

Keijiroo, MD, K. 2000. Diabetes and the effects of ionized water. U.S. Department of Health and Human Services.

Minich, D.M. and J.S. Bland. 2007. Review article: Acid-alkaline balance: Role in chronic disease and detoxification. *Alternative Therapies in Health and Medicine* 13: 62–65.

Naoki, Y., K. Masashi, K. Masaki, O. Akito, S. Osao, H. Toshimasa, H. Katsuaki, N.J. Chen, W. Paul, K. Shoji, M. Arata, and T. Shinichi. 2000. Effect of electrolyzed water on wound healing. *Artificial Organs* 24: 984–987.

Remer, T. and F. Manz. 1995. Potential renal acid load of foods and its influence on urine pH. *Journal of the American Dietetic Association* 95: 791–797.

Rosa, M.C.I., B.J. Kyung, and J.I. Kyu, 2012. clinical effect and mechanism of alkaline reduced water. *Journal of Food and Drug Analysis* 20: 394–397.

Rosborg, I. 2015. "Water without borders: The importance of minerals in drinking water—Reverse Osmosis treated drinking water and health." Paper presented in Stockholm, Sweden.

Schwalfenberg, G. 2012. The alkaline diet: Is there evidence that an alkaline pH diet benefits health? *Journal of Environmental and Public Health* 2012: 727630.

Sequeira, C.A.C. and D.M.F. Santos. 2010. *Polymer Electrolytes: Fundamentals and Applications.* Cambridge: Woodhead Publishing Ltd.

Seung, K.P., W.K. Jae, Y.K. Gwang, S.R. Young, H.K. Geun, C.C. Hyun, K.K. Soo, and W.K. Hyun. 2004. Anticancer effect of alkaline reduced water. *Journal of International Society of Life Information Science* 22: 302–305.

Shirahata, S., S. Kabayama, M. Nakano, T. Miura, K. Kusumoto, M. Gotoh, H. Hayashi, K. Otsubo, S. Morisawa, and Y. Katakura. 1997. Electrolyzed-reduced water scavenges active oxygen species and protects DNA from oxidative damage. *Journal of Cytotechnology* 234: 269–274.

Thomas, D.A. 2003. The drinking water facts—Important issues for buying the best drinking water system. *Bio Natural* 1–10.

Venkitanarayanan, K.S., G.O. Ezeike, Y.C. Hung, and M.P. Doyle. 1999. Inactivation of Escherichia coli and Listeria monocytogenes on plastic kitchen cutting boards by electrolyzed oxidizing water: Use of acid water to clean plastic cutting boards by electrolyzed oxidizing. *Water Journal of Food Protection* 62: 857–860.

Zeng, K. and D. Zhang. 2011. Corrigendum to "Recent progress in alkaline water electrolysis for hydrogen production and applications" [2010. *Progress in Energy and Combustion Science* 36(3): 307–326]. *Progress in Energy and Combustion Science* 37: 631

Zeng, K. and D. Zhang. 2010. Recent progress in alkaline water electrolysis for hydrogen production and applications. *Progress in Energy and Combustion Science* 36: 307–326.

Section IV

Membrane Process Design

12

Mass Transfer Modeling in Hollow Fiber Liquid Membrane Separation Processes

Biswajit Swain, K.K. Singh, and Anil Kumar Pabby

CONTENTS

12.1 Introduction

Separation techniques such as solvent extraction, ion exchange and membrane separation play key roles in many industrial processes. Liquid membrane (LM), which is a combination of membrane separation and solvent extraction, has drawn the attention of many because of its several advantages over conventional solvent extraction processes, such as simultaneous extraction and stripping in a single module, high selectivity and low cost of operation and energy. LM essentially comprises of a solvent layer (carrier) separating a solute containing aqueous phase (feed) and a solute receiving aqueous phase (strip). LMs containing a target-selective carrier have been proposed as alternatives to solvent extraction for selective separation and concentration of solutes from dilute solutions. LMs have found applications in the fields of chemical and pharmaceutical technology, biotechnology, food technology, nuclear and environmental engineering (Boyadzhiev et al., 1995; Klassen et al., 2005; Mohapatra & Manchanda, 2003; Pabby et al., 2015; San Roman et al., 2010). Use of LMs in other fields, such as gas separation, recovery of aroma compounds and applications of ionic liquids as membrane carrier, etc., is increasing rapidly.

An organic phase immobilized in the pores of a micro-porous poly-film (which acts as an inert support) separating the feed and strip phase is called a supported liquid membrane (SLM). Based on the geometry of the support, SLMs are either flat sheet supported liquid membranes (FSSLM) or hollow fiber supported liquid membranes (HFSLM).

A hollow fiber contactor deploying LM represents a promising alternative to conventional contactors because of its several advantages, such as selectivity, very high interfacial area per unit volume, no constraint on density difference between the phases, low cost of operation and energy requirement, ease of scale-up and no problem of stable emulsions and third phase formation (Gabelman & Hwang, 1999). Hydrophobic hollow fiber contactors deploying LMs are used under different configurations such as non-dispersive solvent extraction (NDSX), hollow fiber supported liquid membrane (HFSLM), pseudo emulsion based hollow fiber strip dispersion (PEHFSD), and hollow fiber renewal liquid membranes (HFRLM). These configurations are basically different in the way the liquid phases are contacted (Pabby et al., 2015; San Roman et al., 2010). Figure 12.1 shows representative image of a hollow fiber contactor.

NDSX is simply liquid-liquid extraction, which involves the use of hollow fiber module to contact an aqueous and an organic phase without creating a liquid-liquid dispersion, hence "non-dispersive." In NDSX, the solvent flows either in the lumens or on the shell side and also gets impregnated within the hydrophobic pores of the membrane, which separates the solvent phase and the aqueous phase. Pressure difference is maintained between the aqueous and organic phases to maintain the interface at the mouth of the pores. Scrubbing of loaded organic is possible by NDSX, if required, and this has a wide number of applications in hydrometallurgy, water treatment, recovery of aroma compounds and pharmaceutical applications (Ho & Poddar, 2001; Juang & Huang, 2002; Oritz et al., 2004; Yun et al., 1993). The NDSX mode of operation requires at least two modules to perform simultaneous extraction and stripping operations.

In the HFSLM technique, hollow fiber pores are impregnated with an organic solvent and the feed phase flows on one side of the fiber, and the strip on the other side. The same pressure is maintained on the feed and strip side to avoid organic entrainment from the pores. The technique has the advantage of simultaneous extraction and stripping in a single module with very low solvent inventory but suffers from disadvantages, such as gradual loss of organic phase, progressive wetting of support pores as a consequence of aqueous phase filling, displacement of fluid-fluid interface due to the high differential pressure created, and back extraction due to transfer of H+ ions. This technique is being effectively used for separation of various metal ions and wastewater treatment

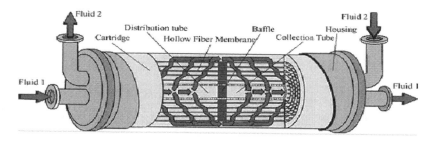

FIGURE 12.1
Schematic of a hollow fiber contactor. (From Liqui-Cel extra flow celgard hollow fiber module. www.liqui-cel .com, accessed on 25th September 2017.)

(Ansari et al., 2009; Kocherginsky et al., 2007; Pabby & Sastre, 2000; Parthasarathy et al., 1997; Vijaylaxmi et al., 2015).

In the PEHFSD technique, a pseudo-emulsion (strip dispersed in organic) flows inside the hollow fiber contactor separated from the aqueous feed solution by the hydrophobic micro-porous fiber for removal and recovery of solutes. This technique combines the advantage of emulsion liquid membranes and the NDSX process in which extraction and stripping occur simultaneously with large specific interfacial area for stripping without membrane stability issues. Pressure difference is maintained between aqueous and emulsion phases. This technique is being used for hydrometallurgical processes (Alguacil et al., 2010; Dixit et al., 2012; Pabby et al., 2015; Pei et al., 2011; Roy et al., 2008; Sonawane et al., 2010).

In the HFRLM technique, hydrophobic hollow fiber pores are filled with the organic phase. The organic in aqueous emulsion prepared with the feed phase (or the strip phase) with high A/O volume ratio is passed through the lumen side of the module. The strip phase (or the feed phase) flows through the shell side. Because of the affinity between the organic phase and hydrophobic wall, a thin organic layer is developed within the lumen side of the fiber. The shear force resulting from the flowing liquid peels off microdroplets from the organic layer, at the same time fresh organic droplets replenish the layer of organic phase. Thus, the liquid membrane is continuously renewed which helps in mass transfer. The pressure difference must be lower than that necessary to displace the organic phase from the pores by the aqueous phase flowing on the shell side. This technique is successfully used in metal extraction and wastewater treatment (Pabby et al., 2015; Ren et al., 2009; Ren et al., 2010; Zhang et al., 2010).

The use of a selective liquid membrane system based on facilitated transport has a great advantage because of its capacity with simultaneous extraction and stripping in one single stage. This ensures maximum driving force and non-equilibrium steady state mass transfer characteristics. This also helps to march toward process intensification by using minimum energy and carrying out multiple operations with low cost. For a successful industrial application of new technology, the availability of a reliable mathematical model and the identification of key parameters are very important. The comprehensive understanding of the process makes the task of optimization of design and operating conditions easier and also helps in scale-up.

Many researchers have reported the use of hollow fiber liquid membrane contactors for various solvent extraction applications with different approaches of mathematical modeling of mass transfer. Danesi (1984), Dixit et al. (2012), Khandalwal et al. (2011), developed the mass transfer model by ignoring the resistance of strip phase, whereas Urtiga et al. (1992), included the strip side resistance with the assumption of instantaneous reaction at the interfaces. As it is difficult to measure the concentrations at interfaces, the overall mass transfer coefficient for mass transfer was represented with the help of bulk concentration of the fluids. Ren et al. (2009; 2010), Urtiga et al. (2005), and Zhang et al. (2010) formulated the mathematical models for HFRLM and SLM techniques by including reaction rate with the resistances of the phases to obtain the overall mass transfer coefficient. In simplified flow models (Bringaset al., 2009; Dixit et al., 2012; Khandalwal et al., 2011; Dansi, 1984–85), one dimensional model describes the mass transfer across the membrane with the assumptions of plug flow, and thus the effect of velocity profile on mass transfer is neglected. In two-dimensional rigorous mass transfers modeling, the radial and axial variations in flow field make the model complex. Very few studies are reported with this modeling approach because of the complexities involved. A numerical analysis of the mass transfer and hydrodynamics that considers multiple phases including membrane phase is required for the modeling of various liquid membrane processes and also for the implementation of different module configurations. Numerical simulation using computational fluid dynamics (CFD) of the hollow fiber membrane contactors may provide interesting insights for the development of membrane technology. However, literature on CFD studies of separation processes in hollow fiber contactor is scare.

Hence, in this study efforts are made to understand, in detail, the different concepts of mass transfer modeling in liquid membrane separation processes using hollow fiber contactors.

12.2 Theory of Solute Transport in Liquid Membranes

Solute transport mechanism in liquid membrane processes is generally described as carrier facilitated transport in which the solute is getting transported after complexion with the organic solvent. This mass transfer process involves the following steps: (i) diffusion of solute through the aqueous phase and its boundary layer, (ii) complexion with organic phase with reversible reactions at feed-membrane interface, (iii) diffusion of solute-solvent complex in organic phase, and (iv) dissociation of complex at the membrane strip-interface. To estimate the quantitative result of the facilitated transport, an understanding of the individual steps mentioned above is essential. Figure 12.2 shows the schematic representation of a solute transport mechanism in liquid membrane with resistances.

In facilitated transport, the solutes get transported through the membrane against their own concentration gradient, which is described as uphill transport. In uphill transport, the driving force is the chemical potential across the membrane. The transport rate is controlled by membrane thickness, pore structures, aqueous diffusion layer thickness, diffusion and distribution coefficients. The transport of the solutes in the organic phase depends upon membrane characteristics such as porosity, tortuosity and viscosity of the

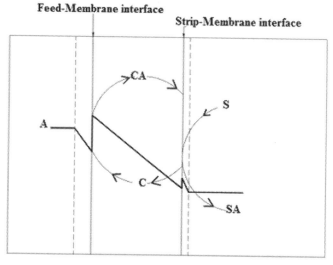

FIGURE 12.2
Schematic of solute transport mechanism in liquid membrane with resistances.

liquid membrane, while in the aqueous phase this depends on flow rates and diffusivity. The bulk diffusion coefficient (D_o) of the solute in the solvent can be estimated from the empirical equation from Walki and Chang (1955):

$$D_o = 7.4 \times 10^{-8} \left(\frac{X^{0.5} M^{0.5} T}{\eta V_m^{0.6}} \right)$$ (12.1)

where, X, M and η are the solvent association parameter, molecular weight and viscosity of the solvent, respectively, V_m is the molecular volume of the carrier and T is temperature. This equation suggests that the diffusion coefficient of the forward transported solute, i.e., ion-organic complex, will be much lower compared to backward transported bare solute (ions) due to a smaller molar volume of the later. This implies a higher concentration of solute at the feed-membrane interface because of faster diffusion of solute ions compared to bulky solute ion-organic complex. Overall mass transport can be also be affected by feed, strip and carrier compositions.

12.3 Mass Transfer Modeling in Hollow Fiber Contactors

Mathematical modeling of mass transfer in hollow fiber liquid membrane contactors has been carried out by various researchers with different assumptions (Bringas et al., 2009). Modeling of mass transfer not only helps in understanding the phenomenon better, but also helps in the scale-up and optimization of design and operational parameters. Modeling incorporates the mathematical formulation of various steps involved in mass transport to predict the overall mass transfer flux followed by solving the formulated equations with the implementation of boundary conditions to obtain spatial and temporal variations of the concentration of the solutes.

Different approaches have been considered for the modeling of mass transfer in hollow fiber liquid membrane processes. These approaches can be broadly classified as (i) diffusive mass transport models, (ii) flow and mass transfer models, and (iii) computational fluid dynamic CFD) models.

12.3.1 Diffusive Mass Transport Model

Mass transfer in liquid membranes occurs broadly by two mechanisms: (i) the diffusion mechanism and (ii) the facilitated transport mechanism.

   i. Diffusion mechanism is useful to explain the transport of solute based on solubility without any chemical reactions where the target solute dissolves in the liquid membrane at the feed-liquid membrane interface, and then the solute diffuses to the liquid membrane-strip interface, where it is recovered in the stripping phase. Figure 12.3 represents the diffusion mechanism of mass transport schematically, where solutes A and B are extracted into membrane phase at feed-membrane interface and recovered in strip phase at strip-membrane interface after diffusing through the membrane.

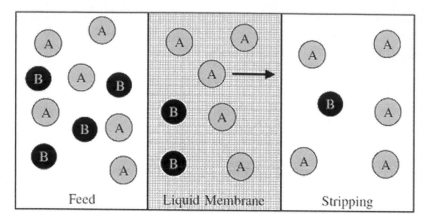

FIGURE 12.3
Schematic of diffusion mass transport mechanism (reprinted with permission from Bringas et al. (2009)).

ii Facilitated transport mechanism occurs when the selective solvent reacts with the target species to formulate solute-organic complex. Facilitated transport is also known as carrier mediated transport and can be classified as of two sorts—i.e., simple facilitated transport and coupled facilitated transport—depending upon the number of species participating with the carrier. Schematic representation of different facilitated transport mechanisms is depicted in Figure 12.4. Figure 12.4 (a) represents a simple facilitated transport mechanism schematically, where a neutral solute (*A*) is extracted selectively and recovered in stripping phase. The coupled facilitated transport occurs when the feed solution contains ionic solute

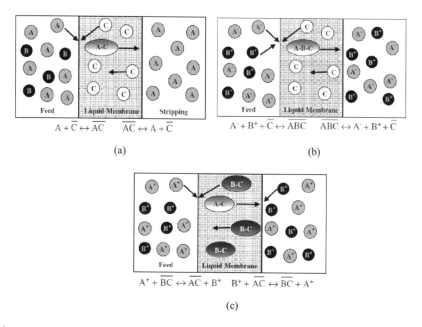

FIGURE 12.4
(a) Simple facilitated transport (reprinted with permission from Bringas E et al. (2009)). (b) Coupled facilitated co-transport (reprinted with permission from Bringas E et al. (2009)). (c) Coupled facilitated counter transport (reprinted with permission from Bringas E et al. (2009)).

and maintains electro-neutrality of the solution. This transport mechanism can be classified as coupled facilitated co-transport and coupled-facilitated counter-transport. In coupled facilitated co-transport mechanism, carrier reacts with both solutes A and B to make a complex (ABC) at the membrane-feed interface that diffuses across the membrane to membrane-strip interface, where both ions are recovered at strip phase, and the membrane is regenerated (Figure 12.4 (b)). In the case of a coupled facilitated counter-transport mechanism, the extractant (BC) reacts with solute (A) at feed-membrane interface to make complex (AC), while at the same time counter-ion (B) releases into the feed solution. The complex (AC) diffuses across the membrane to membrane-strip interface, where the solute (A) releases into back extraction solution (strip phase), and the extractant is regenerated (BC) (Figure 12.4 (c)).

Facilitated transport mechanism can be modeled as (i) diffusion kinematic model and (ii) mixed-kinematic model. In the diffusion kinematic model, the rate determining step is the diffusion across the boundary layers, and in the mixed kinematic model, the rate of interfacial reactions (complexion and decomplexation of metal ion) is considered to be comparable to the rate of diffusion.

12.3.1.1 Diffusional-Kinematic Mass Transport Model

In this model, the rate of interfacial reactions on both sides of the liquid membrane is assumed to be much faster compared to the diffusion of the solute in the aqueous phase boundary layers and through the membrane phase. Therefore, the overall mass transfer of the processes is controlled by diffusion of the solute. Predictions of the model are reported to show promising agreement with experimental results, especially in the case of the viscous liquid membranes, where the diffusion of solute carrier complex through membrane becomes the rate controlling step (Bringas et al., 2009).

In this modeling approach, concentration terms are related to each other by equilibrium relations. Extraction equilibrium can be described by the following reactions:

$$A(aq) + X(aq) + C(org) \xleftrightarrow{K_{eq}} AXC(org) \tag{12.2}$$

where, A, X and C represent the solute ion, anion present with solute in feed solution and organic ligand, respectively. The ratio of concentration of solute ion in the organic phase to that in the aqueous phase is defined as distribution coefficient (K_d). The equilibrium coefficient can be expressed as:

$$K_{eq} = \frac{[AXC]_{org}}{[A]_{aq} \cdot [X]_{aq} \cdot [C]_{org}} = \frac{K_d}{[X]_{aq} \cdot [C]_{org}} \tag{12.3}$$

Under steady state equilibrium, the model developed by Danesi (1984) was one of the initial models developed for mass transfer with simplified assumptions that the concentration gradients are linear and the reactions at the interfaces are instantaneous. The permeability coefficient of the membrane phase was assumed to be constant, irrespective of the metal ion concentration. The model was developed for once through as well as recycle mode of operations in the hollow fiber supported liquid membrane (HFSLM).

This was done by balancing mass flux across the lumen to find the permeability coefficient. Resistances of aqueous and organic phases are to be evaluated to estimate the permeability coefficient.

$$P = \frac{K_d}{K_d \Delta_{aq} + \Delta_{org}}$$

(12.4)

where, (Δ_{aq}) and (Δ_{org}) are the resistances of the aqueous and organic phase, respectively, and can be defined as: $\Delta_{aq} = \left(\frac{r_1}{r_1 - d_{aq}} \right) \frac{d_{aq}}{D_{aq}}$ and $\Delta_{org} = \frac{d_{org} \tau}{D_{org}}$, where (d_{aq}), (d_{org}) and (r_1) represent the aqueous boundary layer, membrane thickness, and inner radius of the lumen. Other parameters; such as, τ is the tortuosity, and (D_{aq}) and (D_{org}) is the diffusivity of metal ions in aqueous and organic phase, respectively. Outlet concentrations in the once through mode and the reservoir concentration in recycle mode of operations can be evaluated with the help of permeability coefficient in Equation 12.4. Dixit et al. (2012) and Khandalwal et al. (2011) developed mathematical models for the prediction of mass transport by calculating the resistances of the aqueous and organic phases by using Fick's first law of diffusion. D'Elia et al. (1986) considered a model, where the overall mass transfer resistance is the sum of individual resistances of feed and membrane phases. It can be written as:

$$\frac{1}{K_{overall}} = \frac{1}{k_{aq}} + \frac{d_{org}}{D_{org} K_d} + \frac{1}{k_{org} K_d}$$

(12.5)

where, d_{org}, D_{org}, k_{aq}, k_{org} represent the thickness of membrane, diffusivity of species in membrane phase and mass transfer coefficient in the aqueous and organic phases, respectively. Including strip phase resistance in addition to the resistances of feed and the membrane phases, Urtiga et al. (1992) modified the above equation as:

$$\frac{1}{K_{overall}} = \frac{1}{k_f} + \frac{r_i}{r_{lm}} \frac{1}{K_d} \frac{1}{k_m} + \frac{r_1}{r_2} \frac{1}{k_s}$$

(12.6)

where, r_{lm} is the log-mean radius of hollow fiber $(r_{lm} = (r_2-r_1)/ln\ (r_2/r_1))$, and k_f, k_m, and k_s, are the mass transfer coefficients of the feed, membrane and strip phases, respectively. The individual mass transfer coefficient can be calculated by using reported correlations. Generally, these coefficients depend on solute diffusivity, geometrical parameters and fluid dynamics. For further details, the reader is referred to the published literature of Gableman and Hwang (1999), Lipnizki & Field (2001), and Skelland (1974).

12.3.1.2 Mixed Kinetic Mass Transport Model

In this modeling approach, the interfacial reactions are not considered instantaneous and thus, the rate of mass transport is affected by the rate of reactions. So the overall resistance comprises of mass transfer resistances at feed-membrane interface and membrane-receiving phase interface, membrane phase and the chemical reaction(s) taking place at interfaces. An expression similar to Equation (12.6) can be obtained if the reaction at interface is considered:

$$\frac{1}{K_{overall}} = \frac{1}{k_f} + \frac{r_1}{r_{lm}}\frac{1}{K_d}\frac{1}{k_m} + \frac{r_1}{r_2}\frac{1}{k_s} + \frac{1}{k_{reaction}}$$ (12.7)

The majority of the applications of liquid membrane processes that have been mathematically modeled using the mixed kinematic approach are selective separations of metals from different aqueous media. Urtiga et al. (2005) and Yoshizuka et al. (1986) studied the mechanism of mass transfer of metals by using this approach, and the kinetics of extraction reactions reported by Harada et al. (1989). Ren et al. (2009; 2010) and Zhang et al. (2010) developed a mathematical model that describes the kinetic behavior of the separation and recovery of metals in hollow fiber liquid membrane contactor.

To predict the overall mass transfer coefficient, individual phase mass transfer coefficients need to be evaluated. Hence, it is required to obtain the theoretical values of k_f, k_m, k_{org}, etc. In most of the cases, solution flows in laminar regime through the hollow fiber module. The mass transfer coefficient for aqueous boundary layer can be calculated by the Leveque equation (Skelland, 1974):

$$Sh = 1.62 Gz^{1/3} \ For \ Gz > 20$$
$$= 3.67, Gz < 10$$ (12.8)

where, Sh and Gz are Sherwood and Graetz numbers, respectively. Kreulen et al. (1993) proposed a correlation valid for the entire range of Gz numbers:

$$Sh = \sqrt[3]{3.67^3 + 1.62^3 Gz}$$ (12.9)

The theoretical values of the membrane mass transfer coefficient (k_m) can be calculated from the following correlation:

$$k_m = \frac{D_{org}\varepsilon}{d_{org}\tau}$$ (12.10)

The mass transfer coefficient depends on the membrane characteristics such as membrane porosity (ε), membrane thickness (d_{org}), and membrane tortuosity (τ) and diffusivity of the species in the membrane phase.

If the organic phase is passed through the shell side, k_{org} can be evaluated by using the empirical correlation of Yang and Cussler (2000):

$$Sh = 1.25\left(Re\frac{d_h}{L}\right)^{0.93} Sc^{0.33}$$ (12.11)

where, Re and Sc represent the Reynolds and Schmidt dimensionless numbers, and d_h and L are hydraulic diameter and effective length of the hollow fiber lumens, respectively. This correlation holds good in the range of $Re \sim 0.5$ to 500 and $SC \sim 500$.

Mass transfer coefficient of interfacial reaction ($k_{reaction}$) can be estimated from the correlation mentioned below (Anil et al., 2001):

$$\frac{1}{k_{reaction}} = \frac{1}{K[X]_{aq}[C]_{org}}$$
(12.12)

where K is forward reaction rate constant. Using Equation (12.3), the above expression becomes:

$$\frac{1}{K_{reaction}} = \frac{1}{K[K_d / K_{eq}]}$$
(12.13)

12.3.2 Flow and Mass Transfer Model

In this approach, mass conservation equation for solute in fluid phases on either side of the membrane phases i.e. tube and shell side are used. Figure 12.5(a) represents the process configuration of NDSX process, where the aqueous feed solution flowing through the tube side is in contact with the organic phase, which is flowing through the shell side and its mass transfer scheme is represented by Figure 12.5(b). The target species (A) and co-species (B) are extracted by the selective extractant if it is assumed as a coupled facilitated co-transport mechanism of mass transfer (ABC). This model involves mathematical complexities because of the coupled equation of fluid flow and mass transfer. This approach is broadly divided into (i) 1-D flow and mass transfer models and (ii) 2-D flow and mass transfer models. In 1-D flow and mass transfer, the models are generally used for quick predictions by assuming average conditions on both lumen side and shell side of hollow fiber membrane, whereas 2-D flow and mass transfer models characterize more realistic variations of conditions and give more accurate results.

Simplified approach to flow and mass transfer modeling is based on average fluid properties with assumptions of plug flow, linear radial and axial concentration gradients. There are several publications based on this modeling approach (Bringas et al., 2009; Dixit et al., 2012; Khandalwal et al., 2011). A highly simplified model which does not consider spatial

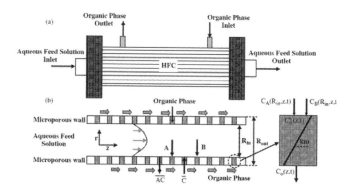

FIGURE 12.5
(a) Process configuration of hollow fiber contactor (b) Mass transfer process in single lumen (reprinted with permission from Bringas E et al. (2009)).

variations of concentration is also possible. The solute mass balance equations for the solutions in the lumen side and shell side derived from the model are as:

$$V_l \frac{dC_{li}}{dt} = -a_{eff} \cdot J \quad \forall t; \forall i \tag{12.14}$$

$$V_{sh} \frac{dC_{shi}}{dt} = a_{eff} \cdot J \quad \forall t \tag{12.15}$$

where, V_l and V_{sh} represent the volume of the lumen side and shell side solutions, respectively, and a_{eff} is the effective mass transfer area. The subscripts 'i', 'l' and 'sh' represent species, lumen and shell, respectively.

Boundary conditions are:

Initial conditions for facilitated co-transport are

$$C_{li} = C_{il}(t=0) \quad \forall i = A, B \tag{12.16}$$

$$C_{shi} = C_{shi}(t=0) \quad \forall i = A, B \tag{12.17}$$

Danesi (1984), Dixit et al. (2012), and Khandalwal et al. (2011), consider the mass transfer model with driving force as the solute concentration in the aqueous feed phase and neglected the organic complex concentration at the membrane-organic interface. The solute mass balance equation expressed in Equation 12.14 can then be rewritten as:

$$V_l \frac{dC_{li}}{dt} = -a_{eff} \cdot \cdot J = -a_{eff} \cdot K_{overall} C_{li} \quad \forall t \tag{12.18}$$

where, $K_{overall}$ is the overall mass transfer coefficient. Integrating the above equation, we get:

$$\ln \left(\frac{C_{li}}{C_{li}^{in}} \right) = -\frac{a_{eff}}{V_l} K_{overallt} t \quad \forall t \tag{12.19}$$

Equation 12.19 can be used to estimate variation of concentration of solute with time in lumen side. A similar equation can be derived for estimating variation of concentration of solute on shell side.

12.3.2.1 1-D Flow and Mass Transfer Models

In this modeling approach, flow on lumen side and shell side is assumed to be plug flow, i.e., velocity variation in the radial direction is neglected. Concentration variation in the radial direction in the lumen and shell side is also neglected. However, concentration variation in axial direction is taken into account. Resistance to mass transfer is described based on film theory. The assumption of plug flow in modeling of liquid membrane separation processes is extensively reported in literatures (Alonso et al., 1999; Bringas et al., 2009).

Solute mass balance in the lumen side fluid can be expressed as:

$$-\frac{\partial C_{li}}{\partial t} = u_l \frac{\partial C_{li}}{\partial z} + \frac{a_{eff}}{S_l L} s..J_i \qquad z \in (0, L); \forall t; \forall i = A, B \tag{12.20}$$

Solute mass balance in the shell side fluid can be expressed as:

$$\frac{\partial C_{shi}}{\partial t} = u_{sh} \frac{\partial C_{shi}}{\partial z} + \frac{a_{eff}}{S_{sh} L} s.J_i \qquad z \in (0, L); \forall t; \forall i = A, B \tag{12.21}$$

Initial conditions, $t = 0$,

$$C_{li} = C_{li}(t = 0) \qquad z \in (0, L), \forall i \tag{12.22}$$

$$C_{shi} = C_{shi}(t = 0) \qquad z \in (0, L), \forall i = A, B \tag{12.23}$$

Inlet boundary condition, $z = 0$,

$$C_{li} = C_{li}^{in} \qquad \forall i = A, B \tag{12.24}$$

$$C_{shi} = C_{shi}^{in} \qquad \forall i = A, B \tag{12.25}$$

The terms expressed in Equation 12.20 and Equation 12.21 are accumulation, convective and effective mass transfer (based on diffusion), respectively. The fiber shape factor(s) for cylindrical geometry based on the inside surface area of lumen can be expressed as (Nobel et al., 1983):

$$s = \frac{(r_2 - r_1)}{r_1 \ln(r_2/r_1)} \tag{12.26}$$

The above equations can be solved with mass transfer flux obtained from individual mass transfer resistances and with concentration in the lumen and shell side.

12.3.2.2 2-D Flow and Mass Transfer Models

This approach considers radial and axial variations of the concentration within lumen and shell side. Also, the radial variation of velocity is also accounted for, by using laminar velocity profiles. Only a few studies have been reported using this modelling approach because of difficulty in solving in the resulting complex governing equations. The mass conservation equation for lumen side is:

$$\frac{\partial C_{li}}{\partial t} = -2u_l\left[1-\left(\frac{r}{r_{in}}\right)^2\right]\frac{\partial C_{li}}{\partial z} + D_{li}\left[\frac{\partial^2 C_{li}}{\partial z^2} + \frac{1}{r}\frac{\partial}{\partial r}\left(r\frac{\partial C_{li}}{\partial r}\right)\right]$$

$$z \in (0,L), r \in (0,r_{in}), \forall t, \forall i = A, B \qquad (12.27)$$

where, $C_{li}(r, z, t)$ is the concentration of solute in lumen side liquid, u_l, the average velocity in the lumen side, D_{li} the diffusivity of the solute in the fluid in the lumen side. The left hand side term in Equation 12.27 represents the accumulation term and the two terms in the right hand side are convective and diffusive terms, respectively. The boundary conditions are:

Initial condition, at t = 0,

$$C_{li} = C_{li}(t=0), \qquad z \in (0,L); r \in (0,r_1)\forall t; \forall l = A, B \qquad (12.28)$$

Inlet boundary condition at z = 0, which is basically a close-open boundary condition in the parlance of classical dispersion model in reaction engineering.

$$2u_l\left[1-\left(\frac{r}{r_1}\right)^2\right]\left(C_{li}^{in} - C_{il}\right) = -D_{li}\frac{\partial C_{li}}{\partial z} \qquad r \in (0,r_1)\forall t; \forall i = A, B \qquad (12.29)$$

Equation 12.29 assumes fully-developed flow before entering into the fiber lumens.
Outlet boundary condition, at z = L

$$\frac{\partial C_{li}}{\partial z} = 0 \qquad r \in (0,r_1)\forall t; \forall i \qquad (12.30)$$

Axisymmetric boundary condition, r = 0,

$$\frac{\partial C_{li}}{\partial r} = 0 \qquad z \in (0,L); \forall t; \forall i = A, B \qquad (12.31)$$

Boundary condition at the feed- liquid membrane interface, $r = r_1$

$$J_i = -D_{li}\frac{\partial C_{li}}{\partial r} \qquad z \in (0,L); \forall t; \forall i = A, B \qquad (12.32)$$

Shell side structural arrangement is complex, which creates difficulties in accurate characterization of radial properties Thus the radial variation of velocity and concentration on shell side is often neglected. The concentration variation in shell side fluid considering counter-current flow can be expressed as:

$$\frac{\partial C_{shi}}{\partial t} = -u_{sh}\frac{\partial C_{shi}}{\partial z} + D_{shi}\frac{\partial^2 C_{shi}}{\partial z^2} + \frac{A}{S_{sh}}s..Ji \quad z \in (0,L);, \forall t; i = A, B \tag{12.33}$$

where, $C_{shi}(z, t)$ represents concentration of the solute in shell side fluid, u_{sh}, the average shell side fluid velocity, D_{sh}, the diffusivity of the solute in shell side fluid, S_{sh}, the flow area, s is the fiber shape factor and J_i is the flux entering into the shell side fluid.

Boundary conditions are:

Initial conditions at, t = 0,

$$C_{shi} = C_{shi}(t = 0) \quad z \in (0,L); i = A, B \tag{12.34}$$

Inlet boundary condition at z = L,

$$u_{sh}\left(C_{shi}^{in} - C_{shi}\right) = -D_{shi}\frac{\partial C_{shi}}{\partial z} \quad \forall t;, i = A, B \tag{12.35}$$

Outlet boundary condition, at z = 0,

$$\frac{\partial C_{shi}}{\partial z} = 0 \quad \forall t \tag{12.36}$$

To solve Equations 12.27 and 12.36 along with initial and boundary conditions, it is required to evaluate the mass transfer flux (Ji) of the target species from the feed to receiving phase by considering diffusion through the organic membrane in the pores. This can be obtained from the overall mass transfer coefficient as discussed in the previous sections.

12.3.3 Computational Fluid Dynamics (CFD) Based Models

Today, more sophisticated complex model are possible. This is because of availability of high speed computers and codes, which enable numerical solution of the complex models. Computational fluid dynamic (CFD) technique is the analysis of systems involving fluid flow, mass and heat transfer, combustion, etc., with the help of computer-based simulations. This technique is being applied for a wide range of applications in different fields.

Broadly, CFD analysis of the system is divided into three steps: pre-processing, processing and post-processing. In pre-processing or physical modeling, the geometry is defined along with the governing equations to be solved with specification of boundary conditions. The whole geometry is discretized into meshes or grids over which numerical solutions of the governing equations are to be obtained. In the processing step, simulation is carried out by solving the equations iteratively over the grid generated in the previous step. In the post-processing step, analysis and visualization of the results obtained from processing step are carried out.

CFD studies of LM and other membrane applications (gas-liquid and gas-gas) by using hollow fiber contactors have been carried out by using finite element methods. COMSOL MultiPhysics, a commercial finite element based solver, is used by Ghadiri et al. (2013), Marjani et al. (2012), and Shirazian et al. (2012; 2011) for numerical simulations of different membrane processes. To understand flow and mass transfer through CFD, a comprehensive two-dimensional model is developed based on the conservation equations of mass and momentum. The developed model for simulation of membrane extraction is capable of predicting radial and axial variation of velocity, pressure and concentration of solutes. The whole geometry of the hollow fiber module is divided into three sub-domains: lumen or tube side, pores of the hollow fiber filled with liquid membrane and lastly, the shell side. Figure 12.6 represents the CFD model subdomains considered in NDSX technique.

Cylindrical polar coordinate is used with the assumptions of asymmetry, r = 0 is the symmetric axis and the radial distance r_1, r_2, and r_3 are lumen inner and outer radii and shell radius. Axial position z = 0 and z = L represent the inlet and outlet positions of lumen side, while shell side inlet and outlet positions can be fixed depending upon the co-current or counter-current flow condition.

The conservation equation for transport of solute ions under steady state condition in the three subdomains is given by Equation 12.37 (Bird et al., 2002):

$$\nabla.(C_i \boldsymbol{v}) + \nabla.(J_i) = 0 \tag{12.37}$$

where, C_i, $\boldsymbol{v}$ and J_i are the concentration of the solute ion, velocity vector and diffusive flux of solute, respectively. The first term in Equation 12.37 represents the convective transport

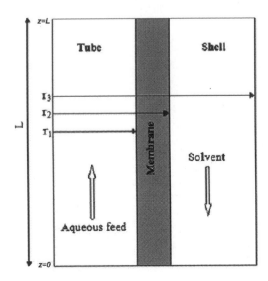

FIGURE 12.6
Model description (reprinted with permission from Shirazian et al. (2012)).

of solute ions and the second term represents diffusive contribution. Diffusive flux and total mass flux can be expressed as:

$$J_i = -D_i \nabla C_i \tag{12.38}$$

$$N_i = D_i \nabla C_i + C_i \boldsymbol{v} \tag{12.39}$$

where, D_i and N_i denote the diffusion coefficient and total mass flux, respectively. Total mass flux in Equation 12.39 is the combination of diffusive and convective mass fluxes. Diffusive flux becomes total mass flux (N_{mi}) when the solute is diffusing through the stagnant membrane phase.

$$N_{mi} = D_i \nabla C_i \tag{12.40}$$

Boundary conditions for mass transport equations for NDSX mode of operation at different boundaries of subdomains considering counter-current flow of fluids are given by Equations 12.41–12.48. Equation 12.41 represents the symmetric axis at ($r = 0$). At feed-membrane interface ($r = r_1$), there is a concentration jump from feed to membrane because of extraction of solute. This solute concentration jump depends upon the distribution coefficient (K_d). The mass flux at this interface remains constant but direction changes. These boundary conditions at feed-membrane interface are expressed by Equations 12.42–12.43. There is no concentration jump at ($r = r_2$) because of the solute transferring in the same phase. Also, at this junction, mass flux remains constant, but direction changes. These boundary conditions are given in Equations 12.44 and 12.45. Equation 12.46 considers the symmetric condition of shell side (at $r = r_3$), and Equations 12.47 and 12.49 represent zero flux conditions in membrane phase at ($z = 0$) and ($z = L$), respectively. Equations 12.48 and 12.50 represent the inlet conditions of lumen and shell side, respectively. It is assumed that flow is counter-current and fresh solvent enters the shell side.

$$\text{At, } r = 0, \frac{\partial C_{li}}{\partial r} = 0 \tag{12.41}$$

$$r = r_1, C_{mi} = C_{li} \times K_d \tag{12.42}$$

$$N_{li} = -N_{mi} \tag{12.43}$$

$$r = r_2, C_{mi} = C_{shi} \tag{12.44}$$

$$N_{mi} = -N_{shi} \tag{12.45}$$

$$r = r_3, \frac{\partial C_{shi}}{\partial r} = 0 \tag{12.46}$$

$$z = 0, \frac{\partial C_{mi}}{\partial r} = \frac{\partial C_{mi}}{\partial z} = 0 \tag{12.47}$$

$$z = 0, C_{li} = C_{li0} \tag{12.48}$$

$$z = L, \frac{\partial C_{mi}}{\partial r} = \frac{\partial C_{mi}}{\partial r} = 0 \tag{12.49}$$

$$z = L, \quad C_{shi} = 0 \tag{12.50}$$

There are two methods to obtain spatial variation of concentrations in the sub-domains by using the governing equations and boundary conditions given above. In the first and simpler method, mass transfer equations are solved without considering the spatial variation of velocity (Shirazian et al., 2012). This method is based on the assumption of uniform velocity in the lumen and on the shell side. In the second method, mass transfer equation is solved considering spatial variations of velocity in the lumen and on the shell side. This necessitates solution of Navier-Stokes equations along with the mass transfer equations. The Navier-Stokes equations are:

If the mass transfer equation is solved with the spatial variations in velocity (lumen and shell side), Navier-Stokes equations are to be solved, coupled with mass transfer equation. In this case, longitudinal diffusion is considered in the model. The Navier-Stokes equations are:

$$\rho(v.\nabla)\boldsymbol{v} = \nabla.\eta(\nabla \boldsymbol{v} + (\nabla \boldsymbol{v})^T) - \nabla p + F \tag{12.51}$$

$$\nabla.\boldsymbol{v} = 0 \tag{12.52}$$

where, ρ, v, η, p, and F denote density of the fluid (kg/m³), velocity vector, dynamic viscosity (kg/m.s), pressure (Pa), and force vector (N), respectively. Equation 12.52 represents the continuity equations for incompressible fluid. Navier-Stokes equation (Equation 12.51) is not solved in the domain of stagnant membrane in the pores of the hollow. Boundary conditions for the solution of the Navier-Stokes equations for lumen and shell side are given in Equations 12.53–12.54.

$$\text{At } z = 0, v_{lz} = v_{lz_{max}}\left(1 - \left(\frac{r}{r_1}\right)^2\right), v_{lr} = 0 \tag{12.53}$$

$$\text{At } z = L, v_{shz} = v_{shz_{max}}\left(1 - \left(\frac{r - r_2}{(r_3 - r_2)}\right)^2\right), v_{shr} = 0 \tag{12.54}$$

where, $v_{lz_{max}}$, $v_{shz_{max}}$ are the maximum velocities of the fluids on the lumen side and shell side respectively. Other boundaries of the domains considered as no-slip boundaries in the hollow fiber module. Table 12.1 contains the required parameters for CFD model of mass transfer.

TABLE 12.1

Parameters Required for CFD Model of Mass Transfer in
Hollow Fiber Contactor for NDSX Mode of Operation

Model Parameters	Unit
Lumen inner radius (r_1)	m
Lumen outer radius (r_2)	m
Lumen length (L)	m
Number of fibers (n)	–
Module diameter (R)	m
Porosity (ε)	–
Tortuosity (τ)	–
Diffusion coefficient lumen side solution (D_{li})	m^2/s
Diffusion coefficient shell side solution (D_{shi})	m^2/s
Effective diffusion coefficient membrane phase ($D_{mi} = D_{shi}*(\varepsilon/\tau)$)	m^2/s
Inlet concentration of solutions (C_{-li0}, C_{-shi0})	mol/m^3
Distribution coefficient (K_d)	–
Maximum velocity in lumen and shell ($v_{lz_{max}}$, $v_{shz_{max}}$)	m/s

12.3.3.1 CFD Case Studies of Solvent Extraction Processes

CFD method elaborated in the previous section is used for simulation of membrane extraction processes using hollow fiber contactor. For this, two different cases reported in literature are considered.

Case study-1

In this case study, CFD method discussed in the previous section is used to simulate extraction of alkali metals by using non-dispersive solvent extraction technique in hollow fiber membrane contactor, for which CFD modeling is reported by Ghadiri et al. (2013), and experimental data have been reported by Haddaoui et al. (2004). The CFD model predicts cesium extraction by dicyclohexano-18-crown-6 in chloroform system. CFD model is based on simultaneous solution of species transport and Navier-Stokes equations using COMSOL Multiphysics.

The model is developed with the following assumptions: complexion reaction takes place at the feed-membrane interface, steady state process, isothermal operation, equilibrium is attained everywhere at the feed-membrane interface, and fully developed laminar flow.

Numerical Simulations Simulation parameters and physical conditions are required to solve the model equations discussed in the previous section. Table 12.2 contains the model parameters used in this case study. COMSOL Multiphysics was used to solve the model equations with the boundary conditions. The software uses finite element method (FEM) for numerical solutions of the model equations. The grid was composed of 5, 55, 584 triangular elements.

Figure 12.7 shows the comparison of the result predicted by the CFD model with the experimental results of Haddaoui et al. (2004). The results of Ghadri et al. (2013) are also compared.

TABLE 12.2

Parameters Used in Case Study-1 using CFD Model

Model Parameters	Value
Lumen inner radius (m)	120×10^{-6}
Lumen outer radius (m)	150×10^{-6}
Lumen length (m)	0.254
Module diameter (m)	0.12
Porosity (ε)	0.3
Tortuosity (τ)	3.3
Aqueous diffusion coefficient of Cs (I) (m²/s)	2.06×10^{-9}
Diffusion coefficient of complex-organic phase (m²/s)	1.29×10^{-9}
Effective diffusion coefficient membrane phase (m²/s)	1.17×10^{-10}
Extractant concentration (mol/m³), DC 18-6	1–0
Inlet concentration of cesium (mol/m³)	0.2–1.2

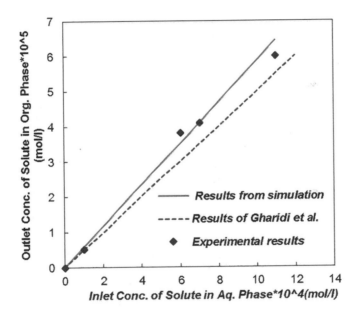

FIGURE 12.7

Validation of CFD model of cesium concentration in hollow fiber contactor operated in NDSX mode. The experimental results are from Haddaoui et al. (2004) and the results of the CFD model of Ghadiri et al. (2013) are given.

The prediction of the CFD model described in the previous section are found to be better than the results of Ghadri et al. (2013).

Figure 12.8 shows the radial concentration profile in the middle of the axial length of the module. It is evident from the graph that the concentration jump is well-implemented at the membrane-feed interface (Equation 12.42).

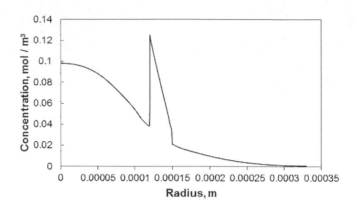

FIGURE 12.8
Radial concentration profile of Cs in the middle of the axial length.

Case Study-2   In this case study, separation of acetone by a supercritical fluid (propane) using hollow fiber contactor is simulated. The results obtained from the CFD model are compared with the results reported in a previous numerical study (Miramini et al., 2013) and experimental data obtained from literature (Bothun et al., 2003).

The assumptions in the CFD model of this case study are same as the assumptions in the CFD model of the previous case study.

Numerical Simulations   Table 12.3 contains the model parameters used in this case study. COMSOL Multiphysics was used to solve the model equations with the boundary conditions. The grid was composed of 291,064 triangular elements.

Figure 12.9 shows the results obtained from the CFD model. The experimental results obtained by Bothun et al. (2003) and the CFD results of Miramini et al. (2013) are also presented. A good agreement between the results of the CFD model and experiments can be observed.

Figure 12.10 shows the radial concentration profile in the middle of the axial length of the module. It is evident from the graph that the concentration jump is well-implemented at membrane-feed interface (Equation 12.42).

TABLE 12.3

Parameters Used in Case Study-2 CFD Using Model

Model Parameters	Value
Lumen inner radius (m)	300×10^{-6}
Lumen inner radius (m)	510×10^{-6}
Shell inner radius (m)	760×10^{-6}
Lumen length (m)	1.067
Porosity (ε)	0.75
Tortuosity (τ)	3.3
Diffusion coefficient of acetone in feed (m^2/s)	1.14×10^{-9}
Diffusion coefficient of acetone in propane (m^2/s)	1.1×10^{-8}
Distribution coefficient (K_d)	1.63
Kinematic viscosity (m^2/s)	2.2×10^{-7}
Solvent to feed molar ratio (-)	3

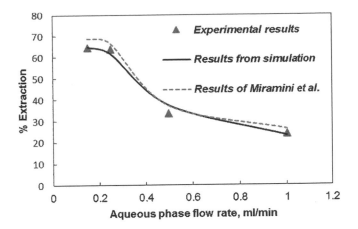

FIGURE 12.9
Comparison of experimental and simulation results of percentage acetone extraction in hollow fiber contactor operated in NDSX mode. Experimental results are from Bothun et al. (2003) and the results of the CFD model of Miramini et al. (2013) are also given.

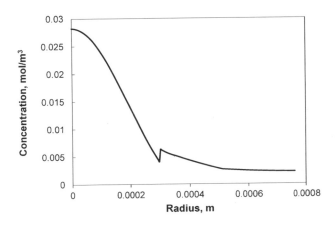

FIGURE 12.10
Radial concentration profile of acetone in the middle of the axial length.

12.4 Conclusions and Future Perspective

This work is an effort to summarize different approaches of mass transfer modeling being practiced in liquid membrane separation processes using hollow fiber contactor. Mass transfer in membrane contactors are modeled in three different ways: (i) diffusive mass transfer model (ii) flow and mass transfer model (iii) CFD based models.

Two different approaches have been used for diffusive mass transport model: diffusion kinematic approach and mixed kinematic approach. The parameters required for both the approaches of modeling are mass transport coefficients and kinetic parameters that can be evaluated from empirical correlations. This makes the results of the diffusive mass transfer modeling system specific.

The flow and mass modeling approaches lead to either simple one dimensional, or rigorous two-dimensional model equations. The assumption of plug flow simplifies the problem and helps in easy and quick calculations. There are only a few publications reported on the two-dimensional rigorous modeling approach because of its complexity.

With the availability of high-speed computers and various commercial computational software, CFD based models are finding more and more applications. The CFD based model solves Navier-Stokes equations, as well as species transport equations. No empirical equation is needed to get the results from CFD based models. In a typical CFD based model hollow fiber module is represented by three sub-domains: lumen, membrane and shell side. Axial and radial variations of concentration can be obtained and rigorously analyzed for important parameters, which affect the performance of the hollow fiber contactor. This understanding can eventually be used for design and process optimization.

Nomenclatures

C	Concentration of solute [mol/m³]
K_d	Distribution coefficient
$K_{eq.}$	Equilibrium constant
D_o	Diffusion coefficient of solute in solvent [m²/s]
D_{aq}	Diffusion coefficient of solute in aqueous [m²/s]
D_{org}	Diffusion coefficient of solute in organic phase [m²/s]
P	Permeability coefficient [m/s]
Δ_{aq}	Aqueous film resistance [s/m]
Δ_{org}	Membrane resistance [s/m]
d_a	Aqueous film thickness [m]
d_{org}	Membrane thickness [m]
$K_{overall}$	Overall mass transfer coefficient [m/s]
k_{aq}	Aqueous phase mass transfer coefficient [m/s]
k_{org}	Organic phase mass transfer coefficient [m/s]
k_f	Feed side mass transfer coefficient [m/s]
k_m	Membrane phase mass transfer coefficient [m/s]
k_s	Strip phase mass transfer coefficient [m/s]
$k_{reaction}$	Reaction mass transfer coefficient [m/s]
Sh	Sherwood number
Gz	Graetz number
Re	Reynolds number
Sc	Schmidt number
d_h	Hydraulic diameter [m]
L	Effective length of lumen [m]
V	Volume of lumen [m³]
a_{eff}	Effective area of mass transfer [m²]
t	Time [s]
v	Velocity [m/s]
v_{lz}	Lumen axial velocity [m/s]
v_r	Lumen radial velocity [m/s]
v_{lzmax}	Maximum axial velocity of lumen [m/s]
v_{shz}	Shell side axial velocity [m/s]
v_{shr}	Shell side radial velocity [m/s]

v_{shzmax}	Maximum axial velocity of shell side [m/s]
S	Cross-sectional flow area [m^2]
s	Shape factor
r_1	Inner radius of lumen [m]
r_2	Outer radius of lumen [m]
r_3	Shell side radius [m]
J	Diffusive mass flux [mol/m^2.s]
N	Total mass flux [mol/m^2.s]
D_{li}	Diffusion coefficient of solute in the lumen side [m^2/s]
D_{shi}	Diffusion coefficient of solute in the shell side [m^2/s]
p	Pressure [Pa]
F	Volumetric body force [N/m^3]

Greek Letters

ε	Porosity
τ	Tortuosity
ρ	Density [kg/m^3]
η	Viscosity [kg/m.s]

Subscripts & Superscripts

l	Lumen
sh	Shell
int	Interface
i	Species
l_i	Species in lumen
sh_i	Species in shell
in	Inlet

Abbreviations

CFD	Computational fluid dynamics
HFSLM	Hollow fiber Supported liquid membrane
HFRLM	Hollow fiber renewal liquid membrane
LM	Liquid membrane
PEHFSD	Pseudo emulsion based hollow fiber strip dispersion
NDSX	Non-dispersive solvent extraction
SLM	Supported liquid membrane
DC 18-6	Dicyclohexano-18-crown-6

References

Alguacil, F.J., M. Alonso, A. Lopez-Delgado, I. Padilla, and H. Tayibi. 2010. Pseudo-emulsion based hollow fiber strip dispersion pertraction of iron (III) using [PJMTH]$_2$$^+SO_4$$^{2-}$ ionic liquid as carrier. *Chemical Engineering Journal* 157: 366–372.

Alonso, A.L., B. Galan, M. Gonzalez, and I. Ortiz. 1999. Experimental and theoretical analysis of a non-dispersive solvent extraction pilot plant for the removal of Cr (VI) from galvanic process waste waters. *Industrial & Engineering Chemistry* 38: 1666–1675.

Ansari, S.A., P.K. Mohapatra, D.R. Raut, T.K. Seshagiri, B. Rajeswari, and V.K. Manchanda. 2009. Performance of actinide partitioning extractants in hollow fiber supported liquid membranes for the transport of actinides and lanthanides from high level nuclear waste. *Journal of Membrane Science* 337: 304–309.

Bird, R.B., W.E. Stewart, and E.N. Lightfoot. 2002. *Transport phenomena, 2nd ed.* New York: John Wiley & Sons.

Bothun, G.D., B.I. Knutson, H.J. Strobel, S.E. Nokes, E.A. Brignole, and S. Diaz. 2003. Compressed solvent for the extraction of fermentation products within a hollow fiber membrane contactor. *Journal of Supercritical Fluid* 25: 119–134.

Boudot, A., J. Floury, and H.E. Smorenburg. 2001. Liquid-liquid extraction of aroma compounds with hollow fiber contactor. *AIChE Journal* 47: 1780–1793.

Boyadzhiev, L., Z. Lazarova In noble RD and Stern SA, 1995. Liquid membranes (Liquid petraction), Membrane separation Technology: Principle and Applications. *Elsevier Science B.V.*: 283–300.

Bringas, E., M.F. San Roman, J.A. Irabien, and A. Ortiz. 2009. An over view of the mathematical modelling of liquid membrane separation processes in hollow fiber contactors. *Journal of Chemical Technology and Biotechnology* 84: 1583–1614.

Danesi, P.R. 1984. Permeation of metal ions through hollow fiber supported liquid membrane: Concentration equations for once through and recycle module arrangement, *Solvent Extraction and Ion Exchange* 2: 115–120

Danesi, P.R. Separation of metal species by supported liquid membranes. 1984-1985. *Separation Science and Technology* 19: 857–894.

Danesi, P.R. 1948. Permeation of metal ions through hollow fiber supported liquid membrane: Concentration equations for once through and recycle module arrangement. *Solvent Extraction and Ion Exchange* 2: 215–220.

D'Elia, N.A., L. Dahuron, L.A. Cussler, and E.L. Cusseler. 1986. Liquid-liquid extraction with micro porous hollow fibers. *Journal of Membrane Science* 29: 309–319.

Dixit, S., Mukhopadhyay, S., G. Smita, K.T. Shenoy, H. Rao, and S.K. Ghosh. 2012. A mathematical model for pertraction of uranium in hollow fiber membrane contactor using TBP. *Desalination and Water Treatment* 38: 195–206.

Gabelman, A. and S.T. Hwang. 1999. Hollow fiber membrane contactors. *Journal of Membrane Science* 159: 61–106.

Ghadiri, M. and S. Shirazian. 2013. Computational simulation of mass transfer in extraction of alkali metals by means of nano porous membrane extraction. *Chemical Engineering & Processing: Process Intensification* 69: 57–62.

Haddaoui, J., D. Trebouet, J.M. Loureiro, and M. Burgard. 2004. Nondispersive solvent extraction of alkali metals with the dicyclohexano 18 crown 6: Evaluation of mass transfer coefficients. *Separation Science and Technology* 39: 3839–3858.

Harada, M., Y. Miyake, and Y. Kayahara. 1989. Kinetics mechanism of metal extraction with hydroxyoximes. *Journal of Chemical Engineering of Japan* 22: 168–176.

Ho, W.S. and T.K. Poddar. 2001. New membrane technology for removal and recovery of chromium from waste waters. *Environmental Progress & Sustainable Energy* 20: 44–52.

Juang, R.S. and H.L. Huang. 2002. Modelling of Non dispersive extraction of binary Zn (II) and Cu (II) with D2EHPA in hollow fiber devices. *Journal of Membrane Science* 208: 31–38.

Khandalwal, P., S. Dixit, S. Mukhopadhyay, and P.K. Mohapatra. 2011. Mass transfer modelling of Cs(I) through hollow fiber supported liquid membrane containing calix-[4]-bis (2,3-naptho)-crown-6 as the mobile carrier. *Chemical Engineering Journal* 174: 110–116.

Klassen, R., P.H.M. Feron, and A.E. Jansen. 2005. Membrane contactors in industrial applications. *Chemical Engineering Research and Design* 83: 234–246.

Kocherginsky, N.M., Q. Yang, and L. Seelam. 2007. Recent advances in supported liquid membrane technology. *Separation and Purification Technology* 53: 171–177.

Kumar, A., R. Haddad, and A.M. Sastre. 2001. Integrated membrane process for gold recovery from hydrometallurgical solutions. *AIChE Journal* 47(2): 328–340.

Lipnizki, F. and R.W. Field. 2001. Mass transfer performance for hollow fiber modules with shell-side axial feed flow: Using an engineering approach to develop a framework. *Journal of Membrane Science* 193: 195–208.

Liqui-Cel extra flow celgard hollow fiber module. www.liqui-cel.com, accessed on 25th September, 2017.

Marjani, A. and S. Shirazian. 2012. Simulation of heavy metal extraction in membrane contactors using computational fluid dynamics. *Desalination* 282: 422–428.

Miramini, S.A., S.M.R. Razavi, M. Ghadiri, and S.Z. Mahadavi. 2013. CFD simulation of acetone separation from an aqueous solution using supercritical fluid in a hollow fiber membrane contactor. *Chemical Engineering & Processing: Process Intensification* 72: 130–136.

Mohapatra, P.K. and V.K. Manchanda. 2003. Liquid membrane based separation of actinides and fission products. *Indian Journal of Chemistry* 42A: 2925–2939.

Nobel, R.D. Shape factors in facilitated transport through membranes. 1983. *Industrial & Engineering Chemistry Fundamentals* 22: 139–144.

Oritz, I., E. Bringas, M.F. San Roman, and A.M. Urtiga. 2004. Selective separation of zinc and iron from spent pickling solutions by membrane based solvent extraction: Process viability. *Separation Science and Technology* 39: 2441–2455.

Ortiz, I., B. Galan, F. San Roman, and R. Ibanez. 2001. Kinetics of separating of multicomponent mixtures by non-dispersive solvent extraction, Ni and Cd. *AIChE* 47: 895–905.

Pabby, A.K., S.H.S. Rizvi, and A.M. Sastre. 2015. *2nd Edition Handbook of Membrane Separations Chemical, Pharmaceutical, Food and Biotechnological Applications*. CRC Press.

Pabby, A.K. and A.M. Sastre. 2000. Hollow fiber supported Liquid Membrane for the separation/concentration of Gold(I) from aqueous cyanide media: Modelling and mass transfer evaluation. *Industrial & Engineering Chemistry Research* 39(1): 146–154.

Parthasarathy, N., M. Pelletier, and J. Buffle. 1997. Hollow fiber based supported liquid membrane: A novel analytical system for trace for metal analysis. *Analytica Chimica Acta* 350: 183–195.

San Roman, M.F., E. Bringas, R. Ibanez, and I. Ortiz. 2010. Liquid membrane technology: Fundamentals and review of its applications. *Journal of Chemical Technology and Biotechnology* 85: 2–10.

Pei, L., L.M. Wang, W. Guo. 2012. Stripping dispersion hollow fiber liquid membrane containing carrier PC-88A and HNO_3 for the extraction of Sm^{3+}. *Chinese Chemical Letters* 23: 101–104.

Pei, L., L.M. Wang, W. Guo, N. Zhao. 2011. Stripping dispersion hollow fiber liquid membrane containing PC-88A as carrier and HCl for transport behaviour of trivalent dysprosium. *Journal of Membrane Science* 378: 520–530.

Ren, Z., W. Zhang, H. Meng, J. Liu, S. Wang. 2010. Extraction separation of Cu (II) and Co(II) from sulphuric solutions by hollow fiber renewal liquid membrane. *Journal of Membrane Science* 365: 260–268.

Ren, Z., H. Meng, W. Zhang, J. Liu, and C. Cui. 2009. The transport of Copper (II) through hollow fiber renewal liquid membrane and hollow fiber supported liquid membrane. *Separation Science and Technology* 44: 1191–1197.

Roy, S.C., J.V. Sonwane, N.S. Rathore, A.K. Pabby, P. Janardan, R.D. Changrani, P.K. Dey, and S.R. Bhardwaj. 2008. Pseudo-emulsion based hollow fiber strip dispersion, technique (PEHFSD): Optimization, modelling and application of PEHFSD for recovery of U(IV) from process effluent. *Separation Science and Technology* 43: 3305–3332.

Shirazian, S.,M. Rezakazemi, and F. Fadai. 2011. Supercritical extraction of organic solutes from aqueous solutions by means of membrane contactors: CFD simulation. *Desalination* 277: 135–140.

Shirazian, S., A. Modhadassi, and S. Moradi. 2011. Numerical simulation of mass transfer in gasliquid hollow fiber membrane contactors for laminar flow conditions. *Simulation Modelling Practice & Theory* 17: 708–718.

Shirazian, S., M. Pishnamazi, M. Rezakazemi, A. Nouri, M. Jafai, S. Noroozi, and A. Marjani. 2012. Implementation of finite element method for simulation of mass transfer in membrane contactors. *Chemical Engineering & Technology* 35: 1077–1084.

Skelland, A.H.P. 1974. Diffusional mass transfer. New York: Wiley.

Sonwane, J.V., A.K. Pabby, A.M. Sastre. 2010. Pseudo emulsion based hollow fiber strip dispersion (PEHFSD) technique for permeation of Cr (VI) using Cyanex-923 as carrier. *Journal of Hazardous Materials* 174: 541–547.

Urtiaga, A.M., M.J. Abellan, J.A. Irabien, and I. Ortiz. 2005. Membrane contactors for the recovery of metallic compounds, modelling of copper recovery from WPO process. *Journal of Membrane Science* 257: 161–170.

Urtiaga, A.M., M.I. Ortiz, E. Salazar, and J.A. Irabien. 1992. Supported liquid membrane for the separation-concentration of phenol. 1. Validity and mass transfer evaluation. *Industrial & Engineering Chemistry Research* 31: 877–886.

Vijaylaxmi, R.,S. Chaudhury, M. Anitha, D.K. Singh, S.K. Agarwal, and H. Singh. 2015. Studies on yttrium permeation through hollow fiber supported liquid membrane from nitrate medium using di-nonyl phenyl phosphoric acid as the carrier phase. *International Journal of Mineral Processing* 135: 52–56.

Walki, C.R. and P. Chang. 1955. Correlation of diffusion coefficients in dilute solutions. *AIChE Journal* 1: 264–270.

Yun, C., R. Prasad, and K. Sarkar. 1993. Hollow fiber solvent extraction removal of toxic heavy metals from aqueous waste streams. *Industrial & Engineering Chemistry Research* 32: 1186–1194.

Zhang, W., C. Cui, Z. Ren, Y. Dai, H. Meng. 2010. Simultaneous removal and recovery of copper (II) from acidic waste water by hollow fiber renewal liquid membrane with LIX984N as carrier. *Chemical Engineering Journal* 157: 230–237.

13

Design of Membrane Systems Using Computational Fluid Dynamics and Molecular Modeling

Siddhartha Moulik, Rehana Anjum Haldi, and Sundergopal Sridhar

CONTENTS

13.1 Introduction

Unit operations involving the transport of molecules by application of membrane as a semi-permeable barrier for selective separation is found to have wide applications dealing with several challenging issues frequently facing industries, in an environmentally safe and cost-effective way (Buonomenna & Golemme, 2012). Membranes have extensive applications in various process industries, such as pharmaceutical (Deegan et al., 2011), textile effluent treatment (Marcucci et al., 2001), food and dairy product concentration (Daufin et al., 2001), etc. Predicting the performance of a particular membrane process is indeed a difficult task due to complex interactions between the dissolved constituents and the membrane, besides the complicated hydrodynamic behavior of the fluid flowing within the e module. Process efficiency varies widely based on feed stream composition and operating conditions. Hence, membrane selection and process scale-up using reliable mathematical models based on molecular modeling and computational fluid dynamics (CFD), appear to be very promising in recent studies. Molecular dynamics (MD) simulation is widely used to analyze the transport coefficient of small molecules through a polymer membrane from a combination of the excess free energy and the diffusion constant (Madhumala et al., 2017). Condensed-phase optimized molecular potentials for atomistic simulation studies (COMPASS) force field has been widely used to optimize and predict the structural, conformational and thermo-physical condensed phase properties of polymers (Moulik et al., 2015).

In the COMPASS force field approach, total energy (E_T) of the system is represented by the sum of the bonding and nonbonding interactions.

$$E_T = E_b + E_o + E_\varnothing + E_{OOP} + E_{PE} + E_{VDW} + E_Q \qquad (13.1)$$

where, E_b, E_o, $E_\varnothing$, E_{opp}, and E_{PE} represent the energy associated with the bond, bond angle bending, torsion angle rotation, out of loop and potential energy, respectively. The last two terms represent non-bonded interactions, which include of van der Waals term (E_{VDE}) and electrostatic force (E_Q). Besides molecular dynamic simulation, CFD analysis is gaining importance for membrane module design. In CFD, partial differential equations of fluid dynamics are solved numerically using appropriate numerical discretization scheme to obtain solutions for problems where analytical solutions to the governing equations are not available (Praneeth et al., 2014). Finite difference method (FDM), finite element method (FEM) and finite volume method (FVM) are three major numerical discretization schemes used in CFD. Out of all three methods, FDM is conceptually simple to formulate. However, FEM is suitable for the accommodation of complex geometries (Wiley et al., 2003). This chapter will cover detailed overview of the application of MD and CFD in different membrane processes.

13.2 Overview of Molecular Modeling and CFD in Membrane Processes

13.2.1 Optimization of Pressure Driven Processes

Pressure driven processes such as reverse osmosis (RO), nanofiltration (NF), ultrafiltration (UF) and microfiltration (MF) are widely used in a number of industrial applications owing to their low cost and high membrane packing density. RO and NF are extensively used in industrial effluent treatment (Dolar et al., 2011), drinking water purification (Pervov et al., 2000), and the recovery of valuable antibiotics from pharmaceutical effluent streams (Kosutic et al., 2007), whereas the UF process is widely used for enzyme concentration (Butterworth et al., 1970), milk protein separation (Ding et al., 2002), fruit juice clarification (Girard et al., 1999), etc. Prior to all such operations, optimization of the membrane is performed on a lab scale flat sheet, whereas scale-up necessitates trials on a spiral wound membrane (SWM) module. The life span of SWM modules currently available on the market is in between one and three years, depending upon the field of utilization (Lee et al., 2011). Since fouling is one of the major operational issues responsible for reducing membrane life, the optimization of spacer design to overcome the fouling problem in the SWM module is a major step forward. Figure 13.1 represents the diamond and ladder configurations, two main types of spacers used to develop a commercial spiral wound membrane module. In the ladder configuration, regularly spaced transverse filaments are interconnected with longitudinal filaments, which create a ladder type structure which reduces the concentration polarization layer by interrupting the momentum and concentration boundary layers (Radu et al., 2014). Figure 13.2 (a) (i) represents a SWM module in a partly unwound state, in which the geometry of the module with spacer was constructed with a large number of flow cells, whereas, Figure 13.2 (a) (ii) and (iii) represent detailed geometry of the computational domain. Figure 13.2 (b) depicts a detailed meshed computational domain used in our earlier study to optimize the spacer configurations for production of

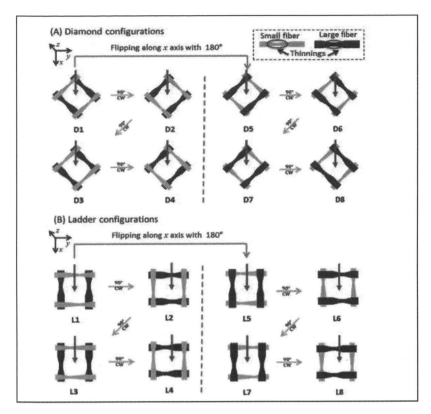

FIGURE 13.1
Schematic representation of (a) diamond and (b) ladder-type feed spacers used for simulation by Radu et al. (2014). (Reproduced with permission from Elsevier.)

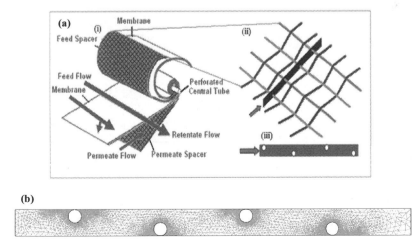

FIGURE 13.2
(a) Development of computational domain and (b) meshed computational domain for CFD study by Moulik et al. (2016). (Reproduced with permission from Elsevier.)

fructose from aqueous solution (Moulik et al., 2015). During the process, the fluid flow was analyzed by solving continuity and momentum balance equations assuming the fluid to be incompressible (Equations 13.2 and 13.3):

$$\nabla . V = 0 \tag{13.2}$$

$$\rho \frac{DV}{Dt} = -\nabla P + \mu \nabla^2 V \tag{13.3}$$

where V is the velocity field (m/s), μ represents fluid viscosity (Pa.s) and ρ is fluid density (kg/m^3). At the inlet (x = 0), a fully developed laminar parabolic velocity profile is specified with a constant average velocity (with respect to time) and the tangential velocity (u_w) (m/s) at the membrane surface is set to zero. The permeate flux (v_w)(m/s) is calculated by using Equation 13.4:

$$v_w = L_p(\Delta P - \Delta \pi) \tag{13.4}$$

where, the TMP (ΔP) is defined as the difference between the local pressure on the feed side of the membrane and constant permeate pressure (Pa), whereas $\Delta \pi$ is the osmotic pressure created by concentration polarization (Pa). The average flux (L/m^2.h) is calculated by integrating the local permeate velocity (v_w) (m/s) over the channel length (L) (m) for both upper and lower membrane leaves, in the spirally wound arrangement.

$$J = \frac{1}{L} \int_0^L v_w \, dx \tag{13.5}$$

The convection and diffusion equation (Equation 13.6) is used to find solute concentration within the membrane channel:

$$u \frac{\partial c}{\partial x} + v \frac{\partial c}{\partial y} = D \left(\frac{\partial^2 C}{\partial x^2} + \frac{\partial^2 C}{\partial y^2} \right) \tag{13.6}$$

where the diffusion coefficient (D)(m^2/s) depends on concentration (c)(mol/m^3) and temperature. At the inlet (x = 0), initial concentration of fructose is specified while average concentration is constant with time. Condition of zero diffusion is applied on the outlet boundary, while insulation boundary condition is applied on the spacer surfaces. The convective flux of solute toward the membrane equals the sum of diffusive backward transport of the solute and the convective flux through the membrane.

$$\left(vc - D \frac{\partial c}{\partial y} \right) = vc(1 - R) \tag{13.7}$$

where, (R) is the rejection coefficient. Figure 13.3 (a) shows the concentration profile inside the channel for functionalized polyamide (FPA) 150 nanofiltration membrane,

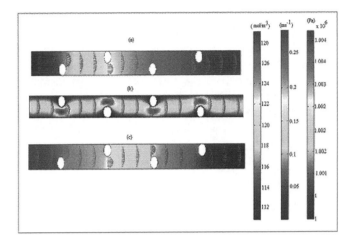

FIGURE 13.3
(a) Concentration, (b) velocity, and (c) pressure profiles inside FPA-150 membrane at 10 bar pressure analyzed by Mouliket al. (2016). (Reproduced with permission from Elsevier.)

FIGURE 13.4
Snapshot of the initial configuration of hydrated polyamide surrounded by two water reservoirs developed by Ding et al., (2014). (Reproduced with permission from Elsevier.)

whereas Figures 13.3 (b) and (c) represent the velocity and pressure profiles inside the membrane channel at 10 bar feed pressure. That major portion of the fluid flow in the SWM module is in the bulk flow direction. However, due to the presence of axial filaments, two distinct zones pertaining to flow attachment and separation are created near the wall. The inlet velocity of the fructose solution was 0.1 ms^{-1}, which increased to 0.26 ms^{-1} near the filament surface due to a vigorous, periodic disturbance in the flow. Velocity fluctuated in the range of 0.12–0.19 ms^{-1} depending on the position of the spacers (Moulik et al., 2015). Studies using molecular dynamics are also performed by few researchers for the development of hydrophilized RO/NF membranes. Ding et al. (2014) developed a molecular dynamics model to investigate both structural and dynamical properties of water trapped within a highly cross-linked polyamide RO membrane. Figure 13.4 depicts the initial configuration of hydrated polyamide surrounded by two water reservoirs. Due to the formation of hydrogen bonding, 90% of the water molecules were embedded in the polyamide membrane. In the central part of the membrane, the translational diffusion coefficient of confined water is less than the bulk value (Ding et al., 2014).

13.2.2 Pervaporation

Pervaporation (PV) is a specific sub-area of membrane separation most suitably applied to separate organic mixtures, heat-sensitive compounds, azeotropic and close boiling mixtures due to its small footprint, scalability, energy saving capability and ease of integration into other unit operations (Sunitha et al., 2013). In the PV process, mass transport through the dense membrane includes three steps: (1) sorption of the targeted component into the membrane at the upstream side, followed by (2) diffusion of the component through the membrane and (3) desorption in a vapor phase at the downstream side (Prasad et al., 2016). The chemical compatibility between the membrane material and the targeted components and the difference in diffusivity of feed components largely influence process efficiency. In the PV process, diffusion occurs due to concentration gradient. Diffusion coefficients were calculated using the Equation 9.8 (Moulik et al., 2016):

$$J_i = P_i[p_{i,f} - p_{i,p}] = \frac{D_i}{h}[C_{i,f} - C_{i,p}] \tag{13.8}$$

where, $C_{i,f}$ and $C_{i,p}$ are the composition of liquids in feed and permeate side, respectively. P and h are the permeate pressure and membrane thickness, correspondingly. Subscript 'i' stands for water or organic component. Diffusion coefficient (D) of molecules can also be calculated using MD simulation. The mobility of molecules within the membrane matrix can be analyzed using mean square displacement (MSD), which can be computed using Equation 13.9, meanwhile the Einstein relationship is used to calculate diffusion coefficient (D) of molecules from the slope of MSD (Equation 13.9):

$$MSD = |r_i(t) - r_i(0)|^2 \tag{13.9}$$

$$D = \lim_{t \to \infty} \frac{\left\langle |r_i(t) - r_i(0)|^2 \right\rangle}{6t} \tag{13.10}$$

where, $r(0)$ and $r_i(t)$ are the initial and final (after time (t)) position coordinates of atom (i), respectively. The Connolly surface method is widely used to calculate the fractional free volume (FFV) within the membrane matrix (Moulik et al., 2015). An amorphous model of poly(vinyl alcohol) membrane for dehydration of ethanol is shown in Figure 13.5 (Moulik et al., 2015). The diffusion coefficient of water and ethanol in the range of 3 to 30 wt% feed water concentration at 303 K was calculated using Einstein relationship (Equation 13.10) and plotted against different feed water concentrations (Figure 13.6). It can be observed that the diffusivity of water increased considerably with increasing water content in the feed mixture. Few experimental and theoretical studies are reported in the literature based on CFD for evaluating the hydrodynamics and mass transfer within the pervaporation process. Mafi et al. (2013) reported a CFD model to predict the mass transfer of aroma compounds through a hydrophobic membrane in PV. The effect of baffles on mass transfer within a slit type PV module is studied by Liu et al. (2005). Moraveji et al. (2013) modeled a pervaporation process with polydimethylsiloxane (PDMS) membrane for ethanol/water separation. In our earlier study, a combined model based on molecular modeling and CFD has been reported for the dehydration of liquid propellants (Moulik et al., 2015) and n-Methyl-2-pyrrolidone (Prasad et al., 2016). Schematic of the computational domain

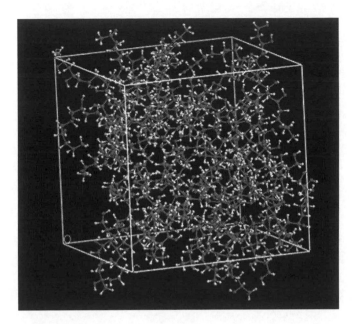

FIGURE 13.5
Amorphous model of Poly(vinyl alcohol).

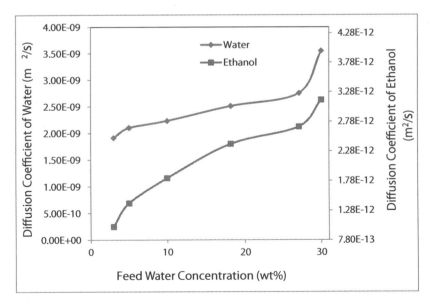

FIGURE 13.6
Diffusion coefficient of water for different feed compositions at 303 K.

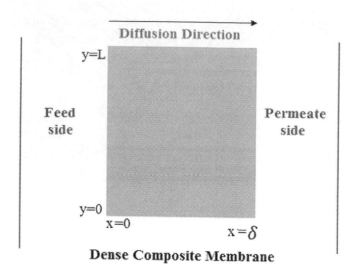

FIGURE 13.7
Schematic of computational domain of the PV process for CFD study.

of the PV process for hydrodynamic study is shown in Figure 13.7. The major assumptions used to develop the model were as follows:

- Thermodynamic equilibrium was considered at feed-membrane interface.
- Mass transfer resistance by the membrane support layer was assumed to be negligible.
- Steady state and isothermal mass transport occur through the membrane.
- The concentration polarization effect was considered negligible.

The continuity equation in the membrane was written as follows:

$$D_{W,membrane}\left[\frac{\partial^2 C_{W,membrane}}{\partial x^2} + \frac{\partial^2 C_{W,membrane}}{\partial y^2}\right] = 0 \tag{13.11}$$

where, $C_{W,membrane}$ is the concentration of permeating molecules through the polymeric membrane and $D_{W,membrane}$ denotes the diffusion coefficient of water.

The boundary conditions for Equation 13.6 across the membrane were:

$C_i(x = 0) = C_i^*$ (Thermodynamic equilibrium)
$C_i(x = \delta) = 0$ (Dry membrane permeate side)

At $y = 0$ and $y = L$, $\dfrac{\partial C_{membrane}}{\partial y} = 0$ (insulation boundary)

Figure 13.8 depicts the concentration distribution of water within the membrane for 10 wt% water concentration in water-ethanol system. At the feed-membrane interface, the water concentration is found to be 55,000 mol/m³, whereas at the permeate-membrane interface the value reduced to 498 mol/m³.

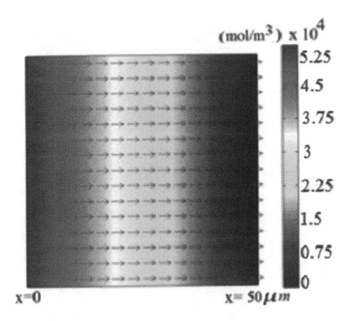

FIGURE 13.8
Concentration distribution of water within the membrane for 10 wt% water concentration in ethanol-water system.

13.2.3 Forward Osmosis

During the last decade, forward osmosis (FO) has made a resurgence in food processing applications, especially in concentrating liquid food products and natural colorants at ambient conditions by retaining sensory and inherent nutritional value (Cath et al., 2006). Due to its intrinsic features of low pressure and ambient temperature requirements, FO has widened its application in the field of desalination, power generation and treatment of industrial wastewater. Unlike pressure-driven membrane processes, FO utilizes the natural osmotic pressure gradient as a driving force for the transport of solvent water molecules. Several researchers have tried FO on a laboratory scale to concentrate beverages and liquid food (Anna et al., 2012) using cellulose acetate, cellulose triacetate and aromatic polyamide membranes. In order to scale-up and commercialize the FO process, reliable mathematical models are needed to predict the effect of certain key operating parameters. In this regard, only a few studies have been reported on modeling of the transport phenomena in the FO process. MD and CFD have emerged as dependable tools in predicting diffusion properties and transport phenomena in specific polymer-permeant systems. A few theoretical studies visualize the mass transfer phenomena across the membrane to optimize FO parameters and enable scale-up. Gai et al. (2014) investigated the performance of nanoporous graphene membranes using molecular modeling to demonstrate its higher efficiency compared with cellulose triacetate. Figure 13.9 illustrates the molecular model for FO used in seawater desalination by fluorinated porous graphene (GF) and nitriding porous graphene membranes (GN). In our earlier study, a couple of MD and CFD models were developed for the concentration of fructose sugar. COMPASS force field was used for molecular dynamics study (Figure 13.10). Details of this force field are mentioned in Section 13.1. Initially, the structure of polyamide membrane was constructed as per the procedure mentioned by Ding et al. (2014). Water (H_2O), fructose, sodium (Na^+) and chloride (Cl^-) ions

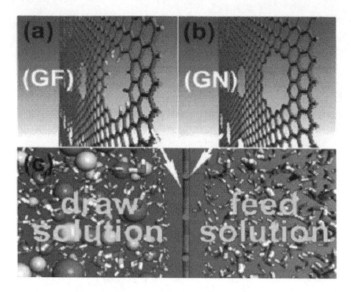

FIGURE 13.9
(a) Fluorinated porous graphene membrane, (b) nitriding porous graphene membrane and (c) molecular model of FO system with Fluorinated porous graphene membrane by Gai et al. (2014). (Reproduced with permission from Royal Society of Chemistry.)

FIGURE 13.10
Molecular model of forward osmosis process for concentration of fructose sugar solution.

were built first at the atomic level and minimized using COMPASS force field. The box was equally partitioned off into feed and draw chambers by a polyamide membrane. Fructose solution of 0.05 M and aqueous NaCl solution of 0.5–2.5 M were added to the feed and draw channels, respectively. The system was equilibrated under 500 ps NPT MD simulation, followed by a 500 ps NVT dynamics study to investigate the diffusion properties of water and Na^+ within the FO process. The mobility of water and Na^+ within the polyamide membrane were investigated using the mean square displacement (MSD), calculated using Equation 13.9. The diffusion coefficient (D) of molecules was calculated from the slope of MSD graphs using Einstein relationship (Equation 13.1). The schematic of the computational domain of the FO process is shown in Figure 13.11. The computational domain

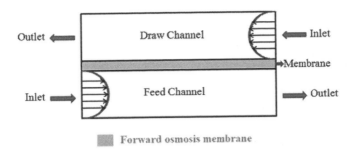

FIGURE 13.11
Schematic of computational domain of forward osmosis.

consists of feed and draw channels separated by a polyamide membrane. Assuming the fluid to be incompressible, stationary and laminar within feed and draw channels, the steady state continuity and momentum balance equations were written as:

$$\frac{\partial u}{\partial x} + \frac{\partial v}{\partial y} = 0 \tag{13.12}$$

$$u\frac{\partial u}{\partial x} + v\frac{\partial u}{\partial y} = -\frac{1}{\rho}\frac{\partial P}{\partial x} + v\left(\frac{\partial^2 u}{\partial x^2} + \frac{\partial^2 u}{\partial y^2}\right) \tag{13.13}$$

$$u\frac{\partial v}{\partial x} + v\frac{\partial v}{\partial y} = -\frac{1}{\rho}\frac{\partial P}{\partial x} + v\left(\frac{\partial^2 v}{\partial x^2} + \frac{\partial^2 v}{\partial y^2}\right) \tag{13.14}$$

where, $\vec{v}$ is the velocity of the fluid, P represents pressure and μ, the dynamic viscosity of the fluid. In both feed and draw channels, laminar inlet velocities were set at the outlet end of the channels and zero change in velocities was assumed.

The continuity equation in the membrane can be written as follows:

$$D_{W,membrane}\left[\frac{\partial^2 C_{W,membrane}}{\partial x^2} + \frac{\partial^2 C_{W,membrane}}{\partial y^2}\right] = 0 \tag{13.15}$$

where, $C_{W,membrane}$ and $D_{W,membrane}$ represent the concentration of water within the membrane and diffusion coefficient of water through the membrane respectively, calculated from molecular simulation.

The boundary conditions for the continuity equation in the membrane are:

$$C_i(x = 0) = C_f \text{ (Inlet feed concentration)} \tag{13.16}$$

$$C_i(x = \delta) = C_p \text{ (Inlet draw concentration)} \tag{13.17}$$

$$At \ y = 0, y = L \text{ and } \frac{\partial C_{membrane}}{\partial y} = 0 \text{ (insulation boundary)} \tag{13.18}$$

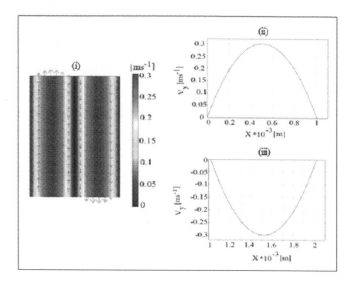

FIGURE 13.12

Surface Velocity Distribution inside the Forward Osmosis Membrane Module for Initial Fructose Concentration of 0.05 M and NaCl Concentration of 1.5 M (Madhumala et al., 2017). (Reproduced with permission of Wiley.)

Figure 13.12 represents the surface velocity profile inside the channels and membrane at constant feed and draw flow rates of 0.25 ms^{-1} on either side of the membrane. Additionally, the arrows clearly establish the direction of flow inside feed and draw channels, respectively. As expected, the velocity in both feed and draw channels were found to be higher at the middle of each channel, marked by the appearance and expansion of red hue, with a characteristic velocity of 0.25 ms^{-1}, whereas the velocity was lower at the wall of each channel due to no slip boundary condition. The y-axis velocity profile inside the feed and draw channels was found to be parabolic, which is similar to the profile attained by fully developed laminar flow in a channel without wall suction or injection. During the CFD study, inlet feed and draw concentrations were assumed as 0.05 and 1.5 M, respectively.

13.2.4 Membrane Bioreactor

In recent years, membrane bioreactor (MBR) technology has become a popular means for treating residential wastewater, to some degree in view of progressively stringent release necessities. To improve process efficiency, the layer units in these frameworks are quite often coupled with high impact reactors. To scale-up MBR in industry, CFD studies are widely used for evaluating the hydrodynamic scenario within the process. Nassehi et al. (1998) coupled the Navier Stokes and Darcy equations together to illustrate the flow field in cross-flow membrane filtration. Wang et al. (2010) reported CFD approach to simulate submerged and airlift MBRs. In our earlier study, we developed a CFD model for a submerged membrane bioreactor (SMBR) for treating dairy industrial effluent using PAN and PVDF hollow fiber membranes (Praneeth et al., 2014). Hollow fiber membranes were submerged within the reactor and a vacuum pump was connected to draw the final permeate. The computational domain used for hydrodynamic simulation is depicted in Figure 13.13, whereby a single fiber surrounded by a fluid layer is used for the geometric domain. At the inlet, perpendicular to the flow direction within the HF, uniform pressure condition is applied. Due to low permeability, the slip velocity at the interface can be neglected.

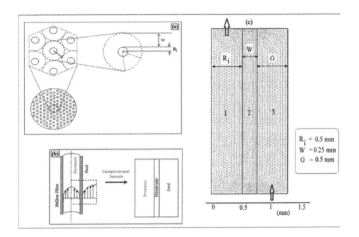

FIGURE 13.13
Computational domain of SMBR for CFD study for the treatment of dairy industrial effluent (Praneeth et al., 2014). (Reproduced with permission of Elsevier.)

The non-uniform mesh conditions are applied at the internal and external channels. The model was developed by assuming that each HF in the module has the same efficiency of filtration. Assuming the fluid to be incompressible, stationary and laminar in the fluid channels of sub-domains 1 and 3, the steady state continuity and momentum balance equations were written as:

$$\nabla . \vec{v} = 0 \tag{13.19}$$

$$\rho(\vec{v}.\nabla)\vec{v} = -\nabla P + \mu \nabla^2 \vec{v} \tag{13.20}$$

where, $\vec{v}$ = velocity of the fluid (m s^{-1}), P = pressure (Pa) and μ = dynamic viscosity (Pa s).

The Darcy Brinkman model was used to solve the flow profile in the porous medium (sub-domain 2). The equation of the Darcy Brinkman model is given below:

$$\nabla \vec{P} = -\frac{\mu}{k}\vec{v} + \mu_{\text{eff}}\nabla^2 \vec{v} \tag{13.21}$$

$$\nabla . \vec{v} = 0 \tag{13.22}$$

where μ_{eff} is effective viscosity (Pa.s) defined as $\mu/\in$, in which $\in$ refers to porosity of the medium in sub-domain 2 and (k) represents intrinsic permeability (m^2) of the HF membrane module. Figure 13.14 shows the surface velocity profiles within the HF membranes at the time of experimental runs. Figure 13.14 (a, c) and (d, f) depict the velocity profiles within the internal and external channels of a unit PAN and PVDF HF membrane for four different sections, respectively. The arrows in Figures 13.14 (b, e) clearly establish the direction of flow inside PAN and PVDF membranes for shell-side filtration. The maximum velocity is found to be at the lumen side in both membranes. The velocity inside PAN is higher than PVDF due to the hydrophilic nature of the PAN polymer. The z-velocity profiles at internal and external channels in the unit cell for both the membranes in all four sections were found to be parabolic.

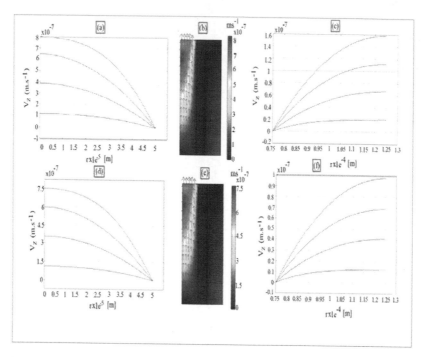

FIGURE 13.14

Velocity profiles in the internal channel and external channel of the PAN (a, c) and PVDF (d, f) HF membranes (arrows are representing the direction of flow of dairy effluent from shell side to tube side for PAN (b) and PVDF (e)) at TMP 0.8 bar (Praneeth et al., 2014) (Reproduced with permission of Elsevier.)

13.2.5 Other Membrane Processes

Gas separation, proton exchange membrane fuel cell, membrane distillation, electrodialysis, etc., are also emerging as viable industrial processes. Gas separation is widely used in natural gas sweetening (Lin et al., 2006), oxygen separation (Dyer et al., 2000), syngas production (Dyer et al., 2000), CO_2/N_2 gas separation (Zhao et al., 2008), etc. Shan et al. (2012) developed a molecular model for analyzing separation performances of porous graphene membranes for CO_2/N_2 separation. Zhou et al. (2006) performed MD simulation to study the diffusion of H_2, O_2, N_2, CO_2, CH_4, and n-C_4H_{10} gases in pure and silica filled poly(1-trimethylsilyl-1-propyne) membrane. The mean square displacement of hydrogen within pure and silica filled PTMSP at 300.0 K is shown in Figure 13.15. Displacement of the gas in pure and silica filled PTMSP are different after 200 ps time interval, after which the gas displacement in the filled PTMSP increased rapidly, indicating higher diffusivity. A CFD model based on Navier-Stokes, continuity, mass balance and Nernst–Planck equations to understand the flow and concentration profiles within an electrodialysis (ED) cell, is developed by Tadimeti et al. (2016). Different types of membrane flow channels were selected for this study. Figure 13.16 shows fluid flow profile inside ED channel with (a) netted spacer geometry, (b) right angle triangular corrugation and (c) rectangular corrugation. It was found that the introduction of corrugation over membrane surfaces inside flow channel enhanced inter-layer mixing and the boundary layer thickness, which improved ED performance. Membrane Distillation (MD), one of the second-generation membrane processes, has become a promising technology not only for seawater or brackish water

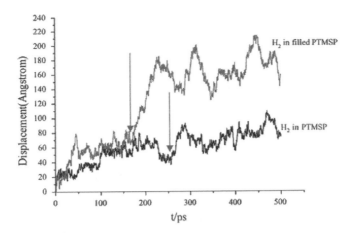

FIGURE 13.15
Displacement of hydrogen within pure and filled PTMSP at 300.0 K reported by Zhou et al. (2006). (Reproduced with permission of Elsevier.)

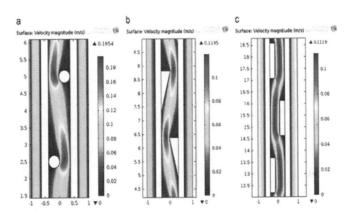

FIGURE 13.16
(a) Fluid flow profile inside ED channel with (a) netted spacer geometry, (b) right angle triangular corrugation and (c) rectangular corrugation analyzed by Tadimeti et al. (2016). (Reproduced with permission of Elsevier.)

desalination but also for absolute purification of products, concentrate/RO reject treatment and recovery of volatile components from aqueous solutions. It involves the transport of vapor molecules through the pores of a hydrophobic membrane. CFD is a useful tool to understand the mass and heat transfer profile within a process to help scale it up. Hayer et al. (2014) developed a CFD based model to predict the heat and mass transfer phenomena within a direct contact membrane distillation set-up for seawater desalination. During the process, the feed and permeate side temperatures were kept at 50 and 17°C, respectively. Temperature distribution within the process is shown in Figure 13.17. This study reveals that permeate gets heated along the fibers due to heat transfer occurring between the feed and permeate across the membrane. No significant change in feed temperature was found as the feed flow rate was much higher than the permeate flow rate and the transferred heat was not sufficient to reduce its temperature.

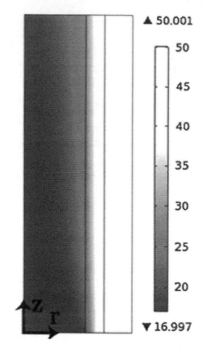

FIGURE 13.17
Temperature distribution during DCMD process by Hayer et al., 2015. (Reproduced with permission of Elsevier.)

13.3 Conclusions

This study covers the current research scenario on computing the complex flow phenomena, membrane structure and interaction of the target components with the membrane material for preferential separation at the molecular level through the application of CFD and molecular modeling. To design efficient membrane systems for challenging industrial separations, the optimization of spacer design for spiral wound membrane module and understanding of transport phenomenon within pervaporation, forward osmosis, membrane bioreactor and other membrane processes could be achieved by CFD. Evaluation of polymer-feed interactions, diffusion coefficient of feed components, glass transition temperature and fractional free volume within the membrane matrix has been carried out by molecular dynamics simulation. Besides the scope of applicability as per the process performance of different membranes, an insight into the molecular structure of the membrane and knowledge about the flow field inside the module would help the researcher to narrow down their research to more specific and target-oriented objectives. Based on such results, one can utilize the same membrane for several applications by observing the hydrodynamic nature and changes brought about at the molecular level to improve membrane-solute interactions for a better performance.

References

Anna, V.S., L.D.F. Marczak, and I.C. Tessaro. 2012. Membrane concentration of liquid foods by forward osmosis: Process and quality view. *Journal of Food Engineering* 111(3): 483–489.

Butterworth, T.A., D.I.C. Wang, and A.J. Sinskey. 1970. Application of ultrafiltration for enzyme. Retention during continuous enzymatic reaction. *Biotechnology and Bioengineering* 12(4): 615–631.

Buonomenna, M.G. and G. Golemme. 2012. Advanced materials for membrane preparation. *Bentham Science Publishers*. doi: 10.2174/97816080530871120101.

Cath, T.Y., A.E. Childress, and M. Elimelech. 2006. Forward osmosis: Principles, applications, and recent developments. *Journal of Membrane Science* 281(1–2): 70–87.

Daufin, G., J.P. Escudier, H. Carrere, S. Berot, L. Fillaudeau, and M. Decloux. 2001. Recent and emerging applications of membrane processes in the food and dairy industry. *Food and Bioproducts Processing* 79(2): 89–102.

Deegan, A.M., B. Shaik, K. Nolan, K. Urell, M. Oelgemöller, J. Tobin, and A. Morrissey. 2011. Treatment options for wastewater effluents. *International Journal of Environmental Science and Technology* 8(3): 649–666.

Ding, L., O.A. Akoum, A. Abraham, and M.Y. Jaffrin. 2002. Milk protein concentration by ultrafiltration with rotating disk modules. *Desalination* 144(1–3): 307–311.

Ding, M., A. Szymczyk, F. Goujon, A. Soldera, and A. Ghoufi. 2014. Structure and dynamics of water confined in a polyamide reverse-osmosis membrane: A molecular-simulation study. *Journal of Membrane Science* 458: 236–244.

Dolar, D., K.K. Kosutic, and B. Vucic. 2011. RO/NF treatment of wastewater from fertilizer factory—Removal of fluoride and phosphate. *Desalination* 265: 237–241.

Dyer, P.N., R.E. Richards, S.L. Russek, and D.M. Taylor. 2000. Ion transport membrane technology for oxygen separation and syngas production. *Solid State Ionics* 134(1–2): 21–33.

Gai, J.G., X.L. Gong, W.W. Wang, X. Zhang, and W.L. Kang. 2014. An ultrafast water transport forward osmosis membrane: Porous graphene. *Journal of Materials Chemistry* A2: 4023–4028.

Girard, B. and L.R. Fukumoto. 1999. Apple juice clarification using microfiltration and ultrafiltration polymeric membranes. *LWT—Food Science and Technology* 32(5): 290–298.

Hayer, H., O. Bakhtiari, and T. Mohammadi. 2015. Simulation of momentum, heat and mass transfer in direct contact membrane distillation: A computational fluid dynamics approach. *Journal of Industrial and Engineering Chemistry* 21: 1379–1382.

Kosutic, K., D. Dolar, D. Asperger, and B. Kunst. 2007. Removal of antibiotics from a model wastewater by RO/NF membranes. *Separation and Purification Technology* 53(3): 244–249.

Lee, K.P., T.C. Arnot, and D. Mattia. 2011. A review of reverse osmosis membrane materials for desalination—Development to date and future potential. *Journal of Membrane Science* 370(1–2): 1–22.

Liu, S.X., M. Peng, and L.M. Vane. 2005. CFD simulation of effect of baffle on mass transfer in a slit-type pervaporation module. *Journal of Membrane Science* 265: 124–136.

Lin, H., E. Van Wagner, R. Raharjo, B.D. Freeman, and I. Roman. 2005. High-performance polymer membranes for natural-gas sweetening. *Advanced Materials* 18(1): 39–44.

Madhumala, M., S. Moulik, T. Sankarshana, and S. Sridhar. 2017. Forward-osmosis-aided concentration of fructose sugar through hydrophilized polyamide membrane: Molecular modeling and economic estimation. *Journal of Applied Polymer Science* 134(13), 44649(1–12).

Mafi, A., A. Raisi, and A. Aroujalian. 2013. Computational fluid dynamics modeling of mass transfer for aroma compounds recovery from aqueous solutions by hydrophobic pervaporation. *Journal of Food Engineering* 119: 46–55.

Marcucci, M., G. Nosenzo, G. Capannelli, I. Ciabattia, D. Corrieri and G. Ciardelli. 2001. Treatment and reuse of textile effluents based on new ultrafiltration and other membrane technologies. *Desalination* 138(1–3): 75–82.

Moraveji, M.K., A. Raisi, S.M. Hosseini, E. Esmaeeli, and G. Pazuki. 2013. CFD modeling of hydrophobic pervaporation process: Ethanol/water separation. *Desalination of Water Treatment* 51: 3445–3453.

Moulik, S., S. Nazia, B. Vani, and S. Sridhar. 2016. Pervaporation separation of acetic acid/water mixtures through sodium alginate/polyaniline polyion complex membrane. *Separation and Purification Technology* 170: 30–39.

Moulik, S., K.P. Kumar, S. Bohra, and S. Sridhar. 2015. Pervaporation performance of ppo membranes in dehydration of highly hazardous MMH and UDMH liquid propellants. *Journal of Hazardous Materials* 288: 69–79.

Moulik, S., P. Vadthya, Y.R. Kalipatnapu, S. Chenna, and S. Sridhar. 2015. Production of fructose sugar from aqueous solutions: Nanofiltration performance and hydrodynamic analysis. *Journal of Cleaner Production* 92(1): 44–53.

Nassehi, V. 1998. Modelling of combined Navier–Stokes and Darcy flows in crossflow membrane filtration. *Chemical Engineering Science* 53: 1253–1265.

Pervov, A.G., E.V. Dudkin, O.A. Sidorenko, V.V. Antipov, S.A. Khakhanov, and R.I. Makarov. 2000. RO and NF membrane systems for drinking water production and their maintenance techniques. *Desalination* 132: 315–321.

Praneeth, K., S. Moulik, V. Pavani, K.B. Bhargava, J. Tardio, and S. Sridhar. 2014. Performance assessment and hydrodynamic analysis of a submerged membrane bioreactor for treating dairy industrial effluent. *Journal of Hazardous Materials* 274: 300–313.

Prasad, N.S., S. Moulik, S. Bohra, K.Y. Rani, and S. Sridhar. 2016. Solvent resistant chitosan/poly(ether-block-amide) composite membranes for pervaporation of n-methyl-2-pyrrolidone/water mixtures. *Carbohydrate Polymers* 136: 1170–1181.

Radu, A.I., M.S.H. van Steen, J.S. Vrouwenvelder, and M.C.M.V. Loosdrecht. 2014. Spacer geometry and particle deposition in spiral wound membrane feed channels. *Water Research* 64: 160–176.

Sunitha, K., K.Y. Rani, S. Moulik, S. Satyanarayana, and S. Sridhar. 2013. Separation of NMP/water mixtures by nanocomposite PEBA membrane: Part I. Membrane synthesis, characterization and pervaporation performance. *Desalination* 330: 1–8.

Tadimeti, J.G.D., V. Kurian, A. Chandra, and A. Chattopadhyay. Corrugated membrane surfaces for effective ion transport in electrodialysis. *Journal of Membrane Science* 499: 418–428.

Wang, Y., M. Brannock, S. Cox, and G. Leslie. 2010. CFD simulations of membrane filtration zone in a submerged hollow fibre membrane bioreactor using a porous media approach. *Journal of Membrane Science* 363: 57–66.

Wiley, D.E. and D.F. Fletcher. 2003. Techniques for computational fluid dynamics modelling of flow in membrane channels. *Journal of Membrane Science* 211(1): 127–137.

Zhao, L., E. Riensche, R. Menzer, L. Blum, and D. Stolten. 2008. A parametric study of CO2/N2 gas separation membrane processes for post-combustion capture. *Journal of Membrane Science* 325(1): 284–294.

Zhou, J.H., R.X. Zhu, J.M. Zhou, and M.B. Chen. 2006. Molecular dynamics simulation of diffusion of gases in pure and silica-filled poly(1-trimethylsilyl-1-propyne) [PTMSP]. *Polymer* 47(14): 5206–5212.

Section V

Energy

14

Carbon-Polymer Nanocomposite Membranes as Electrolytes for Direct Methanol Fuel Cells

Gutru Rambabu and Santoshkumar D. Bhat

CONTENTS

14.1 Introduction to Direct Methanol Fuel Cells

In direct methanol fuel cells (DMFC), the membrane electrode assembly (MEA) is the key component comprising of an anode (Pt-Ru/C), cathode (Pt/C) and an ion conducting polymer sandwiched between anode and cathode (Kumarudin et al., 2009). In a typical DMFC system, aqueous methanol is fed to the anode side of MEA, where the electrochemical oxidation of methanol takes place to generate protons, electrons and carbon dioxide. The protons generated at the anode will pass through the solid polymer electrolyte and reach the cathode, where the electrons arrive through the external circuit and combine with both protons and oxygen, which is fed to the cathode. The following are the overall reactions at the anode and cathode in DMFC (Aricò et al., 2001):

Anode: $CH_3OH + H_2O \rightarrow CO_2 + 6H^+ + 6e^-$

Cathode: $\dfrac{3}{2}O_2 + 6H^+ + 6e^- \rightarrow 3H_2O$

Overall reaction: $CH_3OH + \dfrac{3}{2}O_2 \rightarrow 2H_2O + CO_2$ (E_{cell}: 1.18 V).

14.1.1 Challenges in DMFC

The critical parameter impacting DMFC performance significantly is the methanol crossover from anode to cathode through the polymer electrolyte, leading to secondary reactions and mixed potential at the cathode, thereby reducing the overall cell performance. Methanol crossover is caused by concentration gradient, depending on operational current and molecular transport through electro-osmotic drag (Jiang & Chu, 2002). Methanol crossover during the DMFC operation causes (1) electrode de-polarization, (2) mixed potential, resulting in the open-circuit voltage (OCV) of the DMFC to fall below 0.8 V, (3) consumption of O_2, (4) cathode catalyst poisoning by CO (an intermediate of methanol oxidation), and (5) water accumulation on the cathode (water being produced by methanol oxidation), which limits O_2 access to cathode catalyst sites. Moreover, the overall fuel utilization efficiency is lowered when there is excessive methanol crossover (Zhao et al., 2009).

14.1.2 Membrane Electrolytes in DMFC

The state-of-the-art polymer electrolyte membrane for DMFCs is Nafion®, which has shown promising results due to its remarkable mechanical and chemical stability along with a high proton conductivity. These characteristics are attributed to the presence of a strong hydrophobic fluorinated backbone and the presence of hydrophilic sulfonic acid groups (Samms et al., 1996). The amphiphilic features between hydrophobic and hydrophilic phases enable the sulfonic acid groups to form ionic clusters with inverse micelle structures of a size of approximately 4 nm, which are thus much larger than the diameter of a methanol molecule (0.38 nm). Hence, there may be severe methanol crossover when Nafion® is used as a polymer electrolyte membrane (PEM) in DMFC. On the other hand, the cost of Nafion® is also too high for commercialization (Zhang et al., 2014). Thus, it is important to design a PEM with high proton conductivity and low methanol crossover.

In the past few years, researchers have explored various polymer alternatives to Nafion®, like sulfonated poly(ether ether ketone) (SPEEK) due to its comparable conductivity, superior thermo-chemical properties, lower methanol crossover and greater cost-effectiveness (Xue & Yin, 2006). The differences between the properties of SPEEK and those of Nafion® can be explained by their micro-structure, wherein Nafion® forms well-separated hydrophobic and hydrophilic regions and well-connected hydrophilic domains enabling facile proton and water transport. In the case of SPEEK, hydrophobic and hydrophilic separation is less pronounced due its rigid backbone (Kreuer, 2001). The narrow channels of SPEEK, however, possess the advantage of restricting methanol crossover. Figure 14.1 shows the schematic representation of micro-structures of both Nafion® and SPEEK (Kreuer, 2001).

The properties of SPEEK membranes strongly depend on the sulfonation level. Sulfonation can be performed by the post-treatment of PEEK through an electrophilic

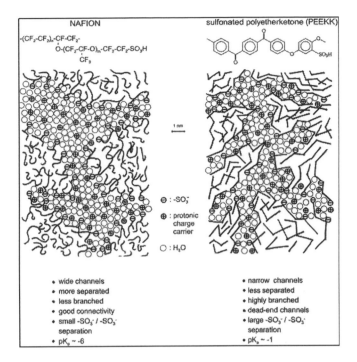

FIGURE 14.1
Micro-structural differences between Nafion® and SPEEK. (Reprinted from Joo, D.J., W.S. Shin, J.H. Choi, S.J. Choi, M.C. Kim, M.H. Han, T.W. Ha, and Y.H. Kim, *Dyes and Pigments*, 73: 59–64, 2007. With permission.)

substitution reaction (Li et al., 2005) or by using sulfonated monomers during polymerization (Wang et al., 2002). However, the mechanical properties of SPEEK membranes could deteriorate progressively at a higher DS. Indeed, according to a DS increase, the long-term stability of highly sulfonated PEEK membranes is considerably affected owing to hydroxyl radical initiated degradation. At a high DS, these membranes are prone to swell in an aqueous methanol solution. To overcome these issues, SPEEK is used in the form of a blend/composite in DMFC. The incorporation of inorganic oxides (SiO_2, ZrO_2, and TiO_2, etc.), phosphates and heteropolyacids in SPEEK were the other methods chosen to reduce methanol crossover and increase ionic conductivity in DMFCs (Nunes et al., 2002; Ruffmann et al., 2003).

Another class of polymers based on sulfonated poly(phthalalizone ether ketone) (SPPEK) is also explored as an alternative membrane due to its excellent chemical, mechanical and thermal stabilities, like SPEEK, and its proton conductivity, which can also be tuned by controlling the degree of sulfonation (Gao et al., 2003). Additionally, the nitrogen moiety present in SPPEK also helps in developing acid-base interactions, which lead to facile proton transport. However, besides these benefits, high swelling and methanol crossover in DMFCs of SPPEK at a high degree of sulfonation is one of the major drawbacks. Hence, literature reports suggest the use of stabilized composite membranes using inorganic materials, such as silica and zirconium phosphate as additives in the SPPEK matrix (Su et al., 2006).

Recently, carbon nanomaterials like MWCNTs, fullerene and graphite nanomaterials are identified as potential nano-additives in alternate polymers to form nanocomposite membrane electrolytes and are evaluated in terms of their physico-chemical properties along with methanol crossover. This chapter focuses on polymer electrolyte membranes

TABLE 14.1

Comparison of Present Results with Literature Reports Based on CNTs, Fullerene and CNF Based Composite Membranes for DMFC

Base matrix	Additive	Proton Conductivity (mS cm⁻¹)	Methanol Permeability (× 10⁻⁷ cm² s⁻¹)	Reference
Nafion®	Im-CNT	150	10.0	Asgari et al. (2013)
SPEI	S-MWCNT	3.98	11.7	Heo et al. (2012)
PVA	s-MWNTs	75	0.03	Yun et al. (2011)
SPESEKK	sCNTs	4.3	0.96	Zhou et al. (2011)
SPEEK	fCNTs	43.1	1.68	Gahlot & Kulshrestha (2015)
Nafion®	S-fullerene	97	8.5	Rambabu et al. (2016)
SPEEK	SCNF	128	5.02	Liu et al. (2017)
SPEEK	PSSA-CNTs	86.6	3.87	Present work
SPEEK	S-fullerene	96.3	3.51	
SPPEK	SGNF	101	3.25	

specifically consisting of the above carbon nanomaterials, which enhance DMFC performance in terms of electrochemical selectivity. Table 14.1 shows recent literature reports on carbon nanomaterial composite membranes and their comparison in terms of proton conductivity and methanol permeability. The performance of these composites are discussed in detail in the following sections.

14.2 Functionalized Carbon Nanoadditives in PEMs

14.2.1 Carbon Nanotube Based Composite Membranes

Multi-walled carbon nanotubes (MWCNTs) are considered to be a promising additive material due to their unique structural and physical properties, such as their significant strength due to the strong sp^2 bond, along with a large surface area and low density (Zhang & Silva, 2011; Kannan, 2008). Recently, several research groups have explored the applications of CNTs in PEMs. CNTs were chosen as additives mainly to address the methanol permeability and mechanical strength issues of PEMs in DMFCs. On the other hand, uniform dispersion of pristine CNTs were difficult as they are held together by Van der Waals forces (Zhang et al., 2011), and the incorporation of pristine CNTs in membranes could cause a decrease in proton conductivity by blocking the ionic channels. To tackle these issues, surface functionalization of CNTs was done with various functional groups (Zhang et al., 2011; Zhou et al., 2011; Joo et al., 2008).

For instance, Zhou et. al (2011) prepared a composite membrane of sulfonated poly(ether sulfone ether ketone ketone) with sulfonated CNTs with improved tensile strength due to the interfacial hydrogen bonding. Sulfonation enables CNTs to exhibit enhanced proton conductivity along with improved dispersion in the polymer matrix, whereas methanol permeability of these composite membranes becomes linearly reduced (Zhou et al., 2011). Gahlot and Kulshrestha (2015) prepared composite membranes of SPEEK with electrically aligned CNTs, wherein carboxylic and sulfonic acid functionalized CNTs were introduced in the SPEEK matrix, and the resultant mixture was subjected to an electric field; a dipole moment was introduced due to the variation of dielectric constant of SPEEK and CNTs as a result of which CNTs were aligned in the direction of the applied electric field. Compared to randomly aligned CNTs, electrically aligned CNTs have shown better properties in terms of proton conductivity and methanol permeability in DMFCs (Gahlot & Kulshrestha, 2015). Further, Joo et al. (2008), used CNTs as additives in a sulfonated polyether sulfone matrix (SPES). In this study, sulfonic acid modified CNTs and Pt-Ru nanoparticle modified CNTs were incorporated in the SPES matrix. These modified CNTs were distributed uniformly in SPES and showed improved tensile strength and proton conductivity with mitigated methanol permeability (Joo et al., 2008).

14.2.2 Fullerene Based Composite Membranes

Buckminsterfullerene (C_{60}) can also be a potential additive for composite membrane electrolytes in fuel cells due to its high electron affinity, volumetric functional group density and radical scavenging property (Tasaki et al., 2007). Wang et al. (2007) investigated composite membranes containing functionalized fullerene as an additive in the Nafion® matrix and studied their effect on ionic conductivity. However, the functionalization routes followed were complex, and real fuel cell polarization studies were not undertaken for these membranes (Wang et al. 2007). Saga et al. (2008) prepared a nanocomposite membrane of sulfonated polystyrene with fullerene as an additive and studied its properties in terms of mechanical and chemical stability, methanol permeability and proton conductivity. In this study, it is proven that fullerene acts as an effective barrier for methanol transport and improves chemical stability due to its radical scavenging property (the alkene group (-C=C-) present in fullerene act as radical scavenger), wherein hydroxyl (•OH) and hydroperoxyl (•OOH) radicals produced during the DMFC operation, causing membrane degradation, are trapped by fullerene. On the other hand, the mechanical stability and proton conductivity of the membrane could not be improved due to poor compatibility and lack of ion conducting groups in fullerene (Saga et al., 2008).

14.2.3 Carbon Nanofiber Based Membranes

In addition to the above-mentioned materials, graphite/carbon nanofibers (GNF/CNF) are also studied as additives in PEMs. Liu et al. (2017) explored the synthesis and sulfonation of carbon nanofibers (CNFs) and their use as additives in SPEEK for DMFC application. The composite membranes were prepared by varying the content of SCNF from 0.5–2 wt% in relation to SPEEK by solution cast method. These membranes reveal higher proton conductivity compared to a pristine membrane, due to their morphological difference. Composite membranes show good phase separation between hydrophobic backbone and self-aggregated hydrophilic regions, resulting in facile proton transport. In addition to this, the composite membrane show reduced methanol permeability, as low as 5.02×10^{-7} cm^2/s, leading to improved electrochemical selectivity.

With this background, we have reported composite membranes of SPEEK using polystyrene sulfonic acid functionalized CNTs and sulfonated fullerene apart from composite membranes of SPPEK using sulfonated GNF to address the challenges associated with polymer electrolytes in DMFCs. This chapter explains the preparation of these composite membranes and their physico-chemical properties along with DMFC polarization studies, in detail.

14.3 Different Functionalization Routes for Additives

Poly(sodium-p-sytrene-sulfonate) (NaPSS) is an amphiphilic polymer which contains both hydrophilic and hydrophobic chains that can modify the surface of CNTs. Pristine CNTs are held together by Van der Waals forces, leading to their poor dispersion in polymer matrices. Because CNTs do not have hydrogen atoms or any other groups on their surface, they are chemically inactive and PSSA grafting to CNTs becomes quite challenging. Hence, prior to the addition of PSSA, CNTs were oxidized (–COOH-CNTs) to induce reactive groups (–COOH) and therefore, hydrophilicity. The modification of CNT with PSSA leads to the formation of π-π interaction of hydrophobic styrene moiety part of with CNT that imparts sufficient dispersion ability. The hydrophilic sulfonic acid groups of PSSA facilitate additional hydrophilicity and electrostatic repulsions to balance the Van der Waals interactions of the CNTs (Zhang et al., 2011).

Oxidation of fullerene was carried out using 6 M H_2SO_4 and HNO_3 (1:1 ratio) at 130°C for 8 h, as reported in the literature of Kanbur and Küçükyavuz (2012), to activate the surface for further functionalization. Oxidized fullerene was sulfonated using 4-benzene diazonium sulfonic acid precursor formed by diazotization reaction route, as reported by Ji et al. (2011). In brief, the typical sulfonation process involves two steps: (i) sulfonating precursor (4-Benzene diazoniumsulfonic acid) prepared from sulfanilic acid via diazotization and then (ii) treatment of fullerene with 4-benzene diazonium sulfonic acid in dilute ethanol and hypophosphorous acid at a reaction temperature of 0–5°C for 2 h to create sulfonated fullerene, which is designated as Sfu in the following sections.

GNF was functionalized with 4-benzene sulfonic acid following the similar procedure adopted for fullerene. The schematic representation of the functionalization procedure for all the three additives is given in Figure 14.2.

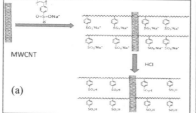
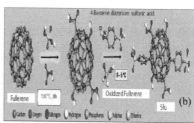

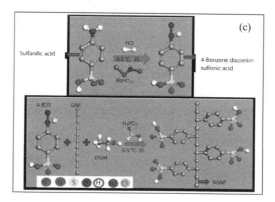

FIGURE 14.2
Schematic representation of functionalization for (a) CNTs, (b) Fullerene and (c) GNF.

14.3.1 Characterization of Functionalized Additives

Functionalization of additives is confirmed by FT-IR spectra, elemental analysis and thermo-gravimetric analysis. Figure 14.3 shows TEM morphology for PSSA-CNTs, Sfu and SGNF. TEM morphology of PSSA-CNTs clearly reveals that the tubular structure of CNTs was retained and not affected by the presence of PSSA. Pristine CNTs (Figure 14.3 (a)) hold together as chains due to Van der Waal interactions between the tubes. The CNTs' holding together as ropes is clearly seen in Figure 14.3 (a). In contrast, PSSA modified CNTs are spread over spaciously due to the electrostatic repulsion, which are imparted by the sulfonic acid groups present in PSSA (Zhang et al., 2008), as shown in Figure 14.3 (b).

TEM also implies detailed structural changes in fullerene. Figure 14.3 (c, d) shows TEM images of pristine fullerene, oxidized fullerene and Sfu. It is noteworthy from Figure 14.3 (c) that pristine fullerene exists as spherical particles with the combination of hexagons. For oxidized fullerene, defects are seen on the surface due to the acid treatment, with the basic bucky ball structure not being disturbed, as shown in Figure 14.3 (d). Sfu molecules are associated together forming intermolecular hydrogen bonding networks and electrostatic interactions leading to the formation of bulky particles. Similar results are observed by Kanbur and Küçükyavuz (2012) in the case of H_2SO_4 and HNO_3 treated fullerene. The bulky nature of Sfu is beneficial in controlling methanol permeability as methanol molecules presumably travel in a torturous path around these bulky particles, to diffuse through the membrane.

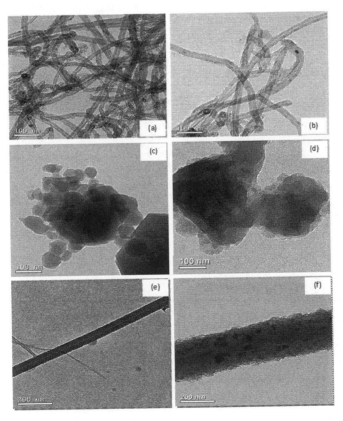

FIGURE 14.3
TEM morphology for (a) CNTs, (b) PSSA-CNTs, (c) Fullerene, (d) Sulfonated fullerene, (e) GNF and (f) SGNF.

TEM analysis of GNF and SGNF is shown in Figure 14.3 (e, f). It is noted that functionalized GNF (Figure 14.3 (f)) is partially unwrapped on the outer surface, forming a layer-like structure and creating defects on the surface due to the electrostatic repulsions introduced by the sulfonic acid groups (Zhang et al., 2011) whereas pristine GNF (Figure 14.3 (e)) exhibits smooth fibrous morphology. This phenomenon is also seen in earlier studies, wherein GNF forms wrinkled graphene sheets when it is functionalized with nitrogen and fluorine due to the electronegativity difference between C, N, and F (Peera et al., 2015). This partial opening of outer layers in GNF contributes to a better distribution of the additive in the polymer matrix to form a composite.

14.4 Fabrication of Composite Membranes

SPEEK/SPPEK was dissolved in dimethyl acetamide (DMAc) to prepare 5 wt% solution. Required weight percentages (0.25, 0.5 and 1 wt%) of the additives (PSSA-CNTs/SFu/SGNF) in relation to polymer were dispersed in DMAc and sonicated for 30 min. The additive dispersion was added to the polymer solution (5 wt% in DMAc) and sonicated for 1 h followed by continuous stirring for 24 h. The resultant solution was cast on a flat Plexiglass plate and dried under vacuum at 80°C for 12 h. The dried membrane was then peeled off the glass plate and washed thoroughly to remove residual solvent.

14.5 Effect of Functionalized Additives on Membrane Properties

14.5.1 Physico-Chemical Properties

Figure 14.4 displays the surface morphology of SPEEK-PSSA-CNT, SPEEK-Sfu and SPPEK-SGNF composite membranes. It is to be noted that functionalized additives are distributed homogeneously throughout the polymer matrix, which can be attributed to the interfacial interactions between sulfonic acid groups of additive and polymer. The composite membranes exhibit dense structures without any micropores, and the incorporation of these functionalized additives in the polymer also improves the surface roughness, whereas the pristine membranes show smooth morphology.

Tensile strength and elongation of the aforesaid membranes were carried out to understand the mechanical properties of the membranes, and the values are presented in Table 14.2. Tensile strength and elongation of the composite membranes increases in relation to the content of functionalized additives. It is noteworthy that the incorporation of these additives to the polymer matrices enhances the mechanical strength of the membrane, probably due to the interfacial structures and electrostatic interaction of sulfonic groups of additive with polymer matrix (Zhou et al., 2011). On the other hand, elongation for composite membranes decreases with the addition of these additives, which may be due to restricted polymer chain flexibility inducing the rigidity in the matrix.

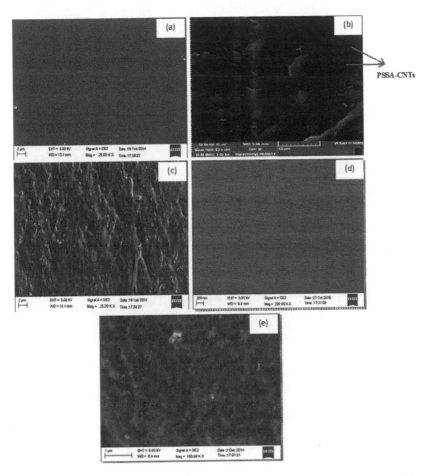

FIGURE 14.4
SEM morphology for (a) pristine SPEEK, (b) SPEEK-PSSA-CNTs, (c) SPEEK-Sfu,(d) pristine SPPEK and (e) SPPEK-SGNF.

Thermo-gravimetric analysis (TGA) for SPEEK-PSSA-CNT, SPEEK-Sfu and SPPEK-SGNF composite membranes was carried out to understand their thermal stability as shown in Figure 14.5 (a). It is to be noted that all of the membranes show similar profiles of TG curves; however the temperatures corresponding to the onset of thermal degradation and the slope of mass loss are different. All of the membranes show three stages of degradation, a first-stage weight loss at 100°C may be attributed to the removal of moisture and physically adsorbed water. The second-stage weight loss observed from 280–400°C is attributed to the splitting of sulfonic acid groups in the composite membranes. The third-stage weight loss observed at 460°C is attributed to the degradation of the polymer backbone (Zhong et al., 2006). In the case of SPEEK-PSSA-CNT, higher mass loss was observed in comparison to other composite membranes, which may be due to the degradation of PSSA grafted on to CNT (Du et al., 2008). However, all membranes are thermally stable up to 300°C, which is vital for PEM in DMFCs.

TABLE 14.2

Properties of Composite Membranes

Membrane	IEC (meq./g)	Water Uptake (%)	Tensile Strength (MPa)	Elongation at Break (%)	Proton Conductivity (mS/cm)		MeOH Permeability (10_{-7}cm2/s)	Electrochemical Selectivity (10_4 S s cm^{-3})
					30°C	60°C		
Pristine SPEEK	1.54	30.4	9.85	9.35	44	62	6.46	6.8
SPEEK-PSSA-CNTs (0.5 wt.%)	1.66	34.2	14.09	7.40	53.1	86.6	3.87	13.7
SPEEK-Sfu (0.5 wt.%)	1.77	39.5	12.80	6.57	55.3	96.3	3.51	15.7
Pristine SPPEK	1.45	23.5	10.70	8.20	28	51.3	5.71	4.9
SPPEK-SGN (0.5 wt.%)	1.74	32.6	14.10	5.70	68	101	3.25	20.9

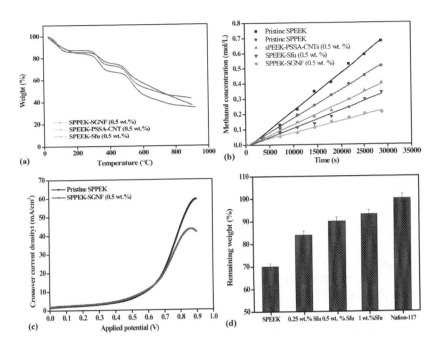

FIGURE 14.5
(a) TGA for SPEEK-PSSA-CNTs, SPEEK-Sfu and SPPEK-SGNF composite membranes, (b) Methanol concentration profile for the membranes, (c) LSV curves for pristine SPPEK and SPPEK-SGNF membranes and (d) Oxidative stability of SPEEK-Sfu composite membranes.

14.5.2 Ion Exchange Capacity and Water Uptake

Ion exchange capacity (IEC) of the membranes was measured by an acid-base titration method using the following relation (Jiang et al., 2012):

$$IEC = \frac{Volume\ of\ NaOH \times Normality\ of\ NaOH}{Dry\ weight\ of\ the\ sample}\ meq./g \qquad (14.1)$$

The IEC of pristine and composite membranes are given in Table 14.2. Inclusion of these functionalized additives improves the IEC due to the presence of additional sulfonic acid groups in addition to the ion exchangeable groups in the polymer matrix. Improvement in IEC depends on the extent of sulfonation of nano-additive—in the present case, IEC values are observed to be in the order of SPEEK-Sfu > SPPEK-SGNF > SPEEK-PSSA-CNTs. The higher IEC for SPEEK-Sfu composite membranes may be due to the presence of additional sulfonic acid groups of Sfu. As fullerene has a high volumetric density for functional groups, a greater number of the same can be grafted on its surface compared to CNTs and GNF (Tasaki et al., 2007).

Water uptake of the membranes was determined by measuring the mass difference between the wet and dry membranes using the following relation (Meenakshi et al., 2013):

$$Water\ uptake(\%) = \frac{W_{wet} - W_{dry}}{W_{wet}} \times 100 \qquad (14.2)$$

Water uptake for the composite membranes along with their pristine membranes is shown in Table 14.2. It is noteworthy that composite membranes show a greater affinity towards water due to the presence of a large number of hydrophilic groups such as carboxylic, hydroxyl and sulfonic acid groups, which enhance the membrane's hydrophilic nature as water molecules form intermolecular hydrogen bonding networks with the sulfonic acid groups of the polymer and the functionalized additive, thereby enhancing the water absorption capacity of the membrane. As a result, composite membranes show a higher water uptake than pristine ones. It is also noteworthy that SPEEK and SPPEK based composite membranes showed a higher water uptake than Nafion-117.

14.5.3 Proton Conductivity and Methanol Permeability

Proton conductivity of the membranes is measured using a four-probe DC method, wherein resistivity of the membranes is measured by considering the probe distance and thickness of the membrane sample to calculate the conductivity. The experimental details can be found in the literature of Meenakshi et al. (2013). Proton conductivity is measured under fully hydrated conditions at 30°C and 60°C, as presented in Table 14.2. Although SPEEK and SPPEK based membranes have a higher IEC and water uptake than Nafion-117, their proton conductivity is suppressed by the presence of more dead-end channels (discontinuous sulfonic acid groups) (Kreuer, 2001). The addition of these functionalized additives improves the connectivity between the proton conducting channels of SPEEK/SPPEK through hydrogen bonding, providing the facile transport of protons through the membrane (Zhou et al., 2011). These hydrogen bonds between the sulfonic acid groups facilitate continuous pathways for proton conduction, and hence a higher conductivity is observed for composites than pristine membranes. However, the ionic conductivity for the aforesaid membranes is higher at 60°C than 30°C, due to the increased mobility of protons, wherein vehicular mechanism may be dominant over structural diffusion (Kreuer, 2001). On the other hand, proton conductivity decreases beyond an optimum loading of additives (0.5 wt%) due to the aggregates formed, which obstruct proton transport (Knauth et al., 2011).

Methanol permeability for the membranes was evaluated by using a two compartment diffusion cell arranged with stirrers. Here, one compartment is filled with 2 M methanol, referred to as compartment A, and second compartment is filled with deionized water, referred as compartment B. The membrane, which is to be subjected to methanol permeability study, is clamped between two compartments. Due to the concentration gradient, methanol from A permeates to B through the membrane. The concentration of methanol, referred as C_2, in B in relation to time is measured by gas chromatography (Thermo scientific Trace GC-700) equipped with a capillary column and flame ionization detector (FID). The methanol permeability through the membranes is calculated from the equation (Heo et al., 2013):

$$P = \frac{k_2 \times V_2 \times L}{(C_1 - C_2) \times A} \tag{14.3}$$

where P is the permeability of the membrane and k_2 is the slope of concentration profiles of compartment B. V_2 is the volume of solution in B and C_1 and C_2 are the methanol concentration in compartments A and B, respectively. L is the thickness, and A is the area of the membrane.

Methanol concentration profiles for the membranes with respect to time are shown in Figure 14.5 (b), and the corresponding permeability values are shown in Table 14.2. It is noteworthy that methanol permeability of composites decreased in comparison to pristine membranes. As the additive content increased from 0.25 to 1 wt%, it reduced the methanol permeability by acting as a barrier for methanol transport through the membrane by forming interfacial interactions with the polymer matrix. The dispersion of these nano-additives makes the membrane surface more rough and denser and offers tortuous pathways for the methanol movement, thereby restricting its crossover. Methanol permeability of SPEEK-PSSA-CNTs, SPEEK-Sfu and SPPEK-SGNF show values of 40%, 46% and 44%, respectively, less than their pristine counterparts. In addition to this, methanol crossover is also evaluated by Linear Sweep Voltammetry (LSV), wherein current density of the MEAs is measured as a function of applied voltage (Rambabu et al., 2016). LSV curves for SPPEK and its composite membranes are shown in Figure 14.5 (c). The crossover current is directly proportional to the amount of methanol permeated through the MEAs. In the present case, the composite membranes show lower crossover current compared to pristine membranes, suggesting its methanol impermeable characteristics.

Electrochemical selectivity is an important parameter to be considered for the membranes, and is determined by considering proton conductivity and methanol permeability (Meenakshi et al., 2013), as expressed here:

$$\text{Electrochemical selectivity}(\beta) = \frac{Proton\ conductivity(\sigma)}{Methanol\ permeability(p)} \tag{14.4}$$

Electrochemical selectivity of the membranes is measured by considering their corresponding proton conductivity and methanol permeability values, as reported in Table 14.2. It is observed that the electrochemical selectivity for the composite membranes is higher in comparison to pristine membranes, which is attributed to improved proton conductivity and a reduced methanol permeability of the composites. SPPEK-SGNF shows higher electrochemical selectivity in comparison to other composites.

14.5.4 Oxidative Stability

During the oxidation of methanol, many side products are formed as reactive intermediates. Among them are hydroxyl and peroxide radicals which initiate membrane degradation, and, as a result of which, pin holes are formed, leading to severe fuel crossover. Hence, membranes must be resistant to these radicals for long-term operation. Oxidative stability of pristine SPEEK and SPEEK-Sfu composite membranes, along with Nafion-117, is represented in Figure 14.5 (d) by measuring remnant weight with respect to different membranes. SPEEK-Sfu composite membranes have better oxidation resistance than pristine SPEEK with a weight loss of only 7% for SPEEK-Sfu (1 wt%). Higher oxidative stability in composite membranes is due to the effective capturing of HO· and HOO· radicals formed by Fenton's solution during experiments (Saga et al., 2008). In the case of pristine SPEEK, after Fenton's test for 2 h, 30% weight loss is observed due to the oxidation of the membrane by the radicals, while, for SPEEK-Sfu composites, the weight loss is lower in relation to different contents of Sfu, thus confirming the radical scavenging properties of fullerene. Interestingly, for Nafion-117, the weight loss is negligible.

14.5.5 Fuel Cell Polarization Studies

DMFC performance of the aforesaid membranes was evaluated by fabricating membrane electrode assemblies (MEAs). In brief, 15 wt% teflonized Toray-TGP-H-120 carbon paper of 0.37 mm thickness was used as the backing layer. To prepare the gas diffusion layer (GDL), carbon paper was teflonized to form a hydrophobic layer on which a Vulcan XC-72R macro-porous carbon layer was coated to have a loading of 1.5 g/cm². For the anode reaction layer, 60 wt% Pt–Ru (1:1 atomic ratio) supported on Vulcan XC-72R carbon mixed with the binder (7 wt% Nafion in relation to Pt/C) coated onto one of the GDLs, constituted the catalyst layer on the anode, while 40 wt% Pt catalyst supported on Vulcan XC-72R carbon mixed with binder (Nafion 10 wt%) coated onto the other GDL, constituted the catalyst layer on the cathode. The catalyst loading on both the anode and cathode was kept at 2 mg cm⁻². The MEAs were prepared by sandwiching the membrane electrolyte between two electrodes using a hot press at 80°C.

The MEAs were evaluated using a conventional fuel cell fixture with parallel serpentine flow field machined on graphite plates and the cells were tested by feeding 2 M aqueous methanol solution at a flow rate of 2 mL min⁻¹ to the anode, and oxygen to the cathode at a flow rate of 300 mL min⁻¹. All the MEAs were tested at a cell temperature of 60°C by using electronic load Model-LCN4-25-24/LCN 50-24 from Bitrode Instruments (U.S.), measuring the cell potential as a function of current density, galvanostatically.

Figure 14.6 (a) shows the cell polarization data for composite membranes along with pristine samples. Peak power density of 93 mW/cm² at load current density of 300 mA/cm²

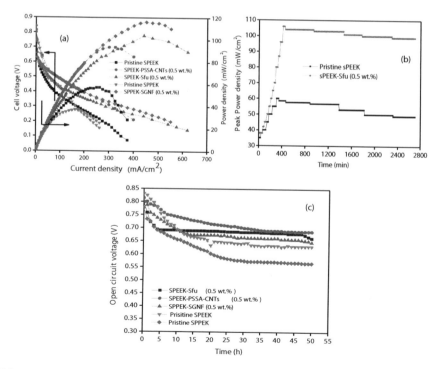

FIGURE 14.6

(a) Polarization studies for pristine SPEEK, SPPEK and their composite membranes, (b) Time evaluation studies for pristine SPEEK and SPEEK-Sfu membranes and (c) Stability test for membranes in terms of OCV in relation to time.

is observed for SPEEK-PSSA-CNT (0.5 wt%), which is 39% higher in comparison to the peak power densities observed for pristine SPEEK, Nafion-117 and recast Nafion membranes. SPEEK-Sfu (0.5 wt%) show peak power density of 103 mW/cm^2, which is 44% higher than pristine SPEEK. SPPEK-SGNF composite membranes show a peak power density of 115 mW/cm^2 at load a current density of 475 mA/cm^2, which is 69% greater than the peak power density observed for pristine SPPEK. The improved power density for the composite membranes is attributed to the higher electrochemical selectivity, which is the cumulative effect of proton conductivity and methanol permeability. The peak power density slightly decreased when the loading of additive reached 1 wt% in the membrane, and these results are consistent with proton conductivity values. Higher ionic conductivity and restricted methanol crossover are the main reasons for the enhanced performance. Higher OCV is observed for composite membranes that correlates with the lower methanol permeability behavior, as shown in Table 14.2. Functionalized additives in the SPEEK/SPPEK matrices enable better proton conduction and act as methanol barriers, leading to the improved performance in DMFCs.

Time evolution studies represented in Figure 14.6 (b) show the maximum (peak) power density of each polarization curve as a function of time for the MEAs comprising pristine SPEEK and SPEEK-Sfu (0.5 wt%). For both membranes, initially, maximum power density increases with time and reaches a steady state value after 400 min. This may be attributed to the activation of all of the catalyst particles and the complete sorption of membranes in MEAs (Yildirim et al., 2008). For the individual MEA, the measurements were done for six days (450 min per day). At the end of sixth day measurement, pristine SPEEK shows a peak power density loss of 11 mW/cm^2 whereas SPEEK-Sfu shows only 6 mW/cm^2 loss, attributable to its low methanol permeability and high oxidative stability, in comparison to pristine SPEEK.

Durability is one of the critical factors for PEMs in DMFCs for its viability in commercial applications. The stability test was performed by measuring change in OCV in relation to time for composite membranes and compared with pristine membranes, as shown in Figure 14.6 (c). In OCV condition, methanol permeability is higher in comparison to load condition, hence OCV condition is chosen to evaluate the stability of the membranes. It is to be noted that the composite membranes show lower OCV loss relative to pristine membranes due to restricted methanol permeability and improved stability. SPEEK-PSSA-CNT exhibits better durability compared to other matrices due to the strong covalent grafting of PSSA to CNT, which ensure that polymer chain flexibility is not reduced drastically due to the addition of PSSA-CNT, as seen in Table 14.2. This provides better durability to the composite with a slight compromise in DMFC performance.

14.6 Conclusions

Carbon nanomaterials have created a significant impact on the development of polymer electrolytes for DMFCs. The performance of composite membrane electrolytes is influenced by the choice of carbon nanomaterial, i.e., CNTs, CNF/GNF, and fullerene, in the present case. Due to the remarkable thermal and mechanical stability, these materials improve the stability of the composite membranes. The methanol barrier effect of these materials mitigate fuel crossover which is one of the major impediments in the commercialization of DMFCs. Structural orientation and composition of these carbon materials in the membrane matrix has a great impact on the properties of membranes. Surface functional-ization of these materials impart a hydrophilic nature, fine-tune its electrical conductivity and provide better interfacial interactions between additives and the polymer matrices. Furthermore, these functional groups enable the facile ion transport, thus improving the ionic conductivity of the composite membranes. Taking into consideration these character-istics and challenges, recent progress in carbon nanomaterial based composite membranes for DMFC application is explained in this chapter. From the cell polarization and durabil-ity data, we can conclude that the composite membranes of SPEEK-PSSA-CNT offers more durability due to covalent grafting and SPPEK-SGNF exhibits a superior DMFC perfor-mance due to its higher electrochemical selectivity, compared to all the subjected compos-ites. Thus, there is a trade-off in the selection of these materials for commercial viability, depending on the application.

Acknowledgments

Mr. Rambabu Gutru would like to thank CSIR for the Senior Research Fellowship and for the research being carried out at the CSIR-Central Electrochemical Research Institute-Madras Unit. Dr. Santoshkumar D. Bhat would like to thank the DU-MLP-0090 (CSIR-YSA) project for the funding to carry out the research presented in this chapter.

References

Aricò, A.S., S. Srinivasan, and V. Antonucci. 2001. DMFCs: From fundamental aspects to technology development. *Fuel Cells* 1: 133–61.

Asgari, M.S., M. Nikazar, P. Molla-abbasi, and M.M. Hasani-Sadrabadi. 2013. Nafion®/histidine functionalized carbon nanotube: High-performance fuel cell membranes. *International Journal of Hydrogen Energy* 38: 5894–5902.

Du, C.Y., T.S. Zhao, and Z.X. Liang. 2008. Sulfonation of carbon-nanotube supported platinum cata-lysts for polymer electrolyte fuel cells. *Journal of Power Sources* 176: 9–15.

Gao, Y., G.P. Robertson, M.D. Guiver et al. 2003. Sulfonation of poly(phthalazinones) with fuming sulfu-ric acid mixtures for proton exchange membrane materials. *Journal of Membrane Science* 227: 39–50.

Gahlot, S. and V. Kulshrestha, 2015. Dramatic improvement in water retention and proton con-ductivity in electrically aligned functionalized CNT/SPEEK nanohybrid PEM. *ACS Applied Materials & Interfaces* 7: 264–272.

Heo, Y., S. Yun, H. Im, and J. Kim. 2012. Low methanol permeable sulfonated poly(ether imide)/sulfonated multiwalled carbon nanotube membrane for direct methanol fuel cell. *Journal of Applied Polymer Science* 126: E467–477.

Heo, Y., H. Im, and J. Kim. 2013. The effect of sulfonated graphene oxide on sulfonated poly(ether ether ketone) membrane for direct methanol fuel cells. *Journal of Membrane Science* 425–426:11–22.

Jiang, R. and D. Chu. 2002. CO$_2$ Crossover Through a Nafion membrane in a direct methanol fuel cell. *Electrochemical and Solid-State Letters* 5: A156–159.

Jiang, Z., X. Zhao, Y. Fu, and A. Manthiram. 2012. Composite membranes based on sulfonated poly(ether ether ketone) and SDBS-adsorbed graphene oxide for direct methanol fuel cells. *Journal of Materials Chemistry* 22: 24862–24869.

Ji, J., G. Zhang, H. Chen, and S. Wang. 2011. Sulfonated graphene as water-tolerant solid acid catalyst. *Chemical Science* 2: 484–487.

Joo, S.H., C. Pak, E.A. Kim, et al. 2008. Functionalized carbon nanotube-poly(arylene sulfone) composite membranes for direct methanol fuel cells with enhanced performance. *Journal of Power Sources* 180: 63–70.

Kamarudin, S.K., F. Achmad, and W.R.W. Daud. 2009. Overview on the application of direct methanol fuel cell (DMFC) for portable electronic devices. *International Journal of Hydrogen Energy* 34: 6902–6916.

Kannan, R., B.A. Kakade, and V.K. Pillai. 2008. Polymer electrolyte fuel cells using Nafion-based composite membranes with functionalized carbon nanotubes. *Angewandte Chemie International Ed.* 47: 2653–2656.

Kanbur, Y. and Z. Küçükyavuz. 2012. Synthesis and characterization of surface modified fullerene. *Fullerenes, Nanotubes, and Carbon Nanostructures* 20: 119–126.

Knauth, P., E. Sgreccia, A. Donnadio, M. Casciola, and M.L. Di Vona. 2011. Water activity coefficient and proton mobility in hydrated acidic polymers. *Journal of the Electrochemical Society* 158: B159–165.

Kreuer, K.D. 2001. On the development of proton conducting membranes for hydrogen and methanol fuel cells. *Journal of Membrane Science* 185: 29–39.

Li, X., Z. Wang, H. Lu, C. Zhao, H. Na, and C. Zhao, 2005. Electrochemical properties of sulfonated PEEK used for ion exchange membranes. *Journal of Membrane Science* 254: 147–155.

Liu, X., Z. Yang, Y. Zhang, C. Li, J. Dong, Y. Liu, and H. Cheng. 2017. Electrospun multifunctional sulfonated carbon nanofibers for design and fabrication of SPEEK composite proton exchange membranes for direct methanol fuel cell application. *International Journal of Hydrogen Energy* 42: 10275–10284.

Nunes, S.P., B. Ruffmann, E. Rikowski, S. Vetter, and K. Richau. 2002. Inorganic modification of proton conductive polymer membranes for direct methanol fuel cells. *Journal of Membrane Science* 203: 215–225.

Peera, S.G., A.K. Sahu, A. Arunchander, K. Nath, and S.D. Bhat. 2015. Deoxyribonucleic acid directed metallization of platinum nanoparticles on graphite nanofibers as a durable oxygen reduction catalyst for polymer electrolyte fuel cells. *Journal of Power Sources* 297: 379–387.

Rambabu, G., N. Nagaraju, S.D. Bhat. 2016. Functionalized fullerene embedded in Nafion matrix: A modified composite membrane electrolyte for direct methanol fuel cells. *Chemical Engineering Journal* 306: 43–52.

Ruffmann, B., H. Silva, B. Schulte, and S.P. Nunes. 2003. Organic/inorganic composite membranes for application in DMFC. *Solid State Ionics* 162/163: 269–275.

Saga, S., H. Matsumoto, K. Saito, M. Minagawa, and A. Tanioka. 2008. Polyelectrolyte membranes based on hydrocarbon polymer containing fullerene. *Journal of Power Sources* 176: 16–22.

Samms, S.R., S. Wasmus, and R.F. Savinell. 1996. Thermal Stability of Nafion® in Simulated Fuel Cell Environments. *Journal of the Electrochemical Society* 143: 1498–1504.

Su, Y.H., Y.L. Liu, Y.M. Sun, J.Y. Lai, M.D. Guiver, and Y. Gao. 2006. Using silica nanoparticles for modifying sulfonated poly(phthalazinone ether ketone) membrane for direct methanol fuel cell: A significant improvement on cell performance. *Journal of Power Sources* 155: 111–117.

Tasaki, K., J. Gasa, H. Wang, and R. Desousa. 2007. Fabrication and characterization of fullerene-Nafion composite membranes. *Polymer* 48: 4438–4446.

Wang, F., M. Hickner, Y.S. Kim, T.A. Zawodzinski, and J.E. McGrath. 2002. Direct polymerization of sulfonated poly(arylene ether sulfone) random (statistical) copolymers: Candidates for new proton exchange membranes. *Journal of Membrane Science* 197: 231–242.

Wang, H., R. DeSousa, and J. Gasa. 2007. Fabrication of new fullerene composite membranes and their application in proton exchange membrane fuel cells. *Journal of Membrane Science* 289: 277–283.

Xue, S. and G. Yin. 2006. Methanol permeability in sulfonated poly(etheretherketone) membranes: A comparison with Nafion membranes. *European Polymer Journal* 42: 776–785.

Yildirim, M.H., A.R. Curo`s, J. Motuzas, A. Julbe, D.F. Stamatialis, and M. Wessling. 2009. Nafion®/H-ZSM-5 composite membranes with superior performance for direct methanol fuel cells. *Journal of Membrane Science* 338: 75–83.

Yun, S., H. Im, Y. Heo, and J. Kim. 2011. Crosslinked sulfonated poly(vinyl alcohol)/sulfonated multi-walled carbon nanotubes nanocomposite membranes for direct methanol fuel cells. *Journal of Membrane Science* 380: 208–215.

Zhang. H., C. Ma, J. Wang, X. Wang, H. Bai, and J. Liu. 2014. Enhancement of proton conductivity of polymer electrolyte membrane enabled by sulfonated nanotubes. *International Journal of Hydrogen Energy* 39: 974–986.

Zhang, W. and S.R.P. Silva. 2011. Application of carbon nanotubes in polymer electrolyte based fuel cells. *Review on Advanced Materials Science* 29: 1–14.

Zhang, X., Q. Tang, D. Yang et al. 2011. Preparation of poly(p-styrenesulfonic acid) grafted multi-walled carbon nanotubes and their application as a solid-acid catalyst. *Materials Chemistry and Physics* 126: 310–313.

Zhou, W., J. Xiao, Y. Chen et al. 2011. Sulfonated carbon nanotubes/sulfonated poly(ether sulfone ether ketone ketone) composites for polymer electrolyte membranes. *Polymers for Advanced Technologies* 22: 1747–1752.

Zhao, T.S., C. Xu, R. Chen, and W.W. Yang. 2009. Mass transport phenomena in direct methanol fuel cells. *Progress in Energy and Combustion Science* 35: 275–292.

Zhong, S., T. Fu, Z. Dou, C. Zhao, and H. Na. 2006. Preparation and evaluation of a proton exchange membrane based on cross linkable sulfonated poly(ether ether ketone)s. *Journal of Power Sources* 162: 51–57.

Zhu, Y., S. Zieren, and A. Manthiram. 2011. Novel crosslinked membranes based on sulfonated poly(ether ether ketone) for direct methanol fuel cells. *Chem Comm* 47: 7410–7412.

15

Bioethanol Production in a Pervaporation Membrane Bioreactor

Anjali Jain, Ravi Dhabhai, Ajay K. Dalai, and Satyendra P. Chaurasia

CONTENTS

15.1 Introduction

In today's society, a steady and reliable source of energy is demanded for human prosperity, development and commercial progress. The energy policy of a country must be focused on attaining optimum and efficient usage of its varied primary resources for energy generation, Conventional fuels are non-renewable, limited and polluting, and therefore need to be used prudently. The world is not facing only progressive exhaustion of non-renewable energy resources, but also atmospheric pollution and climate changes as repercussions of the use of oil-derived fuels. Therefore, alternative resources derived from biomass are becoming potential substitutes to conventional fuels. One of the most important examples of biofuels is bioethanol produced from biomass through fermentation technology. In particular, bioethanol production from non-food biomass resources, such as lignocellulosic feedstock, is of the utmost relevance in countries with large populations and growing gasoline consumption such as Brazil, Egypt, China and India.

Bioethanol has about only two-thirds of the energy content and heat value of petrol; nevertheless ensues in much more efficient combustion. The use of bioethanol as an oxygenate increases the oxygen content and allows better oxidation of hydrocarbons in gasoline. It also reduces the amounts of carbon monoxide and aromatic compounds released into the atmosphere (Prasad et al., 2007). Also, the values of octane number, flame speeds (39 m/s), broader flammability limits (3.3–19), and heats of vaporization are higher for ethanol than for petrol. In an internal combustion engine, these characteristics account for a higher compression ratio, shorter burn time, leaner burn engine, leading to theoretical efficiency benefits for bioethanol, compared to gasoline (Balat, 2007).

The yeast, *Saccharomyces cerevisiae*, is currently the most effectively used bioethanol producing microorganism employed in fermentation (Azhar et al., 2017). However, there is a major issue related to the fermentation process, i.e., inhibition of *S. cerevisiae* by the ethanol being produced. Inhibition of yeast decreases the ethanol yield and overall productivity of the process. Most of the fermenting yeast cannot tolerate more than 10–12 vol% of ethanol, therefore, for achieving complete conversion of sugar, it becomes mandatory to begin with a relatively dilute sugar solution. The large quantities of water consumed in bioethanol production lead to high costs due to large bioreactors used in fermentation as well as downstream processing for ethanol recovery by distillation, which is energy intensive (O'Brien et al., 2004). The quantity of water required for the purpose of dilution in the fermentation process is also a concern for water scarce countries. In addition to water and energy savings in membrane bioreactors (MBR), the continuous removal of ethanol from the fermenter accelerates fermentation and increases throughput.

15.2 Production and Consumption Scenario of Bioethanol

In 2016, the worldwide production of bioethanol remained similar to the previous year's production of 117.7 million m³. Presently, the United States is the world leader in the bioethanol production sector (59.5 million m³), followed by Brazil (27.8 million m³). In 2017, the world's production of bioethanol is expected to remain stable (117.6 million m³) (CropEnergies, 2017). In 2015, the U.S. produced an estimated 14.7 billion gallons (56 billion liters) of ethanol, equivalent to 527 million barrels of crude oil, or 31% of U.S. crude oil imports (Nguyen et al., 2017). In 2015, 144 million tons of biomass (primarily corn) was used within the U.S. to produce biofuels, which provided 5% of the domestic transportation fuel needs (Energy Information Administration, 2016). However, the volume of biomass potentially available for production is far greater. The world fuel ethanol production data for 2016–2012 is shown in Figure 15.1.

India's annual energy consumption growth is about 4.6%, and currently it is the third largest consumer of energy following the United States and China (Global Energy Statistical Yearbook, 2017). Currently, India's bioethanol production is dependent on cane molasses (Aradhey, 2016). Although India is the second largest producer of sugarcane, however it contributes only 1% of the global biofuel production. In 2008, India generated around 2,150 million liters of ethanol, of which 280 million liters were used in blending with gasoline. The production was decreased to 1.07 billion liters in 2009. The ethanol used for blending totaled 100 million liters in 2009 and further decreased to 50 million liters in 2010 (Singh et al., 2017). In 2014, the bioethanol production was allowing only ~1% blending (Biswas & Manocha, 2014). However, in 2017, fuel ethanol was expected to achieve 2% blending throughout the country. But, the production of ethanol is expected to decline by ~8% (1.9 billion liters), due to the second consecutive year of decreased sugarcane production

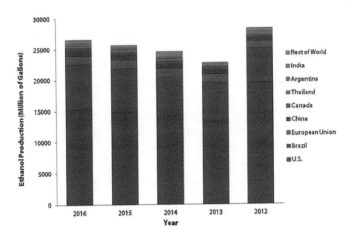

FIGURE 15.1
Global fuel ethanol production data (Modified from: Renewable Fuel Association, 2017).

(Aradhey, 2016). To solve this problem, effective substitutes of biomass sources need to be tested. In India, huge agricultural farmlands generate large quantities of agricultural waste/residues, which can be utilized as alternative sources of green energy (Prasad et al., 2007). India generates about 686 million metric tons of crop biomass annually—out of this ~ 34% surplus biomass may be used for biofuel production (Hiloidhari et al., 2014). For sustainable development, new technologies should also be improvised for better conversion of the biomass.

15.3 Lignocellulose as Feedstock for Bioethanol Production

Lignocellulosic materials have the potential to serve as the greatest source of feedstock in the bioethanol industry and the most plentiful reproducible resource on the planet. It has been estimated that lignocellulosic feedstock can be used to produce ~4.42×10^5 million liters of bioethanol annually. If the waste crop residues are also included, this expands to 4.91×10^5 million liters, ~16 times current production (Sarkar et al., 2012). The variety of feedstocks used in bioethanol production includes straws, rice husk, wheat bran, corn stover, cotton stacks, sugarcane bagasse, low value fruits, sorghum, spruce salix, municipal waste, used paper, soil organic carbon, wheat flour, etc. (Achinas & Euverink, 2016).

15.4 Bioethanol Production from Lignocellulosic Feedstock

The process of ethanol production depends on the types of feedstock used. For lignocellulosic feedstock, generally, there are four major steps in ethanol production: pretreatment, hydrolysis (saccharification), fermentation and product recovery (Behera et al., 2010). Pretreatment methods increase the surface area of the biomass and remove lignin to make it more accessible for hydrolysis. Saccharification can be chemical or enzymatic, and the goal is to release a maximum number of monomeric sugar molecules from complex poly-saccharides. Fermentation converts sugars to ethanol, and the produced alcohol can be recovered by various product recovery operations (Behera et al., 2010).

The structural characteristics of biomass affect the fermentation yield and productivity. Dhabhai et al. (2013) have reported that pretreated wheat straw (PWS) exhibited a larger available surface area and pore volume along with low hemicellulose and lignin contents. Maximum increase in surface area (7.1 m^2/g compared to 4.0 m^2/g for untreated wheat straw) was obtained at pretreatment conditions of a $180°C$ temperature, 0.5% (v/v) acid, and a time of 7 min. SEM imaging of biomass revealed that pore breaking, compression of pores, and partial pore blocking in the case of high temperature ($190°C$) pretreatment conditions may be the reason behind decreased surface area of biomass. FT-IR analysis showed almost complete hemicellulose removal and acid-soluble lignin removal after dilute acid pretreatment, but insufficient removal of acid insoluble lignin (Dhabhai et al., 2012; 2013).

15.5 Fermentation

Fermentation may be defined as an anaerobic biochemical reaction in which an enzyme produced by a microorganism catalyzes the conversion of sugars to alcohol or acetic acid with the evolution of CO_2 (Dhabhai, 2012). The sugars released during pretreatment and saccharification are fermented by bacteria, yeast or fungi to produce ethanol. The reaction of ethanol production from glucose in the presence of microbes and nutrients involves the production of 2 moles each of ethanol and CO_2 from one mole of glucose. Similarly, 3 moles of xylose produce 5 moles each of ethanol and CO_2. According to reaction stoichiometry, 1g of glucose or xylose produces 0.511 g of ethanol and 0.488 g of CO_2. Therefore, the theoretical maximum yield of ethanol is 0.511 g ethanol/g sugar used. But neither glucose nor xylose delivers this much ethanol in practice, and the acceptable ethanol yields are in the range of 0.40–0.50 g/g sugar from both types of sugar (Chen et al., 2010).

15.5.1 Types of Fermentation Processes

The industrial modes of operation of bioethanol production are: (1) batch, (2) continuous, (3) fed batch, and (4) semi-continuous. In batch fermentation process, the substrate, nutrients and microbe culture are fed into the fermenter altogether. The advantages include complete sterilization, easier management of feedstock, and greater flexibility for various product specifications in batch operations, which can be practiced with unskilled labor as it does not need careful control as in the other processes (Ivanova et al., 2011). In the continuous operation, feed containing substrate, microbe culture and nutrients, is charged continuously and product containing ethanol, cells, and residual sugar is taken from the top of the fermentor. The advantages are improved volumetric productivity and, consequently, smaller vessel volumes and lower investment and operational costs (Ivanova et al., 2011). The major disadvantage is that yeasts cultivated for a long time under anaerobic conditions lose their ability to produce ethanol. Additionally, at high dilution rates the substrate is not completely consumed and yields decrease. Moreover, if the system operates near an unstable steady state, any minor disturbance in dilution rate, temperature or substrate concentration could not be off-set by the culture, and the system may show oscillations with time or start operating with lower productivity (Sanchez & Cardona, 2008). In fed-batch operation, the feed solution containing substrate, yeast culture and the nutrients, are charged periodically while product is removed discontinuously. The intermittent inclusion improves process productivity by maintaining a low sugar concentration, reducing the inhibitory effect of substrate. In semi-continuous processes, a fragment of the culture is withdrawn periodically, and fresh medium is added. The advantages include no wastage of time in non-productive idle time for cleaning and resterilization and no requirement for a separate inoculum vessel, except at the initial startup. However, the drawback is a higher investment cost due to the larger fermenter volume and high chances of contamination and mutation due to long cultivation periods and periodic handling (Prasad et al., 2007).

15.5.2 Fermentation with Immobilized Yeast

One of the strategies used for improving ethanolic fermentation is the immobilization of cell that allows the implementation of continuous processes with higher yields and productivities with increased cell concentrations, resulting in shorter residence time and smaller reactor size (Dhabhai, 2012). Other major advantages include prolonged cellular stability, increased ethanol yield, increased tolerance to high substrate concentration, reduced end product inhibition, easier product recovery leading to decreased energy demands and process expenses, regeneration and reuse of cells for extended periods in batch system, feasibility of continuous processing, and reduction of risk of high cell densities (Nikolic et al., 2010).

The most widely used immobilization methods are based on cell entrapment in gels, such as carrageen and Ca-alginate spheres (Azhar et al., 2017). The main drawback of a cell entrapment system is the instability of Ca-alginate against phosphates and the disruption of gel particles due to CO_2 evolution during fermentation (Yu et al., 2007). One other commonly used method for cell immobilization is encapsulation, which encloses cells within a thin semi-permeable membrane. This method has several advantages such as mechanical and chemical stability of the membrane system, possibility of high loading and regulation of the fermentation reaction by selective diffusion of substrate and product. However, it is time-consuming and laborious since capsule disintegration, rupture and cell escape may easily happen due to any simple flaw during preparation or application (Ishola et al., 2015). Other methods are based on passive adhesion to the surfaces, such as glass beads or stainless-steel wire spheres (Yu et al., 2007). Various lignocellulosic materials have also been tried as microbial support, e.g., orange peel, wooden chips or blocks, sorghum bagasse, rice husk, wild sugarcane and wheat starch granules (Azhar et al., 2017). For the adhesion or adsorption system, as the cells attach to the carriers' surface by electrostatic interactions or covalent binding of the cells, the two major drawbacks are the limitation of biomass loading by the carrier's surface, and the effect of various factors that can cause cell desorption, thereby limiting the operational stability. Lignocellulosic support carriers have advantages in terms of cost, safety, and waste treatment but the main drawback is the non-uniform structure, often present in a particulate, powder or chip form. For fermentation of mixed lignocellulosic sugars, a mixed sugar strategy co-culture immobilization was utilized by physical separation of the two microorganisms by means of immobilization of one of the microorganisms (Dhabhai et al., 2012; Dhabhai et al., 2013).

15.6 Membrane Bioreactor (MBR) Systems

For ethanol recovery, distillation is currently the standard downstream process, but being an energy demanding process, accounts for a major part of the total energy utilization in a distillery or biorefinery. The quantity of water required, for the purpose of dilution in the fermentation process, is also a concern. Distillation plants also demand high quantities of cooling water which generally requires the water in a ratio of about 4:1 for bioethanol production (Current Outlook, Profitability, and Weather Information). It has been recognized that if ethanol recovery and the fermentation process were combined, there would be a reduction in the overall cost of the process. If ethanol is recovered directly from the bioreactor, by recycling the broth in a fermenter through a separation unit, it preserves yeast viability and enhances feasibility to achieve complete conversion of a highly concentrated feed (Lewandowska, 2007; Vane 2005). The downstream recovery of ethanol from fermentation broth is an important factor affecting bioethanol production as it accounts for about 30–40% of the total process cost. Various separation techniques, such as extraction, adsorption, membrane processes (microfiltration, nanofiltration, reverse osmosis, pervaporation, and electrodialysis, etc.) are available for the removal and concentration of the target compounds from different feedstock. Among these, membrane processes have attracted greater attention for the following reasons: (i) they improve efficiency and product quality by minimizing ethanol inhibition through continuous removal from the fermentation broth, (ii) they are cost-effective and environmentally friendly, and (iii) they have the flexibility of coupling with other separation processes (Boyaval et al., 1996). Also, the membrane process can reduce thermal heating and cooling loads by separating a significant amount of water from ethanol at ambient temperature conditions. In addition to water and energy savings in a membrane bioreactor (MBR), the continuous removal of ethanol from the fermenter accelerates fermentation and increases throughput ("Current Outlook, Profitability, and Weather Information"). The two configurations in which membrane separation can be integrated with the fermentation process are as follows:

I. When an ethanol selective membrane is externally attached to the fermenter, a significant amount of energy is required for sustaining a continuous flow through the membrane. However, this mode reduces fouling tendency and increases flux.

II. When an ethanol selective membrane is used as internally integrated (submerged/immersed) part of the fermenter, the advantage is that they require less energy to run, compared to external MBR. However, this could pose operational difficulties at high cell concentrations due to fouling, and usually a larger membrane area has to be utilized.

15.7 Theory of Pervaporation (PV)

PV is a membrane process in which a liquid mixture that can be binary or multicomponent, is separated by partial vaporization using a non-porous membrane. The feed mixture is in direct contact with one side of the membrane, whereas permeate is removed in vapor state from the other side by applying vacuum/sweeping gas, then condensed and recovered. The driving force for mass transfer from feed side to permeate side of the membrane

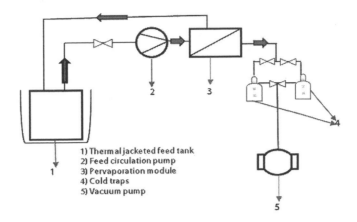

FIGURE 15.2
Schematic diagram of pervaporation setup.

is chemical potential gradient, established by applying a difference in partial pressures across the membrane by using a vacuum pump on the downstream side of the membrane. A schematic of the PV process is shown in Figure 15.2. The performance of a given PV membrane is estimated in terms of permeate flux and selectivity (Kujawski, 2000). Flux is defined as the permeate flow rate per unit membrane area per unit time for a given membrane thickness (Aminabhavi et al., 2005). The selectivity of a membrane can be estimated by using the following two, dimensionless parameters (Kujawski, 2000):

$$\text{Separation factor, } \alpha = \frac{Y_A / Y_B}{X_A / X_B} = \frac{Y_A / (1 - Y_A)}{X_A / (1 - X_A)} \tag{15.1}$$

$$\text{Enrichment factor, } \beta = Y_A / X_A \tag{15.2}$$

where X_A is weight fraction of preferentially permeating species in the feed phase, Y_A is the weight fraction of preferentially permeating species in the permeate phase, with $X_A + X_B = 1$ & $Y_A + Y_B = 1$.

15.8 Pervaporation Membranes for Extraction of Ethanol from Aqueous Solutions

The most commonly used organophilic PV membrane is poly(dimethylsiloxane), PDMS, also known as silicone rubber. The high diffusivity of ethanol in PDMS, due to the free rotation of Si-OH bond, and high contact angle of water on PDMS surface due to the latter's hydrophobic nature, results in high ethanol selectivity (Peng et al., 2010). In a comprehensive review, Beaumelle et al. (1993) reported that the fluxes and separation factors of unmodified PDMS membranes for the removal of ethanol range from 1 to 1000 $g/m^2.h$ and less than 10, respectively. Several attempts have been made to modify PDMS, in order to improve PV performance (Vane, 2004). Another membrane which has received significant attention is PTMSP, poly[1-(trimethylsilyl)-1-propyne]. The flux and separation factors are higher by about threefold and twofold, respectively, as compared to pervaporation by PDMS under similar conditions. Unfortunately, PTMSP shows a declining trend for flux and selectivity over time because of its chemical and/or physical aging (Vane, 2005). Separation factors of composite membranes and zeolites are reported to be in the range of 7–59 and 12–106, respectively, much higher than polymeric ones. However, in some cases, the ethanol-water separation factors might even exceed these ranges. The separation factor of ethanol over water has been reported to be 218 when using a silicate zeolite membrane where high permeate ethanol concentration of 98.2 wt% was obtained from a fermentation broth containing 20 wt% ethanol (Nomura et al., 2002). Zeolite membranes are best in terms of high separation factor and permeability, but are known to be 10–50 times more expensive than their polymer-based counterparts. Zeolite membranes are also more likely to be affected by fermentation by-products (Vane, 2005). Performance data of some polymeric materials reported in the literature, other than PDMS and PTMSP, have also exhibited greater separation factors. However, the commercial unavailability of such constituted membranes restricts their practical application.

15.9 Bioethanol Production in MBR

In the 1980s, ethanol recovery using pervaporation separation technology was a widely researched topic. More recently, the focus of PV studies with ethanol has shifted from ethanol-water mixtures to fermentation broths (Chovau et al., 2011; Mohammadi et al., 2005; Aroujalian et al., 2006). Membrane bioreactors (MBR) with different configurations have been implemented for production of metabolites, with simultaneous separation of fermented products using a pervaporation technique. There are some reports in the literature discussing continuous ethanol production by pervaporation using different cultures, membranes and configurations. Various pervaporation results presenting separation of ethanol from fermentation broth are summarized in Table 15.1. O'Brien and Craig (1996) combined a traditional batch fermentation using *S. cerevisiae* and glucose with a PV system using a PDMS membrane for ethanol extraction. This study was successful in reducing the inhibition effect of ethanol by maintaining a relatively low concentration of ethanol in the fermentation broth. A year later, Schmidt et al. (1997), tried a PTMSP membrane in a system similar to that in the study of O'Brien and Craig (1996). The membrane showed higher flux, higher ethanol selectivity and increased resistance to fouling as compared to PDMS. However, the PTMSP membrane revealed slight deterioration with respect to time, which was because of its being used in unsupported state. In another study on a PTMSP membrane, Fadeev et al. (2003) also observed similar deterioration. They disapproved the reason suggested by Schmidt et al. (1997) and reported that the PTMSP membrane deteriorated because of internal fouling by the by-products formed during fermentation. They found that glycerol, the main by-product of fermentation caused the flux to decline by 30%; however, other by-products (n-propanol, acetone, and butanediol) were absorbed into the membrane but butanediol was not found in the permeate, and therefore the diol was most likely occupying the polymer-free volume and reducing mass transport of components

TABLE 15.1

Recovery of Bioethanol from Fermentation Broth in PVMBR

Membrane	Total Flux (g/m².h)	Max. Selectivity	Conditions	Reference
PDMS	257.4	2.7	30°C, 5 torr	This study
PDMS	350–600	7–9	35°C, pH 4, 30 torr	Fan et al. (2017)
PDMS	385-I run 370-II run	8.8 (I run) 9.5 (II run)	35°C, pH 4, 30 torr	Fan et al. (2014)
PDMS	93.8 (ethanol flux)	5.11	50°C	Jaimes et al. (2014)
PDMS	5.85	10.63	30°C, 3% EtOH	Bello et al. (2014)
PDMS	704	4.8	3wt% EtOH in broth, 30°C, pH of 5	Gaykawad et al. (2013)
PDMS	460	7	32°C	Esfahanian et al. (2012)
Cellulose acetate	3800	9.2	50°C	Kaewkannetra et al. (2012)
Mixed matrix ZSM-5/PDMS	9.35	7.8	50°C	Offeman and Ludvik (2011)
Cellulose acetate	55.3–193.5	9.3–2.2	Between 50°C–70°C for PV	Kaewkannetra et al. (2011)
PDMS–PAN–PV composite membrane	133.6	–	30°C, pH of 4.7	Staniszewski et al. (2009)
PDMS-PAN-PV	119.8–161.2	8	Fermentation at 30°C, PV at 65°C	Lewandowska et al. (2007)
Supported Trioctylamine liquid membrane	0.1	80	54°C	Thongsukmak et al. (2007)
PDMS–PAN–PV composite membrane	161.2 to 119.8	–	–	Staniszewski et al. (2007)
PDMS	~ 600	8–9	40–60°C, 1–40 torr	Aroujalian et al. (2006)
Composite PDMS	26.1 (ethanol flux)	7.9	35°C, pH of 4.5	Wu et al. (2005)
PDMS	5.1	10	34°C, pH of 6.5	O'Brien et al. (2004)
PTMSP	not quantified	not quantified	30°C	Fadeev et al. (2003)
Silicalite coated with silicone rubber	3.7 (ethanol flux)	60	30°C, agitation at 600 rpm	Ikegami et al. (2002)
Silicalite	23–27.6	85.9–218	30–35°C, agitated at 30–60 rpm, pH of 4	Nomura et al. (2002)
Hollow fibre micro-porous polypropylene	50.7–64.5	7.5	30°C, pH of 4–5	Kaseno et al. (1998)
1.PTMSP 2.PDMS	1. 6.7 2. 1.8	1. 5 2. 13	PV at ambient temperature (about 25°C), fermentation at 37°C, 150 rpm	Schmidt et al. (1997)
PDMS	14.3–36.4	1.8–6.5	35°C, agitation at 100 rpm, pH of 5	Brien et al. (1996)
PDMS	180.2	6.78	32°C, pH of 4.5 41g/L EtOH in broth, 4–5 torr	Younan et al. (1992)

across the membrane. Kaseno et al. (1998) conducted a study comparing fermentation combined with PV in fed-batch mode using a micro-porous polypropylene hollow fiber membrane. Kaseno et al. focused on the effect that pervaporation would have on fermentation, rather than on PV membrane performance. They found that fermentation performance was twice as efficient as the conventional fermentation performance and the generation of wastewater during fermentation was also found to have reduced by 38.5% in the coupled fermentation-PV process, which may serve as an important benefit for water scarce countries. Fan et al. (2017) also reported that only 18.4% of the water was necessary for the coupled process as compared to the traditional batch fermentation. Ikegami et al. (1997) have shown that, by a coupled fermentation-pervaporation process using an ethanol selective silicalite membrane, 85 vol% ethanol could be obtained, and the process exhibited about a 20% increase in the average glucose consumption rate. Ikegami et al. (1999) reported interesting data that showed permeation flux to drastically decrease to 33% of that of an ethanol-water solution with increasing glucose concentration, and the decrease in the total flux was attributed to reduction in water flux. However, the amount of ethanol that was recovered was almost constant and independent of the glucose concentration, resulting in an increase in separation factor from 23 to 137. They also found that glucose was not adsorbed by the membrane, suggesting glucose molecules strongly inhibit the adsorption of water into the silicalite membrane. In 2002, Ikegami et al. and Nomura et al. investigated the recovery of ethanol from fermentation broth using silicalite membranes. In these studies, fermentation broth was used as feedstock for pervaporation rather than combining fermentation with PV to determine the effect of fermentation broth on membrane properties. Nomura et al. (2002) reported that a silicalite membrane could be successfully used to separate ethanol from fermentation broth, as the selectivity and flux were found to be higher with the fermentation broth than with the ethanol-water mixture. However, the flux did decrease with respect to time. Ikegami et al. (2002) determined that the decrease in flux was due to the partial coating of membrane by succinic acid and glycerol, which formed as by-products during fermentation. They resolved this problem successfully by coating the silicalite membrane with silicone rubber. In both of these studies, the selectivity of the silicalite membranes was found to be very high but the flux was much lower than observed in previous studies with other membranes. It was suggested that this could be due to the sugar, nutrients, yeast cells, and by-products present in the fermentation broth. Wu et al. (2005) studied the effect of yeast cells on the performance of a composite PDMS membrane. They used inactive yeast cells and observed that the flux and selectivity decreased with an increase in cell weight, but the performance was still better than for the pure ethanol-water mixture, suggesting that the inactive cells in suspension improve ethanol transfer through the membrane. Moreover, fermentation broth containing active yeast cells gave higher flux than with inactive cells, indicating that the metabolism of active cells improves ethanol mass transfer through the membrane. Aroujalian et al. (2006) focused their study on the effect of the presence of glucose on flux and selectivity of a PDMS film on a PVDF-supported membrane. The total flux and ethanol selectivity were found to decrease upon increasing glucose concentration.

With an increase in the popularity of using maize as feedstock to produce bioethanol in the United States, O'Brien et al. (2004) studied a coupled fermentation-pervaporation process using hydrolyzed maize fiber as feedstock and *E. coli* as the fermenting microorganism. O'Brien et al. were successful in maintaining a low ethanol concentration in the fermentation broth. Recently, Offeman and Ludvik (2011) also looked at the effect of fermentation components on PDMS and ZSM-5/PDMS membranes using maize as feedstock. They found that the ZSM-5/PDMS membrane performance was better than the pure

PDMS membrane in PV with ethanol-water mixtures. However, the performance of the membrane was greatly reduced with respect to time when a fermentation broth was used as feed, whereas the PDMS membrane showed no significant reduction in performance with respect to time, indicating that the fermentation broth was deactivating the zeolite component in the membrane. Kaewkannetra et al. (2011) tried experiments with sweet sorghum juice as feedstock and found that cellulose acetate membrane showed high values of flux and selectivity (flux, 5–11.4 kg/m^2.h and selectivity, 7–14.2) for ethanol-water mixture. However, both flux and selectivity of the membrane decreased significantly with the sweet sorghum fermentation broth, as the feed for PV. Chen et al. (2012) have investigated ethanol production by *S. cerevisiae* in a continuous and closed circulating fermentation (CCCF) system using a PDMS pervaporation membrane bioreactor and reported ethanol volumetric productivity of 1.39 g/L.h in the third cycle, with a yield rate of 0.13 h^{-1}. Fan et al. (2014) studied the continuous and complete coupling of ethanol fermentation and pervaporation and reported the average cell concentration, glucose consumption rate, ethanol productivity, ethanol yield, and the total ethanol amount produced reached as 19.8 g/L, 6.06 g/L.h, 2.31 g/L.h, 0.38, and 609.8 g/L, respectively. During the continuous fermentation process, in situ removal of ethanol promoted the yeast cell second growth, but an accumulation of the secondary metabolites in the broth became the main inhibitor against cell growth and fermentation. According to published research, the yeast cell and the byproducts of the yeast cell metabolism (such as glycerol, acetic acid and other carboxylic acids) in the fermentation broth could lead to reduction in membrane flux (Kaewkannetra et al., 2011; García et al., 2009).

Recently, Fan et al. (2017) also reported ethanol fermentation in the coupled process; the total permeation flux of the membrane was in the range of 350 g/m^2.h and 600 g/m^2.h, and the selectivity was in the range of 7.0 and 9.0. However, the flux for the ethanol-water model solution was found to be 1100 g/m^2.h (Fan et al., 2016). The reduction in membrane flux in the coupled process is attributed to the accumulated by-products in the fermentation broth, which do not permeate through the membrane. Fu et al. (2016) studied the integrated fermentation and pervaporation process in batch, fed-batch and continuous mode. In the case of integration in batch mode, after 40 h of fermentation about 2 g/L of glucose remained in the fermentation broth with a cumulative ethanol concentration of 63.8 g/L, as compared to the 63.4 g/L of ethanol after 50 h of fermentation in conventional fermentation. Therefore, when integrated with batch ethanol fermentation, the ethanol productivity was enhanced compared to the conventional process. In fed-batch and continuous fermentation integrated with PV process, a total flux of 396.2–663.7 g/m^2.h and 332.4–548.1 g/m^2.h with corresponding separation factors of 8.6–11.7 and 8–11.6, were observed. At the same time, ethanol production of 417.2 g/L and 446.3 g/L were also achieved in permeate in fed-batch and continuous operations, respectively.

In our study, the performance of traditional batch fermentation with PVMBR using a commercial flat sheet PDMS membrane (purchased from Pervatech, Netherlands) has been compared using *S. cerevisiae* Y35. The membrane has an effective thickness of 3–5 µm PDMS as the top layer, support consisting of polyethylene terephthalate (PET) with a thickness of 130 µm as the sub-layer and an intermediate UF membrane polyimide (PI) as the first membrane layer with thickness of 100 µm. Analysis of produced ethanol in fermentation was done by gas chromatography (Agilent 7890A series) using DB Wax column (Restek Corp., USA) at oven/column temperature of 250°C, FID temperature of 275°C, a sample volume of 3µl and a run time of 4 min. The flow rate of hydrogen, air and helium were 245,400 and 28.598 mL per min., respectively. An analysis of glucose was done by HPLC (Agilent 1100 series) at 80°C using Sugar Pak column, RID detector and Millipore water as eluent at a 0.25 ml/min flow rate and a 5 µl injection volume, for a 35-min. run time. In the PVMBR setup, ethanol yield improved to 0.46 g/g as compared to 0.43 g/g obtained in conventional batch fermentation. The ethanol productivity increased by 60%, relative to the conventional batch fermentation. In conventional system, 100% sugar was utilized in 8 h, compared to just 6 h in PVMBR. The trend of ethanol yield andspecific productivity in fermentation without and with PVMBR are shown in Figures 15.3 and 15.4, respectively. The concentration of ethanol in fermentation broth and in permeate was 7.3 and 13.6 vol%, respectively. Total flux and ethanol flux through the pervaporation unit were 257.4 g/m².h and 36.9 g/m².h, respectively. Selectivity and enrichment factor of the PDMS membrane were found to be 2.8 and 2.5, respectively. Esfahanian et al. (2012) have also reported an increase in ethanol productivity by 26.83% over conventional batch fermentation with ethanol concentration in permeate side approximately 6 to 7 times higher than that of the broth, for the PDMS membrane.

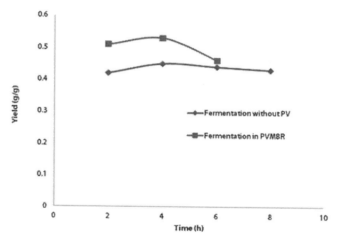

FIGURE 15.3
Trend of ethanol yield in fermentation without and with PVMBR.

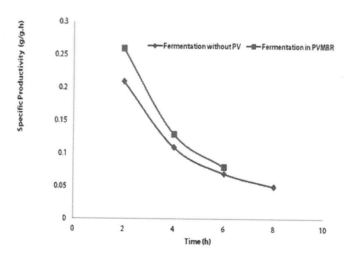

FIGURE 15.4
Trend of specific productivity in fermentation without and with PVMBR.

Most of the current research in the field of fermentation coupled with PV, as discussed above, focuses on the different membranes that can be used, the effect that components in fermentation broth has on membranes and the influence of PV on fermentation. There is, however, a definite lack of research in terms of kinetics of the membrane reactor system where fermentation and pervaporation are combined.

15.10 Economic Assessment of the Integrated Fermentation-Pervaporation Process

In terms of the investment and production cost, several studies have been reported on the economic assessment of the integrated fermentation–pervaporation process for bioethanol production. Most of the studies have given a positive conclusion in terms of high ethanol productivity and low energy consumption. Luccio et al. (2002) have reported that for a complete plant installation, the highest allowable membrane cost to maintain economic feasibility was around \$500/m^2, while the adaptation of an existing plant allowed for a cost of up to \$800/m^2, considering a minimum return rate of 17% on the investment costs. Therefore, improvement in the performance of the membrane is a critical step to reduce production costs. Wasewar et al. (2006) reported the effect of membrane flux and selectivity on the ethanol production cost. A reduction of 22% was found in the direct production cost of ethanol when the flux was increased by 200%. A decrease of 10% was observed when the separation factor increased by 20% and costs decreased by 35% for a 100% increase in separation factor, indicating the need to develop membranes with higher performance and more reliability to reduce the ethanol production cost. In the hybrid process, to reduce the effect of membrane fouling caused by cells and other by-products, microfiltration can be used to retain the remains of cells in the fermentation broth. Groot et al. (1992; 1993) have compared the different process alternatives and concluded that continuous ethanol production with recycling of the microorganism using microfiltration leads to the lowest production costs. In comparison to the batch process, the costs were 15% lower. O'Brien et al. (2000) have also reported that the capital costs for the fermenter in the case of coupled fermentation–PV, were reduced by 75% because of the 12-fold reduction in the required volume. It has been established that, with the development of a high-performance membrane, the hybrid fermentation and PV process will have good potential in ethanol production.

15.11 Conclusions

Coupling pervaporation with the fermentation process decreases ethanol inhibition, reduces water consumption and wastewater generation and increases yeast productivity. Studies have been carried out on ex-situ integration of PV with a fermenter, which is proven to be a viable method to recover ethanol from fermentation broths. The challenge is the availability of a highly ethanol selective membrane on a commercial scale and the assessment of the longer-term trials with fermentation broths to evaluate membrane stability and fouling behavior.

References

Achinas, S. and G.J.W. Euverink. 2016. Consolidated briefing of biochemical ethanol production from lignocellulosic biomass. *Electronic Journal of Biotechnology* 23: 44–53.

Aminabhavi, T.M., B.V.K. Naidu, S. Sridhar, and R. Rangarajan. 2005. pervaporation separation of water-isopropanol mixtures using polymeric membranes: Modeling and simulation aspects. *Journal of Applied Polymer Science* 95(5).

Aradhey, A. 2016. India biofuels annual. In: Global agricultural information network (GAIN) report no IN6088. USDA-Foreign Agricultural Services. https://gain.fas.usda.gov/Recent%20 GAIN%20Publications/Biofuels%20Annual_New%20Delhi_India_6-24-2016.pdf.

Aroujalian, A., K. Belkacemib, S.J. Davids, G. Turcotted, and Y. Pouliot. 2006. Effect of residual sugars in fermentation broth on pervaporation flux and selectivity for ethanol. *Desalination* 193: 103–108.

Azhar, M.S.H., R. Abdulla, S.A. Jambo, H. Marbawi, J.A. Gansau, A.A. Mohd Faik, and K.F. Rodrigues. 2017. Yeasts in sustainable bioethanol production: A review. *Biochemistry and Biophysics Reports* 10: 52–61.

Balat, M. 2007. Global bio-fuel processing and production trends. *Energy Explore Exploit* 25: 195–218.

Beaumelle, D., M. Marin, and H. Gibert. 1993. Pervaporation with organophilic membranes: State of the art. *Food Bioproduction Process* 71(C2): 77.

Behera, S., S. Kar, and R. Chandra. 2010. Comparative study of bio-ethanol production from mahula (*Madhuca latifolia* L.) flowers by *Saccharomyces cerevisiae* cells immobilized in agar agar and Ca-alginate matrices. *Applied Energy* 87: 96–100.

Bello, R.H., P. Linzmeyer, C.M.B. Franco, O. Souza, N. Sellin, S.H.W. Medeiros, and C. Marangoni. 2014. Pervaporation of ethanol produced from banana waste. *Waste Management* 34: 1501–1509.

Biswas, A.K. and N. Manocha. 2014. India's biofuel policies: Progress or boondoggle? *NGI Knowledge Exchange* 5: 6–8.

Boyaval, P., C. Lavenant, G. Gésan, and G. Daufin. 1996. Transient and stationary operating conditions on performance of lactic acid bacteria crossflow microfiltration. *Biotechnology Bioengineering* 49: 78–86.

Chen, C., X. Tang, Z. Xiao, Y. Zhou, Y. Jiang, and S. Fu. 2012. Ethanol fermentation kinetics in a continuous and closed-circulating fermentation system with a pervaporation membrane bioreactor. *Bioresource Technology* 114: 707–710.

Chen, H. and W. Qui. 2010. Key technologies for bioethanol production from lignocellulose. *Biotechnology Advances* 28: 556–562.

Chovau, S., S. Gaykawad, A.J.J. Straathof, and B. Van der Bruggen. 2011. Influence of fermentation by-products on the purification of ethanol from water using pervaporation. *Bioresource Technology* 102: 1669–1674.

CropEnergies. 2017. Site accessed on September 8, 2017: http://www.cropenergies.com/en /Bioethanol/Markt/Dynamisches_Wachstum.

"Current Outlook, Profitability, and Weather Information." Iowa State University, University Extension. Accessed May 17, 2017: www.extension.iastate.edu/agdm/info/outlook.html.

Dhabhai, R. 2012. Ph.D. thesis submitted on "Studies on bioethanol production from lignocellulosic feedstock" to the Malaviya National Institute of Technology. Jaipur, India.

Dhabhai, R., S.P. Chaurasia, and A.K. Dalai. 2012. Efficient bioethanol production from Glucose Xylose mixtures using co-culture of Saccharomyces cerevisiae immobolised on Canadian pine wood chips and free Pichia stipitis. *Journal of Biobased Materials and Bioenergy* 6: 594–600.

Dhabhai, R., S.P. Chaurasia, and A.K. Dalai. 2012. Influence of pretreatment conditions on composition of liquid hydrolysate and subsequent enzymatic saccharification of remaining solids. *The Canadian Journal of Chemical Engineering* 91(7): 1223–1228.

Dhabhai, R., S.P. Chaurasia, and A.K. Dalai. 2013. Effect of pretreatment conditions on structural characteristics of wheat straw. *Chemical Engineering Communications* 200(9): 1251–1259.

Dhabhai, R., S.P. Chaurasia, A.K. Dalai, and A.K. Singh. 2013. Kinetics of bioethanol production employing mono and co culture of Saccharomyces cerevisiae and Pichia stipites. *Chemical Engineering & Technology* 36: 1651–1657.

Di Luccio, M., C.P. Borges, and T.L.M. Alves. 2002. Economic analysis of ethanol and fructose production by selective fermentation coupled to pervaporation: Effect of membrane costs on process economics. *Desalination* 147: 161–166.

Energy Information Administration. 2016. Use of Energy in the United States Explained: Energy Use for Transportation. U.S. Department of Energy. Accessed on September 4, 2017: http://www.eia.gov/energyexplained/?page=us_energy_transportation.

Esfahaniana, M., A.H. Ghorbanfarahi, A.A. Ghoreyshi, G. Najafpoura, H. Younesib, and A.L. Ahmad. 2012. Enhanced bioethanol production in batch fermentation by pervaporation using a PDMS membrane bioreactor. *International Journal of Engineering, Transactions B: Applications* 25(4): 249–258.

Fadeev, A.G., S.S. Kelley, J.D. McMillan, Y.A. Selinskaya, V.S. Khotimsky, and V.V. Volkov. 2003. Effect of yeast fermentation by-products on poly[1-(trimethylsilyl)-1-propyne] pervaporative performance. *Journal of Membrane Science* 214: 229–238.

Fan, S., Z. Xiao, M. Li, S. Li, T. Zhou, and Y. Hu. 2016. Pervaporation performance in PDMS membrane bioreactor for ethanol recovery with running water and air as coolants at room temperature. *Journal of Chemical Technology and Biotechnology* 92: 292–297.

Fan, S., Z. Xiao, M. Li, and S. Li. 2017. Ethanol fermentation coupled with pervaporation by energy efficient mechanical vapor compression. *Energy Procedia* 105: 933–938.

Fan, S., Z. Xiao, Y. Zhang, X. Tang, C. Chen, W. Li, P. Yao, and Q. Deng. 2014. Enhanced ethanol fermentation in a pervaporation membrane bioreactor with the convenient permeate vapor recovery. *Bioresource Technology* 155: 229–234.

Fu, C., D. Cai, S. Hu, Q. Miao, Y. Wang, P. Qin, Z. Wang, and T. Tan. 2016. Ethanol fermentation integrated with PDMS composite membrane: An effective process. *Bioresource Technology* 200: 648–657.

García, M., M.T. Sanz, and S. Beltran. 2009. Separation by pervaporation of ethanol from aqueous solutions and effect of other components present in fermentation broths. *Journal of Chemical Technology and Biotechnology* 84: 1873–1882.

Gaykawad, S.S., Y. Zha, P.J. Punt, L.A.M. vander Wielen, and A.J.J. Straathof. 2013. Pervaporation of ethanol from lignocellulosic fermentation broth. *Bioresource Technology* 129: 469–476.

Global Energy Statistical Yearbook. 2017. Site accessed on October 5, 2016: yearbook.enerdata.net/total-energy/world-consumption-statistics.html

Groot, W.J., M.R. Kraayenbrink, R.H. Waldram, R.G.J.M. Lans, K.C.A.M. Luyben. 1992. Ethanol production in an integrated process of fermentation and ethanol recovery by pervaporation. *Bioprocess Engineering* 8: 99–111.

Groot, W.J., M.R. Kraayenbrink, R.H. Waldram, R.G.J.M. Lans, and K.C.A.M. Luyben. 1993. Ethanol production in an integrated fermentation/membrane system. Process simulations and economics. *Bioprocess Engineering* 8: 189–201.

Hiloidhari, M., D. Das, and D.C. Baruah. 2014. Bioenergy potential from crop residue biomass in India. *Renewable Sustainable Energy Review* 32: 504–512.

Ikegami, T., H. Yanagishita, D. Kitamoto, K. Haraya, T. Nakane, H. Matsuda, N. Koura, and T. Sano. 1997. Production of highly concentrated ethanol in a coupled fermentation/pervaporation process using silicalite membranes. *Biotechnology Techniques* 11: 921–924.

Ikegami, T., H. Yanagishita, D. Kitamoto, K. Haraya, T. Nakane, H. Matsuda, N. Koura, and T. Sano. 1999. Highly concentrated aqueous ethanol solutions by pervaporation using silicalite membrane—Improvement of ethanol selectivity by addition of sugars to ethanol solution. *Biotechnology Letter* 21: 1037–1041.

Ikegami, T., H. Yanagishita, D. Kitamotoa, H. Negishi, K. Haraya, and T. Sano. 2002. Concentration of fermented ethanol by pervaporation using silicalite membranes coated with silicone rubber. *Desalination* 149: 49–54.

Ishola, M.M., T. Brandberg, and M.J. Taherzadeh. 2015. Simultaneous glucose and xylose utilization for improved ethanol production from lignocellulosic biomass through SSFF with encapsulated yeast. *Biomass Bioenergy* 77: 192–199.

Ivanova, V., P. Petrova, and J. Hristov. 2011. Application in the ethanol fermentation of immobilized yeast cells in matrix of alginate/magnetic nanoparticles, on chitosan-magnetite microparticles and cellulose coated magnetic nanoparticles. *International Review of Chemical Engineering* 3: 289–299.

Jaimes, J.H.B., M.E.T. Alvarez, J.V. Rojas, and R.M., Filho. 2014. Pervaporation: Promissory method for the bioethanol separation of fermentation. *Chemical Engineering Transactions* 38.

Kaewkannetra, P., N. Chutinate, S. Moonamart, T. Kamsan, and T.Y. Chiu. 2011. Separation of ethanol from ethanol–water mixture and fermented sweet sorghum juice using pervaporation membrane reactor. *Desalination* 271: 88–91.

Kaewkannetra, P., N. Chutinate, S. Moonamart, T. Kamsan, and T.Y. Chiu. 2012. Experimental study and cost evaluation for ethanol separation from fermentation broth using pervaporation. *Desalination and Water Treatment* 41: 88–94.

Kaseno, I. Miyazawa, and T. Kokugan. 1998. Effect of product removal by a pervaporation on ethanol fermentation. *Journal of Fermentation and Bioengineering* 86(5): 488–493.

Kujawski, W. 2000. Application of pervaporation and vapor permeation in environmental protection. *Polish Journal of Environmental Studies* 9(1): 13–26.

Lewandowska, M. and W. Kujawski. 2007. Ethanol production from lactose in a fermentation/pervaporation system. *Journal of Food Engineering* 79(2): 430–437.

Mohammadi, T., A. Aroujalian, and A. Bakhshia. 2005. Pervaporation of dilute alcoholic mixtures using PDMS membrane. *Chemical Engineering Science* 60: 1875–1880.

Nguyen, Q., J. Bowyer, J. Howe, S. Bratkovich, H. Groot, E. Pepke, and K. Fernholz. 2017. Global production of second generation biofuels: Trends and influences. Report accessed at www.dovetailinc.org/report_pdfs/2017/dovetailbiofuels0117.pdf: 1–15.

Nikolic, S., L. Mojovic, D. Pejin, M. Rakin, and M. Vukasinovic. 2010. Production of bioethanol from corn meal hydrolysates by free and immobilized cells of Saccharomyces cerevisiae var. ellipsoideua. *Biomass and Bioenergy* 85: 1750–1755.

Nomura, M., T. Bin, and S. Nakao. 2002. Selective ethanol extraction from fermentation broth using a silicalite membrane. *Separation and Purification Technology* 27: 59–66.

O'Brien, D.J. and J.C. Craig. 1996. Ethanol production in a continuous fermentation/membrane pervaporation system. *Applied Microbiology Biotechnology* 44: 699–704.

O'Brien, D.J., G.E. Senske, M.J. Kurantz, and J.C. Craig. 2004. Ethanol recovery from corn fiber hydrolysate fermentations by pervaporation. *Bioresource Technology* 92: 15–19.

O'Brien, D.J., L.H. Roth, and A.J. McAloon. 2000. Ethanol production by continuous fermentation–pervaporation: A preliminary economic analysis. *Journal of Membrane Science* 166(1): 105–111.

Offeman, R.D. and C.N. Ludvik. 2011. Poisoning of mixed matrix membranes by fermentation components in pervaporation of ethanol. *Journal of Membrane Science*. 367: 288–295.

Peng, P., B. Shi, and Y. Lan. 2010. A Review of membrane materials for ethanol recovery by pervaporation. *Separation Science and Technology* 46(2): 61–73.

Prasad, S., A. Singh, and H.C. Joshi. 2007. Ethanol as an alternative fuel from agricultural, industrial and urban residues. *Resources, Conservation and Recycling* 50: 1–39.

Sanchez, O. and C.A. Cardona (2008). Trends in biotechnological production of fuel ethanol from different feedstock. *Bioresource Technology* 99: 5270–5295.

Sarkar, N., S.K. Ghosh, S. Bannerjee, and K. Aikat. 2012. Bioethanol production from agricultural wastes: An overview. *Renewable Energy* 37: 19–27.

Schmidt, S., M. Myers, S. Kelley, J. McMillan, and N. Padukone. 1997. Evaluation of PTMSP membranes in achieving enhanced ethanol removal from fermentations by pervaporation. *Applied Biochemistry and Biotechnology* 63: 469–482.

Singh, S., A. Adak, M. Saritha, S. Sharma, R. Tiwari, S. Rana, A. Arora, and L. Nain. 2017. Bioethanol production scenario in India: Potential and policy perspective. *Sustainable Biofuels Development in India*.

Staniszewski, M., W. Kujawski, and M. Lewandowska. 2007. Ethanol production from whey in bio-reactor with co-immobilized enzyme and yeast cells followed by pervaporative recovery of product—Kinetic model predictions. *Journal of Food Engineering* 82: 618–625.

Staniszewski, M., W. Kujawski, and M. Lewandowska. 2009. Semi-continuous ethanol production in bioreactor from whey with co-immobilized enzyme and yeast cells followed by pervapora-tive recovery of product—Kinetic model predictions considering glucose repression. *Journal of Food Engineering* 91: 240–249.

Thongsukmak, A. and K.K. Sirkar. 2007. Pervaporation membranes highly selective for solvents present in fermentation broths. *Journal of Membrane Science* 302(1): 45–58.

Vane, L.M. 2004. "Options for combining pervaporation membrane systems with fermentors for efficient production of alcohols from biomass." U.S. Environmental Protection Agency. AIChE Annual Meeting.

Vane, L.M. 2005. Review: A review of pervaporation for product recovery from biomass fermenta-tion processes. *Journal of Chemical Technology and Biotechnology* 80: 603–629.

Wasewar, K. and V. Pangarkar. 2006. Intensification of recovery of ethanol from fermentation broth using pervaporation: Economical evaluation. *Chemical and Biochemical Engineering Quarterly* 20: 135–145.

Wu, Y., Z. Xiao, W. Huang, and Y. Zhong. 2005. Mass transfer in pervaporation of active fermenta-tion broth with a composite PDMS membrane. *Separation and Purification Technology* 42: 47–53.

Younan, L., C. Zhu, and M. Liu. 1992. Removal of ethanol from continuous fermentation broth by pervaporation. *Journal of Chemical Industry and Engineering (China)* 7(1).

Yu, J., X. Zhang, and T. Tan. 2007. A novel immobilization method of Saccharomyces cerevisiae to sorghum bagasse for ethanol production. *Journal of Biotechnology* 129: 415–420.

16

Recovery of Value-Added Products in Process Industries through Membrane Contactors

M. Madhumala, Rosilda Selvin, and Sundergopal Sridhar

CONTENTS

16.1 Introduction

Separation methods are being applied mainly for the purification and concentration of valuable products besides recovery of reactants. On the whole, separation processes have found to account for nearly 45 to 50 percent of capital and operating costs within industries, every year. Furthermore, the industrial objective to improve cost-effectiveness, boost energy efficiency, increase productivity and prevent pollution necessitates the demand for more efficient, alternative separation processes. In response to these needs, scientific bodies have begun to endorse the development of high-risk, innovative separation technologies. In particular, membrane technology has grown significantly because of increasingly stringent requirements for product purity and environmental norms, which enforce zero liquid discharge.

In the chemical process industry, separation and purification methods assume considerable significance to enable isolation of end products, recovery of intermediates and recycle of reactants. The separation of useful chemical entities from aqueous streams released by various industries is important from both economic and environmental points of view. Recently, there has been a growing demand for production of fermentative products, such as organic acids, due to their wide and varied application in the chemical, food, beverage, textile, plastic, pharmaceutical and cosmetic industries (John et al., 2007). Additionally, they are also being used as primary feedstocks for synthesis of various biodegradable polymers and polyesters. Levulinic acid is an important carboxylic acid that has gained significant potential, particularly in pharmaceutical, agricultural, cosmetic, chemical, and food industries (Erickson et al., 2012). It can be catalytically converted to fuel additives, such as gasoline and biodiesel for which a patent has been granted in 2009 for a process involving permeation enhanced reactive extraction of levulinic acid using a membrane wherein the acid was brought in contact with the liquid alcohol phase (Boestert et al., 2009). Acrylic acid is another beneficial chemical that is widely used for industrial production of polymers, esters, plastics, adhesives, coatings, elastomers, floor polishes and paints. It is mainly produced via the chemical conversion of propylene or fermentative production from sugars. The typical concentration of carboxylic acids produced by the fermentation of sugars and molasses or by chemical conversion methods is usually 10% w/v in the reaction media (Ravishankar, 2016). Purification of an aqueous stream containing low acid concentration is very difficult due to their high solubility in water. Several isolation techniques such as adsorption onto a porous solid surface, evaporation, distillation, solvent extraction and precipitation have been used for separation of organic acids from fermentation broths in the recent past (Li et al., 2016). Recently, the application of liquid-liquid membrane contactor system employing a hydrophobic membrane barrier was found to be a new approach for separating carboxylic acids from aqueous streams. These systems can also be used in extracting biochemical, pharmaceutical and hydrometallurgical products. Due to the advantages of increased productivity, the ease of pH control in a fermenter, the reduction in process wastes and production/recovery costs, the membrane-based reactive extraction method allows continuous separation of carboxylic acids in a single step. Acid extraction in these systems is possible by the use of an organic solvent system on the other side of the membrane wherein an extractant is dissolved in the diluent (Nymeijer et al., 2004). The performance of the process mainly depends on parameters, such as distribution coefficient, extraction efficiency, acid-amine loading ratio, reaction constants, solvent type, pH and temperature of the solution. Extensive research on the development of a cost-effective membrane contactor system with the employment of less expensive solvents for continuous separation of carboxylic acids from fermentation broth is still ongoing.

Hexane, a saturated hydrocarbon is one of the important volatile solvents obtained by the distillation of petroleum (Stellman et al., 1998). Due to its low boiling temperature of 69°C, less sensible heat of 335 kJ/kg and high solubility with oil-bearing materials, hexane has acquired great importance in the chemical industry, mainly in the extraction of vegetable oils from their natural sources (Kumar et al., 2017). The miscella after oil extraction usually contains 20–30% w/w of oil along with a large quantity of solvent. Conventional distillation process as a standalone for concentrating the oil from 20 wt% to 50 wt% or even higher is considered to be disadvantageous as it consumes a large amount of heat load and vitiates the product quality by thermal degradation. The use of solvent-resistant polymeric membranes exhibiting sufficient permeation properties in combination with a conventional distillation process could be advantageous for processing crude oil-solvent miscella. Figure 16.1 shows the schematic of conventional distillation/membrane distillation hybrid process, wherein the mixture of oil-hexane miscella (usually 25–35% w/w oil) from the extractor is initially fed to a conventional distillation unit to remove the maximum amount of hexane and concentrate the miscella to 70–90% w/w oil, with low enthalpy input, owing to a low concentration of the high boiling oil miscella. The concentrated miscella is then pumped into the membrane distillation column operated under downstream vacuum to achieve 99% of oil concentration, by selectively removing hexane present in the feed at ≤ 20%. The hexane recovered in two stages can be recycled back to the reactor for the continuous extraction of oil from seeds. Most researchers have employed ultrafiltration, nanofiltration and RO using membranes made of polyvinylidene fluoride (PVDF), polysulfone, polydimethylsiloxane (PDMS), polyethersulfone (PES), aromatic polyimide (PI), polyamide and cellulose acetate polymers. Tubular ceramic modules have also been used for hexane solvent recovery, due to the intrinsic properties of uniform pore size distribution besides chemical, thermal and mechanical stabilities. In spite of high solvent flux recovery, some solute loss into permeate stream was also encountered when operated at high pressures. Tres et al. (2012) applied a commercial, hollow fiber UF membrane module made of PES/poly (vinylpyrrolidone) blend for the separation of refined *n*-hexane/soybean oil mixture. With an increasing mass ratio of oil in hexane, solute rejection and total permeate flux were found to increase from 10% to 28.7% and 12.2 to 65.3 kg/m^{2}h, respectively. Pagliero et al. (2011) studied separation of sunflower oil from hexane using a flat sheet composite made of PVDF substrate having a silicone rubber or cellulose acetate membrane as the selective top layer, as well as a commercial PI substrate coated with an active PDMS layer. The PVDF-Si membrane exhibited better results with a limiting flux of 12 L/m^{2}h and oil rejection of 46.2% at a trans-membrane pressure of 7.8 bar and a temperature of 30°C. Thus, separation of solvent from miscella by a laboratory-scale, cross-flow filtration cell under ambient temperature and low-pressure conditions offers a promising alternative to distillation. Here, we will be discussing the application of a liquid-liquid membrane contactor and vacuum membrane distillation (VMD) processes employing indigenous solvent resistant polymeric and commercial membranes for recovery of volatile organic solvents and carboxylic acids.

This article intends to explain the application of membrane processes such as liquid-liquid membrane contactors and membrane distillation for the effective separation and

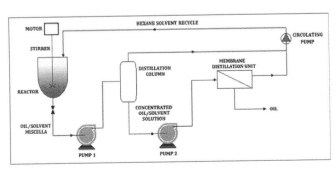

FIGURE 16.1
Hybrid conventional distillation/membrane distillation process for hexane recovery from vegetable oil miscella.

recovery of valuable chemicals, such as levulinic acid, acrylic acid from aqueous stream and hexane solvent from sunflower oil miscella. A brief coverage on potential problems faced by industries during the separation of chemicals such as organic acids and volatile solvents using conventional methods is provided. This is followed by a description of membrane processes and the principles of operation with an emphasis on mass transfer aspects. Details on synthesis of microporous membranes, specifications of commercial membranes, and experimental procedures followed for the separation of chemicals are also mentioned. The influence of membrane type and operating parameters on separation performance is discussed. Finally, outlines of the conclusions drawn while investigating the performance of indigenous membranes for selective separation of valuable chemicals from aqueous solution are presented.

16.2 Introduction to Membrane Contactor Systems

16.2.1 Liquid-Liquid Membrane Contactors

In recent years, liquid-liquid membrane contactors have been playing a major role in enabling an easier recovery of carboxylic acids from various process streams (Drioli et al., 2005; Drioli et al., 2009). Separation of acid in these systems is accomplished by passing the feed solution and organic solvent on either side of a hydrophobic microporous membrane. Polymeric membrane materials exhibiting outstanding properties of a high contact angle ($\theta > 90°$), chemical and thermal stability are typically used for contactor applications. Figure 16.2 (a) shows schematic of liquid water droplet on a hydrophobic membrane surface. Commercial membranes made up of hydrophobic materials, such as polytetrafluoroethylene (PTFE), polypropylene (PP), polyethylene (PE) and PVDF, have been used so far for contactor applications (Pabby et al., 2008). Table 16.1 shows the details of contact angle values for different commercial and modified hydrophobic membrane materials in an air-water system at ambient temperature. The modification of hydrophobic membranes can be done using radiation graft polymerization (Khayet & Matsuura 1992), plasma polymerization (Kong et al., 1992), surface coating techniques (Jin et al., 2008) and even modifying the properties of hydrophilic polymer material by adding tetraethoxysilane (TEOS) crosslinker into the doping solution during membrane preparation by phase inversion technique (Heba et al., 2015). These devices are very compact and exhibit easy diffusion of acid molecules at liquid-membrane interface, which thereby creates a large contact area for efficient mass transfer. Additionally, they are modular, enabling easy scale-up and can be efficiently integrated into new or pre-existing production lines. Figure 16.2 (b) shows a typical liquid-liquid membrane contactor system of two liquid phases separated by a porous hydrophobic barrier coupled with simultaneous regeneration of organic solvent. The hydrophobic nature of the membrane avoids direct contact of fermentative product with toxic organic solvent. During forward extraction of acid, the aqueous stream flowing on the feed side will not wet the membrane, whereas the organic solvent used at the permeate side *will* wet the membrane, implying the formation of an interphase at the feed/membrane side. The acid extracted from the aqueous medium into the organic solvent is stripped using a second membrane contactor unit. Further, the process enables continuous

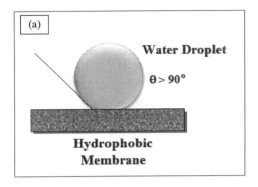

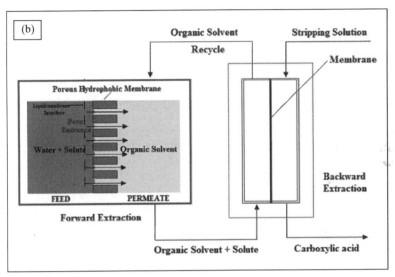

FIGURE 16.2
(a) Schematic of water droplet on hydrophobic membrane surface (b) membrane contactor system for continuous acid extraction coupled with simultaneous solvent regeneration.

regeneration of organic solvent with long-term stability and regulates fermenter pH without the addition of base. So far, this technology has been successfully implemented for the extraction of several organic acids, including lactic acid, propionic acid, acetic acid, butyric acid and formic acid, etc.

Transport of the solute through the membrane during forward extraction process takes place as a result of concentration gradient (Δc) acting on the components in the feed. A component is transferred from the feed phase to the permeate phase in these three steps:

1. Transport of component from the upstream boundary layer in the feed phase to the membrane surface.
2. Diffusion of component through the membrane pore.
3. Transfer of component from the downstream membrane-liquid interface to the bulk permeate phase.

TABLE 16.1

Contact Angle Values of Various Hydrophobic Polymeric Membranes for Contactor Applications

Membrane Material	Structure	Contact Angle (θ)	Reference
Polyvinylidene fluoride	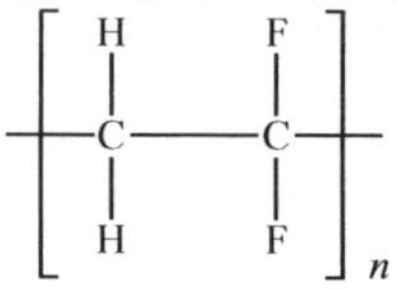	89°	Sengupta and Pittman (2008)
Polyethylene	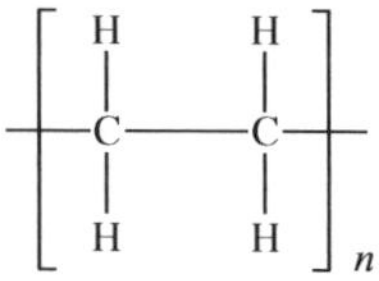	96°	Sengupta and Pittman (2008)
Cellulose Acetate (Modified)	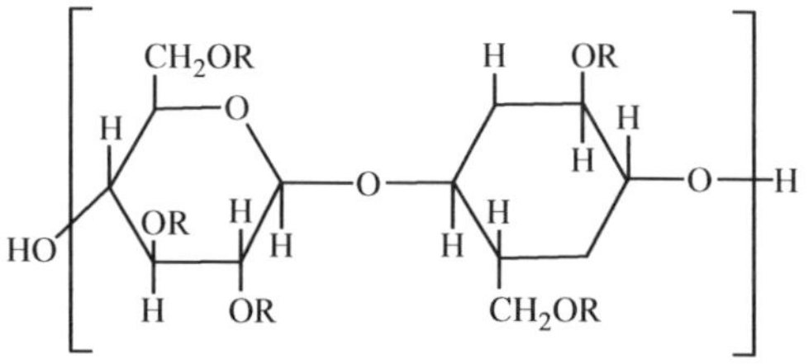 R: CH₃CO	96°	Khayet and Matsuura (2011)
Polyethersulfone	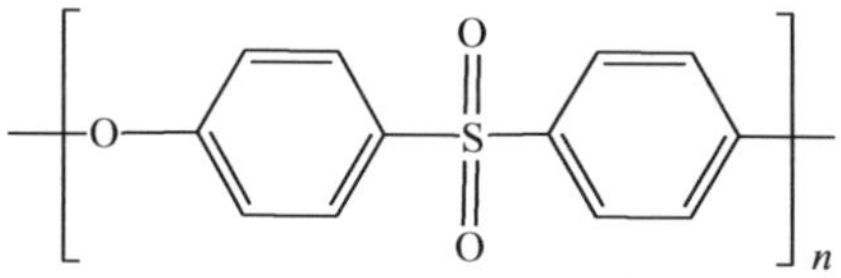	98 ± 2°	Heba et al. (2015)
Polypropylene	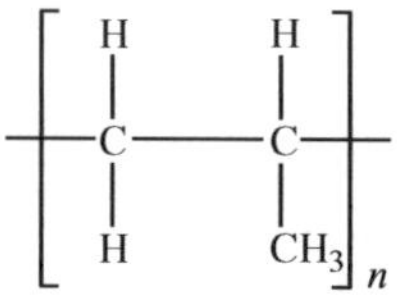	108°	Sengupta and Pittman (2008)
Poly(phthalazinone ether sulfone ketone)	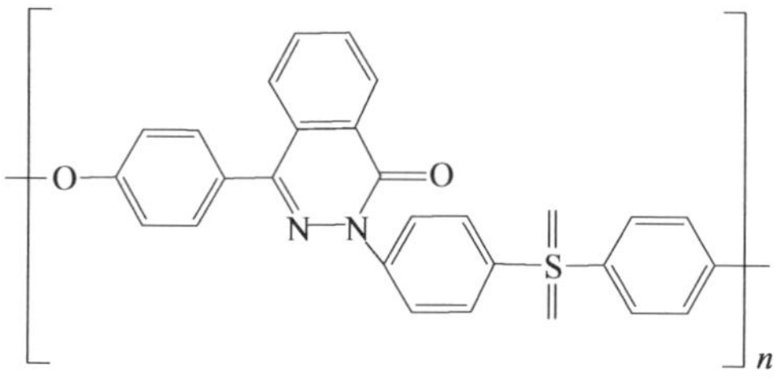	110°	Jin et al. (2008)
Polytetrafluoroethylene	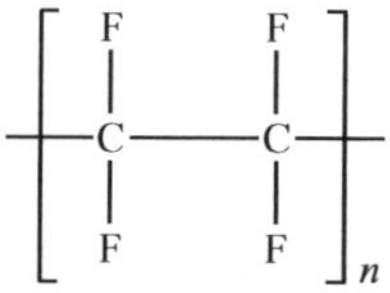	112°	Johnson and Nguyen (2017)

(Continued)

TABLE 16.1 (CONTINUED)

Contact Angle Values of Various Hydrophobic Polymeric Membranes for Contactor Applications

Membrane Material	Structure	Contact Angle (θ)	Reference
Polyvinyl chloride		116.4°	Demirci et al. (2014)
Cellulose Nitrate (Modified)		120°	Kong et al. (1992)

The flux (J) of the component can be expressed in terms of overall mass transfer coefficient (K_{ov}):

$$J = K_{ov} \times \Delta c \tag{16.1}$$

$$\frac{1}{K_{ov}} = \frac{2}{k_f\left(feed\ phase\right)} + \frac{1}{k_m\left(membrane\right)} + \frac{1}{k_p\left(permeate\ phase\right)} \tag{16.2}$$

If the mass transfer resistance ($1/k$) is completely dominated by the membrane phase, then Equation 16.1 reduces to:

$$J = \frac{D * k_d}{l} \times \Delta c \tag{16.3}$$

To prevent the organic solvent from permeating into the feed solution, the pressure of the latter is maintained slightly higher. Furthermore, the feed solution does not displace the fluid in the pores as long as the pressure on the feed side is kept below a critical value known as the breakthrough pressure. For liquid/liquid extraction, the breakthrough pressure is given by the Young-Laplace equation:

$$\Delta p = \frac{2\gamma \cos\theta}{r} \tag{16.4}$$

16.2.2 Membrane Distillation

Membrane distillation (MD) is another emerging membrane contactor technology that has gained worldwide importance due to its inherent features of process safety, low cost and separation capability when compared to conventional processes, such as distillation and evaporation (Alkhudhiri et al., 2012). The key advantages of MD over conventional separation processes include its relatively lower energy consumption compared to distillation, a considerable separation of dissolved and non-volatile species, its reduced vapor space as compared to conventional multi-stage flash distillation, a lower operating pressure than pressure-driven membrane processes, a lower operating temperature and corrosion-related problems compared to evaporation (Sharmiza et al., 2012). The large vapor space required by a conventional distillation column is replaced in MD by the pore volume of a microporous membrane of around 100 μm thickness. While conventional distillation relies on extensive vapor-liquid contact, MD employs a hydrophobic microporous membrane to permit a vapor-liquid interface.

Membrane fouling is a smaller problem in MD than in other membrane processes because the pores are relatively large compared to the 'pores' or diffusional pathways in nanofiltration or ultrafiltration, which do not get easily clogged since vacuum is applied at the downstream side in MD rather than pressure at the upstream surface (Lawson & Lloyd, 1997). Its applications include dehydration of high boiling solvents, separation of volatile inorganic acids from effluents, recovery of volatile organic solvents and useful chemical entities, concentration of agro-based and organic solutions, desalination of brackish water and concentration of sugar solutions (Pangarkar et al., 2016; Kiai et al., 2014; Alves & Coelhoso, 2006).

16.2.2.1 Flux and Selectivity Calculations

Flux J (kg/m²h), is determined as the amount of liquid transported though the membrane per unit time, per unit membrane area.

$$J = \frac{W}{A_m \times t}$$

(16.5)

Membrane selectivity (α) is defined as:

$$\alpha = \frac{Y(1-X)}{X(1-Y)}$$

(16.6)

16.3 Indigenous Membranes and Commercial Modules Investigated for Separation of Chemical Entities

Hydrophobic membranes for liquid-liquid membrane contactor and membrane distillation applications were prepared by phase inversion technique. PVDF/PVP polymer solution was prepared by dissolving 12 g of PVDF polymer in 88 mL of dimethylformamide solvent containing 6 g of PVP as an additive. The dope solution was then cast over a supporting macroporous polyester nonwoven fabric of 100 μm thickness using a doctor blade. Immediately after casting the solution, the non-woven backing with the plate was sumerged for 10 min. in a non-solvent bath of ice-cold water to obtain an ultra-porous membrane. The ultraporous PVC membrane for MD application was prepared using a homogenous polymer solution containing 4.75 g of PVC polymer and 20.25 mL of N-Methyl-2-pyrrolidone (NMP) solvent. The property of PVC membrane for MC application was altered using additives such as $LiCl_2$, $ZnCl_2$, polyethylene glycol (PEG)-6000 and glycerol. Four different polymer dope solutions containing 20 wt% PVC were prepared using NMP as solvent with additive in each dope constituting 20% of polymer weight. A ceramic tubular module with 0.5 cm inner diameter (ID), 1 cm outer diameter (OD), 0.6 m length and 3 kDa pore size was procured from TAMI Industries, France and operated under cross-flow filtration mode for hexane solvent recovery from sunflower oil miscella. The feed solution flows tangential to the direction of the membrane surface producing two streams including liquid passing through the membrane termed as permeate, whereas the remaining fraction is called retentate.

16.4 Results and Discussion

16.4.1 Case Study 1: Reactive Extraction of Carboxylic Acids Using Indigenous Liquid-Liquid Membrane Contactor System

Studies on the reactive extraction of carboxylic acids were performed using indigenous membrane-based stirred assembly (Madhumala et al., 2014). Equal volumes (89 mL) of aqueous acid solution and organic solvent comprising tri-*n*-octylamine (TOA) extract-ant dissolved in 1-octanol were charged into two glass chambers separating a porous hydrophobic membrane. The solutions present in both the chambers were kept under stirring at a constant rate of 400 revolutions/min (rpm) using overhead and magnetic stirrer. The extractant present in the organic solution facilitated the transport of acid molecules from bulk of aqueous solution through the membrane into a bottom glass chamber containing organic solution. The concentration of acid present in aqueous and organic solutions was determined using acid-base titration method and material bal-ance, respectively.

16.4.1.1 *Reactive Extraction of Levulinic Acid from Industrial Effluent Using Microporous Polyvinyl Chloride Membrane*

a. Influence of Additives on Extraction of Levulinic Acid

Figure 16.3 shows the performance of an indigenous PVC membrane on extraction efficiency with respect to different additives for initial levulinic acid content of 575 mol/m^3 and TOA concentration of 10% (v/v in 1-octanol). The membrane pre-pared with glycerol as additive exhibited a higher extraction efficiency (21.74%) as compared to other formulations containing $LiCl_2$, $ZnCl_2$ or PEG as additives. This observation is attributed to a higher pore size formed in the PVC membrane dur-ing phase inversion, as revealed by SEM analysis (Figure 16.4). During the phase inversion process, the solvent and glycerol molecules are replaced with water in the precipitation bath.

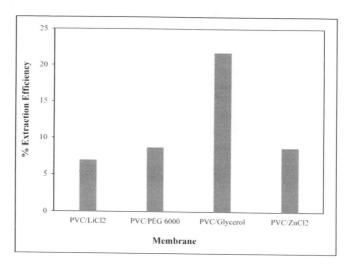

FIGURE 16.3
Influence of additive loading on extraction efficiency of levulinic acid for initial acid concentration of 575 mol/m³ and tri-*n*-octylamine concentration of 10% (v/v) in 1-octanol.

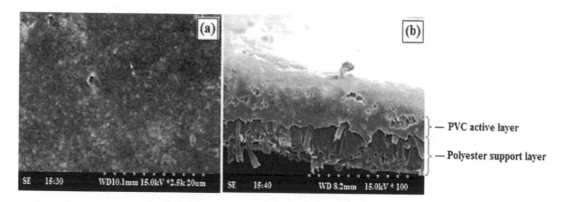

FIGURE 16.4
SEM image of PVC/glycerol membrane (a) surface and (b) cross-section.

16.4.1.2 *Reactive Extraction of Acrylic Acid from Synthetic Solution Using Ultraporous Polyvinylidene Fluoride (PVDF)/Polyvinylpyrrolidone (PVP) Blend Membrane*

a. Effect of Amine Concentration on Extraction Efficiency

Figure 16.5 shows the effect of initial TOA concentration on % extraction efficiency of acrylic acid. The initial concentrations of acrylic acid in water and tri-*n*-octylamine in 1-octanol solution varied from 100–500 mol/m³ and 206–620 mol/ m³, respectively. It was observed that increasing content of the amine extractant in organic solvent resulted in enhancement of the extent of acrylic acid recovery in the permeate side. For an acrylic acid concentration of 100 mol/m³, PVDF/ PVP blend membrane exhibited significant improvement in extraction efficiency from 30.64% to 54.74% within a processing time of 1 h.

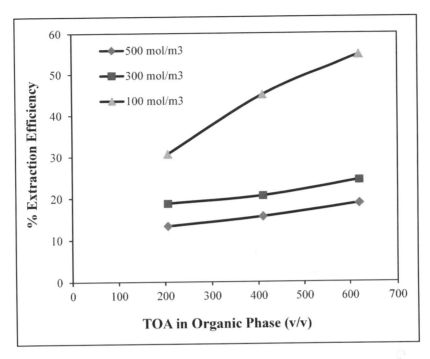

FIGURE 16.5
Effect of tri-*n*-octylamine concentration on extraction efficiency at different initial acrylic acid concentrations.

16.4.2 Case Study 2: Recovery of Hexane Volatile Solvent from Sunflower Oil Miscella Using Ultraporous PVC Membrane by Vacuum Membrane Distillation

Bench scale experiments were performed using indigenous a laboratory MD system equipped with MD manifold assembly, an overhead stirrer, vapor condensation trap and a vacuum pump (Madhumala et al., 2014). The synthesized porous PVC membrane was placed over a porous SS 316 metal support separating two bell-shaped glass pipe reducers. Hexane-oil mixture of a known composition was charged into the upper compartment, whereas the downstream side of the cell was connected to a vacuum pump. A condenser placed in between the vacuum pump and lower compartment of the cell condensed the hexane vapors present in the feed mixture. Permeate obtained was further analyzed using a refractometer to determine the purity of hexane solvent.

16.4.2.1 Effect of Downstream Pressure on Membrane Selectivity and Total Flux

Figure 16.6 shows the effect of downstream pressure on membrane selectivity and total flux under a downstream pressure ranging from 2.5 mmHg to 8 mmHg at constant hexane-oil feed composition (80% v/v), using a porous PVC membrane. The increment in permeate pressure resulted in a reduction in total flux, from 0.156 to 0.147 kg/m²h, while selectivity toward hexane was found to increase from 8.46 to 22.47. The reduction in permeate flux is a consequence of the decrease in vapor pressure gradient. Another set of trials were carried out using commercial tubular ceramic membrane module, with observations shown in Table 16.2. By increasing the concentration of hexane in oil from 40 to 80% (v/v), the flux was found to increase from 0.062 to 0.32 kg/m²h. The study revealed

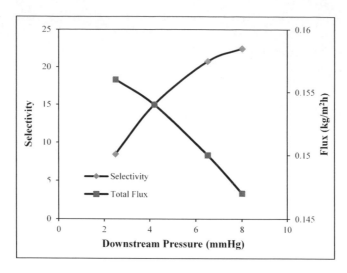

FIGURE 16.6
Effect of downstream pressure on membrane selectivity and total flux for initial hexane concentration of 80% (v/v) in sunflower oil.

TABLE 16.2

Performance of Commercial Tubular Ceramic Membrane Module at Constant Vacuum of 20 mmHg

S. No.	Concentration of hexane in oil (%volume)	RI value at 20°C		Flux (kg/m²h)
		Feed	Permeate	
1.	20	1.4623	1.3799	0.04
2.	40	1.4474	1.3805	0.06
3.	50	1.4382	1.3812	0.07
4.	60	1.4292	1.3833	0.16
5.	70	1.4188	1.3842	0.25
6.	80	1.4017	1.3827	0.32

*RI value of pure hexane = 1.3794.

that both indigenously synthesized PVC and commercial ceramic membranes exhibited hexane purity of > 95% in permeate, owing to a low boiling point (68°C) and high vapor pressure of the solvent.

16.5 Conclusions and Future Prospects

The application of membrane technology has been increasing in various sectors of process industries due to such advantages as process simplicity, reliability, low energy requirement and high contact area per unit volume. Membrane contactor systems have become one of the most advancing technologies for the separation of both gaseous and liquid mixtures in various industries. It is very important to find a sustainable and economically feasible process for separation of carboxylic acids from the fermentation broth. Amine-based reactive extraction using a hydrophobic porous barrier offers a large mass transfer area for continuous extraction of acid from fermentation broths at reasonably low production costs. Membrane distillation is an emerging technology that has attracted research interest in solving separation problems in many areas as it can be operated at a temperature lower than that required in conventional evaporation or distillation. A hybrid process combining conventional distillation with membrane distillation shows great potential for oil concentration, due to reduction in operating cost and improvement in quality of oil. Further studies could focus on scale-up of these membranes for commercialization of membrane distillation technology in industries.

Nomenclature

Δp	Pressure difference (N/m^2)
r	Membrane pore radius (m)
l	Membrane thickness (mm)
J	Flux of permeating component (kg/m^2h)
Δc	Concentration gradient (mol/m^3)
K	Overall mass transfer coefficient (m/s)
k	Individual mass transfer coefficient (m/s)
D	Diffusion coefficient of species (m^2/s)
k_d	Distribution coefficient of species between the two phases
t	Time (s)
γ	Interfacial tension between aqueous and organic phases (N/m)
A	Cross-sectional area (m^2)
W	Mass of permeate collected (kg)
θ	Angle of contact between air, membrane, and water ($°$)
X	Mass fraction of preferentially permeating component in feed
Y	Mass fraction of preferentially permeating component in permeate
α	Membrane selectivity

Subscripts

m	Membrane
ov	Overall
f	Feed
p	Permeate
m	Membrane

References

Alkhudhiri, A., N. Darwish, and N. Hilal. 2012. Membrane distillation: A comprehensive review. *Desalination* 287: 2–18.

Alves, V.D. and I.M. Coelhoso. 2006. Orange juice concentration by osmotic evaporation and membrane distillation: a comparative study. *Journal of Food Engineering* 74: 125–133.

Boestert, J.L.W.C.D., J. Pieter Haan, and A. Nijmeijer. 2009. *Process for permeation enhanced reactive extraction of levulinic acid.* U.S. Patent 7,501,062 B2.

Demirci, N., M. Demirel, and N. Dilsiz. 2014. Surface Modification of PVC Film with Allylamine Plasma Polymers. *Advances in Polymer Technology* 33: 1–8.

Drioli, E. and L. Giorno. 2009. *Membrane operations: Innovative Separations and Transformations.* Weinheim: Wiley-VCH Verlag GmbH & Co. KGaA.

Drioli, E., E. Curcio, and G. Di Profio. 2005. State of the Art and Recent Progresses in Membrane Contactors. *Chemical Engineering Research and Design* 83: 223–233.

Erickson, B., J.E. Nelson, and P. Winters. 2012. Perspective on opportunities in industrial biotechnology in renewable chemicals. *Biotechnology Journal* 7: 176–185.

Heba, A., E. Ayman, K. Maaly, and E. Elham. 2015. Hydrophobic polyethersulfone porous membranes for membrane distillation. *Frontiers in Chemical Science Engineering.* DOI 10.1007/s11705 -015-1508-4: 1–10.

Jeevan Kumar, S.P., S. Rajendra Prasad, R. Banerjee, D.K. Agarwal, K.S. Kulkarni, and K.V. Ramesh. 2017. Green solvents and technologies for oil extraction from oilseeds. *Chemistry Central Journal* 11: 1–7.

Jin, Z., D.L. Yang, S.H. Zhang, and X.G. Jain. 2008. Hydrophobic modification of poly(phthalazinone ether sulfone ketone) hollow fiber membrane for vacuum membrane distillation. *Journal of Membrane Science* 310: 20–27.

John, R.P., K.M. Nampoothiri, and A. Pandey. 2007. Fermentative production of lactic acid from biomass: an overview on process developments and future perspectives. *Applied Microbiology and Biotechnology* 74: 524–534.

Johnson, R.A. and M.H. Nguyen. 2017. *Understanding Membrane Distillation and Osmotic Distillation.* U.S.: John Wiley & Sons, Inc.

Khayet, M. and T. Matsuura. 2011. *Membrane Distillation: Principles and Applications.* U.K.: Elsevier.

Kiai, H., M.C. García-Payo, A. Hafidi, and M. Khayet. 2014. Application of membrane distillation technology in the treatment of table olive wastewaters for phenolic compounds concentration and high quality water production. *Chemical Engineering and Processing: Process Intensification* 86: 153–161.

Kong, Y., X. Lin, Y. Wu, J. Cheng, and J. Xu. 1992. Plasma polymerization of octafluorocyclobutane and hydrophobic microporous composite membranes for membrane distillation. *Journal of Applied Polymer Science* 46: 191–199.

Lawson, K.W. and D.R. Lloyd. 1997. Membrane Distillation. *Journal of Membrane Science* 124: 25–1.

Li, Q.Z., X.L. Jiang, X.J. Feng, J.M. Wang, C. Sun, H.B. Zhang, M. Xian, and H.Z. Liu. 2016. Recovery processes of organic acids from fermentation broths in the biomass-based industry. *Journal of Microbiology and Biotechnology* 26: 1–8.

Madhumala, M., D. Madhavi, T. Sankarshana, and S. Sridhar. 2014. Recovery of hydrochloric acid and glycerol from aqueous solutions in chloralkali and chemical process industries by membrane distillation technique. *Journal of Taiwan Institute of Chemical Engineers* 45: 1249–1259.

Madhumala, M., D. Satyasri, T. Sankarshana, and S. Sridhar. 2014. Selective extraction of lactic acid from aqueous media through a hydrophobic H-Beta Zeolite/PVDF Mixed Matrix Membrane Contactor. *Industrial & Engineering Chemistry Research* 53: 17770–17781.

Nymeijer, K., T. Visser, R. Assen, and M. Wessling. 2004. Super selective membranes in gas–liquid membrane contactors for olefin/paraffin separation. *Journal of Membrane Science* 232: 107–114.

Pagliero, C., N.A. Ochoa, P. Martino, and J. Marchese. 2011. Separation of sunflower oil from hexane by use of composite polymeric membranes. *Journal of the American Oil Chemists' Society* 88: 1813–1819.

Pangarkar, B.L., S.K. Deshmukh, V.S. Sapkal, and R.S. Sapkal. 2016. Review of membrane distillation process for water purification. *Desalination Water and Treatment* 57: 2959–2981.

Ravishankar, R.V. 2016. *Advances in Food Biotechnology*. U.K.: John Wiley & Sons.

Sengupta, R.A. and R.A. Pittman. 2008. Application of membrane contactors as mass transfer devices. In *Handbook of Membrane Separations: Chemical, Pharmaceutical, Food, and Biotechnology Applications*. A.K. Pabby, S.S.H. Rizvi, and A.M. Sastre, eds. 7–24. New York: CRC Press.

Sharmiza, A., H. Manh, W. Huanting, and X. Zangli. 2012. Commercial PTFE membranes for membrane distillation application: effect of microstructure and support material. *Desalination* 284: 308–297.

Stellman, J.M. 1998. *Encyclopaedia of Occupational Health and Safety: Guides, indexes, directory*. Geneva, Switzerland: International Labour Organization.

Tres, M.V. et al. 2012. Separation of soybean oil/*n*-hexane miscella using polymeric membranes. *Journal of Food Science and Engineering* 2: 616–624.

17

Recent Research Trends in Polymer Nanocomposite Proton Exchange Membranes for Electrochemical Energy Conversion and Storage Devices

A. Muthumeenal, M. Sri Abirami Saraswathi, D. Rana, and A. Nagendran

CONTENTS

17.1 Introduction

The era of a fossil-fuel based economy is coming to an end. The world's foremost geologists believe that the peak in world oil production will occur within the next 10 years. In fact, whether or not the oil peak occurs in 10 years or in 40 years makes little difference since, once it happens, oil demand will quickly outstrip supply and prices will rapidly increase. The main source of our energy is fossil fuels, which causes brutal pollution and cannot provide long-term use. Nuclear energy is very costly and causes issues with its disposal. Other energy sources, such as tidal and wind schemes, are inadequate. Solar, thermal, and hydro energy sources are feasible, but entail an immense amount of capital. Hence, changes toward new, alternate energy systems with different energy sources, which would greatly impact human society, have recently gained attention. Among the various alternate energy systems, electrochemical energy conversion systems, such as fuel cells, are increasingly important in the field of automobiles, cogeneration and portable power systems. Storage systems, such as redox flow batteries (RFBs), are suitable for use in large power storage applications. This chapter focuses on fuel cells, particularly polymer electrolyte membrane fuel cells (PEMFC), direct methanol fuel cells (DMFC) and Vanadium redox flow batteries (VRFBs) (Figures 17.1 through 17.3). These energy conversion and storage devices have gained much attention due to their high power density and environmental benignity, i.e., fewer greenhouse emissions, a crucial factor of air and water pollution in this current scenario (Gulder & Gunes, 2011).

Fuel cells bear similarities both to batteries, which they share, the electrochemical nature of the power generation process, and to engines, which, unlike batteries, will work

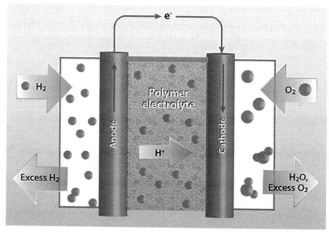

FIGURE 17.1
Schematic representation of Proton Exchange Membrane Fuel Cell (PEMFC).

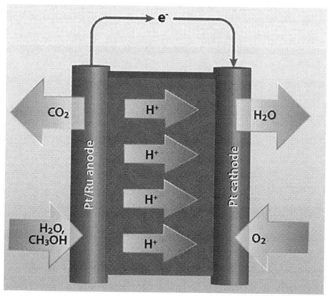

FIGURE 17.2
Schematic representation of Direct Methanol Fuel Cell (DMFC).

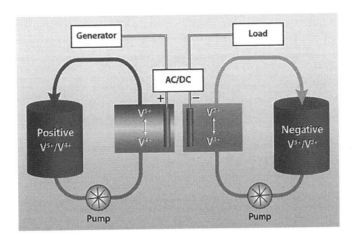

FIGURE 17.3
Schematic representation of Vanadium Redox Flow Battery (VRFB).

continuously, consuming a fuel of some sort. This is where the analogies stop, however. Unlike engines or batteries, a fuel cell does require recharging, it operates quietly and efficiently, and when hydrogen is used as fuel (PEMFCs), it generates only power and water. Thus, it is a so-called zero emission engine. The DMFC is a liquid- or vapor-fed PEM fuel cell operating on a methanol/water mix and air, and therefore deserves careful consideration. One of the most elegant solutions to the fueling problem would be to make fuel cells operate on a liquid fuel.

Among the energy storage technologies, RFB is considered as the best option for medium to large scale storage due to an excellent combination of energy efficiency, capital cost and life cycle expenditures, without specific site requirements. Flow batteries store chemical energy and generate electricity by a redox reaction between vanadium ions dissolved in the electrolytes. The RFB is a secondary battery; it is termed redox flow because the chemical species are stored outside the battery and supplied to it by pumps. Vanadium redox flow batteries (VRFBs) are widely recognized as promising candidates for large-scale energy storage systems (Joerissen et al., 2004). The batteries are further characterized by their longevity, in the range of several years, easy maintenance and a high overall energy efficiency of up to 90%.

The above-mentioned electrochemical energy conversion and storage devices share a common, vital component: the solid polymer electrolyte membrane, which acts as an electrolyte as well as a separator between the anode and cathode. The necessary criteria of the membranes for the energy conversion applications are: (i) high specific ionic conductivity (0.1 S/cm^2), (ii) good water retention, (iii) durability in desirable operating conditions, (iv) extremely low permeability to reactant species to maximize coulombic efficiency, (v) high electrolyte transport to maintain uniform electrolyte content and to prevent localized drying and (vi) a low cost (Sridhar et al., 2001).

The real technical breakthrough for membrane technology occurred when DuPont registered a trademark for a perfluorinated ion exchange membrane, Nafion®, which has commonly been used as an ion exchange membrane in fuel cell and in VRFBs (Hwang & Ohya, 1996; Luo et al., 2008). This material shows an excellent proton conducting property and substantial durability in fuel cells and acidic vanadium electrolytes in VRFB environments.

17.2 Veracities of Nafion in Fuel Cell and VRFB Environments

Although Nafion membranes find extensive applications in fuel cells and VRFB environments, they do have limitations. One of the major problems with the perfluorosulfonic acid membranes has been, and still is, their high cost. Thus, for a PEMFC operating at the desired power density of about 0.6 Wcm^{-2}, the cost of membrane alone will be about US$ 120 kW^{-1}. The glass transition temperature of Nafion limits the upper temperature ranges to about 100°C. The useful operating temperature is closer to 80°C because the membrane loses water, and hence the ionic conductivity reduces at a high temperature. It functions as proton conductors only in a highly hydrated state, so humidification is necessary (Sridhar et al., 2001).

On the other hand, the presence of electrolytes crossing over the membranes during the electrochemical processes can be responsible for a decreasing cathode potential. Consequently, during DMFC operation, methanol species are readily transported across perfluorinated sulfonic acid (PFSA) membranes (mostly methanol and water) (Aricò et al., 2001; Kalhammer et al., 1998). This results in the drawback of methanol crossover from anode to cathode, which is mostly performed by: (i) diffusion through the water-filled channels within the Nafion structure and (ii) active transport together with protons and their solvated water molecules during DMFC operation (electro-osmotic drag). The crossed-over methanol is chemically oxidized to CO_2 and H_2O at the cathode, decreasing the fuel utilization efficiency and depolarizing the cathode. Apart from this, it can also adversely affect the cathode performance due to the consumption of oxygen by the parasitic methanol oxidation at the cathode catalyst layer, lowering its partial pressure (Chu & Gilrnam, 1994). It is believed that the methanol crossover from the anode to the cathode causes a reduction in efficiency of DMFC to 35% (Kalhammer et al., 1998). On the other hand, the high water permeability in perfluorinated membranes can also cause cathode flooding and thus lower cathode performance due to mass transport limitations (Aricò et al., 2001).

In RFBs, the ideal ion exchange membrane only transfers H^+ or other non-reacting ions (such as Na^+, HSO_4^-, etc.) to complete the internal circuit while separating the catholyte and anolyte to avoid capacity loss and self-discharge. H^+ ions generated on the cathode during the charging process pass through the cation exchange membranes. Electrolyte ions passing through the membrane are mainly H^+ ions. The function of the cation exchange membrane is to allow the H^+ ions and reject vanadium ion transport. The transfer of H^+ ions is accelerated by the De Grotthuss mechanism for charge transfer (De Grotthuss, 1806). Vanadium ion transfer is caused by migration and diffusion in the sulfuric acid, which dissociates in the solution in two steps as follows:

$$H_2SO_4 \rightleftharpoons H^+ + HSO_4^- \qquad k_{a1} = 10^5$$

$$HSO_4^- \rightleftharpoons H^+ + SO_4^{2-} \qquad k_{a2} = 1.02 \times 10^{-2}$$

The first dissociation constant ka_1 is predominantly larger than the second dissociation constant k_{a2}; thus, sulfuric acid ions are dissolved in the form of HSO_4^-. In the cation exchange membrane, HSO_4^- ions are not allowed to permeate as a result of the Donnan exclusion. In this circumstance, the H^+ ion transport number is estimated to be considerably high but not equal to 1. A small number of vanadium ions transferred by migration and diffusion cannot be ignored; the transport is related to the mobility of vanadium ions in the membrane.

However, over long-term operation, the ions in the catholyte and anolyte tend to diffuse across the membrane until they reach equal concentrations in both half-cells. Unequal diffusion rates of vanadium ions across the membrane can cause a disparity between the state of charge (SOC) of the anolyte and catholyte, thereby leading to capacity fade. A small number of vanadium ions transferred by migration and diffusion cannot be ignored; the transport is related to the mobility of vanadium ions in the membrane. This phenomenon indicates that vanadium ions are not completely excluded from the membrane, thus allowing for vanadium ion crossover (Xi et al., 2007), which can cause self-discharge of the battery (Weidmann et al., 1998). The transport of vanadium ions across the membrane is very important for evaluating the capacity loss in VRFB. In redox flow cells, one must pay attention to water transport across the membrane due to electro-osmosis and volume osmosis. Electro-osmosis is caused by water molecules hydrated to H$^+$ ions; the driving force of electro-osmosis is the potential gradient generated in the membrane. In the ion exchange membrane, water molecules are transferred with H$^+$ ions, from the anode half-cell toward the cathode half-cell. This phenomenon induces an increase and decrease in the vanadium ion concentration in the anode half-cell and cathode half-cell, respectively. Thus, excessive water transfer between two half-cells can lead to precipitation of vanadium salt in the cell (Sukkar & Skyllas-Kazakos, 2003). Volume osmosis is caused by the concentration (osmotic pressure) difference between the two half-cells, and thus the phenomenon depends on the state of charge (SOC). Nevertheless, the diffusion of vanadium ions across the membrane will determine the coulombic efficiency of the cell and capacity loss with a continuous charge/discharge cycle and still remains an important factor in determining cell performance of VRFB.

The Nafion membrane accounts for about 10–15% of the total cost of the battery system and 40% of the total cost of a VRFB cell stack (Eckroad, 2007). The high cost and relatively poor selectivity of Nafion membranes for vanadium ions, have however, precluded their use in commercial systems. With Nafion, there is a crossover of methanol when used in DMFC and vanadium ions, in the case of VRFB, which results in a decrease in energy efficiency.

To overcome the above shortcomings associated with Nafion, extensive research studies have been already made. Therefore, there exits an urgent need to develop new ionomeric membranes that will provide improved and economical proton exchange membranes (PEMs) for fuel cell applications since the development of membranes for redox flow cells has tended to mirror the work already carried out in the area of fuel cell membranes. The research and development of novel PEMs is known to be one of the most challenging issues regarding the PEM fuel cell and VRFB technology. The chemical, physiochemical, economical, and more or less interconnected requirements of the PEM material are quite extensive in many ways. The key to PEM research is the development of alternate new materials or modifying existing ones which possess promising properties as ion exchange membranes.

17.3 Need for Developing Polymer Nanocomposite Membranes

The Nafion membrane functions as a proton conductor only in a highly hydrated state, so humidification is necessary. Its conductivity plummets when the cell temperature goes above 80°C. Thus, it is crucial for an ion exchange membrane to maintain its water content at elevated temperatures and for that we should develop materials that do not require water to maintain their proton conductivity. In DMFC, higher temperatures (130°C–200°C) could improve the kinetics of the methanol oxidation reaction at the anode and reduce methanol crossover, which would enhance cell efficiency significantly. In a H_2/O_2 fuel cell, high temperatures (150°C) will allow the fuel cell to tolerate higher levels of carbon monoxide (produced as a byproduct in the fuel reformer). Thus, there is a strong driving force to operate PEMFCs at higher temperatures. In VRFB with a Nafion membrane, coulombic efficiency is mainly affected by the cross-mixed vanadium ion, indicating the capacity loss and reduction in the overall efficiency of the system. One promising solution to the above-mentioned challenges in PEMs is to combine polymers with inorganic materials to incorporate the advantages of both, otherwise known as nanocomposite membranes. This approach of developing composite membranes has two objectives—one is to improve the mechanical properties of the composite membranes, i.e., the polymers give the flexibility to the composite, and the inorganic component present in the composite enhances the stability (thermal, mechanical, hydrolytic). The other objective is to make the composite membrane with the value-adding properties, such as high proton conductivity, low methanol crossover, etc. (Uchida et al., 2003). It has also been suggested that the size of the particles (nano or micro), surface properties (acid or basic), and the functionalization determine whether the filler, besides acting as a reinforcing component as mentioned above, can impart a significant improvement in proton conductivity (Croce et al., 2001).

Although some inorganic materials are not intrinsic proton conductors, they can be functionalized with proton-conducting groups. There are several extensive reviews that discuss and summarize PEM advancements for fuel cell applications (Li et al., 2003; Laberty-Robert et al., 2011). These recent reviews discuss in detail the design, synthesis, and properties of inorganic–organic hybrid or nanocomposite membranes for fuel cell and VRFB applications. Much attention is given to initial polymer matrices and the inorganic phase, particularly those based on advanced inorganic materials, such as (i) ZrO_2, SiO_2, TiO_2, P_2O_5 and Zeolite nanoparticles, (ii) TiO_2 nanotubes and nanowires, and (iii) 2-D layered graphene oxide (GO), MoS_2 (Transition metal dichalcogenides) were specifically discussed. The remarkable interest in polymer nanocomposite membranes has stemmed from the pursuit of PEMs with a lower cost and a high performance. This chapter focuses on recent progress in developing various nanocomposite PEMs based on modified perfluorosulfonic acid (PFSA) membranes with an additional inorganic phase that can retain water and/or conduct protons at elevated temperatures, and non-fluorinated hydrocarbons (including aliphatic or aromatic structures) to replace PFSA that make them attractive for fuel cell and VRFB applications.

17.3.1 Polymer Nanocomposite Membranes for PEMFC and DMFC Applications

17.3.1.1 PFSA Based Nanocomposite Membranes

In order to eliminate the inadequacy of Nafion and enhance its properties, several researchers have endeavored to configure it with varying approaches.

A synthesized nano TiO_2 powder was used by Sacca et al. (2005) to form a membrane for medium temperature PEMFC. A standardized and reproducible method for film preparation based on the doctor blade technique was developed. The incorporation of inorganic filler in the composite membrane provides good mechanical and thermal resistance and additionally improves water uptake and IEC values better than the commercial Nafion membrane. The membrane was tested in a single cell from 80 to 130°C in humidified H_2/air. The results obtained were compared with commercial Nafion 115 and a home-made recast Nafion membrane. Power density values of 0.514 and 0.256 Wcm^{-2} at 0.56 V were obtained at 110 and 130°C, respectively, for the composite Nafion–Titania membrane.

Nanocomposite Nafion/metal oxide (MO_2; where M = Zr, Si, Ti) membranes were synthesized by in-situ sol–gel method and characterized for high temperature operation of PEM fuel cells by Jalani et al. (2005). Nafion-ZrO_2 sol–gel membranes, in particular, demonstrated a higher water uptake and conductivity than unmodified Nafion membranes. Also, Nafion-MO_2 (M = Si, Ti) showed promising water uptake properties. The degradation temperatures and T_g improved for all nanocomposites. This shows that these membranes are tolerant to temperatures above 120°C. Thus, both chemical and physical properties were modified by incorporating nano-sized, inorganic additives having higher acidity and water uptake properties.

Park et al. (2008) synthesized fine particles of ZrO_2–SiO_2 binary oxide with different Zr:Si ratios, and composite membranes were prepared by their using the recasting procedure. The binary oxides were synthesized from sodium silicate and a carbonate complex of zirconium by sol–gel technique. The composite membranes were then prepared by blending a 10% (w/w) Nafion-water dispersion with the inorganic compound. Scanning electron micrographs confirmed good distribution of inorganic fillers in the polymer matrix. All composite membranes showed a higher water uptake than unmodified membranes, and the proton conductivity increased with increasing zirconia content at 80°C. By contrast, the proton conductivity decreased with zirconia content for the composite membranes containing binary oxides at 120°C. The composite membranes were tested in a 9 cm^2 commercial single cell at both 80°C and 120°C in humidified H_2/air under different relative humidity (RH) conditions. Composite membranes containing ZrO_2–SiO_2 binary oxide (Zr/Si = 0.5) gave the best performance of 610 mWcm^{-1} under conditions of 0.6 V, 120°C, 50% RH and 2 atm. Consequently, the binary oxide was found to be effective in improving proton conductivity and water uptake at high temperatures and low humidity.

A new hybrid membrane, based on incorporation of SiO_2–P_2O_5 glass electrolytes into hydrophilic channels of Nafion 115 by in situ sol–gel synthesis of tetraethoxysilane (TEOS) and trimethylphosphate (TMP), was developed by So et al. (2010). The contribution of SiO_2–P_2O_5 glass electrolytes to the proton becomes more noticeable at the extreme condition of 100°C/dry out state wherein the De Grotthus mechanism predominantly governs proton conduction (De Grotthuss, 1806). The higher SiO_2–P_2O_5 content in the hybrid membranes allows well-connected/enlarged hydrophilic channels along with an increase in the number of free/bound water molecules. These interesting features are expected to enable the hybrid membranes to present higher proton conductivity than the pristine Nafion 115, which becomes more noticeable at high-temperature/low-humidity conditions, and also to provide an in-depth understanding of the proton transport mechanism in terms of hydrophilic-channel structure and state of water.

A series of Nafion/Pd-silica fiber composite membranes were introduced to reduce methanol permeability without compromising on the essential proton conductivity of ionomer membranes by Thiam et al. (2013) for DMFC applications. The silica-supported palladium nanofibers had diameters ranging from 100 nm to 200 nm and were synthesized by a facile electro-spinning method. Nafion/Palladium-silica nanofiber (N/Pd-SiO_2) composite membranes with various fiber loadings were prepared by a solution casting method. The morphology evaluations of the composite membranes indicated that the inorganic composite fibers were evenly distributed throughout the membrane matrix and formed an adhered sub-layer with the Nafion bulk. Addition of composite fibers during membrane fabrication resulted in a permeable barrier for methanol. The composite membranes with optimum fiber content (3 wt.%) showed an improved proton conductivity of 0.1292 S cm^{-1} and a reduced methanol permeability of 8.36 × 10^{-7} cm^{2}s^{-1}. The moderate composite fiber loading of 3 wt.% resulted in better membrane performance, and this set of composite membranes will continue to be studied as a potential electrolyte for use in DMFCs.

An organically-modified ceramic material (TiO_2-RSO_3H) to be used as filler in Nafion-based composite membranes was synthesized by covalently grafting propylsulfonic acid groups on the surface of TiO_2 nanoparticles (Cozzi et al., 2014). The introduction of covalently bound sulfonic acid groups led to increased IEC and improved capacity of the material to retain more ore strongly bound water. The highest conductivity value was obtained for the composite membrane containing 10 wt.% TiO_2-RSO_3H (0.08 S cm^{-1} at 140°C). The presence of the filler resulted in a general enhancement in cell response in terms of both high power density and low methanol crossover, with respect to an unfilled Nafion membrane. The authors suggested that the propylsulfonic functionalized Titania be used as a valid proton conducting filler to be used in Nafion-based composite membranes operating at temperatures higher than 100°C.

A GO laminated Nafion membrane for DMFC was originally proposed by Lin and Lu (2013). Compared to a GO-dispersed polymer composite membrane, a GO laminated proton-conducting membrane is advantageous for DMFC operating at higher methanol feed concentrations. The novel laminate membrane consists of highly ordered GO paper with parallel orientation prepared by using a vacuum filtration method. This dual-layer membrane is fabricated by laminating the Nafion 115 membrane with a highly orientated GO paper through transfer printing followed by hot pressing. The GO laminated Nafion 115 membrane with a GO layer thickness of 0.8 mm had an IEC value of 0.99 meq. g^{-1} and a proton conductivity of 2.35×10^{-2} S cm^{-1}, and these values are close to those of the pristine Nafion 115 membrane. The methanol permeability of the GO paper was 70% lower than that of the Nafion 115 membrane, which is probable because the parallel orientation of the GO paper effectively decreases methanol crossover. The GO laminated Nafion membrane is far superior to the pristine Nafion membrane in DMFC performance when operating at an 8-M methanol feed concentration.

GO-Nafion nanohybrid membranes have been prepared by Peng et al. (2016). Nafion chains have been reacted onto GO through atom transfer radical addition (ATRA) reaction between the C–F groups of Nafion and the C=C groups of GO. Incorporation of Nafion chains to GO sheets enhances their interfacial compatibility with Nafion matrix. The Nafion chains attached on GO sheet surface might induce the aggregation of sulfonic acid fraction of Nafion chains so as to form proton-conducting domains. Compared to the neat recast Nafion membrane, addition of GO-Nafion results in 1.6-fold rise in proton conductivity, relative to the corresponding nanocomposite PEM. The single cell tests on the nanocomposite PEM also exhibit about 35–40% increase in fuel cell performance.

Similarly, phosphonic acid functionalized graphene oxide (PGO) is synthesized and doped into Nafion to prepare nanohybrid membranes (Zhang et al., 2016). The Nafion/PGO nanohybrid membranes exhibited improved proton conductivity at both 100% RH and low RHs compared to recast Nafion membrane. At 100% RH, the phosphonic acid groups on GO nanosheets form hydrogen bond networks with water molecules, which act as well-organized pathways for proton hopping to enhance the proton conducting capacity. The incorporation of PGO into Nafion matrix generates additional proton conducting sites and enhances water adsorption and retention ability of nanohybrid membranes. More significantly, the distribution of phosphonic acid groups in the membrane is controlled by the unique structure of GO nanosheets, which benefits the formation of new pathways for proton hopping at low humidity conditions. Consequently, the hybrid membranes show improved proton conducting capacity, particularly under high temperatures or low relative humidity. The hybrid membrane with 2 wt% PGO displayed a proton conductivity of 0.277 S cm^{-1} at 100°C and 100% RH, and 0.0441 S cm^{-1} at 80°C and 40% RH, which are 1.2 and 6.6 times higher than that of the pure Nafion membrane.

A novel strategy of in-situ growth of MoS_2 to prepare MoS_2/Nafion composite membranes for highly selective PEM, was employed by Feng et al. (2013). The strong interactions between the Mo precursor (($NH_4)_2MoS_4$) and Nafion's sulfonic acid groups in dimethyl formamide (DMF) environment probably lead to a selective growth of MoS_2 flakes mainly around the ionic clusters of the resultant MoS_2/Nafion composite membrane. This process significantly promotes aggregation and thus leads to better connectivity of these ionic clusters, which favors the increase in proton conductivity of the resultant PEMs. Meanwhile, the existence of MoS_2 flakes in those ionic channels could effectively increase the tortuosity of the membrane and prevent methanol transport through the membrane, contributing to a dramatic decrease in methanol crossover. Consequently, the MoS_2/Nafion composite membranes exhibit greatly increased selectivity. Under severe conditions, such as 50°C with 80 v/v% methanol concentration, an increase in the membrane selectivity by nearly 2 orders of magnitude compared to that of the recast Nafion membrane could be achieved, which proved this method to be a promising way to prepare high performance PEMs. Table 17.1 summarizes the physicochemical and electrochemical characteristics of the modified Nafion nanocomposite membranes for PEMFC applications.

TABLE 17.1

Summary of Modified Nafion Nanocomposite Membranes for PEMFC Applications

Polymer Nanocomposite Membranes	Comments	References
Nafion/nanometer TiO_2 powder	Improves water uptake and IEC. Good mechanical and thermal resistance. Power density values of 0.514 and 0.256 Wcm^{-2} at 0.56 V at 110 and 130°C.	Sacca et al. (2005)
Nafion/MO_2 (M = Zr, Si, Ti)	Higher water uptake and ionic conductivity. Tolerant to high temperature above 120°C.	Jalani et al. (2005)
Nafion/ZrO_2–SiO_2 binary oxide	Effective in improving proton conductivity and water uptake at high temperatures and low humidity. (Zr/Si = 0.5) gives the 610 $mWcm^{-1}$ at 0.6 V, 120°C, 50% RH and 2 atm.	Park et al. (2008)
Nafion/SiO_2–P_2O_5	Grotthus mechanism predominantly governs the proton conduction at 100°C/dry out state. Higher proton conductivity than the pristine Nafion 115.	So et al. (2010)
Nafion/Pd/ SiO_2	Improved proton conductivity of 0.1292 S cm^{-1} and a reduced methanol permeability of 8.36×10^{-7} cm^2 s^{-1}.	Thiam et al. (2013)
Nafion/TiO_2-RSO_3H (R-propyl)	Highest conductivity value of 0.08 S cm^{-1} at 140°C. Covalently bound sulfonic acid groups led to increased IEC and water retention properties.	Cozzi et al. (2014)
GO laminated Nafion	Methanol permeability is 70% lower than Nafion 115. IEC value of 0.99 meq. g^{-1} and a proton conductivity of 2.35×10^{-2} S cm^{-1} are close to pristine Nafion.	Lin and Lu, (2013)
Nafion/GO	1.6-folds of proton conductivity than pristine Nafion. 35–40% increase in the fuel cell performance.	Peng et al. (2016)
Nafion/PGO	2 wt.% PGO has proton conductivity of 0.277 S cm^{-1} at 100°C and 100% RH, and 0.0441 S cm^{-1} at 80°C and 40% RH, which are 1.2 and 6.6 times higher than that of pristine Nafion membrane.	Zhang et al. (2016)
Nafion/MoS_2	Higher proton conductivity than pristine Nafion. The existence of MoS_2 flakes increases the tortuosity and decreases methanol crossover. Two orders of magnitude higher membrane selectivity. than Nafion.	Feng et al. (2013)

17.3.1.2 Sulfonated Hydrocarbon Polymer Based Nanocomposite Membranes with Inorganic Oxides

A series of sulfonated poly(ether ether ketone) (SPEEK of degree of sulfonation, DS = 87%) based nanocomposite membranes were prepared by in-situ zirconium oxide formation (Silva et al., 2005). In terms of morphology, it was observed that in-situ zirconium alkoxide hydrolysis enables the fabrication of homogeneous membranes that offer a promising adhesion between inorganic domains and the polymer matrix. It was found that the increase of zirconium oxide content leads to a decrease in water, methanol, carbon dioxide and oxygen permeability coefficients, which prevented reactants' loss in DMFC operation and lead to long-term stability. It increased the water/methanol selectivity and reduced the carbon dioxide/nitrogen and oxygen/nitrogen selectivities. The obtained results showed that the inorganic oxide network decreased the proton conductivity and water swelling. Furthermore, the various amounts of zirconium oxide in the SPEEK matrix enabled the fabrication of composite membranes with a wide range of properties concerning proton conductivity, water uptake, and methanol and water permeation.

Composite membranes of SPEEK with three different ion exchange capacities (1.35, 1.75 and 2.1 meq. g^{-1}) and blended to 5 and 10% additive loadings were prepared individually by solution casting (Sambandam & Ramani, 2007). The additives were silica, and sulfonic acid functionalized silica synthesized by condensation of appropriate precursors through a sol–gel approach. The agglomerate size of the additives was found to be between 2 and 5 µm. Water uptake of the composite increased with IEC of SPEEK. Composite membranes displayed lower water uptake in comparison to pristine SPEEK. The membranes prepared using sulfonic acid functionalized silica offered proton conductivities up to 0.05 S cm^{-1} at 80°C and 75% RH and 0.02 S cm^{-1} at 80°C and 50% RH. Hydrogen crossover current densities of the membranes were in the order of 1–2 mA/cm^2. MEA tests showed acceptable performance at 80°C and RH in the range of 50–75%.

Hydrated tin oxide (SnO$_2$·nH$_2$O) has been used as an inorganic filler in SPEEK and composite membranes were prepared and characterized (Mecheri et al., 2008). The durability and the electrochemical performance of the composite membranes are substantially enhanced by the addition of tin oxide. The SPEEK composite with 50 wt% tin oxide showed decent proton transport characteristics, reduced methanol uptake and increased stability with respect to the plain SPEEK membrane, making it apt for application as an electrolyte in DMFCs.

The authors reported a novel method for fabricating composite membranes with SPEEK (DS = 61%), metal oxides MO$_2$ (SiO$_2$, TiO$_2$ and ZrO$_2$) and PANI (Tripathi & Shahi, 2009). With the aid of redox polymerization process, SPEEK–metal oxide membrane surfaces were modified with PANI. The synergetic effect of metal oxides and surface modification with PANI increased the water retention capacity and reduced the methanol permeability of the composite membranes. These composite membranes exhibited very low methanol permeability (1.9–1.3 × 10^{-7} cm^2 s^{-1}), which was lower than values reported for either SPEEK–metal oxide or SPEEK/PANI membranes. Moreover, high selectivity parameter values at 70°C, especially in the case of S–SiO$_2$–PANI and S–TiO$_2$–PANI, indicated their great advantages over the Nafion 117 membrane for targeting moderate temperature fuel cell applications.

The functionalized silica powder with sulfonic acid groups (SiO^x-S) was added into the SPEEK matrix (DS = 55.1%) to prepare SPEEK/SiO^x-S composite membranes in order to avoid excessive swelling and even dissolution at high temperatures (Gao et al., 2009). The rate of decrease in both the swelling degree and methanol permeability of the membranes depended on the amount (in wt.%) of the addition of SiO^x-S powder. The SPEEK/SiO^x-S

(15%, by mass) membrane swelled only 27.3% at 80°C, whereas pure SPEEK membrane swelled up to 52.6% at the same temperature. All SPEEK/SiO^x-S composite membranes had much lower methanol permeability than that of Nafion 115. The addition of the SiO^x-S powder not only leads to higher proton conductivity, but also increases the dimensional stability at higher temperatures. The SPEEK/SiO^x-S (20%, by mass) membrane could endure temperature up to 145°C, at which in 100% RH, its proton conductivity surpassed that of the Nafion 115 membrane and reached 0.17 S cm^{-1}, whereas the pure SPEEK membrane dissolved at 90°C.

Composite membranes comprising of sulfonated polysulfone (SPSU) and titanium dioxide (TiO$_2$) (SPSU/TiO$_2$) were developed by Devrim et al. (2009) for fuel cell applications. The sulfonation of PSU was carried out with trimethylsilyl chlorosulfonate in 1,2 dichloroethane at room temperature. The incorporation of TiO$_2$ augmented the thermal stability but higher filler concentrations reduced the miscibility of the composite component and resulted in brittle membranes. The composite membranes exhibited conductivity values in the range of 10^{-3}–10^{-2} S cm^{-1}, which increased with operating temperature. The maximum current density of 300 mA/cm^2 was obtained for SPSU/TiO$_2$ membrane at 0.6 V for an H$_2$–O$_2$ fuel cell working at 1 atm and 85°C. Single cell tests performed at different operating temperatures indicated that the SPSU/TiO$_2$ composite membrane is more hydrodynamically stable and also performed better than SPSU membranes.

Liu et al. (2011) developed novel nanocomposite proton exchange membranes using sulfonated mesoporous silica nanoparticles (SMSNs) as inorganic fillers through direct blending with sulfonated polyimides (SPIs). As compared with SPI, the thermal stability, water uptake and methanol permeability of the resulting nanocomposite membranes were increased with the introduction of sulfonated mesoporous silica. The highest water uptake value of 54.2% was obtained for the composite membrane with 3 wt.% SMSNs. The hybrid membrane showed the lowest methanol permeability value of 5.23 × 10^{-6} cm^2 s^{-1}. Due to the aggregation of SMSNs particles for higher than 3 wt.%, water uptake, as well as the methanol permeability values, decreased slightly. All the composite membranes exhibited excellent proton conductivity values when compared to Nafion 117. The membranes displayed maximum proton conductivity at different testing temperatures, when 7 wt.% SMSNs were incorporated into the SPI, which indicates that the addition of SMSNs can also improve the proton conductivity of composite membranes.

The multilayered (sandwich like) membranes were fabricated using sulfonated polysulfone (SPSU) and SiO$_2$, and tested as a PEM in fuel cell environment (Padmavathi et al., 2012). The addition of a SiO$_2$ layer between the two layers of SPSU to form the multilayered composite membrane enhanced its dimensional stability, but slightly decreased the proton conductivity parameter when compared to the conventional SPSU/SiO$_2$ composite membrane. In addition to that, higher water absorption, lower methanol permeability and higher flame retardant nature were also observed for the fabricated multilayered membrane. When the content of silica is low, the silica cross-linked network in the SPSU matrix is not so perfect. Increasing the amount of silica additive plays an important role in the membrane matrix, resulting in enhanced conductivity by the multilayered membrane. The hydrophilic silica may mainly exist near the hydrophilic ion clusters and ion channels, thus changing the microstructure of SPSU matrix, which would facilitate the holding of methanol. In the parent membrane, the SO$_3$H groups facilitate methanol permeation via hydrogen bonding. In 2–6 wt% silica loaded multilayer membranes, the content of sulfonic acid groups decreased, and, as a result, their assistance to methanol permeation also decreased. Additionally, interactions between silica OH and SO$_3$H groups suppressed the formation of methanol permeation paths. In other words, the methanol permeation path was highly tortuous in the presence of silica. In the case of 8 and 10 wt% silica loaded multilayer membranes, a slight increase in methanol permeability was attributed to poor dispersion of silica (because of agglomeration at higher concentration of silica), due to which the interaction between the silica and the matrix was not sufficient to inhibit methanol permeation. The 2 wt.% SiO_2 loaded multilayered membrane in DMFC showed a peak power density of 86.25 mW cm^{-2} at a current density of 200 mA/cm^{-2} with a methanol flow rate of 20 mL min^{-1} at the anode, an O_2 flow rate of 40 ml min^{-1} at the cathode, for anode loading of 0.25 mg Pt:Ru cm^{-2}, cathode loading of 0.375 mg Pt cm^{-2} and methanol concentration of 1 M at room temperature. This performance was higher than that obtained for Nafion 117 membrane (52.8 mW cm^{-2}) in the same single cell test assembly. Hence, due to the enhanced dimensional stability, reduced methanol permeability and higher maximum power density, the SPSU/SiO$_2$/SPSU multilayered membrane can be a promising candidate for use as an electrolyte membrane in DMFC applications.

Sakamoto et al. (2014) have developed a new composite membrane based on sulfonated polyimide containing triazole groups (SPI-8) as a matrix ionomer and SiO_2 nanoparticles. During fuel cell operation at 53% RH and 80°C, the inclusion of SiO_2 nanoparticles confers a positive effect on enhancing the fuel cell performance. The single cell with 10 wt% SiO_2/SPI-8 was found to display the highest I–E performance, with the highest mass activity at 0.85 V and the smallest oxygen-transport over potential (O_2-gain), as well as the lowest ohmic resistance. Reduction of ohmic resistance led to an improvement in cell performance. This greatly denotes that SiO_2 nanoparticles were able to promote the back-diffusion of water produced in the cathode catalyst layer to the anode catalyst layer, retaining high water content in the membrane during the fuel cell operation. If water generated at the cathode was predominantly exhausted through the microporous layers (MPL), it would certainly cause the O_2 gas diffusion to decrease. In contrast, when a large fraction of the water generated at the cathode back-diffuses toward the anode, the O_2 gas diffusion rate into the cathode/PEM interface should increase, accompanied by increased water content in the PEM, resulting in higher performance. One of the possible reasons for such an improvement in O_2 gas diffusivity by the use of the SiO_2/SPI-8 is an enhanced back-diffusion of water from the cathode toward the anode. This indicates that both protons and oxygen were sufficiently supplied to the utilized Pt catalysts through the Nafion binder by the addition of SiO_2 in the SPI-8 membrane. Thus, the improvement of the IR-free I–E performance for the SiO_2/SPI-8 cells can be ascribed to an enhanced cathode performance. It was seen that the cell with a bilayer SPI-8 membrane having 10 wt% SiO_2 in the anode side layer and 3 wt.% SiO_2 in the cathode side layer displayed performance greater than that with a uniform dispersion of 10 wt% SiO_2, particularly in the higher current density region at low RH, which can be attributed with reliability to the fact that the concentration gradient of SiO_2 in the SPI-8 led to improvement of the back diffusion of water through the membrane.

Hybrid membranes based on sulfonated polyphenyl sulfone (SPPA), tungstophosphoric acid (TPA) and SiO_2 (SPPS/TPA/SiO$_2$) were fabricated and investigated for their properties for fuel cell applications at high temperature and low humidity conditions (Devrim, 2014). The sulfonation reaction of polyphenyl sulfone polymer (PPS) was carried out with trimethylsilyl chlorosulfonate in 1,2 dichloroethane, at ambient temperature. The hybrid membrane was composed of a mixture of SPPS solution, TPA and SiO_2 particles. The composite membranes showed good thermal resistance with the incorporation of TPA/SiO$_2$. The existence of TPA/SiO$_2$ improved the water uptake, proton conductivity and oxidative stability of the hybrid membranes. The membranes were tested in a single cell with an active area of 5 cm^2. The operating temperature range of 70 to 120°C and 100% and 30%

RH conditions and with the catalyst loading of 0.4 mg Pt/cm^2 for both anode and cathode sides indicated that the SPPS/TPA/SiO$_2$ hybrid membrane was more stable and performed better than pure SPPS membrane.

The authors evaluated the proton conductivity of the novel composite membranes based on sulfonated polysulfone (SPSU) and SiO$_2$ doped with phosphomolybdic acid (PMA) (Martínez-Morlanes et al., 2015). The composite membranes were prepared by solution casting technique. The addition of inorganic particles alters the thermal, mechanical properties, as well as proton conductivity. From the results obtained with thermal analysis, the composite membranes behave well at temperatures up to 200°C. At higher temperatures, the proton conductivity of the nanocomposite membranes increased, which is greater than the SPSU/SiO$_2$ membranes. This enhancement was more pronounced when the PMA weight percent was increased from 10% to 20%, due to the surface functional sites for proton transfer catered by PMA particles. The hybrid membrane with 2% SiO$_2$ and 20% PMA seems to be the most promising candidate for fuel cell applications.

The introduction of silicon dioxide (SiO$_2$) as inorganic filler in the PBI membrane has positive effect, and thus hybrid membranes were prepared and characterized as alternative materials for fuel cell applications (Devrim et al., 2016). The PBI/SiO$_2$ hybrid membrane shows higher proton conductivity and acid retention properties than the PBI membrane. The PBI/SiO$_2$ (5 wt.%) hybrid membrane exhibits a maximum proton conductivity of 0.103 S cm^{-1} at 180°C. Maximum current density of 0.24 A/cm^2 was observed at 165°C and a cell voltage of 0.6 V. A single slice test cell operating at 140°C–180°C with PBI/SiO$_2$ hybrid membrane was more stable and also performed better than the pristine PBI membrane.

In order to improve the physicochemical properties of SPEEK membranes, sulphated zirconia (SZ) nanoparticles were used as inorganic additives and nanocomposite membranes were fabricated (Mossayebi et al., 2016) and characterized for fuel cell application at intermediate temperatures. Various loadings of SZ were introduced into different SPEEK membranes. The membrane preparation conditions were optimized by surface response method. The optimum content of SZ was 5.94 wt.% that represented proton conductivity of 3.88 mS cm^{-1} (at 100°C and 100% RH) and oxidative stability of 102 min. Furthermore, the addition of SZ nanoparticles enhanced mechanical, oxidative and chemical stabilities, as well as proton conductivity. The composite membranes based on poly(ethylene oxide)/GO were fabricated by a novel method, which presents electrolyte membrane without polymer modifications and characterized for its applications (Cao et al., 2011). The GO content in the membrane promoted high proton conductivity of about 0.134 S cm^{-1} at 60°C, as well as adequate mechanical properties, which improved resistance to methanol crossover. The PEO/GO composite membrane exhibited a maximum tensile strength of 52.22 MPa and Young's modulus 3.21 GPa, and the fractured elongation was about 5%. The single cell with the composite membrane provides a maximum power density of 53 mW cm^{-2} at 60°C.

Due to the lack of proton conductive groups (carboxylic acid and other oxygen containing groups are not sufficient proton conductive groups), the incorporation of the pristine GOs could lead to a decrease in the proton conductivity of the membranes and therefore a decrease in the fuel cell performance. In this regard, the incorporation of sulfonated GOs could be a better choice. On the one hand, the incorporation of the sulfonated GOs not only improves the proton conductivity of the obtained membranes but also facilitates the formation of homogeneous membranes due to the high compatibility between the sulfonated GOs and -SO$_3$H. On the other hand, the strong interfacial interactions between the high surface areas of graphene framework and the host membranes could improve the mechanical stability of the membranes. One could therefore, expect the incorporation of sulfonated GOs into host PEMs to greatly improve the electrochemical performance of DMFCs. Sulfonated organosilane functionalized graphene oxides (SSi-GO) have been used as a filler in sulfonated poly(ether ether ketone) (SPEEK) membranes and character-ized for fuel cell applications (Jiang et al., 2013). These SSi-GOs can remain exfoliated and are homogeneously distributed in the SPEEK matrices after the evaporation of the DMF solvents due to their good compatibility and the strong interfacial interactions between SSi-GOs and SPEEK, enabling the fabrication of membranes with a uniform structure. The inclusion of SSi-GOs enhances the ion-exchange capacity (IEC), water uptake, and proton conductivity of the membrane. The composite membranes with well-controlled contents of SSi-GOs, display high proton conductivity and low methanol permeability than Nafion 112 and Nafion 115, making them particularly attractive as PEMs for DMFCs. The com-posite membrane with optimal SSi-GOs content exhibit over 38% and 17% higher power densities, respectively, than Nafion 112 and Nafion 115 membranes in DMFCs, offering the possibility to cut down the DMFC membrane's cost significantly while maintaining high performance.

Nanohybrid membranes using polymer and a SGO nanosheet layer were fabricated and uniformly dispersed in a chitosan-based membrane (Liu et al., 2014). Due to the strong electrostatic attractions between $-SO_3H$ of SGO and $-NH_2$ of chitosan, compared with chi-tosan control and GO-filled membranes, SGO-filled membranes attained enhanced ther-mal and mechanical stabilities. These strong interactions inhibit the mobility of chitosan chains and reduce the area swellings of SGO-filled membranes, reinforcing their structural stabilities. The production of acid–base pairs along chitosan–SGO interface, which work as facile proton hopping sites, thus creates continuous and wide proton transfer pathways, yielding enhanced proton conductivities under both hydrated and anhydrous conditions. A 122.5% increase in hydrated conductivity and a 90.7% increase in anhydrous conduc-tivity are obtained by incorporating 2% SGO. The membranes have afforded acceptable PEMFC performance under anhydrous conditions.

The phosphorylated graphene oxide (PGO) nanosheets are incorporated into chitosan matrix to prepare nanohybrid membranes, since phosphonic acid (PA) groups are uti-lized as a feasible proton carrier for possessing distinct intrinsic proton conduction (Bai et al., 2015). It is found to be about 26 wt% of PGO on chitosan matrix, which considerably increases the IEC from 0.44 mmol g^{-1} of GO to 0.79 mmol g^{-1}. Due to the strong electrostatic interactions between $-PO_3H$ and $-NH_2$, PGO-filled membranes achieve higher thermal and mechanical stabilities when compared to chitosan control and GO-filled membranes. PGO caters efficient hopping sites ($-PO_3H$, $PO_3^- \cdots {}^+_3 HN-$), which allow the formation of highly conductive channels along its surface. These channels are found to ease proton con-duction under both hydrated and anhydrous conditions. The nanohybrid membrane with 2.5% PGO acquires a 22.2 time increase in conductivity from 0.25 mS cm^{-1} to 5.79 mS cm^{-1} (160°C, 0% RH). Due to this advantage, the hydrogen fuel cell comprising PGO-filled mem-branes displays much improved cell performance than those using chitosan control and GO-filled membranes.

A sequence of composite membranes, based on PBI with various weight ratios of Zwitter ion-coated graphene oxide (ZC-GO) containing ammonium and sulfonic acid groups, was synthesized and characterized for DMFC applications (Chu et al., 2015). Even though there is a decrease in thermal stability of composite membranes with increasing ZC-GO con-tent, the other properties like methanol permeability, proton conductivity, water uptake, and swelling ratio of the composites increased. The maximum proton conductivity of 4.12×10^{-2} S cm^{-1} at 90°C and RH 100% was obtained with the composite membrane that contained 25 wt% ZC-GO and exhibited a low methanol permeability of 1.38×10^{-7} cm^2 s^{-1} at 25°C. The performance obtained with composite membranes is much lower than that of Nafion 117 under the same conditions. The excellent properties of the composite membranes can be ascribed to the homogeneous dispersion of ZC-GO and the formation of proton-transport channels in the membranes.

A series of ionically cross-linked PEMs based on SPES and quaternized graphene oxide (QGO) was successfully prepared using QGO as the inorganic filler and crosslinking agents (Zhao et al., 2016). The ionic interactions take place between the sulfonic acid groups of SPES and the quaternary ammonium groups of QGO. The addition of QGO enhanced the mechanical properties and oxidative stabilities of the composite membranes. Meanwhile, the swelling ratio, water uptake and methanol permeability of the ionic cross-linked SPES/QGO composite membranes were reduced to much lower values than that of the pristine SPES membrane. SPES-10-QGO composite membrane revealed less water uptake and swelling ratio, but yielded a high proton conductivity of about 0.08 S cm^{-1} at 80°C.

The hybrid PEMs for DMFCs were prepared by histidine functionalized GO sheets incorporated into SPEEK as base matrix (Yin et al., 2016). The acid-base pairs formed between $-SO_3H$ groups (proton donors) in SPEEK and imidazole groups (proton acceptors) in histidine molecules transport protons synergistically, which yielded efficient proton conduction channels inside the hybrid membranes. The hybrid membranes showed maximum proton conductivity at 100% RH and room temperature. The low methanol permeability in the range of 1.32–3.91 × 10^{-7} cm^2 s^{-1} of hybrid membranes was due to the incorporation of functionalized GO flakes. The 4 wt% functionalized GO hybrid membrane exhibits a higher selectivity of 5.14 × 10^5 S s cm^{-3} and a maximum power density of 43.0 mW cm^{-2} for single slice DMFC cell, which is 80.7% higher than that of bare SPEEK membrane.

The polymer nanocomposites based on phosphoric acid PBI/GO were prepared by dispersion of various amounts of GO in PBI polymer matrix followed by phosphoric acid doping for high temperature fuel cell applications (Üregen et al., 2016). The inclusion of GO in the PBI matrix helps to enhance the acid doping, proton conductivity and acid retention properties. The single test cell with PBI/GO composite membrane (2 wt.% GO) performed better than pure PBI membrane at non-humidified conditions. The maximum power density of PBI/GO-1 membrane can reach 0.38 W/cm^2 at ambient pressure and 165°C, and the current density at 0.6 V reaches up to 0.252 A/cm^2, with H_2/air. The results indicate that the PBI/GO nanocomposites could be utilized as a PEM for high temperature fuel cell applications.

Polypyrrole (PPy) is doped (*via* potentiostatic synthesis), and its electrochemical behavior as membrane material has been studied (Molina et al., 2016). The ionic conductivity was superior (1.6 × 10^{-3} S cm^{-1}) due to the high porosity of the material as demonstrated by EIS and cyclic voltammetry measurements, where the PPy/GO film was in use as a freestanding membrane.

Chitosan nanocomposite membranes were fabricated using sulfonated chitosan (SCS) and sulfonated graphene oxide (SGO) to evaluate the electrochemical properties of the membrane (Shirdast et al., 2016). 10 wt% SCS and different amounts of SGO nanosheets were added into the chitosan matrix. At 5 wt.%, SGO content, the permeability reduced by 23%, the proton conductivity increased by 454%, and therefore the selectivity enhanced by 650%, compared to that of pure chitosan. The enhancement of proton conductivity is mainly due to the synergistic effect of SCS.

One-step solvo-thermal method was used to prepare PEMs by the dispersion of iron titanate (Fe_2TiO_5) nanoparticles onto SGO nanosheets (Beydaghi et al., 2016). SGO/Fe_2TiO_5 hybrid material is dispersed into PVA matrix so as to fabricate thin film membranes by using solution casting method. The nanocomposite PEM with 5 wt.% SGO/Fe_2TiO_5 yielded a high proton conductivity of about 0.061 S cm^{-1} and low methanol permeability of about 4.78 × 10^{-6} cm^2 s^{-1} to deliver a power density of about 20.47 mW cm^{-2} at 30°C. The electrochemical and physicochemical properties of the composite membranes demonstrated the role of Fe$_2$TiO$_5$/SGO nanosheets as an effective hybrid material in the development of novel nanocomposite PEMs.

17.3.2 Polymer Nanocomposite Membranes for VRFB Applications

The idea of nanocomposite membranes for VRFBs was inspired by the extensive research reported to develop Nafion composite membranes for DMFC applications (Shao et al., 2006; Tang et al., 2007), which generally took advantage of interactions between organic or inorganic components and the sulfonic acid groups of ion exchange membranes. The introduction of nanofillers leads to a reduction in methanol crossover by blocking the hydrophilic clusters of Nafion membranes. This principle has also been applied in VRFB applications, and thus decreased vanadium ion permeability with minimized proton conductivity loss, leading to an improvement in VRFB performance. Similarly, ion exchange membranes for VRFB application have been developed by the pore-filling method (Kim et al., 2014), polymer blending (Ling et al., 2012) and the inorganic nanofiller doping technique (Mai et al., 2011; Wang et al., 2012). However, the commercialization of VRFB is still limited due to the vanadium ion permeability and high cost. To improve the performance of VRFB, by reducing the permeation of the vanadium ions, various Nafion based nanocomposite and hydrocarbon based nanocomposite membranes have been proposed.

17.3.2.1 Tailored Nafion Based Nanocomposite Membranes for VRFB Applications

To reduce methanol permeability, sol–gel derived Nafion/SiO$_2$ hybrid membranes have been successfully used by Mauritz and Warren (1989) and Mauritz and Stefanithis (1990). Nafion/SiO$_2$ hybrid membranes can reduce the crossover of methanol because of presence of polar clusters (pores) of Nafion being filled with SiO$_2$. The specific nanostructure of Nafion/SiO$_2$ hybrid membranes inspired the Xi et al. (2007), and they found that the same kind of membrane could reduce vanadium ion permeability, in comparison with the Nafion membrane. The Nafion/SiO$_2$ hybrid membrane was prepared by in-situ sol–gel method and exhibited nearly the same IEC and proton conductivity as that of pristine Nafion 117 membrane. Nafion/SiO$_2$ hybrid membrane displayed very low vanadium ion permeability compared to the Nafion membrane, for its polar clusters were filled with SiO$_2$ nanoparticles. The VRFB single cell with Nafion/SiO$_2$ hybrid membrane showed higher coulombic and energy efficiencies and a lower self-discharge rate than that of the Nafion system. The results from the cycling tests revealed that Nafion/SiO$_2$ hybrid membrane has good chemical stability in vanadium and acid solutions. The experimental results confirmed that the Nafion/SiO$_2$ hybrid membrane approach is a promising strategy to overcome the vanadium ions crossover in VRFB.

Shul and Chu (2014) fabricated a Nafion/GO layered membrane by coating the pristine Nafion 117 membrane with a Nafion/GO emulsion. The two-dimensional layered GO structure can act as an effective blockade to the transport of vanadium ions due to the significant increase in tortuosity (Dai et al., 2014a). GO/Nafion membrane showed slightly higher water uptake and IEC than pristine Nafion membrane due to the hydrophilic properties of GO and the existence of oxygen functional groups on its surface. Although the IEC did not change significantly, the permeability of the VO^{2+} ion decreased considerably from 20.5 × 10^{-7} to 6.1 × 10^{-7} cm^2 min^{-1} by reducing vanadium ions crossover. The VRFB single cell performance of the GO/Nafion membrane displayed higher coulombic efficiency compared to the Nafion membrane, while the voltage efficiency of the GO/Nafion membrane was lower. This is due to the rise of internal resistance and the decrease in vanadium ion permeability by applying another layer on the Nafion membrane. As a result, the energy efficiency of the GO/Nafion membrane was higher than that of the pristine Nafion membrane at low and medium current densities, except at 80 mAcm^{-2}.

The colloidal zeolite-Nafion composite membrane (ZNM) was synthesized by Yang et al. (2015), with a two-layer structure consisting of a top layer of colloidal silicalite in Nafion matrix and a base layer of pure Nafion. The colloidal silicalite-Nafion composite membranes with overall zeolite nanoparticle contents of 5 and 15 wt.%, which were denoted by ZNM-5 and ZNM-15. Due to the intracrystalline pores of silicalite (0.56 nm diameter, defined by 10-member rings), which were impervious to the multivalent hydrated metal ions but permeable to H_3O^+, the composite membrane displayed higher H^+/VO^{2+} ion transport selectivity (defined as the ratio of the slope for H^+ to that for VO^{2+} in the permeation curves) than the Nafion membrane. The ZNM-5 showed a lower H^+/VO^{2+} ion transport selectivity than ZNM-15 with a more compact colloidal silicalite layer. But the VRFB with ZNM-15 exhibited a much higher resistance of 3.40 Ω than the battery composed of ZNM-5 (1.72 Ω) and Nafion 117 (1.98 Ω). This is because ZNM-15 had a much higher content of silicalite nanoparticles, and ZNM-5 indicated lower electrical resistance than the Nafion 117 because ZNM-5 was much thinner with a lower silicalite content. The VRFB consisting of ZNM-5 membrane achieved higher coulombic efficiency, voltage and energy efficiencies than Nafion 117 and ZNM-15 membranes. A VRFB equipped with ZNM-5 provided quite stable energy efficiency on a 30-day continuous cyclic operation test, which confirmed the good chemical stability of the silicalite-Nafion composite membrane.

17.3.2.2 Sulfonated Hydrocarbon Polymer Based Nanocomposite Membranes for VRFB Applications

Several non-fluorinated aromatic polymer membranes, such as sulfonated polyimide (SPI) (Yue et al., 2011), sulfonated polysulfone (SPSU) (Chen & Hickner, 2013), sulfonated poly(fluorenyl ether ketone) (SPFEK) (Pan et al., 2013) and SPEEK (Winardi et al., 2014), have been widely explored as candidates for proton conductive membranes in VRFB environments. These aromatic polymer membranes commonly show lower vanadium ion permeability at a lower price than Nafion membranes, but their proton conductivity and stability need to be improved, while their vanadium ion permeability is expected to be further reduced. Adding the inorganic filler to aromatic polymers to prepare organic–inorganic composite membranes is seen as a useful strategy to achieve such aims.

Because the mesoporous TiO_2 particles have sufficient stability, their easily availability and lower price render them promising for use as inorganic fillers in SPI polymers to fabricate SPI/TiO_2 composite membranes (Li et al., 2014a). Due to their less hydrophilic TiO_2, SPI/TiO_2 composite membranes exhibited lower water uptake and swelling ratio than the pure SPI membrane. The room temperature proton conductivity of the SPI/TiO_2 membrane was 3.12×10^{-2} Scm^{-1}, which was higher than that of the bare SPI membrane. The synergistic effect between the hydrated sulfonic acid group and the hydrated mesoporous titanium dioxide particles aids the proton transport behavior through the mesoporous structure. Compared to the Nafion 117 membrane, the proton conductivity of the SPI/TiO_2 membrane was slightly lower at 5.82×10^{-2} Scm^{-1}, due to the exclusive hydrophilic-hydrophobic structure of Nafion. However, the ion selectivity of the SPI/TiO_2 composite membrane was the highest on account of its low vanadium ion permeability. Also, the vanadium resistance of SPI/TiO$_2$ membrane was higher than pure SPI membrane and much lower than that of Nafion 117 membrane, due to the superior size stability of the composite membrane, as represented by its physicochemical properties. As a result, the VRFB with SPI/TiO$_2$ composite membrane was always higher in coulombic and energy efficiencies than that with Nafion 117 membrane especially at low current densities, and the OCV performance of the VRFB operated with the SPI/TiO$_2$ composite membrane showed a superior performance to the Nafion 117 membrane.

A series of SPEEK/GO nanocomposite membranes with various amounts of GO loadings were fabricated Dai et al. (2014b). The incorporation of GO into SPEEK membranes for VRFB can form a hydrophobic/hydrophilic phase separation structure because of the hydrophilic surface of GO, which originates from the oxygen containing functional groups of GO. The hydrophilic functional groups of GO were prone to form hydrogen bonds with polymer chains and consequently enhance ion selectivity. The two-dimensional layered structure of GO nanosheets effectively prevents the transport of vanadium ions. The SPEEK/GO composite membranes were prepared with GO and denoted by SPEEK/GO-X, where X (1, 2, 3 or 5 wt.%) is the weight ratio of GO. The water uptake, swelling ratio and IEC of SPEEK/GO composite membrane increased with increasing GO content due to the hydrophilic nature of GO, whereas a decreasing trend was observed in proton conductivity values due to the blocking effect of the GO. The permeation of VO^{2+} across the composite membranes decreased with increasing GO weight ratio because the impermeable 2-D layered GO nanosheets can serve as effective barriers, while the interfacial interaction between GO and SPEEK matrix restricts the formation of hydrophilic channels used for the transport of vanadium ions. The coulombic and energy efficiencies of VRB single cell performance with SPEEK/GO-2 membrane (96.9% and 84.2%, respectively) were much higher than those with the Nafion 117 membrane (92.8% and 79.5%, respectively) at a current density of 80 mA cm^{-2}. Based on all the above results, the SPEEK/GO-2 composite membrane displayed the highest selectivity of 1.88×10^4 S min cm^{-3}, which was about two times higher than that of the Nafion 117 membrane (0.91×10^4 S min cm^{-3}).

Sulfonated polyimide (SPI) and ZrO$_2$ were blended to prepare a series of novel SPI/ZrO$_2$ composite membranes for VRFB application (Li et al., 2014b). Results of AFM and XRD reveal that ZrO$_2$ has sufficient compatibility with SPI and there are no micro defects in the SPI/ZrO$_2$ membrane. The crystalline structure of the composite membrane can be altered by controlling the quantity of ZrO$_2$. Water uptake, proton conductivity and chemical stability are enhanced with increase in ZrO$_2$ content. All SPI/ZrO$_2$ membranes possess high proton conductivity (2.96–3.72×10^{-2} S cm^{-1}) and low VO^{2+} permeability (2.18–4.04×10^{-7} cm^2 min^{-1}). However, when ZrO$_2$ content is raised beyond 15 wt.%, vanadium ion permeability unexpectedly increased, causing proton selectivity to reduce. Therefore, the optimum weight content of ZrO$_2$ is stipulated as 15 wt.%. VRFB incorporated with SPI/ZrO$_2$-15 membrane showed higher coulombic and energy efficiencies than Nafion 117. Cycling charge-discharge tests indicate that the SPI/ZrO$_2$-15% membrane has promising operational stability in the VRFB system. All results indicate that use of the SPI/ZrO$_2$-15% composite membrane is very promising for VRFB.

Composite membranes of SPI and sulfonated molybdenum disulfide (s-MoS$_2$) were prepared by Li et al. (2015). The SPI/s-MoS$_2$ membranes possess very high proton conductivity of up to 2.75×10^{-2} S cm^{-1} than pristine SPI due to their unique 2-D structure together with sulfonated acid groups of s-MoS$_2$. The blocking effect of inorganic component s-MoS$_2$ led to a decrease in the vanadium ion permeability of SPI/s-MoS$_2$ membrane to 1.23×10^{-7} cm^2 min^{-1}. The VRFB cell with SPI/MoS$_2$ membrane exhibited higher coulombic efficiency and energy efficiency when compared to Nafion 117, at a current density of 20 to 80 mA cm^{-2}. The self-discharge rate of VRFB with SPI/s-MoS$_2$ membrane is much slower than that of Nafion 117.

TABLE 17.2

Summary of Nanocomposite Membranes for VRFB Applications

Nanocomposite Membranes	Comments	References
Nafion/ SiO$_2$	Displayed lower vanadium ion permeability than Nafion. Higher coulombic and energy efficiencies than Nafion. Lower self-discharge rate than Nafion. Has good chemical stability in vanadium and acid solutions.	Xi et al. (2007)
Nafion/GO	Higher water uptake and ionic conductivity than pristine Nafion. VO^{2+} ion permeability decreased dramatically from 20.5×10^{-7} to 6.1×10^{-7} cm^2 min^{-1} by reducing vanadium ion crossover. Exhibit higher coulombic and energy efficiencies than Nafion.	Dai et al. (2014a)
SPI/TiO$_2$	Exhibited lower water uptake and swelling ratio than pure SPI membrane. Exhibit proton conductivity of 3.12×10^{-2} S cm^{-1}, at RT which was higher than the plain SPI membrane. Higher vanadium resistance than pure SPI membrane but lower than Nafion. Higher coulombic and energy efficiencies than Nafion.	Li et al. (2014a)
SPEEK/GO	IEC, water uptake and swelling ratio increased with GO content. Lower proton conductivity than pristine Nafion. Higher coulombic and energy efficiencies (96.9% and 84.2%,) than Nafion 117 membrane (92.8% and 79.5%, resp.) at a current density of 80 mAcm^{-2}.	Dai et al. (2014b)
SPI/ZrO$_2$	Water uptake, proton conductivity and chemical stability increased with ZrO$_2$ content. High proton conductivity (2.96–3.72×10^{-2} S cm^{-1}) and low VO^{2+} permeability (2.18–4.04×10^{-7} cm^2 min^{-1}). Threshold weight ratio of ZrO$_2$ content is 15%.	Li et al. (2014b)
SPI/s-MoS$_2$	High proton conductivity of up to 2.75×10^{-2} S cm^{-1}. Less vanadium ion permeability of 1.23×10^{-7} cm^2 min^{-1}. Higher coulombic and energy efficiencies than Nafion at 20 to 80 mA cm^{-2}. Slower self-discharge rate than Nafion 117.	Li et al. (2015)

Moreover, the better operational stability of SPI/s-MoS$_2$ has excellent potential for long-life VRFB systems. The physicochemical and electrochemical characteristics of the various nanocomposite membranes for VRFB applications are summarized and shown in Table 17.2.

17.4 Conclusions

The present chapter provides a snapshot of the efforts that have been made to summarize the various polymer nanocomposite membranes particularly those based on advanced inorganic nanoparticles to replace the currently used Nafion. They exhibited a significant improvement in thermomechanical and thermal stability, as well as proton conductivity and good barrier resistance at very low filler contents. The main reason for these improved properties in polymer nanocomposite is the strong interfacial interactions between the matrix and the filler surface, as opposed to conventional composites. Although a significant amount of work has already been done on various aspects of Nafion-based and hydrocarbon-based nanocomposite membranes, much research still remains in order to understand the complex structure-property relationships in various nanocomposite membranes, particularly to find the link between the polymer morphology and properties, such as transport and thermo-mechanical properties, relevant to the use of this polymer electrolyte in fuel cells. Another major challenge that remains is balancing the ratio between the proton conductivity and the methanol permeability, which, to date, has not been attained. This concludes that these nanocomposite membranes should be commercially available once this ratio is maximized, and their appropriate utilization will open a new direction in alternative energy sources.

Membranes for VRFB applications have also been extensively studied. Nafion continues to be one of the most widely studied membranes in VRFB systems; however, its high cost, high level of water transfer and high diffusivity values for vanadium ions limit its use in commercial applications. Even though many membranes provide potential results, meeting all the requirements for commercialization of an economically viable system is still a challenge, which requires further research in order to achieve the required cost structure for large-scale grid connected applications. Membranes for VRFB must withstand the highly oxidative pentavalent vanadium (V^{5+}) ions in the case of the original VRFB technology. New, improved perfluorinated and hydrocarbon-based membranes possess sufficient resistance to the oxidizing V^{5+} ions and robust mechanical properties, and they seem to be suitable for use in VRFB applications. An extensive study on nanocomposite membranes is ongoing to find low-cost alternatives to Nafion membranes. Considering their ion selectivity and energy efficiency, functionalized nanocomposite membranes show better performance than pre-functionalized nanocomposite membranes. Nevertheless, research on nanocomposite membranes for application in VRFB is still in its infancy, and more efforts based on new strategies to design and prepare the next generation membranes with outstanding VRFB performance are necessary.

17.5 Future Outlook

Among most battery systems that have been developed and commercialized thus far, the PEMFC and VRFB configurations offer several advantages that render them suitable for a variety of applications. Developing cost-effective materials and optimizing the operational parameters will further expand the scope of application of these systems. Various groups around the world are now focusing their research on low-cost membrane development with long-term stability, which is now being manufactured commercially by several companies. However, an even greater cost reduction and durability are expected with some of the novel membrane materials, which are currently under development, to make fuel cell technology reliable and affordable.

References

Aricò, A.S., S. Srinivasan, and V. Antonucci. 2001. DMFCs: From fundamental aspects to technology development. *Fuel Cells* 1(2): 133–161.

Bai, H., Y. Li, H. Zhang, H. Chen, W. Wu, J. Wang, and J. Liu. 2015. Anhydrous proton exchange membranes comprising of chitosan and phosphorylated graphene oxide for elevated temperature fuel cells. *Journal of Membrane Science* 495: 48–60.

Beydaghi, H., M. Javanbakht, and E. Kowsari. 2016. Preparation and physicochemical performance study of proton exchange membranes based on phenyl sulfonated graphene oxide nanosheets decorated with iron titanate nanoparticles. *Polymer* 87: 26–37.

Cao, Y.C., C. Xu, X. Wu, X. Wang, L. Xing, and K. Scott. 2011. A poly(ethylene oxide)/graphene oxide electrolyte membrane for low temperature polymer fuel cells. *Journal of Power Sources* 196(20): 8377–8382.

Chen, D.Y. and M.A. Hickner. 2013. V^{5+} degradation of sulfonated Radel membranes for vanadium redox flow batteries. *Physical Chemistry Chemical Physics* 15(27): 11299–11305.

Chu, D. and S. Gilrnan. 1994. The influence of methanol on O_2 electroreduction at a rotating Pt disk electrode in acid electrolyte. *Journal of the Electrochemical Society* 141(7): 1770–1773.

Chu, F., B. Lin, T. Feng, C. Wang, S. Zhang, N. Yuan, Z. Liu, and J. Ding. 2015. Zwitterion-coated graphene-oxide-doped composite membranes for proton exchange membrane applications. *Journal of Membrane Science* 496: 31–38.

Cozzi, D., C. de Bonis, A. D'Epifanio, B. Mecheri, A.C. Tavares, and S. Licoccia. 2014. Organically functionalized titanium oxide/Nafion composite proton exchange membranes for fuel cells applications. *Journal of Power Sources* 248: 1127–1132.

Croce, F., L. Persi, B. Scrosati, F. Serraino-Fiory, E. Plichta, and M.A. Hendrickson. 2001. Role of the ceramic fillers in enhancing the transport properties of composite polymer electrolytes. *Electrochimica Acta* 46(16): 2457–2461.

Dai, W., Y. Shen, Z. Li, L. Yu, J. Xi, and X. Qiu. 2014a. SPEEK/Graphene oxide nanocomposite membranes with superior cyclability for highly efficient vanadium redox flow battery. *Journal of Materials Chemistry A* 2(31): 12423–12432.

Dai, W., L. Yu, Z. Li, J. Yan, L. Liu, J. Xi, and X. Qiu. 2014b. Sulfonated Poly(Ether Ether Ketone)/Graphene composite membrane for vanadium redox flow battery. *Electrochimica Acta* 132: 200–207.

De Grotthuss, C.J.T. 1806. On the decomposition of water and the bodies it holds in dissolution using galavanic electricity. *Annales de Chimie Paris* 58: 54–73.

Devrim, Y. 2014. Fabrication and performance evaluation of hybrid membrane based on a sulfonated polyphenyl sulfone/phosphotungstic acid/silica for proton exchange membrane fuel cell at low humidity conditions. *Electrochimica Acta* 146: 741–751.

Devrim, Y., S. Erkan, N. Bac, and I. Eroğlu. 2009. Preparation and characterization of sulfonated polysulfone/titanium dioxide composite membranes for proton exchange membrane fuel cells. *International Journal of Hydrogen Energy* 34(8): 3467–3475.

Devrim, Y., H. Devrim, and I. Eroglu. 2016. Polybenzimidazole/SiO_2 hybrid membranes for high temperature proton exchange membrane fuel cells. *International Journal of Hydrogen Energy* 41(23): 10044–10052.

Eckroad, S. 2007. Vanadium redox flow batteries: An in-depth analysis. *Technical Update EPRI-1014836.* Electric Power Research Institute, Palo Alto, CA.

Feng, K., B. Tang, and P. Wu. 2013. "Evaporating" Graphene Oxide Sheets (GOSs) for rolled up GOSs and its applications in proton exchange membrane fuel cell. *ACS Applied Materials & Interfaces* 5(4): 13042–13049.

Gao, Q., Y. Wang, L. Xu, G. Wei, and Z. Wang. 2009. Proton-exchange sulfonated poly (ether ether ketone) (SPEEK)/SiO_x-S composite membranes in direct methanol fuel cells. *Chinese Journal of Chemical Engineering* 17(2): 207–213.

Gulder, C. and S. Gunes. 2011. Carbon supported Pt-based ternary catalysts for oxygen reduction in PEM fuel cells. *Catalysis Communications* 12(8): 707–711.

Hwang, G.J. and H. Ohya. 1996. Preparation of cation exchange membrane as a separator for the all-vanadium redox flow battery. *Journal of Membrane Science* 120(1): 55–67.

Jalani, N.H., K. Dunn, and R. Datta. 2005. Synthesis and characterization of Nafion-MO_2 (M = Zr, Si, Ti) nanocomposite membranes for higher temperature PEM fuel cells. *Electrochimica Acta* 51(3): 553–560.

Jiang, Z., X. Zhao, and A. Manthiram. 2013. Sulfonated poly(ether ether ketone) membranes with sulfonated graphene oxide fillers for direct methanol fuel cells. *International Journal of Hydrogen Energy* 38(14): 5875–5884.

Joerissen, L., J. Garche, and C. Fabjan. 2004. Possible use of vanadium redox-flow batteries for energy storage in small grids and stand-alone photovoltaic systems. *Journal of Power Sources* 127(1–2): 98–104.

Kalhammer, F.R., P.R. Prokopius, and V.P. Voecks. 1998. Status and prospects of fuel cells as automobile engines. State of California Air Resources Board, Sacramento, CA.

Kim, J., J.-D. Jeon, and S.-Y. Kwak. 2014. Nafion-based composite membrane with a permselective layered silicate layer for vanadium redox flow battery. *Electrochemistry Communications* 38: 68–70.

Laberty-Robert, C., K. Valle, F. Peoreira, and C. Sanchez. 2011. Design and properties of functional hybrid organic–inorganic membranes for fuel cells. *Chemical Society Reviews* 40(2): 961–1005.

Li, Q., R. He, J.O. Jensen, and N.J. Bjerrum. 2003. Approaches and recent development of polymer electrolyte membranes for fuel cells operating above 100°C. *Chemistry of Materials* 15(26): 4896–4915.

Li, J., Y. Zhang, and L. Wang. 2014a. Preparation and characterization of sulfonated polyimide/TiO$_2$ composite membrane for vanadium redox flow battery. *Journal of Solid State Electrochemistry* 18(3): 729–737.

Li, J., Y. Zhang, S. Zhang, X. Huang, and L. Wang. 2014b. Novel sulfonated polyimide/ZrO$_2$ composite membrane as a separator of vanadium redox flow battery. *Polymers for Advanced Technologies* 25(12): 1610–1615.

Li, J., Y. Zhang, S. Zhang, and X. Huang. 2015. Sulfonated polyimide/s-MoS$_2$ composite membrane with high proton selectivity and good stability for vanadium redox flow battery. *Journal of Membrane Science* 490: 179–189.

Lin, C.W. and Y.S. Lu. 2013. Highly ordered graphene oxide paper laminated with a Nafion membrane for direct methanol fuel cells. *Journal of Power Sources* 237: 187–194.

Ling, X., C. Jia, J. Liu, and C. Yan. 2012. Preparation and characterization of sulfonated poly(ether sulfone)/sulfonated poly(ether ether ketone) blend membrane for vanadium redox flow battery. *Journal of Membrane Science* 415–416: 306–312.

Liu, D., L. Geng, Y. Fu, X. Dai, and C. Lü. 2011. Novel nanocomposite membranes based on sulfonated mesoporous silica nanoparticles modified sulfonated polyimides for direct methanol fuel cells. *Journal of Membrane Science* 366(1–2): 251–257.

Liu, Y., J. Wang, H. Zhang, C. Ma, J. Liu, S. Cao, and X. Zhang. 2014. Enhancement of proton conductivity of chitosan membrane enabled by sulfonated graphene oxide under both hydrated and anhydrous conditions. *Journal of Power Sources* 269: 898–911.

Luo, Q., H. Zhang, J. Chen, P. Qian, and Y. Zhai. 2008. Modification of Nafion membrane using interfacial polymerization for vanadium redox flow battery applications. *Journal of Membrane Science* 311(1–2): 98–103.

Mai, Z., H. Zhang, X. Li, S. Xiao, and H. Zhang. 2011. Nafion/polyvinylidene fluoride blend membranes with improved ion selectivity for vanadium redox flow battery application. *Journal of Power Sources* 196(13): 5737–5741.

Martínez-Morlanes, M.J., A.M. Martos, A. Várez, and B. Levenfeld. 2015. Synthesis and characterization of novel hybrid polysulfone/silica membranes doped with phosphomolybdic acid for fuel cell applications. *Journal of Membrane Science* 492: 371–379.

Mauritz, K.A. and I.D. Stefanithis. 1990. Microstructural evolution of a silicon oxide phase in a perfluorosulfonic acid ionomer by an in situ sol-gel reaction. 3. Thermal analysis studies. *Macromolecules* 23(8): 1380–1388.

Mauritz, K.A. and R.M. Warren. 1989. Microstructural evolution of a silicon oxide phase in a perfluorosulfonic acid ionomer by an in situ sol-gel reaction. 1. Infrared spectroscopic studies. *Macromolecules* 22(4): 1730–1734.

Mecheri, B., A. D'Epifanio, E. Traversa, and S. Licoccia. 2008. Sulfonated polyether ether ketone and hydrated tin oxide proton conducting composites for direct methanol fuel cell applications. *Journal of Power Sources* 178(2): 554–560.

Molina, J., J. Bonastre, J. Fernández, A.I. del Río, and F. Cases. 2016. Electrochemical synthesis of polypyrrole doped with graphene oxide and its electrochemical characterization as membrane material. *Synthetic Metals* 220: 300–310.

Mossayebi, Z., T. Saririchi, S. Rowshanzamir, and M.J. Parnian. 2016. Investigation and optimization of physicochemical properties of sulfated zirconia/sulfonated poly (ether ether ketone) nanocomposite membranes for medium temperature proton exchange membrane fuel cells. *International Journal of Hydrogen Energy* 41(28): 12293–12306.

Padmavathi, R., R. Karthikumar, and D. Sangeetha. 2012. Multilayered sulphonated polysulfone/silica composite membranes for fuel cell applications. *Electrochimica Acta* 71: 283–293.

Pan, J.J., S.J. Wang, M. Xiao, M. Hickner, and Y.Z. Meng. 2013. Layered zirconium phosphate sulfophenylphosphonates reinforced sulfonated poly (fluorenyl ether ketone) hybrid membranes with high proton conductivity and low vanadium ion permeability. *Journal of Membrane Science* 443: 19–27.

Park, K.T., U.H. Jung, D.W. Choi, K. Chun, H.M. Lee, and S.H. Kim. 2008. ZrO_2– SiO_2/Nafion composite membrane for polymer electrolyte membrane fuel cells operation at high temperature and low humidity. *Journal of Power Sources* 177(2): 247–253.

Peng, K.J., J.Y. Lai, and Y.L. Liu. 2016. Nanohybrids of graphene oxide chemically-bonded with Nafion: Preparation and application for proton exchange membrane fuel cells. *Journal of Membrane Science* 514: 86–94.

Sacca, A., A. Carbone, E. Passalacqua, A. D'epifanio, S. Licoccia, E. Traversa, E. Sala, F. Traini and R. Ornelas. 2005. Nafion–TiO_2 hybrid membranes for medium temperature polymer electrolyte fuel cells (PEFCs). *Journal of Power Sources* 152: 16–21.

Sakamoto, M., S. Nohara, K. Miyatake, M. Uchida, M. Watanabe, and H. Uchida. 2014. Effects of incorporation of SiO_2 nanoparticles into sulfonated polyimide electrolyte membranes on fuel cell performance under low humidity conditions. *Electrochimica Acta* 137: 213–218.

Sambandam, S. and V. Ramani. 2007. SPEEK/functionalized silica composite membranes for polymer electrolyte fuel cells. *Journal of Power Sources* 170(2): 259–267.

Shao, Z.-G., H. Xu, M.Q. Li, and I.-M. Hsing. 2006. Hybrid Nafion–inorganic oxides membrane doped with heteropolyacids for high temperature operation of proton exchange membrane fuel cell. *Solid State Ionics* 177(7–8): 779–785.

Shirdast, A., A. Sharif, and M. Abdollahi. 2016. Effect of the incorporation of sulfonated chitosan/sulfonated graphene oxide on the proton conductivity of chitosan membranes. *Journal of Power Sources* 306: 541–551.

Shul, Y.G. and Y.H. Chu. 2014. Nafion/Graphene oxide layered structure membrane for the vanadium redox flow battery. *Science of Advanced Materials* 6(7): 1445–1452.

Silva, V.S., B. Ruffmann, H. Silva, Y.A. Gallego, A. Mendes, L.M. Madeira, and S.P. Nunes. 2005. Proton electrolyte membrane properties and direct methanol fuel cell performance: I. Characterization of hybrid sulfonated poly(ether ether ketone)/zirconium oxide membranes. *Journal of Power Sources* 140(1): 34–40.

So, S.Y., Kim, S.C. and Lee, S.Y. (2010). *In situ* hybrid Nafion/SiO_2–P_2O_5 proton conductors for high-temperature and low-humidity proton exchange membrane fuel cells. *Journal of Membrane Science* 360(1–2): 210–216.

Sridhar, P., R. Perumal, N. Rajalakshmi, M. Raja, and K.S. Dhathathreyan. 2001. Humidification studies on polymer electrolyte membrane fuel cell. *Journal of Power Sources* 101(1): 72–78.

Sukkar, T. and M. Skyllas-Kazakos. 2003. Water transfer behaviour across cation exchange membranes in the vanadium redox battery. *Journal of Membrane Science* 222(1–2): 235–247.

Tang, H., Z. Wan, M. Pan, and S.P. Jiang. 2007. Self-assembled Nafion–silica nanoparticles for elevated-high temperature polymer electrolyte membrane fuel cells. *Electrochemistry Communications* 9(8): 2003–2008.

Thiam, H.S., W.R.W. Daud, S.K. Kamarudin, A.B. Mohamad, A.A.H. Kadhum, K.S. Loh, and E.H. Majlan. 2013. Nafion/Pd–SiO_2 nanofiber composite membranes for direct methanol fuel cell applications. *International Journal of Hydrogen Energy* 38(22): 9474–9483.

Tripathi, B.P. and V.K. Shahi. 2009. Surface redox polymerized SPEEK–MO_2–PANI (M = Si, Zr and Ti) composite polyelectrolyte membranes impervious to methanol. *Colloids and Surfaces A: Physicochemical and Engineering Aspects* 340(1–3): 10–19.

Uchida, H., Y. Ueno, H. Hagihara, and M. Watanabe. 2003. Self-humidifying electrolyte membranes for fuel cells: Preparation of highly dispersed TiO_2 particles in Nafion 112. *Journal of Electrochemical Society* 150(1): A57–A62.

Üregen, N., K. Pehlivanoğlu, Y. Özdemir, and Y. Devrim. 2016. Polybenzimidazole/SiO_2 hybrid membranes for high temperature proton exchange membrane fuel cells. *International Journal of Hydrogen Energy* 41(23): 10044–10052.

Wang, N., S. Peng, H. Wang, Y. Li, S. Liu, and Y. Liu. 2012. SPPEK/WO_3 hybrid membrane fabricated via hydrothermal method for vanadium redox flow battery. *Electrochemistry Communications* 17: 30–33.

Weidmann, E., E. Heintz, and R.N. Lichtenthaler. 1998. Sorption isotherms of vanadium with H_3O^+ ions in cation exchange membranes. *Journal of Membrane Science* 141(2): 207–213.

Winardi, S., S.C. Raghu, M.O. Ohnmar, Q.Y. Yan, N. Wai, T.M. Lim, and M. Skyllas-Kazacos. 2014. Sulfonated poly (ether ether ketone)-based proton exchange membranes for vanadium redox battery applications. *Journal of Membrane Science* 450: 313–322.

Xi, J., Z. Wu, X. Qiu, and L. Chen. 2007. Nafion/SiO_2 hybrid membrane for vanadium redox flow battery. *Journal of Power Sources* 166(2): 531–536.

Yang, R., Z. Cao, S. Yang, I. Michos, Z. Xu, and J.H. Dong. 2015. Colloidal silicalite-nafion composite ion exchange membrane for vanadium redox-flow battery. *Journal of Membrane Science* 484: 1–9.

Yin, Y., H. Wang, L. Cao, Z. Li, Z. Li, M. Gang, C. Wang, H. Wu, Z. Jiang, and P. Zhang (2016). Sulfonated poly(ether ether ketone)-based hybrid membranes containing graphene oxide with acid-base pairs for direct methanol fuel cells. *Electrochimica Acta* 203: 178–188.

Yue, M.Z., Y.P. Zhang, and Y. Chen. 2011. Preparation and properties of sulfonated polyimide proton conductive membrane for vanadium redox flow battery. *Advanced Materials Research* 239–242: 2779–2784.

Zhang, B., Y. Cao, S. Jiang, Z. Li, G. He, and H. Wu. 2016. Enhanced proton conductivity of Nafion nanohybrid membrane incorporated with phosphonic acid functionalized graphene oxide at elevated temperature and low humidity. *Journal of Membrane Science* 518: 243–253.

Zhao, Y., Y. Fu, B. Hu, and C. Lü. 2016. Quaternized graphene oxide modified ionic cross-linked sulfonated polymer electrolyte composite proton exchange membranes with enhanced properties. *Solid State Ionics* 294: 43–53.

18

Polyion Complex Membranes for Polymer Electrolyte Membrane and Direct Methanol Fuel Cell Applications

F. Dileep Kumar, Harsha Nagar, and Sundergopal Sridhar

CONTENTS

18.1 Introduction

The shortage of energy resources, a tremendous population growth, and the negatively impacted environment are matters of great concern worldwide. Therefore, an unearthing of promising alternative energy sources that mitigate these issues is a major priority. Fuel cells have garnered attention by successfully generating electricity in an eco-friendly manner, with its only by-products being water and heat (Ergun et al., 2012). The absence of noise and air pollution in fuel cells, renders them to cleaner energy technology, and can be considered foundational to a future sustainable energy system. Polymer-based fuel cells are more intensified in nature due to their compact size, reliability and high power density. The primary components of a fuel cell consist of two electrodes and the electrolyte, wherein the fuel is sent to the anode side of the electrode, where the oxidation of fuel occurs to generate electrons and protons. The oxidant is in the form of either pure oxygen or air, which is supplied to the cathode side, whereas the electrolyte transports the ions from anode to cathode compartment, as shown in Figure 18.1. The concentration gradient present between the electrode and electrolyte accelerates the movement of ions. Though the function of the fuel cell is similar to a battery, there exist several differences. Batteries operate in continuous fuel consumption mode and require replenishment as they store electrical energy in a closed system. Fuel cells do not require recharging, provide greater operating life, are lighter or equal in weight per unit of power output and they deliver a much higher power density, making them potential replacements for batteries. There are no moving parts in fuel cells—an added advantage making them noise-free devices. They do not work on the principle of Carnot's cycle; therefore, their operation is simple and more efficient compared to internal combustion (IC) engines (Person et al., 2011). Amongst the various type of FCs, the more robust ones, based on solid oxides or highly portable and affordable versions, such as proton exchange membrane fuel cell (PEMFC) and direct methanol fuel cell (DMFC), assume greater significance (Srinivasan et al., 2011). The proton exchange membrane (PEM) is the foremost component (along with Pt based catalysts) of the fuel cell as it plays an important role for the conduction of protons and separating of fuel from oxidant. The prerequisite characteristics of PEM membranes are low-cost, with a high proton conductivity and low electron conductivity, with sufficient mechanical properties, oxidative and chemical integrity, low permeability to gases and limited swelling

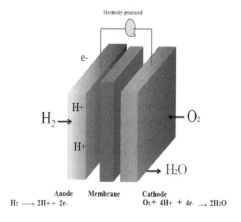

FIGURE 18.1
Fuel cell set-up.

in the presence of water (Silva et al., 2005). It is also imperative to possess a film forming property and flexibility for fabrication into MEAs. The perfluorinated membrane (Nafion®) manufactured by DuPont is widely used in fuel cells; however, problems like methanol crossover, carbon monoxide poisoning of non-reusable expensive Pt and Pt-based electrocatalysts at low temperatures, the high cost of the membrane (US\$ 600–1,200/m^2), and environmentally hazardous disposal of toxic perfluorinate waste, have obstructed the widespread commercialization of fuel cells (Vladimir et al., 2007). To mitigate these issues, the synthesis of an alternative membrane with desired properties is an important area of research. To address these aspects, the present chapter describes the fundamentals and important characteristics of modified acid-base blend membranes for PEMFC and DMFC applications, in detail.

18.2 Challenges Facing Fuel Cells

The major hurdles imposing setbacks to the commercialization of PEMFC and DMFC are as follows:

- The hydrogen fuel used in PEMFC contains a small quantity of carbon monoxide (CO) contaminant that affects the efficiency of anode. Apart from this, the handling and storage of hydrogen is also a difficult task due to their explosive nature.
- Methanol is used as an alternative fuel in place of hydrogen, but its use endures the crossover issue through the membrane. At low operating temperatures, methanol oxidation also occurs, affecting fuel cell efficiency.
- The major barrier that restricts FC applicability, on a large scale, is the high cost of components, such as platinum-based catalyst, the membrane (Nafion), and the problems associated with them, e.g., fuel loss, low thermal and chemical stability, poor moisture retention at high temperatures, etc.

To address these issues, the development of various new polymeric materials and catalyst types, apart from the modification of current membranes with different catalyst loading, has been investigated by researchers. Technological advancements in this field could enable wider commercialization of fuel cells as major sources of energy for the automotive industry and for portable electronics.

18.3 Research Trends in Polymer Electrolyte
Membranes for PEMFC and DMFC

The membrane is a critical component of the fuel cell, as it influences the performance in terms of power density. Therefore, the selection of membrane material is an important criterion, as it is the epicenter of the heart of any fuel cell, which is the MEA. In terms of cost, roughly 75% of the overall price of a PEMFC is associated with its MEA. State-of-the-art perfluorinated ionomer membranes exhibit different limitations that hinder fuel cell commercialization, such as:

- High cost (US$ 500–800.00 m^{-2}) and low proton conductivity nature at elevated temperature (T > 100°C), due to low humidity.
- In the case of nonfluorinated ionomer materials, high water uptake is exhibited even after achieving sufficient proton conductivity, leading to fuel bypass and flooding of cathode.
- Current membranes exhibit high methanol crossover problem, especially in DMFC, which induces reduction in power density through poisoning of cathode catalyst.
- The low stability of ionomers at elevated temperatures leads to polymer degradation.

Therefore, the synthesis of alternative cost-effective membranes is necessary for FC development. Common approaches that have been followed worldwide to synthesize alternative membranes are as follows:

- Modification of perfluorinated ionomer membranes.
- Functionalization of aromatic hydrocarbon polymers/membranes.
- Development of composite membranes based on solid inorganic proton conducting material incorporated within organic polymer matrix or acid-base blends, which improve water retention properties.

Compared to other methods, the synthesis of composite membranes by blending cationic polymers with anionic ones (acid-base blends), followed by the addition of organic/inorganic fillers, are investigated widely. These composite membranes exhibit the following advantages:

- The polyion complex membrane is mechanically more stable compared to the individual sulfonated polymer.
- The interaction between water-soluble polymeric sulfonic acid moieties with the basic amine group reduces the excess water swelling property with an enhanced retention of organic molecules, such as methanol, thus limiting fuel bypass.
- Acid-base blends prepared from water-soluble polymeric sulfonated salts or sulfonic acids exhibit a surprisingly high ion exchange capacity of up to 2.7 meq SO$_3$H/g of polymer, leading to substantial ionic conductance.
- The prepared blends exhibit an excellent performance for both hydrogen–PEMFC and methanol–DMFC systems (Zicheng et al., 2012).

Different types of interactions present in the blend membranes are discussed in the next section.

18.4 Types of Interactions in Acid-Base Blend Membranes

Within blend membranes, different types of interactive forces exist and are classified in the following sub-sections.

18.4.1 Dipole–Dipole/van der Waals Interactions

The modification of polymers through sulfonation or other reactions reduces their mechanical strength. Blending of high electron density groups present in sulfonic acid macromolecule with high dipole moment functional groups present in quaternary phosphonium or ammonium polymer, results in improved mechanical strength. Enhanced interaction between polymer chains (Javaid, 2009), as well as van der Waals forces (VDF) improves mechanical strength. VDF is a distance-dependent interactions force between atoms or molecules, and, compared to ionic, covalent bonds, it is weaker.

18.4.2 Hydrogen-Bonding Interaction Blend Membranes

Many polymers have a problem of swelling after their modification through sulfonation, which also deprives their mechanical strength. Therefore, the blending of these sulfonated polymers with other macromolecules that form hydrogen bonds can enhance their chemical and mechanical stability and reduce swelling. For example, the blending of SPEEK with polyamide (PA) or PEI polymers induces hydrogen bonding interactions between the sulfonic group of SPEEK and imide and amide groups of PEI and PA, respectively. Consequently, the thermal stability of the blend increases and further improves the glass transition temperature (T_g) by 5–15 K, as compared to pristine SPEEK. However, the blend could exhibit phase separation and hydrolytic stability problems at high temperatures, which means that hydrogen bond interactions must be substantiated with other types of interactions to satisfy the desired properties for fuel cell application (Cui et al., 1998).

18.4.3 Ionically Cross-Linked Acid-Base Blends and Acid-Base Ionomers

Interactions occurring in such blends through Vander Waals forces and hydrogen bonds do not meet the desired criteria for a fuel cell membrane, and therefore an alternative approach by mixing polysulfonates and polybases were investigated. These blend membranes consist of electrostatic and hydrogen bridge interactions that enhance their mechanical, chemical and thermal stabilities (Kerres et al., 2004). Moreover, both cationic and anionic polymers can be prepared from the same polymer and blended to carry both sulfonic acid and basic groups in the same backbone (Kerres et al., 2004).

18.4.4 Covalently Cross-Linked Blends

Covalent cross-linking between two uncharged polymers can mitigate swelling and chemical stability issues, but a limited number of covalently cross-linked ionomer membranes are reported (Mikhailenko et al., 2004) as heavy branching combined with an absence of ionic bonds can reduce proton conduction rate.

18.4.5 Covalently and Ionically Cross-Linked Blends

Synthesis of blend membranes that exhibits both covalent and ionic links can overcome the problem of brittleness in membranes, which can otherwise occur in only ionically or covalently cross-linked membranes. Splitting of ionic bonds and bleeding of sulfonated macromolecules in the case of covalently cross-linked blend membranes at high temperatures have been observed. Kerres et al. (2009) blended a polysulfonate with a polysulfinate and a polybase, under the addition of a dihalogeno cross-linker, which is capable

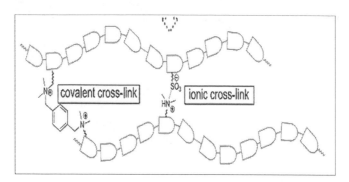

FIGURE 18.2
Membrane casting unit.

of reacting with both sulfinate groups and tertiary amino groups during alkylation. Lei et al. 2017 synthesized the highly stable ionic-covalent cross-linked sulfonated poly(ether ether ketone) membrane for DMFC. The interactions associated between sulfonic acid and pendant tertiary amine groups present in SPEEK membrane are explained in Figure 18.2.

18.5 Membrane Synthesis by Solution Casting and Solvent Evaporation Method

Solution casting and solvent evaporation is the usual method for preparation of dense blend membranes for FCs. In this procedure, appropriate polymer powders are dissolved separately in suitable solvents to make two viscous solutions. The prepared solutions are then mixed together in appropriate ratios and stirred to homogeneity for some time before casting on a glass plate using a casting machine, as shown in Figure 18.3. The thickness of the membrane can be adjusted in the casting machine using an adjustable metallic bar resting on metallic supports on either side to form a precise gap between the bar and the

Legends: A, B: Pressure gauges, C: Start end of knife edge, D: Polymer solution, E: Glass plate, F: Finish end of knife edge

FIGURE 18.3
Covalent-ionic cross-links between tertiary amine and sulfonic acid groups of SPEEK membrane (Adapted from Lei et al., 2017).

plate on which the film is cast. The cast film is kept aside to undergo controlled evaporation of the solvent, which leaves behind a thin, uniform polymer film (Harsha et al., 2015). Solvent evaporation is initially allowed at ambient temperature and later at an elevated temperature (60–80°C) to obtain dense non-porous membranes.

18.6 Membrane Characterization

18.6.1 Analytical Characterization

Fourier transform infrared (FTIR) spectroscopy analysis is useful in identifying the functional groups present in a polymer sample to monitor the vibrational energy levels of the molecules. The analysis was done using a Nicolet-740, Perkin-Elmer-283B (Boston, U.S.) in the wave number range of 400–4000 cm^{-1}. A digital scanning electron microscope (SEM) sourced from JEOL, JSM 5410, Japan was used to examine the microstructure morphology. Thermal degradation of the synthesized membrane samples was examined by a Seiko 220TG/DTA Thermo Analyzer purchased from Tokyo, Japan. Heating was maintained in the range of 25 to 600°C, with a heating rate of 10°C min^{-1}, under constant flushing with inert N$_2$ gas flowing at 200 mL min^{-1}. A Siemens D 5000 powder X-ray Diffractometer, (New Jersey, U.S.) was used to study crystalline nature and inter-segmental spacing. It generates X-rays of 1.5 Å wavelength through a CuK-alpha source across a diffraction angle of 0–65°. The tensile strength of the membranes was evaluated using a universal testing machine (UTM), model AGS-10kNG, made by Shimadzu, in Japan (Harsha et al., 2017).

18.6.2 Physico-Chemical Characterization

18.6.2.1 Ion Exchange Capacity (IEC) and Sorption Studies

IEC was measured to estimate the number of mill equivalents (meq) of ions present per unit weight of the dry polymer whereas sorption study was performed to determine the affinity of the polymer membrane toward water and methanol.

18.6.2.2 Proton Conductivity

Proton conductivities of polymeric blend membranes were measured by electrochemical impedance spectroscopy (EIS) through alternating current (AC) method. The membrane samples were sandwiched between two gold-coated electrodes in a cell and kept inside a temperature- and humidity-controlled oven. These electrodes were used to measure the voltage drop across a known distance and provide the impedance value as a function of frequency at different temperatures and humidity conditions. By putting the impedance value in the below equation, the conductivity of membranes was determined.

$$\sigma = \frac{L}{R_2 \times A} \tag{18.1}$$

where, L is the distance between the electrodes (cm), R_1, the impedance of the membrane and A is cross-sectional area of the membrane (cm^2).

18.6.2.3 Methanol Permeability

Methanol permeability of the polyelectrolyte films was determined by a two-compartment cell system in which the first compartment contained aqueous methanol, and the second compartment was filled with water. The membrane was placed as a barrier between these compartments. Before starting the experiment, the membrane was dipped in deionized water for at least 24 h. Both compartments were kept under stirring to impart uniform concentration gradient as a driving force, for diffusion of methanol from the first to the second compartment. The concentration of methanol in the second compartment was measured by refractive index. Methanol permeability was calculated using Equation 18.2:

$$P = \frac{V_2 \times L \times C_2(t)}{A \times C_1(t - t_i)} \tag{18.2}$$

where, C_1 and C_2 are concentrations of methanol in the first and second compartment, respectively, A is membrane area, L is membrane thickness, V_2 is volume of liquid in the second compartment, t is time, t_i is initial time and P is methanol permeability.

The ratio of proton conductivity to methanol permeability defines the selectivity of the membrane useful for DMFC.

18.7 Recent Trends in Acid-Base Blend Membranes

The first novel acid-base complexes were investigated by Kerres et al. (2001) using acidic polymers, such as sulfonated polyetheretherketone (SPEEK) or sulfonated poly(ether sulfone) (SPSU), and basic polymers, such as poly(4-vinylpyridine) (P4VP), poly(ethyleneimine) (PEI) and polybenzimidazole (PBI). The polyion complex was formed by ionic interactions between the acidic –SO$_3$H groups and the basic –N= groups. These ionic interactions enhance PEMFC performance with high thermal stability. Even in the case of DMFC, the introduction of ionic interactions through polymer blending narrows down the channels in the polymer matrix to reduce methanol permeation and thereby, fuel loss. Aromatic hydrocarbon polymers are effective materials for synthesizing PEM as they are robust and can be modified to increase polar group content, which contributes to enhancement in water uptake capacity, even at elevated temperatures by restricting the evaporation of absorbed water. These polymers exhibit easy recyclability with high decomposition stability and are available at a low cost. Researchers have investigated different aromatic hydrocarbon polymers, such as poly(etheretherketone) (PEEK) (Jinhua et al., 2010), polysulfone (Psf) (Francesco et al., 2006), Polybenzimidazole (PBI) (Jochen et al., 2001), polyethersulfone (PES) (Luca et al., 2018), etc., which can be employed as the polymer backbone for developing proton-conducting polymers. These aromatic polymers are easily sulfonated by concentrated sulfuric, acetyl sulfate, chlorosulfonic acid or sulfur trioxide (Smitha et al., 2003). Among the various hydrocarbon polymers, PES, PEEK and PBI exhibit properties necessary for alternative fuel cell membranes, especially acid-base blends.

18.7.1 SPES-Based Blends

Polyethersulfone (PES) has attracted considerable attention due to its easy preparation procedure and film forming ability, supported by high mechanical, chemical and thermal stabilities. Additionally, it is less expensive than perfluorinated ionomers and can easily undergo sulfonation (Harsha et al., 2016). Kim et al. (2013) reported blend membranes made up of sulfonated poly(arylene ether sulfone) (PAES) with hydrophobic polymers like PES and poly(vinylidene fluoride-cohecafluoropropylene) (PVdF-HFP) for DMFC application. The blend membrane exhibits requisite mechanical properties due to the interactions between -SO_3H groups of SPAES with sulfone groups of PES. Moreover, it reduced methanol crossover and facilitated proton conductivity. Harsha et al. (2015) reported ionically cross-linked blend membranes made by combining SPES with PEI and aniline treated PEI that exhibited high proton conductivity with low methanol permeability, due to the ionic interaction between sulfonic acid groups of SPES and imide and amine groups of PEI and APEI, as shown in Figures 18.4 (a) and 18.4 (b), respectively. Introduction of sulfonated and aminated hydrophilic sites into hydrophobic polymers having rigid aryl backbones appears to produce promising, inexpensive alternative electrolytes for DMFC applications.

Another novel membrane was prepared by dissolving 15% (w/v) SPES polymer powder in NMP solvent under continuous stirring to make a homogeneous SPES polymer solution. Post-degasification, this solution was poured on a Petri dish to the desired thickness

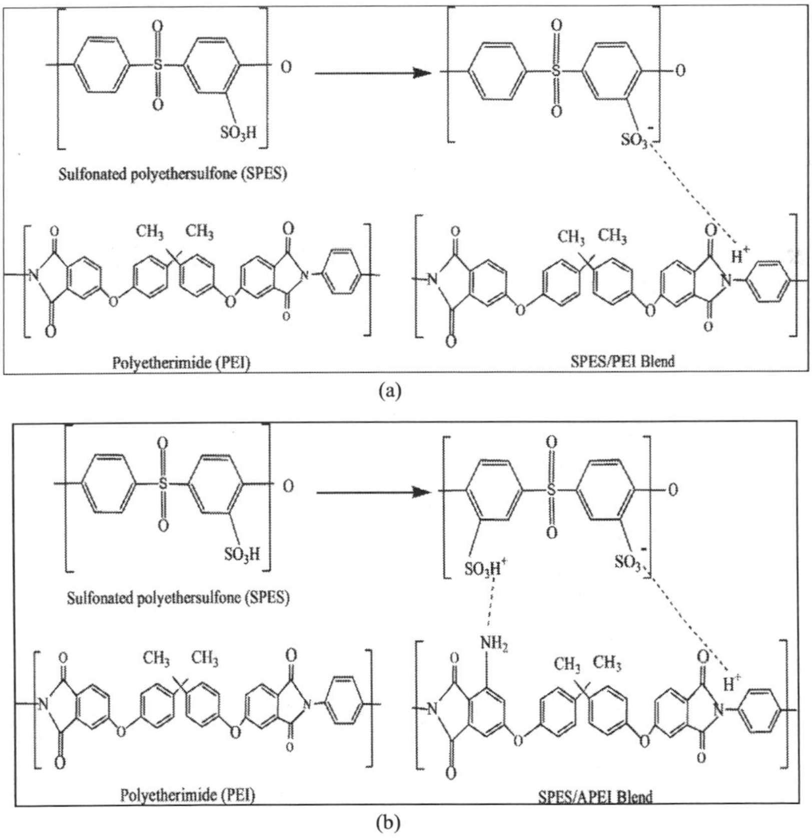

(a)

(b)

FIGURE 18.4
Ionically cross-linked (a) SPES/PEI and (b) SPES/APEI blend membrane (Harsha et al., 2015).

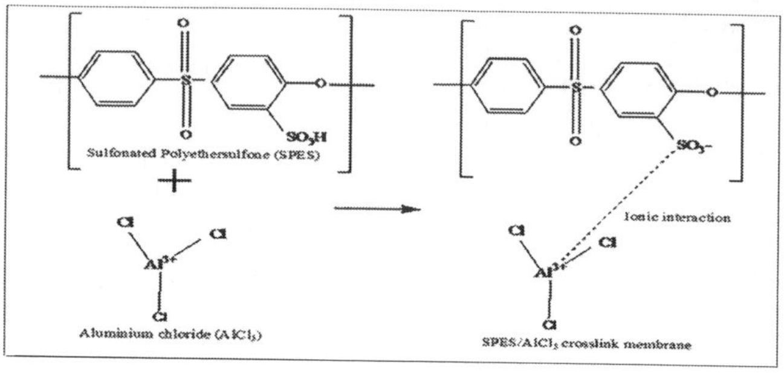

FIGURE 18.5
Ionically cross-linked SPES/AlCl$_3$ membrane.

TABLE 18.1

Methanol Permeability, Proton Conductivity and Selectivity of SPES/AlCl$_3$ Ionically Cross-Linked Membrane

Membrane	Selectivity ×10^5 (Ss^{-1}cm^{-3})	Methanol Permeability (cm^2 s^{-1})	Proton Conductivity (S cm^{-1})			
			30°C	40°C	60°C	80°C
SPES	2.46	1.26×10^{-7}	0.031	0.049	0.068	0.074
SPES/AlCl$_3$	13.6	5.21×10^{-8}	0.071	0.10	0.13	0.17
Nafion®	0.51	1.52×10^{-6}	0.077	0.08	0.103	0.135

and subsequently dried in atmosphere at room temperature, followed by vacuum drying at 80°C for 7 h to obtain a dense non-porous morphology. Furthermore, the membrane was treated with aqueous aluminum chloride salt solution of 15 wt% concentration to form ionic cross-linking between the trivalent aluminum ion and the monovalent sulfonic acid groups, as shown in Figure 18.5. The prepared membrane was investigated in terms of proton conductivity (0.071 Scm^{-1}), methanol permeability (5.21×10^{-8} cm^2 s^{-1}) and selectivity (13.6×10^5 Ss cm^{-3}), as shown in Table 18.1.

18.7.2 SPEEK-Based Blend Membrane

PEEK is an aromatic thermostable polymer with a non-fluorinated backbone with phenyl groups that are separated by ether (–O–) and carbonyl (–CO–) linkages. PEEK can be functionalized by sulfonation, and the degree of sulfonation (DS) can be controlled by manipulating reaction time and temperature. Sulfonation can be performed using concentrated sulfuric acid or oleum. SPEEK exhibits high thermal stability, mechanical strength and adequate proton conductivity that makes it ideal for FC application. However, excess swelling and low stability at high DS associated in SPEEK can be overcome by blending with other polymers (Adolfo et al., 2012).

Liang et al. (2015) synthesized novel ionically cross-linked SPEEK membranes containing diazafluorene functional groups, as shown in Figure 18.6. Ionic interactions occur between the sulfonic acid groups and pyridyl in diazafluorene, which reduces the swelling ratio and methanol permeability for DMFC operation. The ionically cross-linked membranes exhibit low methanol permeability in the range between 0.56×10^{-7} cm^2 s^{-1} and 1.8×10^{-7} cm^2 s^{-1}, which is lower than that of Nafion 117, apart from providing higher selectivity (4.82×10^4 Ss cm^{-3}).

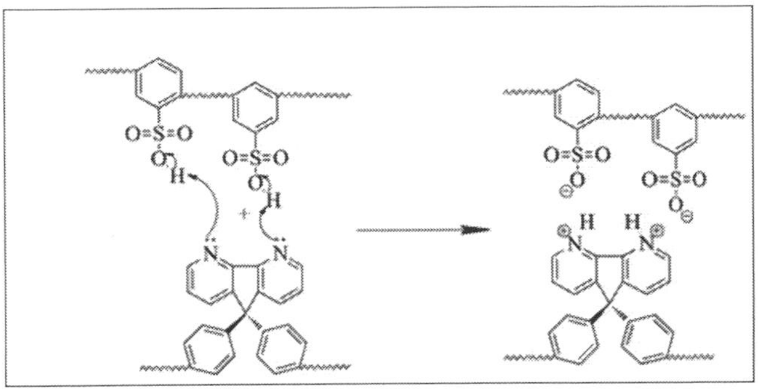

FIGURE 18.6
Mechanism of ionic cross-linking between sulfonic acid moeities and amine groups. (Adapted from Yu, L., G. Chenliang, Q. Zhigang, L. Hui, W. Zhongying, Z. Yakui, Z. Shujiang, and L. Yanfeng, *Journal of Power Sources*, 284: 86–94, 2015.)

Che et al. (2016) reported PEMs based on polyurethane (PU) matrix for operability at high temperatures. The thermal stability of PA doped PU membrane was found to have improved with the introduction of SPEEK into PU to form SPEEK/PU/PA membrane. The electrostatic interactions between PU matrix and SPEEK ensured SPEEK/PU/PA membranes to be homogeneous with compact structure. This interaction induced high thermal stability, proton conductivity and satisfactory mechanical property.

Luu et al. (2011) reported ionically cross-linked composite of SPEEK with strontium (Sr) earth metal by accommodation via ionic bonding. The incorporation of a small amount of Sr led to a considerable increment in thermal and mechanical stabilities due to the ionic interactions developed between anionic sulfonic acid groups and cationic Sr. Moreover, it reduced the methanol crossover and provided high selectivity. The formation of ionic bridge between the metal and sulfonic acid group enables satisfactory performance at elevated temperatures. Pasupathi et al. (2008) investigated an acid–base polymer blend membrane based on SPEEK and PBI for DMFC. An et al. (2016) reported a new approach to synthesize a gradient cross-linked structure with SPEEK membrane through simply post-treatment with NaBH$_4$ and H$_2$SO$_4$ solution at ambient temperature. This treatment causes the conversion of certain benzophenone moieties on SPEEK into benzhydrol groups that further react with sulfonic acid groups of SPEEK to form cross-links. These induced gradient cross-links effectively enhance membrane strength with adequate proton conductivity.

18.7.3 PBI-Based Blends

PBI is an aromatic polymer synthesized from bis-o-diamines and decarboxylates (acids, esters, amides), in the form of a molten state or a solution. The concentration of the tetramine and dicarboxylic acids in the PBI decides their thermal properties (Li et al., 2004). Florian et al. (2015) synthesized novel acid–base blend membranes from polybenzimidazoles F$_6$PBI, PBIOO and PBI HOZOL for application in HT-PEMFCs. Ionic cross-linking between acidic and basic polymers, shown in Figure 18.7, enhanced chemical stability and integrity of the membranes in hot PA. This resulted in higher conductivity with reduced swelling. This study proves that acid–base blends are more suitable alternatives to pure PBI and AB-PBI as membranes for HT-PEMFCs. Future work involves the optimization of electrode nature and properties for MEAs based on blend membranes to achieve higher cell performance than individual PBI or AB-PBI based MEAs.

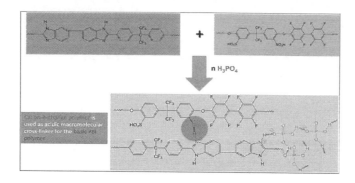

FIGURE 18.7

Ionic cross-linking and formation of hydrogen bridge network by PA doping in the MJK1845 membrane. (Adapted from Florian, M., A. Karin, E. Corina, J. Kerres, and Z. Roswitha, *Journal of Materials Chemistry A*, 3:10864–10874, 2015.)

Mingfeng et al. (2016) synthesized and characterized highly sulfonated SPEEK/PBI ionically cross-linked blend membranes, followed by acidification. SEM pictographs, shown in Figure 18.8, confirm the compatibility between the polymers that constituted a compact structure without any phase separation. The strong ionic interaction and hydrogen bonds between –SO$_3$H and N-basic group, as shown in Figure 18.9, enhance mechanical

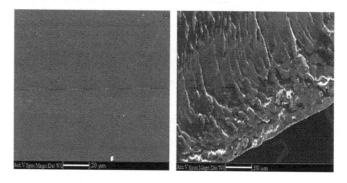

FIGURE 18.8

SEM images of SPEEK/PBI membrane (a) Surface and (b) Cross-section. (Adapted from Mingfeng, S., L. Xuewei, L. Zhongfang, L. Guohong, Y. Xiaoyan, and W. Yuxin, *International Journal of Hydrogen Energy*, 41: 12069–12081, 2016.)

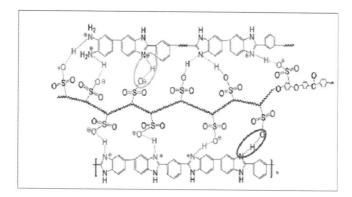

FIGURE 18.9

Structure of SPEEK/PBI composite membranes. (Adapted from Mingfeng, S., L. Xuewei, L. Zhongfang, L. Guohong, Y. Xiaoyan, and W. Yuxin, *International Journal of Hydrogen Energy*, 41: 12069–12081, 2016.)

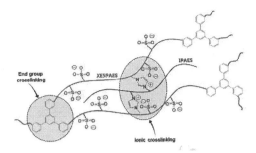

FIGURE 18.10
Formation of polymer chain between ESPAES and IPAES incorporating both ionic and end group cross-linking (Adapted from Lee et al., 2015).

and thermal stabilities with high reduction in methanol permeation, to exhibit high proton conductivity (Mingfeng et al., 2016). Apart from the above-described methods, Lee et al. (2015) synthesized a dually cross-linked membrane made up of end groups that consisted of cross-linkable sulfonated poly(arylene ether sulfone) copolymer (ESPAES) and imidazolium poly(arylene ether sulfone) copolymer (IPAES), as shown in Figure 18.10. The blending of ESPAES and IPAES with thermal treatment enhanced thermal and mechanical stabilities and induced excellent fuel barrier properties, due to the synergistic effects of end groups and ionic cross-links. The performance of various acid-base blend membranes reported in literature is displayed in Table 18.2.

TABLE 18.2

Acid-Base Blend Membranes Reported in Literature and Their Performance

Blend Membranes	Description	References
SPEEK-SPOP	The ionic cross-linking between sulfonic acid groups of SPEEK with basic character of SPOP induces the formation of $SO_3^- \cdots NH^+ \cdots$ bridges that leads to easy dissociation and transportation of protons which enhances proton conductivity (0.120 Scm^{-1}) with a maximum power density of 935 mW cm^{-2} in PEMFCs.	Rambabu et al. (2016)
SPES in SPVdF-co-HFP	The blending of SPES with SPVdF-co-HFP exhibits high proton conductivity (7.2 Scm^{-1}) with low methanol permeability (3.26 cm^{2}s^{-1}) for DMFC applications.	Uma et al. (2017)
cross-linked SPEEK	A novel ionically cross-linked SPEEK containing equal contents of sulfonic acid and pendant tertiary amine groups (TA-SPEEK) has been reported for DMFC application. Along with activation of sulfonic acid groups, grafting of tertiary amine groups to the side chains was introduced that induced ionic and ionic-covalent cross-linking. This improved mechanical properties, dimensional and oxidative stability. Moreover, the dominance of ionic-covalent cross-linking relative to ionic cross-linking resulted in higher selectivity and better single-cell performance.	Linfeng et al. (2017)
dsPFES/imPES	Ionic interactions between sulfonate group in dsPFES and imidazolium group in imPES post blending enhances their physical and chemical stability with a high proton conductivity of 0.24 Scm^{-1} at 80°C.	Sohyun et al. (2017)
PES/PVP	The doping of phosphoric acid (PA) in the blend membrane enhanced proton conductivity (0.21 Scm^{-1}) with a high power density of 850 mW/cm^2 and outstanding stability without extra humidification.	Xu et al. (2015)

Abbreviations: dsPFES: densely sulfonated poly(fluorenyl ether sulfone), imPES: imidazolium-functionalized poly(ether sulfone), PVP: poly(vinyl pyrrolidone), SPOP: sulfonated poly(bis(phenoxy)phosphazene), SPVdF-co-HFP: Sulfonated poly (vinylidene fluoride-co-hexafluoropropylene).

18.8 Molecular Dynamics Simulation Study for Acid-Base Blend Membranes

Experimental studies provide performance data of indigenously synthesized membranes under laboratory conditions. In order to understand the static and dynamic properties of membranes at the molecular level, computation based molecular modeling (MD) could prove to be useful. Computational resources and technologies are cost-effective and consume less time through hypothetical development and virtual evaluation of membranes. MD provides insights into various properties, and the design of chemical structures of the newly developed membrane can be further modified to enhance performance. The initial MD methods were formulated by the theoretical physics community, in 1957, with the use of a hard-sphere model by Alder and Wainwright. Subsequently, researchers reported various membrane related aspects using MD simulation for different separation process (Christian et al., 2007). In the case of FCs, authors reported MD simulation of new types of membranes, such as Kim et al. (2008), who used density functional theory (DFT) calculations and *ab initio* molecular simulations for sulfonated aromatic polyarylene membranes to explain that the presence of strong aryl-SO_3H bond in the electron deficient aromatic ring caused restriction of desulfonation at high temperatures, indicating thermodynamic stability. Moreover, Kim et al. (2008) also explained the significance of the Grotthus-type mechanism, with inter-conversion between Eigen ($H_9O_3^+$) and Zundel cations ($H_5O_2^+$) as limiting structures for hydrated proton transport in the vicinity of the sulfonic acid groups. Choe et al. (2010) investigated the impact of hydration levels on the performance of SPES membrane, viz., Nafion®. At low hydration levels, protons are not completely dissociated from the sulfonic acid group, which induces inadequate solvation and poor conduction. Moreover, the negative interactions associated with SO_2 thwart the effective hydration around SO_3. This effective removal of SO_2 group contributes to the enhancement of FC performance. Fatemeh et al. (2015) investigated the performance of poly(arylene ether sulfone) with different degrees of monomer carboxylation at 353 K and 1 atmosphere pressure with 10 water molecules associated per each sulfonic acid group. The static and dynamics properties of membranes were evaluated by radial distribution function, water cluster size and mean square displacement. The simulation study revealed that hydronium ions have a strong interaction with carboxylic acid groups compared to sulfonic acid groups, which result in low proton conduction, whereas increased amounts of carboxylic acid in the membrane enhance hydrophilicity and, ultimately, proton conduction. The addition of carboxylic acid introduces hydrogen bonds and shortens the distances between acidic groups that accelerate proton conduction due to higher H_2 dissociation at the anode, as shown in Figure 18.11. Blend membranes were also investigated by researchers like Ghasem et al. (2013), who studied the impact of different hydration levels on SPEEK/SPES based blends using MD simulations. Increased hydration level enhances the average distances between sulfur–sulfur groups, sulfonic acid groups and also reduces the coordination number of hydronium ion present near the sulfonic acid groups. Through cluster size distribution analysis of water molecules, it was found that at high hydration levels the large spanning water clusters encompassing all molecules were formed, which improved water and hydronium ion transport properties as shown in Figure 18.12 (Ghasem et al., 2013).

Chetan et al. (2013) deduced water and methanol diffusivities in acid–base polymer blend membranes consisting of SPEEK and polysulfone tethered with different bases: 2-amino-benzimidazole, 5-amino-benzotriazole and 1H-perimidine, through ab-initio MD simulation. The simulation results revealed that hydrogen bonding interactions between the

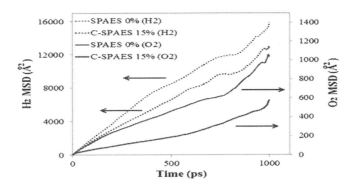

FIGURE 18.11

MSD of H_2 and O_2 gases through SPAES and C-SPAES membranes. (Adapted from Fatemeh, K.K., A.I. Sepideh, and M. Hamid, *International Journal of Hydrogen Energy*, 40: 15690–15703, 2015.)

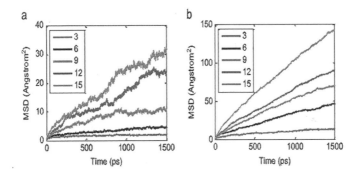

FIGURE 18.12

MSD of (a) Hydronium ion and (b) water molecule in the hydrated SPEEK-SPES blend membranes at different hydration levels. (Adapted from Ghasem, B., Manouchehr, N., and Mohammad, M.H.S., *Journal of Membrane Science*, 429: 384–395, 2013.)

sulfone oxygen of polysulfone and H atom of N base influenced methanol diffusivity by reducing pore size (free volume) of the membrane.

18.8.1 Construction of MD Simulation

MD simulations were performed by constructing a system (polymer chain, water, etc.) using an amorphous builder module, which is minimized and dynamically analyzed by discover module. The minimization happened by selecting an appropriate method with desired iteration. After the minimization and optimization of the system, suitable equations that described the parameters needed to calculate the potential energy were defined as the force field. Further minimizations were performed to optimize the cell with a convergence level of 0.01 Kcal/mol/Å and the dynamics' run was made using a suitable algorithm that enabled calculation of dynamic and static properties, as shown in Figure 18.13.

18.8.1.1 Diffusion Coefficient and Ion-Conductivity

The following governing equations are used to calculate static and dynamic properties of membranes. The diffusivity of molecules through the polymeric membrane matrix

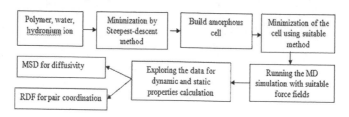

FIGURE 18.13
Steps for MD simulation construction.

was investigated using mean square displacement (MSD) computed by the following equation:

$$MSD = \left| r_i(ts) - r_i(0) \right|^2 \tag{18.3}$$

$$D = \lim_{t \to \infty} \frac{\left(\left| r_i(t) - r_i(0) \right|^2 \right)}{6t} \tag{18.4}$$

where, ri (0) and *and* ri (t) are the initial and final position (after time t) coordinates of atom i, whereas the diffusion coefficient (D) of molecules was calculated from the slope of the MSD curve by Einstein relationship. The ion conductivity (σ) was evaluated by using the Einstein equation:

$$\sigma = \frac{Nz^2 \varepsilon^2 D}{VKT} \tag{18.5}$$

where, σ is the ion conductivity, N is the number of counter ions in the simulation cell, V is the simulation cell volume, k is Boltzmann's constant, z represents total charge per unit of electron with e being the elemental charge (Thongchai et al., 2012)

18.8.1.2 Radial Distribution Function (RDF)

The structural characterization of synthesized membranes at the molecular level were investigated by radial distribution function (RDF) (Thongchai et al., 2012), which depends upon the spherically averaged distribution of interatomic distances (r) between two species, A and B, by totaling N_A and N_B, within a volume (V). Through RDF, the coordination number can be calculated using Equation 18.7, which describes the transport mechanism of the ion:

$$g_{AB}(r) = \frac{V \left[\Sigma \left(r - \left| r_{Ai} - r_{Bi} \right| \right) \right]}{(N_A N_B - N_{AB}) 4\pi r^2 dr} \tag{18.6}$$

$$N_{AB} = \int 4\pi \rho r^2 g_{AB}(r) \, dr \tag{18.7}$$

18.9 Conclusions

The rapid development of fuel cells for energy generation increases their production in terms of the number of pilot and commercial installations for domestic, portable and transportation applications. Despite advancements in the development of FCs, certain drawbacks restrict their large-scale applicability. The commercial Nafion® membrane is the most dominant material for PEMFC and DMFC; however, its high cost, fuel crossover, catalyst poisoning and low stability necessitate an alternative. In this aspect, the development of acid-base polymer blends has attained much attention. Ionic interactions arising from cross-linking between acidic and basic groups enhance proton conductivity at high temperatures and suppress fuel crossover, particularly methanol permeability, but boost mechanical and thermal stabilities of the membranes. Aromatic hydrocarbon polymers, such as PES, PEEK and PBI, exhibit properties that render them potential alternative materials—provided they can be modified suitably and combined to give acid-base blends. Molecular dynamic (MD) simulations provide information of static and dynamic properties of the acid-base blend at different operating conditions, to enable optimization. MD simulation provides a new route for researchers to understand membrane properties, modification routes and feasibility of scale-up to commercial fuel cell systems, which can produce clean and green energy to safeguard the ecosystem.

References

Adolfo, I. and B. Angelo. 2012. Sulfonated PEEK-based polymers in PEMFC and DMFC applications: A review. *International Journal of Hydrogen Energy* 37: 15241–15255.

Bahlakeh, G., M. Nikazar, and M.M.H. Sadrabadi. 2013. Understanding structure and transport characteristics in hydrated sulfonated poly(ether ether ketone)–sulfonated poly(ether sulfone) blend membranes using molecular dynamics simulations. *Journal of Membrane Science* 429: 384–395.

Chetan, V.M. and G. Venkat. 2013. Influence of hydrogen bonding effects on methanol and water diffusivities in acid–base polymer blend membranes of sulfonated poly(ether ether ketone) and base tethered polysulfone. *Journal of Physical Chemistry B*, 117: 5315–5329.

Christian, K., L.A. Walter, and D.P. Tieleman. 2007. Setting up and running molecular dynamics simulations of membrane proteins. *Methods* 41: 475–488.

Cui, W., J. Kerres, and G. Eigenberger. 1998. Development and characterization of ion-exchange polymer blend membranes. *Separation and Purification Technology* 14: 145–154.

De, A., W. Bin, Z. Genlei, Z. Wen, and W. Yuxin. 2016. Gradiently crosslinked polymer electrolyte membranes in fuel cells. *Journal of Power Sources* 301: 204–209.

Deuk, J.K., J.L. Hye, and Y.N. Sang. 2013. Sulfonated poly(arylene ether sulfone) membranes blended with hydrophobic polymers for direct methanol fuel cell applications. *International Journal of Hydrogen Energy* 39: 17524–17532.

Dinh, X.L. and K. Dukjoon. 2011. Strontium cross–linked sPEEK proton exchange membranes for fuel cell. *Solid State Ionics* 192: 627–631.

Ergun, D., Y. Devrim, B. Nurcan, and I. Eroglu. 2012. Phosphoric acid doped polybenzimidazole membrane for high temperature PEM fuel cell. *Journal of Applied Polymer Science* 124: 267–277.

Eunja, K., F. Philippe, W. Naduvalath, and B. Chulsung. 2008. Nanoscale building blocks for the development of novel proton exchange membrane fuel cells. *Journal of Physical Chemistry B* 112: 3283–3286.

Fatemeh, K.K., A.I. Sepideh, and M. Hamid. 2015. Molecular dynamics simulation study of car-boxylated and sulfonated poly(arylene ether sulfone) membranes for fuel cell applications. *International Journal of Hydrogen Energy* 40: 15690–15703.

Florian, M., A. Karin, E. Corina, J. Kerres, and Z. Roswitha. 2015. Novel phosphoric acid-doped PBI-blends as membranes for high-temperature PEM fuel cells. *Journal of Materials Chemistry A.* 3: 10864–10874.

Francesco, L., B. Vincenzo, S. Pietro, S.A. Antonino, and A. Vincenzo. 2006. Development and char-acterization of sulfonated polysulfone membranes for direct methanol fuel cells. *Desalination* 199: 283–285.

Harsha, N., S. Kalyani, V.V.B. Rao, and S. Sridhar. 2015. Synthesis and characterization of polyion complex membranes made of aminated polyetherimide and sulfonated polyethersulfone for fuel cell applications. *Journal of Fuel Cell Science and Technology* 12(6): 061004.

Harsha, N., V.V.B. Rao, and S. Sridhar. 2017. Synthesis and characterization of Torlon-based polyion complex for direct methanol and polymer electrolyte membrane fuel cells. *Journal of Materials Science* 52: 8052–8069.

Harsha, N., C. Sumana, V.V.B. Rao, and S. Sridhar. 2017. Performance evaluation of sodium alginate–Pebax polyion complex membranes for application in direct methanol fuel cells. *Journal of Applied Polymer Science* 134, 44485: 1–11.

Jinhua, C., L. Dengrong, A. Masaharu, K. Hiroshi, and M. Yasunari. 2010. Crosslinking and graft-ing of polyetheretherketone film by radiation techniques for application in fuel cells. *Journal of Membrane Science* 362: 488–494

Juhana, J., A.F. Ismail, T. Matsuura, and M.N.A.M. Norddin. 2013. Stability of SPEEK-Triaminopyrimide polymer electrolyte membrane for direct methanol fuel cell application *Sains Malaysiana* 42: 1671–1677.

Kerres, J., W. Cui. EP 1,073,690, 14 January, 2004.

Kerres, J. 2009. Blend concepts for fuel cell membranes. *Polymer Membranes for Fuel Cells.* U.S.: Springer

Kerres, J.A. Development of ionomer membranes for fuel cells. 2001. *Journal of Membrane Science* 2001, 185: 3–27.

Li, Q., R. He, J.O. Jensen, and N.J. Bjerrum. PBI-Based Polymer Membranes for High Temperature Fuel Cells – Preparation, Characterization and Fuel Cell Demonstration. 2004. *Fuel Cells* 4: 147–159.

Linfeng, L., Z. Xingye, X. Jianfeng, Q. Huidong, Z. Zhiqing, and Y. Hui. Highly stable ionic-covalent cross-linked sulfonated poly(ether ether ketone) for direct methanol fuel cells. 2017. *Journal of Power Sources* 350: 41–48.

Luca, A., I. Cristina, A.A. Gulsen, C. Laure, and S. Jean-Yves. Polyethersulfone containing sulfon-imide groups as proton exchange membrane fuel cells. 2014. *International Journal of Hydrogen Energy* 39: 2740–2750.

Meenakshi, S., D. Bhat, A.K. Sahu, P. Sridhar, S. Pitchumani, and A.K. Shukla. 2011. Chitosan-polyvinyl alcohol-sulfonated polyethersulfone mixed-matrix membranes as methanol-barrier electrolytes for DMFCs. *Journal of Applied Polymer Science* 124: 73–82.

Mikhailenko, S.D., K. Wang, S. Kaliaguine, P. Xing, G.P. Robertson, and M.D. Guiver. Proton con-ducting membranes based on cross-linked sulfonated poly(ether ether ketone) (SPEEK). 2004. *Journal of Membrane Science* 233: 93–99.

Mingfeng, S., L. Xuewei, L. Zhongfang, L. Guohong, Y. Xiaoyan, and W. Yuxin. 2016. Compatible ionic crosslinking composite membranes based on SPEEK and PBI for high temperature pro-ton exchange membranes. *International Journal of Hydrogen Energy* 41: 12069–12081.

Pearson, G., M. Leary, A. Subic, and J. Wellnitz. 2011. Performance comparison of hydrogen fuel cell and hydrogen combustion engine racing cars. *Sustainable Automotive Technologies*: 85–91.

Quantong, C., C. Ning, Y. Jinming, and C. Shicheng. 2016. Sulfonated poly(ether ether) ketone/polyurethane composites doped with phosphoric acids for proton exchange membranes. *Solid State Ionics* 289: 199–206.

Rambabu, G., P.S. Gouse, D.B. Santoshkumar, K.S. Akhila. 2016. Synthesis of sulfonated poly(bis(phenoxy)phosphazene) based blend membranes and its effect as electrolyte in fuel cells. *Solid State Ionics* 296: 127–136.

Silva, V.S., J. Schirmer, R. Reissner, B. Ruffmann, H. Silva, A. Mendes, L.M. Madeira, and S.P. Nunes. 2005. Proton electrolyte membrane properties and direct methanol fuel cell performance: II. Fuel cell performance and membrane properties effects. *Journal of Power Sources* 140: 41–49.

Sivakumar, P., J. Shan, J.B. Bernard, and L. Vladimir. 2008. High DMFC performance output using modified acid–base polymer blend. *International Journal of Hydrogen Energy* 33: 3132–3136.

Smitha, B., S. Sridhar, and A.A. Khan. 2003. Synthesis and characterization of proton conducting polymer membranes for fuel cells. *Journal of Membrane Science* 225: 63–76.

Smitha, B., S. Sridhar, and A.A. Khan. 2004. Synthesis and characterization of sulfonated PEEK membranes for fuel cell application. *International Journal of Polymeric Materials* 21: 99–106.

Sohyun, K., L. Boryeon, and K. Tae-Hyun. 2017. High performance blend membranes based on densely sulfonated poly(fluorenyl ether sulfone) block copolymer and imidazolium functionalized poly(ether sulfone). *International Journal of Hydrogen Energy* 42: 20176–20186.

Srinivasan, G., M. Rethinasabapathy, and S. Dharmalingam. 2011. Development of a solid polymer electrolyte membrane based on sulfonated poly(ether ether)ketone and polysulfone for fuel cell applications. *Canadian Journal of Chemistry* 90: 205–213.

Suxiang, D., K.H. Moham-mad, A.M. Kenneth, and W.M. Jimmy. 2015. Hydrocarbon-based fuel cell membranes: Sulfonated crosslinked poly(1,3-cyclohexadiene) membranes for high temperature polymer electrolyte fuel cells. *Polymer* 73: 17–24.

Thongchai, S., and M. Suwalee. 2012. Ionic conductivity in a chitosan membrane for a PEM fuel cell using molecular dynamics simulation. *Carbohydrate Polymers* 88: 194–200.

Uma, D.A., A. Muthumeenal, R.M. Sabarathinam, and A. Nagendran. 2017. Fabrication and electrochemical properties of SPVdF-co-HFP/SPES blend proton exchange membranes for direct methanol fuel cells. *Renewable Energy* 102: 258–265.

Vladimir, N., M. Jonathan, W. Haijiang, and Z. Jiujun. 2007. A review of polymer electrolyte membranes for direct methanol fuel cells. *Journal of Power Sources* 169: 221–238.

Won, H.L., H.L. Kang, W.S. Dong, S.H. Doo, R.K. Na, H.C. Doo, H.K. Ji, M.L. Young. 2015. Dually cross-linked polymer electrolyte membranes for direct methanol fuel cells. *Journal of Power Sources* 282: 211–222.

Xin, X., W. Haining, L. Shanfu, G. Zhibin, R. Siyuan, X. Ruijie, and X. Yan. 2015. A novel phosphoric acid doped poly(ethersulphone)-poly(vinyl pyrrolidone) blend membrane for high-temperature proton exchange membrane fuel cells. *Journal of Power Sources* 286: 458–463.

Yoong-Kee, C., T. Eiji, I. Tamio, O. Akihiro, and K. Koh. 2010. An *Ab Initio* modeling study on a modeled hydrated polymer electrolyte membrane, sulfonated polyethersulfone (SPES). *Journal of Physical Chemistry B* 114: 2411–2421.

Yu, L., G. Chenliang, Q. Zhigang, L. Hui, W. Zhongying, Z. Yakui, Z. Shujiang, and L. Yanfeng. Intermolecular ionic cross-linked sulfonated poly(ether ether ketone) membranes containing diazafluorene for direct methanol fuel cell applications. 2015. *Journal of Power Sources* 284: 86–94.

Section VI

Environment

19

Integrated Membrane Technology for Promoting Zero Liquid Discharge in Process Industries

R. Saranya, P. Anand, and Sundergopal Sridhar

CONTENTS

19.1 Introduction

Membrane technology has emerged as a robust and advanced option for performing separations in a variety of unit operations. In this technology, membranes are used as filters in separation processes in various applications. Membrane technology is widely used in the processing of raw materials, product recovery, separation of unreacted compounds and wastewater treatment to facilitate sustainability in several process industries. It has the unique advantage of enhancing process efficiency and product quality by retrofitting into an existing process without modifying the process. Membrane processes are largely considered, today, as the consolidated processes that could ensure energy efficiency, sustainability and environmental safety toward industrial growth. Some widely established membrane processes have already been applied in industries on a large scale. Such uses include solute concentration by ultrafiltration (UF), tertiary effluent treatment and desalination by reverse osmosis (RO), as well as wastewater treatment by membrane reactors (MBR).

To validate the sustainability ranking of membrane technology, it aims to practice sustainable membrane manufacturing processes and systems by means of technology development to perform holistic operation using renewable energy with neither greenhouse gas emission nor the generation of toxic materials or waste. In the present scenario, membrane technology is playing an increasing role in sustaining water supplies by reclaiming various wastewater streams, brackish water and seawater, and so on. On the other hand, many of the membrane technologies—those which are driven by osmotic pressure, electric gradient, salinity gradient and so on—need improvement in terms of techno-economic feasibility, cost and affordability, energy consumption and expertise. To achieve these improvements, advances in the design of membrane materials and processes, either by process integration with conventional methods or by overcoming the current limitations, is needed. This chapter aims to provide an overview on the process know-how of emerging membrane technologies and the advancements in membrane processes to tackle water and energy sustainability. The entire sections of this chapter are arranged so as to provide membrane researchers and process engineers with greater clarity about recent possibilities of practicing novel membrane technology and membrane process integration. Moreover, industrial end-users will also learn about emerging membrane technologies, reinforcing the strengthening linkage between research and application aspects.

19.2 Current Status of Membrane Processes for Industrial Growth

Water and energy have always been crucial for the world's social and economic growth. There is a vast scope for membrane technologies in many of the current process industries to assure water and energy sustainability. Based on the analysis of current status and potential applications of well-known membrane processes, this chapter encompasses the various key aspects of widely used membrane systems and also provides the scope for emerging membrane technologies by scrutinizing research and development challenges to enhance the sustainability of integrated membrane processes. Chemical process industries around the world are in need of sustainable technology in the backdrop of ever-increasing world population, demand for employment, industrialization and a concern for the environment (Pal & Dey, 2012). Over the past few decades, membrane technology has emerged as a viable alternative to conventional separation techniques, owing to its inherent sustainability. Considerable improvements, both in terms of process and product, have been witnessed in the present scenario by traversing the boundary of efficiency and reliability. With present membrane technology, there is nevertheless a great need for improved cost-effectiveness, energy conservation and sustainability.

Some of the membrane processes that are widely applied in industrial processes are in the area of effluent treatment. For instance, RO is presently playing an indispensable role, owing to its efficiency in producing reusable water. Membrane processes are also lately finding applications in the area of treating secondary and tertiary municipal wastewater and also produced water (Giwa & Ogunribido, 2012). UF process has been in use for decades in several process industries, such as dairy, paper and pulp, sugarcane, agro-food, which is attributable to its advantages of no chemical/thermal inputs and simple operation principle. Membranes play a significant role in water and energy conservation and have, for a long time, been used as sustainable solution for seawater desalination. The successful use of membranes in other applications requires new materials and tailored separation characteristics. The scope of membrane technology in several applications helps to broaden the requirement of its significant innovation in both processes and products. Despite its viability as a sustainable technology for process industries, major uses of this technology are water filtration (including desalination) and purification (including groundwater and wastewater), as well as the food and beverage and biotechnology industries (Kurt et al., 2012; Ozgun et al., 2013). RO, UF, MF and MBR have seen significant growth over the past 20 years.

This chapter aims to cover the aspects concerning membrane processes, such as RO for desalination, forward osmosis (FO), various types of electrodialysis (ED), pressure-retarded osmosis (PRO), integrated membrane systems toward energy sustainability and zero liquid discharge (ZLD) in industries.

19.3 Membrane Process Integration for Industrial Sustainability

Industrial applications mostly rely on single integrated process involving fewer processing steps, low energy consumption and minimum chemical requirements, conforming to flexibility, compactness and environmentally benign operation. Membrane processes, today, are largely considered to be the consolidated processes that could ensure energy savings, environmental protection, as well as sustainable industrial growth (Drioli et al., 2012). Integrated processes employing membrane modular design offer greater flexibility in system operation, which is in high demand by the modern manufacturing sector. The industrial importance of membrane technology is mainly due to its indispensable role in several unit operations, such as concentration, fractionation, clarification, and molecular separations, in many chemical and process engineering industries. Additionally, the possibility of combining different membrane operation units within an integrated membrane system along with conventional separation technologies offers interesting new perspectives for redesigning traditional flow sheets of the agro-food industry within the context of process intensification strategies, which is one such way toward sustainability. The utilization of membrane operations as hybrid systems, in combination with other conventional techniques or even other membrane operations is considered the way forward to more rational applications. The synergic interactions derived from hybrid systems render these systems advantageous in many industrial processes (Drioli & Romano, 2001; Daufin et al., 2001). The possibility of redesigning industrial sectors by combining various membrane operations has been studied and in some case realized with minor environmental impact and low energy consumption.

The integration of membrane processes based on the most promising combinations, enables to set higher standards for practicing sustainability in industries and also provide sustainable solutions for end-users (Table 19.1). Process integration can either be in the form of retrofitting the membrane processes with conventional unit operations or by installing new integrated membrane process with a relatively smaller footprint. The outcomes of

TABLE 19.1

Integration of Membrane Processes for Various Industrial Applications

Process I	Process II	Application	Reference
Electrodialysis	Ultrafiltration	Advanced separation of peptides	Suwal et al. (2014)
Electrodialysis	Reverse osmosis	Concentration of brine solutions from RO	Jiang et al. (2014)
Membrane crystallization	Reverse osmosis	Efficient brine reuse	McCutcheon and Elimelech (2006)
Membrane distillation	Membrane bioreactor	Bioethanol production Removal of Trace organic contaminants from industrial effluent	Wijekoon et al. (2014)
Membrane desalination	Membrane crystallization	Treatment of NF retentate and RO brines	Jensen et al. (2016)
Reverse osmosis	Membrane distillation	Concentration of heavy metals and nitrogen compounds	Hayrynen et al. (2013)
Ultrafiltration	Electrodialysis	Fractionation of dairy components	Kelly et al. (2000)

efficient membrane process integration will not only benefit the chemical sector but also adopts as greener, eco-friendly processes in the near future, to secure sustainability status of industry and society. Coupling membrane systems with several other biological, chemical or physical treatments could improve separation efficiency, manage difficult contaminants, cut costs and energy consumption and boost ecological performance. As part of the solution, membranes for water applications are arousing great interest as potentially cost-effective responses to a growing range of purification and separation needs. Some membrane systems incur higher capital or operational costs than conventional processes. For instance, as shown in Figure 19.1, the integration of UF and RO for several water treatment solutions, especially desalination, is already in the process of achieving superior product water quality, while imposing a smaller footprint. Integration and scale-up of these advances will be essential elements of a thorough evaluation, intended to make this a key technology for the treatment of highly saline waste streams stemming from a broad range of industrial production. The value lies in the possibility of exploiting the salt content of waste streams by transforming them into valuable, recoverable products, while also conferring the environmental benefits of water recycling and minimal effluent discharge. But the feasibility of such recent membrane technologies for large-scale applications has so far been constrained by shortcomings in the available membrane and process technologies.

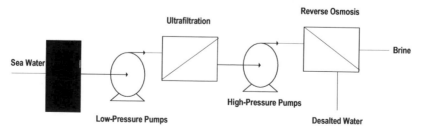

FIGURE 19.1
Conventional UF and RO for seawater desalination.

19.4 Breakthrough Advancements of Membrane Processes in Industries

During the past 40 years, the implementation of membrane technology in unit operations has improved process performance and reliability with lower operating costs, making membranes the most preferred technology in the water treatment industry. In recapping the technological advancements of the twentieth century, with respect to the treatment of fluids with membranes, one can see a significant shift from the traditional aspects of membrane treatment to a more technologically refined process, whereby one can maximize resources and achieve higher quality and performance, all at lower energy inputs than what had been previously required. The research and development efforts put forth in the twentieth century have paved the way for many novel membrane processes, a few of which are summarized in Table 19.2. This chapter will concentrate on four distinct developments in the field of membrane technology, and how each development will contribute and redefine the use and application diversity of membranes in water and wastewater industries. The technology to transform membrane properties has only just begun evolving and, in the future, will be applied to a variety of membrane surfaces, therefore changing completely the system design and operating guidelines.

A variety of industries are finding that it makes sense to re-evaluate the way they treat industrial processes, both to improve the quality of their products and increase the efficiency of their processes. The ongoing evolution of membrane technology allows greater flexibility in designing systems that function under a variety of operating conditions. The development of new membranes continues to expand both the range of chemical compatibilities and physical operating conditions (including pressure, temperature and pH) in membrane systems. From the application point of view, approximately 40% of membrane

TABLE 19.2

Improvised Membrane Systems for Novel Applications

Process	Membrane Type	Application	Reference
Electrodialysis	Dense Ion exchange membranes	Fractionation of amino acids	Readi et al. (2013)
Bipolar membranes based electrodialysis	Bipolar membranes	Removal of salts for production of acids and bases	Zhang et al. (2012)
Electrophoretic membrane contactor	Porous membrane	Separation of whey proteins	Galier and Balmann. (2011)
Reverse electrodialysis (RED) Diffusion dialysis	Anion/cation exchange membrane	Desalination of mining waters and energy production from concentrated brine or seawater	Daniilidis et al. (2014) Hong et al (2015)
NF precipitation/ concentrator system	Slurry precipitation and recycle reverse osmosis (SPARRO)	Treatment of mine drainage, gypsum precipitation	Juby et al. (1996)
Pressure-retarded reverse osmosis	Salinity gradient	Energy generation from pressurized brine	Lee et al. (1981)
Forward osmosis	Osmotic pressure	High salinity induced water treatment. Seawater desalination	Chung et al. (2012); McCutcheon and Elimelech (2006)
Continuous electrodeionization	Ion-exchange resins and bipolar membranes; electro-osmotic gradient	Removal of metal ions and production of ultrapure water	Arar et al. (2014)

sales are destined for water and wastewater applications, while food and beverage processing combined with pharmaceuticals and medical applications account for another 40% of sales—all while membrane use in chemical and industrial gas production continues to grow (Wiesner & Chellam 1999). The ultimate objectives are to advance cost-effectiveness, through energy savings in various operational water treatment contexts, and to appeal to public and private customers.

19.4.1 Electrodialysis

Electrodialysis (ED) is a well-known membrane process and traditional unit operation, originating from the 1950s, that operates by means of electric potential as the driving force. ED possesses inherent advantages of economic feasibility, safety, modular compactness and environmental benignity, making it potentially viable for wider adoption, particularly in pharmaceutical and chemical industries. The process's sustainability can be achieved by determining the electrical energy requirement in ED, by operating it with renewable energy source for faster payback in future (Bruggen et al., 2004). The role of ED in separating charged compounds is applicable in various chemical, biochemical, food, and pharmaceutical industries (Xu & Huang, 2008). Apart from salt separation, ED also finds application in recycling bio nutrients, removing micro pollutants, solvent separation, etc. There have also been recent advances in ED integration leading to the advent of process such as (a) Reverse ED-RED, (b) Ionic liquid assisted ED-ILED, Bipolar membrane electrodialysis-BPMED, and (d) Continuous electrodeionization-CEDI.

The application of RED enables renewable energy capture from salt and water mixtures (Vermaas et al., 2013), as illustrated in Figure 19.2. ILED operates to selectively remove lithium from charged solutes and organic salts from paper and textile industries (Hoshino, 2013; Lopez & Hestekin, 2013). The integration of ED with several other membrane processes, such as nanofiltration and membrane distillation, is also gaining more interest as an efficient tertiary operation in effluent treatment plants (ETPs). With the advent of bipolar membranes, processes related to production, fractionation and conversion of solutes

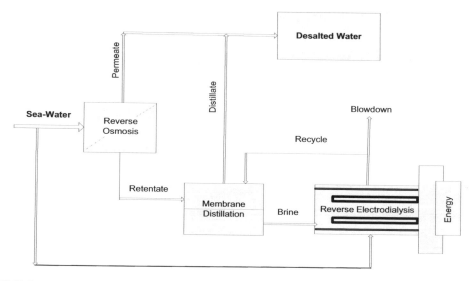

FIGURE 19.2
Electrodialysis reversal for continuous seawater desalination and energy production.

of the same charge and size is being established in BPMED for water-splitting and other niche steps of downstream purification (Tsukahara et al., 2013).

ED holds a breakthrough potential in seawater desalination, and membrane processes for desalination are constantly evolving and still there is a need for further advances by integrating any new process with forward or reverse osmosis, thereby satisfying the factors such as cost, capacity, viability, adaptability, quality and quantity of product yield. Other applications in the selective removal of one or more ions may become attractive when membrane materials are developed with better selectivity. ED in conjugation with other membrane processes helps in large scale adaptability of separation systems for several applications. For instance, BPMED offers a new range of applications, such as nickel and cobalt recovery (Iizuka et al., 2013; Yang et al., 2014), tetrapropyl ammonium hydroxide and vitamin C production (Shen et al., 2013; Song et al., 2012), and so on, thus delineating the scope of facilitate zero discharge of wastewater in a large scale.

19.4.2 Pressure-Retarded Osmosis

Pressure-retarded osmosis (PRO) was first reported in 1976 (Leob et al., 1976), and the first PRO concept-based power plant prototype was opened in 2009 to prove its concept in generating electricity. The schematic representation of typical PRO is shown in Figure 19.3. It is a well-known fact that 97% of our planet is encompassed with seawater and, thus, there is a large energy potential from concentrated streams of seawater and brine, which remain unexploited. PRO technology employs salinity gradient for energy production and is driven by the osmotic pressure so as to enable not only the production of renewable energy, but also to eliminate the hazard of brine disposal. This technology generates power by the difference in the osmotic pressure between aqueous solutions of different salinities (fresh and salt water). With constant hydraulic pressure, seawater or brine is pumped from one side of the membrane, whereas the compartment on the other side of the membrane is concurrently fed by fresh or wastewater. Water passes through the membrane from the fresh water to the seawater side, which, in turn, rotates the turbine for electricity generation. PRO is operated by salinity gradient energy, which implies the power generated by the difference in osmotic pressure between aqueous solutions of different salinities. PRO is more efficient when concentrated brine is employed, whereas RED is more favorable with seawater. PRO is a renewable energy source with high environmental advantages. The implementation of PRO is limited only by cost factor and geographical behavior when compared to other salinity-driven electro-chemical processes.

An important component of a PRO set-up is a membrane by which solutions with different salinities are separated so as to maintain the chemical potential difference for electricity generation. The membrane performance in PRO is characterized by power density,

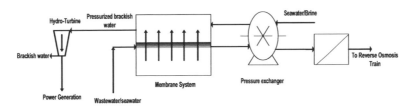

FIGURE 19.3
PRO technology for energy production from seawater.

which is defined as the power output per unit membrane area, and it is critical because it directly affects the cost of the generated power. The requirements for PRO membranes are quite different from those for RO. The thick, dense and highly resistant support layer of RO membranes, which is essential to tolerate high pressures, causes severe concentration polarization in PRO, and thus, a thinner or less dense support layer is highly preferred. Compared to currently available renewable energy resources like wind and solar power PRO would be cost-effective, if capital payback is justified. The scope for reducing the payback time is possible by achieving high power density. There is also a great amount of upcoming research potential for improving the effective membrane area of hosted membranes in PRO, which will enhance the power density and also reduce the membrane fouling issue, thus to reducing maintenance and operation costs (only 30% of the total cost).

19.4.3 Continuous Electro-Deionization

Continuous electro-deionization (CEDI) is a mature water purification system, like the RO process, that ensures reliable and long-term operation and focuses on process design and proper integration among other unit operations involving the elements of module and systems. CEDI (Semmens et al., 2001) is recently gaining importance in nuclear power plants and other electroplating industries for the removal of heavy metals and radioactive elements from rinse electroplating and nuclear wastewaters. CEDI is a hybrid of electrodialysis (ED) and ion-exchange membranes in the diluting compartments that provide a medium of transport for the removal of ions. The CEDI system employs RO pretreatment, which may allow the device to make mixed-bed quality water, thereby providing more reliable operations, divided into two basic categories. The feed water hardness limits of CEDI systems can be met with four main system configurations, such as antiscalant RO/CEDI, softening RO/CEDI, antiscalant/softening/RO/CEDI and antiscalant/RO/RO/CEDI systems that depend on factors like raw water hardness.

19.4.4 Forward Osmosis

Forward osmosis (FO) is another membrane process used to desalinate saline water at a notably reduced cost. It involves the use of a draw solution of aluminum sulfate or ammonium sulfate to transport water from saline water. These draw solutions on chemical treatment or heating result in fresh water (Cath et al., 2006; Chanukya et al., 2013). This new membrane process has been developed as a possible alternative to desalination. It employs a semipermeable, dense, hydrophilic membrane that separates two aqueous solutions (feed and osmotic agent solution) with different osmotic pressures. The difference in osmotic pressure acts as a driving force. An osmotic pressure-driven process operates on the principle of osmotic transport of water across a semi-permeable hydrophilic membrane from a dilute feed solution into a concentrated osmotic agent or draw solution (Nayak & Rastogi, 2010).

   Membranes and distillation processes equally share current desalination production capacity worldwide. RO has emerged as the leader in future desalination installations and will be the key to increasing water supplies for drinking water production throughout the world. Although wealthy Middle Eastern countries have been able to afford distillation processes, RO technology can now produce freshwater (from seawater) at one-half to one-third of the cost of distillation (Miller, 2003). Brackish water desalination is even less expensive than seawater desalination. Membranes have risen to the challenge of supplying desalinated water and continue to perform efficiently and effectively.

Various ongoing developments may lead to the emergence of cost-effective and energy efficient membrane technologies that can transform municipal wastewater treatment practices for high effluent quality production with a high potential of water reuse. The most interesting developments for industrial membrane technologies are related to the possibility of integrating several of these membrane operations in the same industrial cycle, with overall significant benefits in product quality, plant compactness, environmental impact, and energetic aspects. It is also realistic to affirm that new wide perspectives of membrane technologies and integrated membrane solutions for sustainable industrial growth are possible. A variety of technical challenges must be overcome to permit the successful industrial application of new membrane solutions. The intrinsic multidisciplinary character of membrane science has been and still is, today, one of the major obstacles to the further exploitation of its possibilities. The new logic of membrane engineering (Drioli & Romano, 2001), based on a drastic rationalization of the existing process design and not on the more traditional approach of adding one more, eventually innovative, unit at the end of the existing pipe (which has been another obstacle to the growth of membrane units), will also contribute to the exploitation of this technology.

19.5 Membrane Technology for Facilitating Zero Liquid Discharge (ZLD) in Industries

Stringent environmental regulations have necessitated ZLD systems in industries as they offer holistic effluent management along with simultaneous recovery of useful salts, solutes and other chemicals from the spent and brine solutions. There are also other factors triggering the significance of imparting ZLD systems in industries:

- Depleting water resources creating water stress worldwide and an increasing need for water necessitates high throughput wastewater recycling.
- Recycled water is more economical than the water supply from conventional sources.
- Growing awareness and social responsibility on environmental issues due to untreated discharge from industries.
- Compared to the cost incurred for transporting waste in large volumes over long distances, the cost of implementing ZLD for wastewater management is economical.

The two basic ZLD technologies in industries are conventional and hybrid ZLDs, respectively.

19.5.1 Conventional Zero Liquid Discharge Systems

Conventional zero liquid discharge aims at more than 90% water recovery based on evaporation and crystallization techniques. The feed water undergoes a pretreatment step that reduces the scaling potential and is then concentrated by a brine concentrator and a brine crystallizer. The wastes and water (as distillate) are separated in this brine concentrator. From the crystallizer unit, reusable clean water is obtained as the product, while the wastes

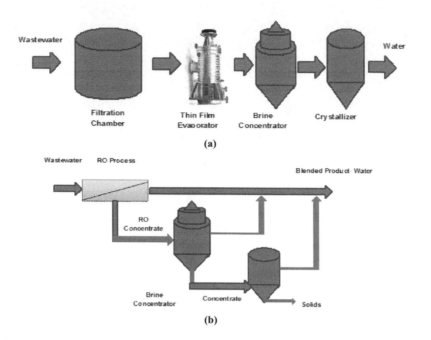

FIGURE 19.4

(a) and (b): Schematic illustration of thermal conventional ZLD system using thin film evaporator and hybrid ZLD system based on RO process.

are stored in the crystallizer, itself. Figure 19.4 (a) illustrates a conventional ZLD system used for recovery of wastewater treatment plants in textile industries. Brine concentrators commonly use mechanical vapor compression for water evaporation. The feed water is previously heated by utilizing the sensible heat from the distillate.

The conventional thermal ZLD system shown in Figure 19.4 (a) includes a thin film evaporator with which evaporation occurs when the wastewater is flowing through it to provide preferential separation in the form of a distillate, thus making water recovery more efficient than the previous method. In this method, calcium sulfate is usually added into the recirculating evaporators, which keep the salts in suspension for preventing scale formation on the heat transfer tubes such that the superheated vapor condenses during water recovery. This thin film evaporator enhances the heat transfer rate by reducing the compression ratio and required energy for the compressor. Thus, the use of energy for the heating and water recovery is also reduced.

19.5.2 Hybrid Zero Liquid Discharge Systems

The recent boom in wastewater treatment technologies has given a promising niche for hybrid ZLD systems as they have overcome the shortcomings of conventional ZLD systems. The conventional ZLD systems, attached either with an evaporator or crystallizer, are encountering difficulty in handling complex streams, such as discharge from petrochemical industries, nuclear power plants, etc. The high energy consumption, attributed to high operational and capital costs, compelled the development of membrane retrofitted hybrid ZLD systems. Most of the hybrid ZLD systems require tertiary stage membrane processes such as RO, nanofiltration or ED to curb the cost, as well as to achieve efficient recovery. Various membrane based processes have emerged as a promising alternative for implementing full-scale ZLD in industries.

19.5.2.1 ZLD Combined with Reverse Osmosis

The method of combining RO with ZLD is one of the most energy-saving systems that help in minimizing waste discharge and maximizing water recovery. The schematic representation of ZLD with RO is shown in Figure 19.4 (b). RO receives maximum water from the stream before it reaches the mechanical vapor compression (MVC) evaporator, and this rate of water recovery eventually decreases the operational cost. The rate of water recovery is, however, limited in the RO process due to three main limitations: (i) very high osmotic pressure for recovering high TDS wastewater, (ii) scaling due to high saline streams requiring the use of anti-scalants, and (iii) RO prone to fouling due to both organics and biofilm formation. Hence, recent research is highly focused on resolving scaling and fouling issues for overall reduction in pressure requirements.

19.5.2.2 ZLD Combined with Electrodialysis

In this hybrid ZLD, electric potential acts as the driving force for the removal of ions by means of the ion exchange membranes. The main advantage of this process is improved product water quality with no chemical addition. As the ion removal takes place by charged membranes, this ZLD combined with electrodialysis is not limited by high osmotic pressure and thus could attain larger water recovery. The ion exchange membranes employed in this process selectively permit the transport of ions. The anions tend to migrate through the membranes toward the negatively charged cathode and vice versa for the cations. In this modified process, the electrodialysis reversal approach is used to reverse the polarity of the electrodes to eliminate the fouling and scaling problems. These methods can be easily used for the treatment of water having a TDS of greater than 1,00,000 mg/L, and the water quality obtained as the product from this process represent usable water. However, when compared to a ZLD system hybridized with RO, the desalting cost is relatively higher, but lower when compared to conventional ZLD without RO. The efficiency of this system is likely to be rated between RO and conventional MVC evaporation.

19.5.2.3 ZLD Combined with Membrane Distillation

This method of hybrid ZLD constitutes membrane distillation, a process similar to pervaporation. In a thermal membrane desalination process, the partial pressure drives the recovered water vapor to move across a hydrophobic microporous membrane. This process requires the liquid- vapor phase transition, thus avoiding the need for high pressure. It can be operated at low temperatures and pressures, such that the passage of toxic compounds and heavy wastes can be avoided. The use of anti-scalant chemical salts is also not required. Compared to other processes, this system will ensure high energy, as well as water recovery. Though it is beneficial to treat high salinity feed water, a limitation of this system exists in obtaining low TDS water.

19.6 Potential of Membrane Technology Toward ZLD

ZLD systems are gaining potential among industries that are interested in recovering useful chemicals from the spent and brine solution obtained from process industries. It is well-known that ZLD systems play an important role in reclaiming process waste. ZLD strives to employ the most advanced effluent treatment technologies to recycle and reclaim virtually almost all wastewater produced. The concept of ZLD focuses to eliminate all of the available discharge to the environment, thereby resulting in (i) the recovery of valuable products from wastewater and (ii) a minimization of polluting substances into the environment away from the wastewater treatment facility, so that zero discharge can be achieved.

The need for state-of-the-art technology for processing reject streams like reverse osmosis, fluidization and evaporation are still burgeoning and have not achieved its final stage for running a ZLD assured ETP in industries. MBR has emerged as cutting-edge technology for treating waste streams in its secondary and tertiary stages to intensify the benefits of ZLD concept. Hence, recovery of all elemental pollutants from the final liquid discharge assures the effective implementation of ZLD in industries. MBRs can thus provide a higher level of treatment leading to ZLD, as they strive to maintain resistance to fluctuations in effluent stream capacity and composition, which many industries consider to be a major setback.

19.7 Advancements and Scope of Membrane Technology for ZLD

Enormous investment in the effluent treatment sector will surely be reciprocated by efficient ZLD, achievable through incorporation of the MBR step. With growing interests toward cost reduction in membrane processes, economic constraints can be eliminated for making MBR a practical solution for full-scale industrial and municipal wastewater treatment plants. Submerged MBR systems have recently been reducing operating cost by 30-fold, owing to membrane life, design and process variation, with further cost reduction expected in forthcoming years. As MBR technology tends to bring together the domains of process design, hydrodynamics, microbiology, polymer technology, chemistry, and so on, it creates a scope for multidisciplinary research, offering valuable advancements in the ETP sector. The bottlenecks of MBR can be overcome and the scope of MBR can be broadened by designing research programs on finding economically suitable anti-fouling additives, analyzing biomass and bio-fouling features, predicting an anti-fouling model, determining the operating factors affecting membrane filtration performance, among other solutions.

19.8 Emerging Trends in State-of-the-Art ZLD Systems

There are many emerging state-of-the-art ZLD technologies other than the conventional and hybrid ZLD systems. This is due to unintended environmental impacts of conventional

ZLDs, specifically raising concerns regarding storing sludge wastes in evaporation ponds, leakage risk, and so on. The few alternative ZLD technologies that have recently found implementation in industries include:

- Seeded RO and high efficiency RO process: These enable treatment of high pH waste streams from nuclear power and mining industries. Further, fouling problems can be overcome as alkaline pH mitigates fouling and biofilm formation.
- Forward osmosis: A promising alternative to RO as it is not limited by the problem of high osmotic pressure. The lower operating pressure and higher cross-flow velocity of FO both deliver greater performance for reducing TDS in high fouling streams.
- Molecular distillation: Vacuum distillation combined with salt-crystallizing spray dryers rather than mechanical vapor compression evaporation equipment helps in recycling process waste from various water streams.
- Wind-assisted intensified evaporation: The driving force, in this case, being high-speed winds that could ensure cost-effective evaporation ponds as part of ZLD.

In addition to the emerging ZLD technologies, it is mandatory to monitor the disposed solid wastes in landfills as it might result in the leaching of chemicals into groundwater. Accordingly, impervious liners and reliable monitoring systems are typically required to prevent potential contamination from solid waste. These safety measures, along with robust ZLD systems, could eventually create a breakthrough in wastewater management in industries.

19.9 Computational Aspects in Membrane Processes

To render membrane processes energy efficient, proper design of pilot and industrial scale plants by modelling and simulation is necessary. A wide range of mathematical models for understanding, evaluating and enhancing the performance of the membrane systems can be formulated, developed and tested. The optimal operating conditions and critical parameters for the design of membrane-based equipment can be established through the simulation of these models. The solution of a membrane module requires local mass transport equations involving conservation and continuity equations for representing the process and flow of material through the module. Formulation of models is tedious, computationally expensive and involves many complexities in flow modeling. These models are formed on the basis of various assumptions that are system-dependent. However, several researchers have attempted and formulated models in certain cases based on many restrictive assumptions, such as:

- Physical properties are assumed to be constant, although changes in concentrations, temperatures and pressures affect the thermodynamic fluid properties, like viscosity and density.
- In seawater desalination, the models are built on the assumption of a binary mixture feed, even though the membranes are often used to separate multicomponent mixtures (seawater itself is a multi-component mixture).

- In gas separations, density is assumed to be constant even though significant pressure changes are likely to be observed on a membrane.

- Isothermal conditions are assumed in RO, whereas, in pervaporation, the heat supplied to vaporze the permeating material often results in a significant feed stream temperature drop.

- In flow models, one-dimensional plug flow is assumed, which means that the concentration variations perpendicular to the bulk flow direction are neglected.

The partial differential equations are formulated from mass, momentum and energy balances based on these assumptions. Algebraic equations are incorporated to represent fluid properties and thermodynamic relationships. The models are then solved by numerical integration of these partial differential equations.

Membrane models are categorized as approximate or simulation models, depending on their utility. The approximate models (Evangelista, 1985; Malek et al., 1994) are used for quick design calculations and typically assume average conditions on either side of the membrane. Simulation models, also known as distributed models (El-Haiwagi et al., 1996; Ben-Boudinar et al., 1992), are used for more accurate calculations, as these models consider spatial variations and are well represented by a mixed set of partial differential and algebraic equations.

Distributed models are solved by discretizing the spatial domain using finite difference (Ben-Boudinar et al., 1992) or orthogonal collocation (Tessendorf et al., 1996) techniques. Finite difference methods divide the spatial domain into a number of elements for which the derivatives are approximated by linear approximations derived from the Taylor series. The error in finite difference methods is a function of the neglected second order derivative terms and the number of elements used. Orthogonal collocation methods assume spatial variation of distributed variables and are approximated by Lagrangian polynomials. The approximations are enforced at discrete points within the spatial domain and orthogonal collocation is carried out for each element independently with continuity of variables and derivatives at element boundaries. This results in a set of non-linear algebraic equivalents, which are then solved using efficient newton type methods.

In general, all of the membrane models assume steady-state conditions and the dynamics of membrane processes are rarely considered. In order to utilize a model for investigating control related aspects and dynamic optimization strategies, it is important that steady-state conditions are not assumed.

19.9.1 Simulation of Spiral Wound Membranes

Spiral wound modules (Bhattacharyya & Williams, 1992) were used for RO as they offer higher permeation rates with easier cleaning. RO process modelling involves complex models, and common approaches include the solution-diffusion model and the Kedem-Katchaisky model (Ohya & Taniguchi, 1975). The classic approach is to neglect the curvature of the channels and consider flow through two flat, spacer-filled channels on either side of the membrane (Figure 19.5). A one-dimensional plug flow model that assumes constant values on either the feed or the permeate side of the membrane and two-dimensional models that describe the true cross-flow nature of the flow are used in the literature.

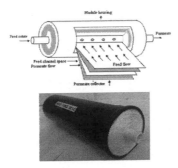

FIGURE 19.5
Flow through a spiral wound membrane.

19.9.2 Simulation of Hollow Fiber Membranes

Hollow fiber modules are commonly used for pervaporation, gas separation and RO (Rautenbach & Albrecht, 1989). Hollow fiber geometry is designed to accommodate a large number of membrane fibers housed in a module shell, and feed is introduced on either the fiber or the shell side assuming plug flow (Figure 19.6). Permeate is usually withdrawn in a co-current or counter-current manner. The fibres are usually considered to be long, thin cylindrical tubes with material injection or removal at the tube walls. The shell side is treated as a continuous phase with axial and/or radial flow through the porous fiber bundle. Concentration and velocity distributions along the length of the fibre are then obtained by the solution of an axial mass balance and each fiber in the module is assumed to have identical specifications. In pervaporation (Tsuyumoto et al., 1997) and RO (Ben-Boudinar et al., 1992), there are significant radial concentration variations

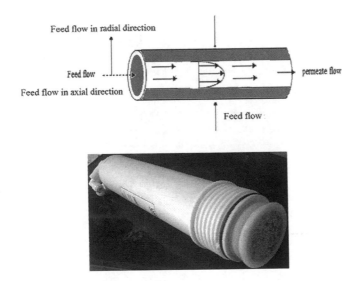

FIGURE 19.6
Flow through a hollow fiber membrane.

inside each fiber. Pressure build-up inside the fiber pore can be significant and will affect the mean driving force for mass transport through the membrane. The pressure change inside the fibers is calculated from the axial momentum balance with a Hagen-Poiseuille relationship. However, the flow is not laminar due to fluid injection or removal rates at the fiber walls. Radial distributions are neglected when describing parallel flow hollow fibre modules.

19.9.3 Design of Membrane Systems

The design of a general membrane separation process requires several factors, such as membrane area, type of module, size of module, process configuration, positioning of additional equipment and optimal operating conditions. In water treatment, the factor that has the greatest influence on the membrane system design is the fouling tendency of the feed water, which is caused by particles or colloidal material present in the feed water and concentrated at the membrane surface. The silt density index (SDI) value of the pretreated feed water correlates fairly well with the amount of fouling material present. The concentration of fouling materials at the membrane surface increases with increasing flux. A system with a high flux rate is therefore likely to experience greater fouling rates and more frequent chemical cleaning. Thus, a membrane system should be designed such that the elements of the system operate within a frame of recommended operating conditions to minimize fouling and exclude mechanical damage. These operating conditions are limited by the maximum recovery, the maximum permeate flow rate, the minimum concentrate flow rate and the maximum feed flow rate, per element. The higher the fouling tendency of the feed water, the stricter the limits of these parameters become. The average flux of the entire system is related to the total active membrane area of the system, which is a characteristic number of a design. Systems operating on high quality feed water are typically designed at high flux values, whereas systems operating on poor quality feed water are designed at low flux values.

In order to establish optimum operating parameters, approximate models and process simulation software can be used, and selection of the best design values involves many complicated and interacting choices, which depend on the type of application and end-user requirements.

19.10 Conclusions

Emerging research and development implicates membrane technology as a potential alternative for complementing or replacing conventional, cumbersome unit operations in industrial processes. However, to achieve sustainability, developmental efforts to commercialize the promising membranes and integrate the improved membrane processes are important. Moreover, basic as well as applied research strategy must be practiced for developing novel membranes and also to deploy new membrane processes. Increasing competition, growing industrial requirements, shift of technologies and other rapid market changes necessitate the need for alternate ways to assess and forecast the behavior of a chemical process. Incorporating computational aspects to understand and forecast the behavior of the process, its components and their relationships, by means of modeling and simulation, is crucial for any membrane process. More advancements in membrane materials, module configurations, and processes are becoming more important for improving system performance, particularly in terms of minimal energy requirements for sustainable industrial growth, which will have a critical role in deciding the future scope of membrane application.

References

Arar, O., U. Yuksel, N. Kabay, and M. Yuksel. 2012. Various applications of electrodeionization (EDI) method for water treatment-A short review. *Desalination* 342: 16–22.

Ben-Boudinar, M., W.T. Hanbury, and S. Avlonitis. 1992. Numerical simulation and optimization of spiral wound modules. *Desalination* 86: 273–290.

Bhattacharyya, D., M.E. Williams, R.J. Ray, and S.B. McCray. 1992. Reverse osmosis: Design. *Membrane Handbook*. Van Nostrand Reinhold.

Bruggen, B.V., A. Koninckx, and C. Vandecasteele. 2004. Separation of monovalent and divalent ions from aqueous solution by electrodialysis and nanofiltration. *Water Research* 38: 1347–1353.

Cath, T.Y., A.E. Childress, and M. Elimelech. 2006. Forward osmosis: Principles, applications, and recent developments. *Journal of Membrane Science* 281: 70–87.

Chanukya, B.S., S. Patil, and N.K. Rastogi. 2013. Influence of concentration polarization on flux behavior in forward osmosis during desalination using ammonium bicarbonate. *Desalination* 312: 39–44.

Chung, T.S., S. Zhang, K.Y. Wang, J. Su, and M.M. Ling. 2012. Forward osmosis processes: Yesterday, today and tomorrow. *Desalination* 287: 78–81.

Daniilidis, A., R. Herber, and D.A. Vermaas. 2014. Upscale potential and financial feasibility of a reverse electrodialysis power plant. *Applied Energy* 119: 257–265.

Daufin, G., J.E. Escudier, H. Carrere. 2001. Recent and emerging applications of membrane processes in the food and dairy industry. *Trans IChemE* 79: 89–102.

Drioli, E., A. Brunetti, G. Di Profio, and G. Barbieri. 2012. Process intensification strategies and membrane engineering. *Green Chemistry* 14: 1561–1572.

Drioli, E. and M. Romano. 2001. Progress and new perspectives on integrated membrane operations for sustainable industrial growth. *Industrial & Engineering Chemistry Research* 40: 1277–1300.

El-Halwagi, A.M., V. Manousiouthakis, and M.M. Haiwagi. 1996. Analysis and simulation of hollowfiber reverse osmosis modules. *Separation Science and Technology* 31: 2505–2529.

Evangelista, F. 1985. A short cut method for the design of reverse osmosis desalination plants. *Industrial and Engineering Chemical Process Design and Development.* 24: 211–223.

Galier, S. and H.R. Balmann. 2011. The electrophoretic membrane contactor: A mass transfer-based methodology applied to the separation of whey proteins. *Separation and Purification Technology* 77: 237–244.

Giwa, A. and A. Ogunribido. 2012. The applications of membrane operations in the textile industry: A review. *British Journal of Applied Science & Technology* 2: 296–310.

Hayrynen, P., I. Galambos, K. Szekrenyes, R.L. Keiski, and G. Vatai. 2013. Concentration of nitrogen compounds and heavy metals by a reverse osmosis-membrane distillation hybrid process. *Proceedings: The 6th Membrane Conference of Visegrad Countries PERMEA 2013*, Poland.

Hong, J.G., B. Zhang, S. Glabman, N. Uzal, X. Dou, H. Zhang, X. Wei, and Y. Chen. 2015. Potential ion exchange membranes and system performance in reverse electrodialysis for power generation: A review. *Journal of Membrane Science* 486: 71–88.

Hoshino, T. 2013. Preliminary studies of lithium recovery technology from seawater by electrodialysis using ionic liquid membrane. *Desalination* 317: 11–16.

Iizuka, A., Y. Yamashita, H. Nagasawa, A. Yamasaki, and Y. Yanagisawa. 2013. Separation of lithium and cobalt from waste lithium-ion batteries via bipolar membrane electrodialysis coupled with chelation. *Separation and Purification Technology* 113: 33–41.

Jensen, C.A.Q., F. Macedonio, and E. Drioli. 2016. Membrane crystallization for salts recovery from brine—An experimental and theoretical analysis. *Desalination and Water Treatment* 57: 7593–7603.

Jiang, C., Y. Wang, Z. Zhang, and T. Xu. 2014. Electrodialysis of concentrated brine from RO plant to produce coarse salt and freshwater. *Journal of Membrane Science* 450: 323–330.

Juby, G., C. Schutte, and J. van Leeuwen. 1996. Desalination of calcium sulphate scaling mine water: design and operation of the SPARRO process. *Water SA* 22: 161–172.

Kelly, P., J. Kelly, R. Mehra, and B.T.O' Kennedy. 2000. Implementation of integrated membrane processes for pilot scale development of fractionated milk components. *Dairy Science & Technology* 80: 139–153.

Kurt, E., D.Y. Koseoglu-Imer, N. Dizge, S. Chellam, and I. Koyuncu. 2012. Pilot-scale evaluation of nanofiltration and reverse osmosis for process reuse of segregated textile dyewash wastewater. *Desalination* 302: 24–32.

Lee, K., R. Baker, and H. Lonsdale. 1981. Membranes for power generation by pressure retarded osmosis. *Journal of Membrane Science* 8: 141–171.

Loeb, S., F. Van Hessen, and D. Shahaf. 1976. Production of energy from concentrated brines by pressure-retarded osmosis: II. Experimental results and projected energy costs. *Journal of Membrane Science* 1: 249–269.

Lopez, A.M. and J.A. Hestekin. 2013. Separation of organic acids from water using ionic liquid assisted electrodialysis. *Separation and Purification Technology* 116: 162–169.

Malek, A., M.N.A. Hawlader and J.C. Ho. 1994. A lumped transport parameter approach in predicting BIO RO permeator performance. *Desalination* 99: 19.

McCutcheon, J.R. and M. Elimelech. 2006. Influence of concentrative and dilutive internal concentration polarization on flux behavior in forward osmosis. *Journal of Membrane Science* 284: 237–247.

Miller, J.E. 2003. Review of water resources and desalination technologies. Sandia National Laboratories. http://www.prod.sandia.gov/cgi-bin/techlib/access-control.pl/2003/030800.pdf

Nayak, C.A. and N.K. Rastogi. 2010. Forward osmosis for concentration of anthocyanin from Kokum (Garcinia indica Choisy). *Separation and Purification Technology* 71: 144–151.

Ohya, H. and Y. Taniguchi. 1975. Analysis of reverse osmosis characteristics of Roga-400 spiral-wound modules. *Desalination* 16: 359–373.

Ozgun, H., R.K. Dereli, M.E. Ersahin, C. Kinaci, H. Spanjers, J.B. van Lier. 2013. A review of anaerobic membrane bioreactors for municipal wastewater treatment: Integration options, limitations and expectations. *Separation and Purification Technology* 118: 89–104.

Pal, P. and P. Dey. 2012. Developing sustainability technology for chemical process industry: Lactic acid by membrane integrated hybrid process. *International Journal of Biological Ecological and Environmental Sciences* 1: 25–41.

Rautenbach, R. and R. Albrecht. 1989. *Membrane Processes.* Chichester, U.K.: John Wiley & Sons.

Readi, O.M.K., M. Girones, W. Wiratha, and K. Nijmeijer. 2013. On the isolation of single basic amino acids with electrodialysis for the production of biobased chemicals. *Industrial and Engineering Chemistry Research* 52: 1069–1078.

Semmens, M.J., C.D. Dillon, and C. Riley. 2001. An evaluation of CEDI as an in-line process for plating rinsewater recovery. *Environmental Process & Sustainable Energy* 20: 251–260.

Shen, J., C. Yu, J. Huang, and B.V. Bruggen. 2013. Preparation of highly purified tetrapropyl ammonium hydroxide using continuous bipolar membrane electrodialysis. *Chemical Engineering Journal* 220: 311–319.

Song, S., Y.F. Tao, H.Q. Shen, B.Q. Chen, P.Y. Qin, and T.W. Tan. 2012. Use of bipolar membrane electrodialysis (BME) in purification of L-ascorbyl-2-monophosphate. *Separation and Purification Technology* 98: 158–164.

Suwal, S., C. Roblet, J. Amiot, A. Doyen, L. Beaulieu, J. Legault, and L. Bazinet. 2014. Recovery of valuable peptides from marine protein hydrolysate by electrodialysis with ultrafiltration membrane: impact of ionic strength. *Food Research International* 65: 407–415.

Tessendorf, S., R. Gani, and M.L. Michelsen. 1996. Aspects of modelling and operation of membrane-based separation processes for gaseous mixtures. *Computers and Chemical Engineering* 20 Suppl.: S653–S658.

Tsukahara, S., B. Nanzai, and M. Igawa, 2013. Selective transport of amino acids across a double membrane system composed of a cation- and an anion-exchange membrane. *Journal of Membrane Science* 448: 300–307.

Tsuyumoto, M., A. Teramoto, and P. Meares. 1997. Dehydration of ethanol on a pilot plant scale, using a new type of hollow-fibre membrane. *Journal of Membrane Science* 133: 83–94.

Vermaas, D.A., J. Veerman, N.Y. Yip, M. Elimelech, M. Saakes, and K. Nijmeijer. 2013. High efficiency in energy generation from salinity gradients with reverse electrodialysis. *ACS Sustainable Chemistry and Engineering* 1: 1295–1302.

Wiesner, M.R. and S. Chellam. 1999. The promise of membrane technology: An expanding understanding of membrane technology is fostering new environmental applications. *Environmental Science & Technology* 33: 360A–366A.

Wijekoon, K.C., F. Hai, J. Kang, W.E. Price, W. Guo, H.H. Ngo, T.Y. Cath, and L.D. Nghiem. 2014. A novel membrane distillation-thermophilic bioreactor system: biological stability and trace organic compound removal. *Bioresource Technology* 159:334–341.

Xu, T.W. and C. Huang. 2008. Electrodialysis-based separation technologies: A critical review. *AIChE Journal* 54: 3147–3159.

Yang, Y., X.L. Gao, A.Y. Fan, L.L. Fu, and C.J. Gao. 2014. An innovative beneficial reuse of seawater concentrate using bipolar membrane electrodialysis. *Journal of Membrane Science* 449: 119–126.

Zhang, Y., S. Paepen, L. Pinoy, B. Meesschaert, and B.V. Bruggen. 2012. Selectrodialysis: Fractionation of divalent ions from monovalent ions in a novel electrodialysis stack. *Separation and Purification Technology* 88: 191–201.

20

Electromembrane Processes in Water Purification and Energy Generation

Sujay Chattopadhyay, Jogi Ganesh Dattatreya Tadimeti,
Anusha Chandra, and E. Bhuvanesh

CONTENTS

20.1 Introduction

Water shortage is a worldwide problem that is increasing day by day and the global population could fall victim to it sooner or later. Increasing urbanization, deforestation, thrust for energy to match higher standards of living have resulted in rapid depletion of water resources and climate change. Several regions of the world are dangerously lacking fresh water reserves. The bitter truth is that the number of available fresh water sources like ponds, lakes, streams, reservoirs, springs, wells and tube-wells are not adequate to meet the demands of growing industrialization. Therefore, means need to be devised to ensure contaminated water is recycled after appropriate purification wherever alternative sources of water are unavailable. Exploration of alternative resources of water and economics of the purification techniques employed need to be thoroughly checked before any recommendation is made.

Water collected from fresh water sources is mostly treated through ultrafiltration (UF) membranes to remove, among other things, suspended solids, turbidity and pathogens. Thus, UF treated water becomes suitable for its subsequent treatment by reverse osmosis (RO) to remove dissolved impurities. Usually, UF/RO treatment is sufficient for fresh water, but its limited availability compels scientists and engineers to explore various other treatment methodologies to make available water from other sources, e.g., (i) brackish water, (ii) municipal wastewater, (iii) sea water and (iv) wastewater from various process industries.

Water from most of these sources is rich in a wide range of metal salts. Seawater is a huge source of saline-rich (≥ 30000 ppm) water, readily available for people living in coastal areas but not potable. Excess salinity forces it to be treated through a series of conventional pretreatment steps, e.g., settling, polyelectrolyte coagulation, hydrogen peroxide treatment, charcoal bed filtration, followed by membrane-based techniques such as MF/UF/RO. After RO treatment, the water becomes completely free from all electrolytes; therefore, a portion of the untreated water is added to make it worthy of drinking. The pressure applied in an RO unit also becomes very high and consumes extra energy. Thus, water obtained from these treatment steps becomes comparatively expensive. Among other processes, Electrodialysis (ED) can compete with RO as an alternative option. Replacing RO by ED will reduce fouling issues but operation cost will rise. No single membrane technique is able to provide water of desired quality due to existence of wide range of impurities and contaminants. Integrated approaches involving suitable separation techniques become essential to meet the desired performance.

Here, we take up issues with electrodialysis technique, which is popular mostly in coastal belts and countries where sea is the major source of water, e.g., Japan, Singapore and Middle Eastern countries. Over the years, ED technique has got enriched with several robust systems and has diversified into various classes, e.g., desalination, capacitive deionization, bipolar membrane electrodialysis and reverse electrodialysis. In the following sections we will take up each of these techniques in detail.

20.2 Electrodialysis

This is a technique to separate dissolved electrolytes from their aqueous solution with the help of oppositely charged polar membranes and applied external potential. The electrolytes undergo hydrolysis and produce ions in solution. Strong electrolytes are fully hydrolyzed while weak electrolytes get partially hydrolyzed. The physical properties (concentration, charge, size) of these hydrated ionic species and their interaction in solution/membrane phases under the influence of applied potential cause each ion to behave distinctly and be selectively transported through an ion exchange membrane (IEM) bearing counter-ionic fixed charges on its surface/matrix. The complete process of electrolytes' separation from solution under applied electric potential is executed with the help of an ED setup. Figure 20.1 shows a basic ED cell consisting of different electrodes, ion exchange membranes (CEM and AEM) and different flow streams, namely (i) diluate/feed solution

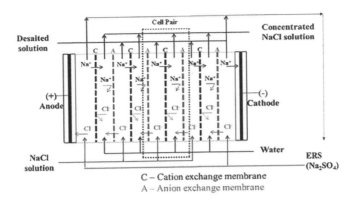

FIGURE 20.1
Schematic diagram of an electrodialysis stack.

(where electrolyte removal takes place), (ii) concentrate stream (gets slowly enriched with electrolytes) and (iii) Electrode rinsing solution (ERS). Each CEM/AEM membrane is separated from its counter-ionic membrane with the help of spacers. Ions reaching electrodes are either oxidized or reduced, depending on the electrode they reach, and ERS ensures ion conductivity inside ED cell and prevents any influence of electrode reactions reaching the diluate/concentrate streams. Electro-neutrality condition is assumed to prevail in all chambers (streams), and only equivalent amount of cations/anions migrate from diluate to concentrate (Geraldes and Afonso 2010; Strathmann 2004) compartment. Large quantity process streams in industrial operations are managed by increasing the number of cell pairs in a single assembly (called ED stack) placed between a pair of electrodes. Even larger volumes of feed are tackled by employing more ED stacks and connecting them in parallel. Multiple channels enable handling such capacities to reduce recurring and fixed costs of maintenance.

ED has been used in: production of potable water; pretreatment of boiler feed water, preparation of table salt from seawater (in Japan); desalination of seawater; removal of heavy metals from brackish water; demineralization of food products; production of organic acids such as formic acid, acetic acid, propionic acid, lactic acid, gluconic acid, oxalic acid, malic acid, itaconic acid, tartaric acid, salicylic acid, citric acid and ascorbic acid; recovery of important chemicals from fermentation broth in pharmaceutical industries; separation of whey protein; de-acidification of fruit juice; power generation (using reverse ED); desalting of crude glycerol in bio diesel production; production of sodium methoxide, acetoacetic ester and methyl methoxy acetate and more (Vadthya et al. 2015; Q. Li, Huang, and Xu 2009; Strathmann 2004).

ED has been applied in both small scale (batch mode) and large scale (continuous mode) operations. Commonly, small volumes are processed in a batch ED process either in (i) constant current mode or (ii) constant potential mode. In batch ED, a small volume of feed is continuously recirculated, which results in time-dependent composition variation in both streams (diluate/concentrate). In continuous mode, the feed stream passes through the ED stack only once, which makes its composition time-invariant. Separation of monovalent ions (KCl, NaCl, etc.) was easier than separation of multivalent ions ($CaCl_2$, $MgCl_2$ etc.) and higher potential drastically reduced the time of ED operation to achieve a desired percentage removal, but flow rate variation did not register any appreciable change in efficiency (Kabay et al. 2006). Caustic recovery from industrial black liquor using a batch ED

(Mishra and Bhattacharya 1984) is facilitated at lower velocity and higher current density once ED is carried out below a limiting current. Batch ED is an unsteady process, and cell resistance continues to rise due to continuous removal of electrolytes from diluate stream under a constant potential that alters current density. The appearance of concentration polarization and the unsteady nature of the solution-membrane interface results in varying potential difference. In a batch ED, the ion removal efficiency mostly passes through an optima where maximum ion flux is recorded. Gradual drop in diluate composition slowly increases solution resistance and reduces efficiency. Nearly uninterrupted process efficiency could be achieved with continuous adjustments of both voltage and current. Parulekar (Parulekar 1998) created a theoretical model to estimate current and potential values' variation with time.

Continuous ED to recover caustic from black liquor using different combinations of membranes were reported by Mishra and Bhattacharya (1987) where lignin fouling over IEM was substantially lowered by introduction of membranes of reduced polarity. Vadthya et al. (2015) reported desalination of crude glycerol during production of biodiesel from oil feed stock. In continuous ED operation, a steady ion transport is achieved due to constant composition of concentrate/diluate streams at a given location inside a flow channel.

20.3 Ion Exchange Membranes

Ion exchange membranes (IEMs) are the backbone of ED and they selectively separate counter-ions based on their charge and hydrodynamic volume. Figure 20.2 shows a graphical representation of a typical ion exchange membrane and its structure. A wide variety of ion exchange membranes are being synthesized on the basis of specific process application (Xu, 2005). IEMs are traditionally made from polymer molecules bearing charged moieties. Thick or dense layer membranes are commonly used in ED. Based on charge moieties of

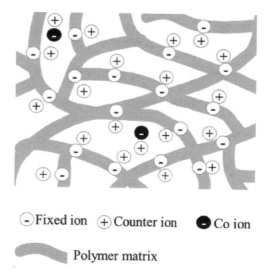

FIGURE 20.2
Structure of an ion exchange membrane adapted from Strathmann. (Adapted from Strathmann, H., *Ion-Exchange Membrane Separation Processes, Membrane Science and Technology Series*, 9, 2004.)

polymer matrix and their arrangement, these IEMs are classified mainly into the following three categories:

1. Cation exchange membranes (CEMs) are made up of polymers containing functional groups like $-SO_3^-$, $-COO^-$, $-PO_3^{2-}$, etc. (e.g., Neosepta CMX and Nafion®). Ideally, these CEMs are selective to cations only.

2. Anion exchange membranes (AEMs) are made up of polymers containing functional groups like $-NH_3^+$, $-NRH_2^+$, $-NR_2H^+$, etc. (e.g., Neosepta AMX and Neosepta AM-1). Ideal AEMs are selective to anions only.

3. Bipolar membranes are made up of a catalytic intermediate layer sandwiched between one anion and one cation exchange layer. No ions can actually cross through these membranes from one side to the other side. Instead, the intermediate layer promotes electrolysis of water and generates H^+ and OH^-, thus forming separate acid and base streams during ED.

Desirable properties for ion-exchange membranes are (Xu, 2005; Strathmann, 2004):

1. *High perm-selectivity*: the ion-exchange membrane should be highly permeable to counter-ions, and impermeable to co-ions.

2. *Low electrical resistance*: electrical resistance of the membrane after equilibration in an electrolyte solution is important for any ED application. If the electrical resistance is lowered, the current registered in the ED circuit for a fixed potential will be higher.

3. *Good mechanical and form stability*: a membrane should have enough mechanical strength and negligible influence in its shape due to swelling or shrinking during its exposure to solutions of varying ionic strength (diluate or concentrated solutions).

4. *High chemical stability*: membranes need to be both chemically and physically stable, preferably over a wide pH range and in the presence of oxidizing agents.

These membranes are further classified into homogeneous and heterogeneous categories based on their chemical composition and nature of their interaction of constituents with the base polymer bearing charged moieties. Homogeneous membranes are made of a single layer of base polymer/copolymer molecules bearing charged moieties and possess adequate physical strength required for process conditions. These membranes are generally synthesized by solution casting or hot melt molding (e.g., Neosepta CMX and Nafion) techniques. Heterogeneous membranes consist of various substance(s) in addition to base polymer molecules bearing charged moieties. Series of materials (e.g., metal or nonmetal oxides, halides, sulfates, carbonates, acids etc.) possessing special features (e.g., size, shape, porosity distribution, ion selectivity, charge, flexibility, mechanical and thermal stability, acid or base resistance etc.) can be blended with the base polymer material in an appropriate quantity to obtain a heterogeneous membrane of desired properties.

CEM blended with potassium perchlorate and magnetic iron/nickel nanoparticles to create heterogeneous membranes were reported in earlier articles (Hosseini et al., 2012; Kim & Van Der Bruggen, 2010). Incorporation of these particles caused noticeable improvement in membrane properties, e.g., water uptake, transport number, perm-selectivity, ion exchange capacity, membrane-resistance and surface charge density. Various dopants and techniques of their incorporation are routinely used to tailor properties of these IEMs according to process requirements. Commonly mechanical, thermal and chemical properties along with ion selectivity of membranes have been reported earlier. Sulfonated silica, metal oxide nanoparticles and phosphonated carbon nanotubes are commonly used dopants in IEMs (Klaysom et al., 2010; Kumar et al., 2009; Ng et al., 2013; Kannan et al., 2011).

20.4 Mathematical Representation of Various Fluxes in Electrodialysis Process

Ions need to overcome various resistances, e.g., hydrodynamic, diffusive and solution, during their transport through IEMs. A series of fundamental laws, including Ohm's law, Fick's law and the Nernst-Planck equation (Strathmann, 2004) and semi-empirical equations to represent useful parameters are commonly used to quantitatively calculate transport process phenomena. The Nernst-Planck equation accounts for both diffusive and convective transport of ions. According to this equation, total flux in ED process is sum total of fluxes due to ionic diffusion ($N_{j,diff}$), electric potential ($N_{j,el}$) and convection ($N_{j,con}$).

$$N_j = N_{j,diff} + N_{j,el} + N_{j,con} \tag{20.1}$$

The overall ionic flux N_j based on the Nernst-Planck equation can be expressed as (Lee et al., 2006; Strathmann, 2004):

$$N_j = -D_j \frac{\partial C_j}{\partial x} - \frac{z_j C_j F D_j}{TR} \frac{\partial V}{\partial x} + v_k C_j \tag{20.2}$$

where D_j is diffusivity of ion j (m²·s⁻¹), C_j is concentration of ion j (mol·m⁻³), z_j is charge of ion j, F is Faraday's constant, 96500 C·geq⁻¹, R is the Universal gas constant, 8.314 J·mol⁻¹·K⁻¹, T is temperature (K), V is Potential difference and v_k is the velocity of component k (m·s⁻¹).

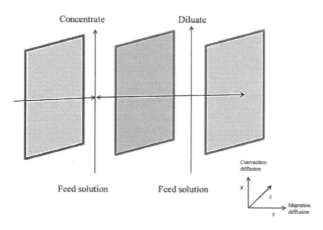

FIGURE 20.3
Direction of ionic flow in ED process.

In Equation 20.2 $-D_j \dfrac{\partial C_j}{\partial x}$ is the diffusive component, $-\dfrac{z_j C_j F D_j}{RT}\dfrac{\partial V}{\partial x}$ is the migration component arising out of applied electric potential and $v_k C_j$ is the convective component of the flux. In general, the direction of fluid flow is parallel to the surface of the membrane in ED (Figure 20.3), and an almost negligible transport of ions take place through the membrane.

20.5 Current-Voltage Characteristics

Current voltage (*I-V*) characteristics of an ED stack are highly crucial in understanding the ion transport phenomena. Briefly, the properties of membrane, the solution (including the type of electrolytes, concentration, charge of ionic species, solubility, flow velocity and hydrodynamic profile) and applied electric potential all influence *I-V* characteristics. In an ideal homogeneous conductor, the flow of electrons increases with applied potential while its resistance remains unchanged, i.e., Ohm's law is followed. In an ED stack, current is carried by hydrated ionic species instead of electrons. Therefore, all properties of ionic species would influence the *I-V* characteristic. At lower potential, a linear behavior of *I-V* plot indicates constancy in the overall cell resistance, i.e., Ohm's law behavior is followed. With increase in potential a boundary layer is formed adjacent to membrane surfaces where a concentration gradient is observed. Figure 20.4 (a) shows a typical *I-V* plot containing three zones in it which are discussed separately in the following sections.

20.5.1 Ohmic Region

Figure 20.4 (a) demonstrates the initial linear portion of the *I-V* plot where ion transport through IEM is linearly dependent on applied potential. For a given electrolyte concentration, flow rate and a given ED stack, once the concentration gradient/polarization is fully

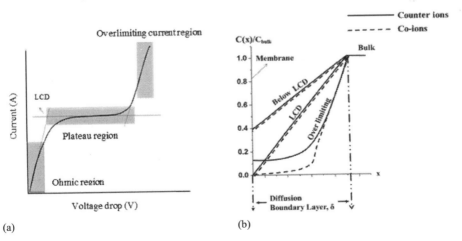

(a) (b)

FIGURE 20.4
(a) Voltage-current characteristics in electrodialysis. (Adapted from Balster, J., M.H. Yildirim, D.F. Stamatialis, R. Ibanez, R.G.H. Lammertink, V. Jordan, and M..Wessling, *The Journal of Physical Chemistry. B*, 111 (9): 2152–2165, 2007.) (b) Different stages of concentration polarization that occur near an ion exchange membrane on the diluate side during ED operation, an idea modified suitably to explain ion transport. (Adapted from Mishchuk, N.A., *Desalination*, 117: 283–295, 1998.)

developed, ion transport attains maximum, i.e., it limits the value of current. At this stage, counter-ion concentration close to the membrane surface approaches zero, and no further rise in current is observed. This current or current density is defined as limiting current (LC) or limiting current density (LCD). LC is a critical parameter and all ED processes should be operated below LC.

20.5.2 Plateau Region

The region beyond LCD is called the plateau region (Figure 20.4 (a)). In this section, current density does not change appreciably even with a considerable rise in potential. This indicates that in this zone the membrane resistance increases non-linearly to accommodate increased potential.

20.5.3 Over Limiting Region

The onset of the over limiting zone starts where the plateau region nears its end. In this region, current density shows a steady rise with applied potential. A number of different phenomena like water splitting, coupled convection, electro convection, gravitational convection, electro osmosis and more have reported to occur (Balster et al., 2007) in this section of *I-V* plot. Mainly electro splitting of water generating large number of H^+/OH^- ions causes this steady rise in current density in this section.

20.6 Concentration Polarization

Electrodialysis is a complex physico-chemical process and is influenced by physical, chemical and ionic properties of (i) membrane, (ii) solution (e.g., molar concentration, viscosity, density, and ion conductivity), (iii) solute (e.g., solute type, charge, size, concentration, diffusivity, extent of ionization) and (iv) flow hydrodynamics (velocity profile, pressure gradient, mixing patern, boundary layer thickness etc. of the flow channel) (Balster, 2006).

The physical properties of an ion, e.g., charge, hydrodynamic radius, diffusivity (solution/membrane) and more, decide its current carrying capacity, which is commonly expressed by its transport number. Differences in mobility of ionic species between solution and charged membrane phases cause a concentration difference at the solution-membrane interface. Ion transport results when an ion can overcome these concentration gradients (Figure 20.4 (b)) (Mishchuk, 1998) on either side of membrane and pass through it. With higher applied potential, the concentration gradient (polarization) also increases. Besides flow characteristics, the nature of solute and its interaction with membrane (solute-solute and solute-membrane interactions) also influence concentration polarization. The thickness of this concentration polararization layer depends on flow velocity, channel geometry and membrane roughness (Lee et al., 2006).

On the concentrate side, solute concentration is higher at membrane surface and drops in bulk, but, in the diluate compartment, the bulk concentration is higher than the concentration at membrane surface. Once concentration of salt present in a solution exceeds the solubility product value, the salt precipitates over the membrane surface which increases membrane resistance due to fouling. Thus, membrane performance (F. Li et al., 2005; Balster, 2006; Lee et al., 2006) drops. For industrial application, fouling is eliminated through precipitation and is commonly addressed by periodic reversing of the electrode polarities with consequent alteration of concentrate and diluate flow channels.

20.7 LCD Determination and Parameters Influencing LCD

At the onset of plateau region in *I-V* plot on an ED process, where the counter-ion concentration at solution-membrane interface approaches zero (Figure 20.4 (b)), we find the LCD. Usually, a graph of *I-V* or *V/I-1/I* (Cowan and Brown method) is used to estimate LCD (Strathmann, 2004). Figures 20.5 (a) and 20.5 (b) show experimental measurements of LCD for 20 mol·m^{-3} CaCl$_2$ solution at two different velocities in an ED setup (Reynolds numbers, *Re* = 69 and 207). A mathematical expression of LCD derived from Nernst Planck equation with the assumption of zero ion concentration at membrane surface (Lee et al., 2006) is given in Equation 20.3:

$$i_{\lim} = \frac{CDzF}{\delta(t_m - t_s)}$$ (20.3)

where t_m and t_s are transport numbers of ion in membrane and solution respectively, i_{lim} is limiting current density (A·m^{-2}), z is the charge of transporting ion, δ is diffusion boundary layer thickness, *F* is Faraday's constant, *D* is diffusivity of ionic species in solution (m^2·s^{-1}) and *C* and C_m are concentrations of ion in bulk and at the membrane phase (mol·L^{-1}).

Experimentally measured LCD values (Tanaka, 2002; Lee et al., 2006) indicated a strong influence of electrolyte concentration and velocity of the fluid. Lee et al. (2006) proposed an empirical equation ($i_{lim} = a \cdot C \cdot u^b$) based on electrolyte concentration (*C*) and flow velocity (*u*). This equation performs satisfactorily when applied to solutions containing a single electrolyte, but was less effective for solutions containing multiple electrolytes. Geraldes and Afonso (2010) theoretically developed LCD expression based on linearized Nernst-Plank equation containing equivalent fractions of each electrolyte, ionic charge and diffusivity. This was experimentally verified with multiple electrolytes. The concept of effective

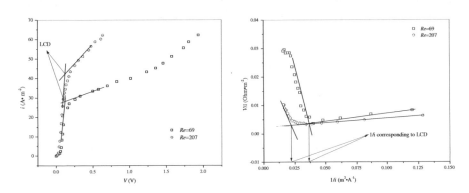

FIGURE 20.5
(a) Determination of LCD by plotting current density (*i*) against potential drop (*V*) across the membrane (*V*) for 20 mol·m^{-3} of CaCl$_2$ feed solution at two different flow rates (*Re* = 69, 207); (b) Determination of LCD from *V/i* vs. *1/i* data for 20 mol·m^{-3} of CaCl$_2$ feed solution at two different flow rates (*Re* = 69, 207).

diffusivity (the total of product of ionic diffusivity and its mole fraction) could effectively predict LCD for multi-ionic solutions (Geraldes & Afonso, 2010). ED process operating at current density beyond its LCD value will result in unwanted electrode reactions and water splitting, thus causing the process to become uneconomical (Lee et al., 2006). The universally accepted limit to current density is 80% of LCD value.

20.8 Mass Transfer Enhancement in Electrodialysis

Ion transport during ED is mainly governed by principles of mass transfer in a flow channel. Concentration polarization is an unavoidable phenomenon and works to limit ion transport. Minimization of this concentration polarization will facilitate ion transport or mass transfer rate. Modification of the flow channel geometry between membranes that minimize boundary layer thickness will reduce concentration polarization by promoting ion transport (Balster et al., 2006; F. Li et al., 2002). Application of rod-type promoters with appropriate spacing has been reported to effectively reduce concentration polarization by increasing mass-transfer coefficient at a specific flow rate. Additionally, Balster et al. (2006) reported that commercial netted spacers and multilayered spacers could effectively reduce concentration polarization. Unconventional techniques such as air sparging, ultrasound application in the diluate channel and membrane surface modification have also been used to reduce concentration polarization. Surface modification of membranes can also reduce pressure drop and minimize concentration polarization (Balster et al., 2007; Balster, 2006; Tadimeti et al., 2016).

Design of appropriate spacer geometry is undoubtedly a complicated technique and should always be verified by experimentation. This demands huge effort, fabrication cost and experimentation time without knowing the right geometry. Therefore, computational fluid dynamics (CFD) tools have often been applied to carry out initial screening process for the appropriate geometry that might facilitate ion transport. F. Li et al. (2002) used CFD simulations to design an improved flow promoter that offers almost 40% rise in Sherwood number, Sh. Ion transport facilitation with reduced power consumption compared to commercial netted spacers could be achieved by modification of promoter geometry. Subsequent studies with modified spacer geometries showed formation of both longitudinal and transverse vortices near the membrane surface that improved ion transport (F. Li et al., 2005). Sherwood number (Sh) and Reynolds number (Re) information for any flow channel geometry in ED are the most commonly used quantities for determining optimum spacer design, using simulation results. Shaposhnik et al. (1998) developed mathematical models based on the Navier-Stokes equation, the equation of continuity and the Nernst-Planck equation to simulate flow and concentration profiles inside ED cell. They verified simulation results through laser interferometric experiments. As well as reducing the time, cost and risks involved in running repeated experiments, CFD simulation can provide crucial information regarding process parameters (flow, concentration and pressure drop) at any location of interest inside the flow channel geometry (Fimbres-Weihs & Wiley, 2010).

20.9 Facilitation of Electrodialysis

Usually microfiltration/nanofiltration/reverse osmosis based deionization of highly concentrated electrolytes solution (>30000 ppm), e.g., a reject stream from RO plant, seawater or brackish water, demands high maintenance (membrane fouling) and energy. ED efficiency drastically drops at electrolyte concentrations below 10000 ppm. Therefore, ED is becoming increasingly popular to desalinate highly concentrated streams, e.g., seawater, to remove organic acids and amino acids from fermentation broth and to recover salts from cheese and whey.

The most convenient route to improve the efficiency of an ED setup is to improve the flow hydrodynamics between membrane layers. As discussed in Section 20.8, flow promoters of different geometries can be explored that are able to facilitate effective transport of eddies in both longitudinal and transverse directions and reduce the boundary layer thickness. The utmost care is necessary to minimize pumping cost, if any arise out of geometry modification; otherwise, any geometric changes would prohibit its feasibility. Therefore, a comparative study of Sherwood Number (Sh) and the Power number (Pn) is most informative for selecting the appropriate spacer geometry that can facilitate ion transport without altering the pumping cost too much. In this regard, a case study performed in our laboratory on citric acid recovery from its solution using an ED setup with different spacer arrangements is presented below.

Organic acids have low mobility through the membranes due to large molecular size and have low conductivities (high electrical resistance due to poor dissociation), which affect the ion transport through membranes and induce the need for high energy to transport such ions (Chandra et al., 2017). Operating an ED setup with highest possible current density, i.e., with minimum concentration polarization, ensures its maximum efficiency (Krol, 1997). The application of a wide range of flow spacers (or promoters) help in improving ion transport. Flow promoters act as mechanical stabilizers and facilitate mixing (Strathmann, 2010; Shaposhnik et al., 1998).

Different categories of flow spacer arrangements were explored with different flow rates (5–20 LPH) and 0.1 mol·L⁻¹ citric acid solution was chosen as the feed stream. Flow promoters of (i) rod type, (ii) twisted tape (TT) and (iii) mesh type TT were analyzed for an ED set up with rectangular flow channel of fixed aspect ratio and detailed geometrical specifications of flow promoters are listed in Table 20.1 (Tadimeti & Chattopadhyay, 2016a).

TABLE 20.1

Geometrical Specifications of Various Flow Promoters

Spacer	No. of Twists per cm Length	Spacing, d in cm	α	β	l_1 in cm	l_2 in cm
A (Rod)	–	2	90	–	8	–
B (Rod)	–	1.33	0	–	4	–
C (TT)	0.5	2	90	–	8	–
D (TT)	0.5	1.33	0	–	4	–
E (Mesh)	0.5	–	0	90	2	1.33

Note: Rod with a 0.25 cm diameter.

Five different orientations of flow promoters of varying geometries were tried, e.g., A and C are rod and twisted tape (TT) type flow promoters oriented horizontal to flow direction while B and D are rod and twisted tape (TT) type flow promoters oriented in perpendicular direction and E is double layer spacer or mesh type TT. 0.1 mol·L^{-1} citric acid as feed, 0.3 mol·L^{-1} citric acid as shielding solution and 0.1 mol·L^{-1}HCl were chosen as ERS for conducting LCD measurements.

LCD, mass transfer coefficient (k), Sherwood number (Sh) and power number (Pn) were estimated from Equations 20.3–20.7. Boundary layer thickness (δ) is dependent on flow profile, physical properties of the fluids, cell geometry and surface morphology of membranes used in an ED cell.

$$\delta = \frac{D}{k} \tag{20.4}$$

where D is diffusivity and k is mass transfer coefficient of the diffusing species in solution.

Equations 20.3 and 20.4 provide a relation between limiting current density, boundary layer thickness (δ) and mass transfer coefficient (m·s^{-1}).

The performance of each spacer was estimated from LCD, mass transfer coefficient and Sh number. So, by using LCD, the boundary layer thickness and mass transfer coefficient can be obtained.

$$k = \frac{i_{lim}\left(t_m - t_s\right)}{CzF} \tag{20.5}$$

$$Sh = \frac{kh}{D} \tag{20.6}$$

where h is inter-membrane distance (m).

The power consumption for pumping fluid through each flow geometry is estimated from power number (Pn):

$$Pn = \frac{\Delta Pu\rho^2 h^4}{L\mu^3} \tag{20.7}$$

where ΔP is pressure drop (Pa), L is the characteristic length of flow channel (m), ρ is solution density (kg·m^{-3}) and μ is solution viscosity (N·m^{-2}·s^{-1}).

We found that with flow rate variation from 5 to 20 LPH (i.e., Re 67 to 267), LCD was doubled (Figure 20.6 (a)). Lowering of diffusion boundary layer thickness at higher flow rates resulted in higher LCD and facilitated the mass transfer rate. Figure 20.6 (a) shows LCD and mass transfer coefficient, k variations with Re. At Re < 134, performance of flow promoter B (rod type oriented perpendicular to flow direction) is nearly similar to an empty channel, while slightly higher LCD values were recorded with flow promoters D (TT oriented perpendicular to flow direction). With $134 \leq Re \leq 267$, higher LCD was recorded with spacer B compared to empty channel. The rod type spacer A (oriented horizontal to flow direction) showed poorer performance (low LCD) than even empty channel in all flow rates measured. Out of all the flow promoters and their arrangements used, mesh and TT

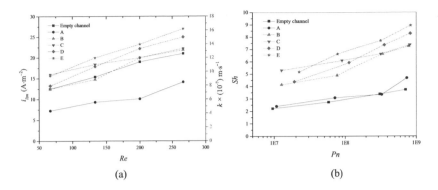

FIGURE 20.6
(a) Plot of i_{lim}, A·m^{-2} and k, m·s^{-1} vs. Re for various flow promoters (A, B, C, D and E) and empty channel; (b) Plot of *Sh* vs. *Pn* for different flow promoter geometries (A–E) and empty channel.

reduced the concentration polarization more effectively than rod type promoters. Twisted tapes, TT, create swirling motion due to longitudinal and transverse vortices in the fluid that move towards boundaries and lower the concentration polarization by mixing various fluid layers (Tadimeti & Chattopadhyay 2016a). Spacers E (double layer spacer or mesh type TT) and D (TT oriented perpendicular to flow direction) showed better performance at *Re* > 134 than other flow promoters and empty channel (Figure 20.6 (a)). While the highest LCD was recorded with spacer C (TT oriented horizontal to flow direction) at *Re* < 134, lower LCD was observed for *Re* ≥ 134. At low *Re* < 134, spacer C showed higher LCD due to a swirling motion created by alignment of twists in the flow direction which reduced diffusion boundary layer thickness and thus enhanced ion transport. At higher velocities (*Re* > 134), the tangential components dominated over the swirling components of C geometry, whereby the reduction in boundary layer thickness is not adequate compared to that caused by longitudinal vortices of geometry E and D (Figure 20.6 (a)).

Figure 20.6 (b) shows a non-linear behavior of *Sh* compared to *Pn*. Spacers B, C, D and E showed higher *Sh* when compared to empty channel, whereas spacer A showed slightly higher *Sh* number than empty channel. At higher flow rates where *Re* > 134, spacer E performed well with highest *Sh* value. Similarly, high *Sh* and higher *Pn* values were recorded with B, C, D and E due to higher pressure drop. Even though mesh type spacer E showed higher LCD, higher pressure drop values rule out this type of promoter as a sustainable solution. At low flow rates (*Re* ≤ 134), spacer C offered a reasonable ion transport (i.e., LCD) at the cost of a small trade off in pressure drop, while at higher *Re*, spacer D performed well with higher LCD and low *Pn* (Figure 20.6 (b)). Overall, the flow promoter C (TT oriented horizontal to flow direction) performed well with reasonable LCD and low pressure drop as the flow was negligibly obstructed by the spacer arrangement to provide swirling motion for increased ion transport.

20.10 Role of Mathematical Modeling in Electrodialysis

Mathematical modeling has significantly helped in development of ED from the stage of an idea to a massive desalination technique. This modeling offers the flexibility of verifying the efficiency of ED process and its feasibility for scale-up from lab/pilot scale to an industrial unit, especially with new feed streams (Rohman et al., 2010; Ortiz et al., 2005). It has thoroughly contributed to improving the design of existing spacers/flow promoters and even to exploring the efficiency of new designs of the ED stack (Tadimeti et al., 2016; Balster et al., 2006; F. Li et al., 2005).

20.11 Challenges and Proposed Remedies in Electrodialysis

Challenges with ED system arises due to the difficulties experienced by ions during their transport through electrolyte solution under a given potential. Other than flow hydrodynamics, physical properties of ions, solution, membrane and solution-membrane interface decide the ion transport mechanism. The following statements appear to be relevant:

1. Flow hydrodynamics influence boundary layer thickness and concentration polarization which is the major resistance to ion transport. Minimization of boundary layer and concentration polarization can be achieved through various means: (i) introducing spacers of different geometries and (ii) membrane surface modification, which is extensively reported in literature (Vasil'eva et al., 2001; Balster, 2006; Tadimeti et al., 2016). Non-conducting spacers cause a shadow effect on ion transport, and they are often replaced by conducting spacers that facilitate ion transport. Tadimetti and Chattopadhyay (2016a) recently established that uninterrupted circular motion of fluid inside the flow channel facilitates ion transport.

2. The nature of electrolytes (strong or weak) and the properties associated with the ions (size, diffusivity, transport number, ion-membrane interactions, etc.) that have liberated from electrolytes (Tadimeti & Chattopadhyay, 2016b) pose challenges to ED separation. A strong electrolyte fully dissociates and increases solution conductivity, while a weak electrolyte dissociates partially, resulting in poor solution conductivity, and ED stack has no control on it. Strong resistance from weakly dissociating organic acids or amino acids drastically reduces efficiency (Ferrer et al., 2006). Separation of large size organic acid molecules using ED is not a feasible option. This problem is overcome by electrodialysis using bipolar membranes a detailed discussion of the same is presented in Section 20.13.

3. Solute-membrane interactions and the resulting surface charge over polar membranes (once equilibrated with the solution) decide membrane resistance. Ions with higher charges get strongly adsorbed over counter-ionic fixed charges through membranes that are then weakly desorbed during ion transport. Thus, membrane resistance increases with a bulky, higher charge-bearing species and its interaction with membranes (Tadimeti & Chattopadhyay, 2016b). Reasonable transport is achieved only with application of higher potential.

20.12 Applications of Electrodialysis

Based on the basic principles of ED and various parameters that might influence the ion transport, we will now take up a few applications of this technology in water purification in the following discussions:

i. Nitrate and hardness removal from water.

ii. Treatment of solutions of heavy metals ions Cu, Zn, Ni, Cr, Fe.

iii. Application of ED to recover organic acids.

iv. Electro-deionization for production of ultrapure water.

i. *Nitrate and hardness removal from water.*

Nitrate is an essential nutrient for human health and mostly comes from the water we drink or the food we consume. Nitrates in water mainly originate from excessive use of fertilizers and manures in agriculture, while hardness appears due to presence of calcium and iron salts. Ground water available through bore wells is also a major source of nitrates and calcium salts, e.g., carbonates and sulfates. However, an excess of either of these salts leads to human health issues. Therefore, their concentration must be lowered below the tolerance limit (Suss et al., 2015). The processes to separate nitrate from drinking water are ion exchange, RO and ED. Ion exchange is an adsorption equilibrium based technique that is extremely slow. The nitrate anion is exchanged with Cl^-, or HCO_3^- ions of the anion exchange resins; the exchange results in water rich in Cl^-/HCO_3^- anions instead of nitrates. In RO, under a pressure above that of the osmotic pressure of the salt solution, the contaminated water is passed through a tight membrane module and almost every ion is removed from water. Therefore, treated water from RO unit is often mixed with a small stream of contaminated water, by a process known as blending. ED is relatively fast and capable of handling large quantity of water (Elyanow and Persechino, 2005).

ii. *Treatment of solutions of heavy metals ions Cu, Zn, Ni, Cr, Fe.*

Most of the water coming out of mining, electroplating and tanning industries consists of heavy metal ions, e.g., Fe, Ni, Sb, Zn and Cu, in a strong H_2SO_4 solution. These streams are very hazardous to aquatic and human health. A common practice is to bleed out a small portion of the stream to keep the concentration unaffected. The concentrated bleed part is then subsequently treated for isolation of Cu through crystallization. A decrease of H_2SO_4 concentration through ED will allow retrieval of useful elements with high efficiency and is a viable alternative to commonly used methods. Experiments with ED showed over 99% of Zn, Mg and Mn retention in feed stream until the acid removal is apprecible. Tests indicated that As and Cu had lower retention rate compared to other metal ions, e.g., Ni, Sb and Fe. With an increase in current density there is no appreciable change in current efficiency (Dubrawski et al., 2014).

iii. *Application of ED to recover organic acids.*

Mixtures of organic acids are obtained from fermentation process in pharmaceutical industries. A series of steps, e.g., pH adjustment, precipitation, re-dissolution, crystallization, drying etc., are employed to recover organic acids of commercial value. Application of ED can drastically reduce the number of intermediate steps to enhance percentage of recovery. Simple ED could successfully recover small acid molecules, e.g., acetic, maleic, citric, oxalic and tartaric, but recovery becomes increasingly difficult with larger size organic molecules because of poor dissociation and stronger adsorption over AEM surface (Zhang et al. 2011). Separation efficiency is dependent on the interaction of the diffusing acid molecules and membrane. Parulekar (1998) analyzed the optimum operating conditions for organic acids recovery with minimum energy input in a batch ED. He reported time-dependent adjustment of applied potential and current through ED cell to be energy efficient. Wang et al. (Wang et al., 2006) reported only AEM to be capable of effectively removing organic acids from waste. They investigated the role of some key parameters, such as voltage, current density, pH, current efficiency and specific energy consumption during ED. Numerous reports (Huang et al., 2007; Wang et al., 2006; Zhang et al., 2011) are available on the recovery of organic acids by ED.

iv. *Electro-deionization to produce ultrapure water.*

Ultrapure water is usually recommended for electronic and pharmaceutical industries. Electrodeionization (EDI) is a unique technique that can provide ultrapure water at very low cost. This is a hybrid separation process involving the ion exchange resins and ion exchange membranes. EDI technique has gained tremendous attention these days because it can deliver ultrapure water as well as recover some precious ionic species from contaminated water. The technique is independent of the nature of the electrolytes that are present in the feed solution (Arar et al., 2014). Usually, weakly-ionized species, such as CO_2 and boron, pose difficulties during separation via RO and ED. EDI is capable of continuous removal of these species to a very high degree of removal. Crucial process parameters that influence the efficiency of an EDI module are current density, flow velocity both in feed and concentrate streams, temperature and TDS of the feed solution. Laktionov et al. (1999) introduced ion exchange textile as conducting spacers in ordinary electrodialysis and experimented on a pilot scale to prepare ultrapure water. The pressure drop was found to be much lower with ion exchange spacers relative to commonly used resin beads. Thus, it becomes cost effective. Strathmann (2010) gave a detailed description of ED technique and the appropriate modifications required to make ED process a continuous operation instead of a batch process. In some cases, the continuous electro-deionization unit contains mixed bed resins in the diluate chamber, while in another set of design, cation and anion exchange resins are placed in separate chambers with a partition of bipolar membrane between them. Detailed description of this technique may be found elsewhere (Strathmann, 2010; Laktionov et al., 1999).

20.13 Bipolar Membrane Electrodialysis (EDBPM)

In this section, we provide an overview of bipolar membrane (BPM), as well as application, advantages, limitations and techniques to improve its performance. This is a different class of polar membrane consisting of two different membranes of opposite polarity placed one over the other after incorporation of a neutral and hydroscopic intermediate layer. With a suitable arrangement of this membrane, along with CEM/AEM, we can simultaneously produce acid and alkali from an electrolyte solution without adding any extra chemical (Figure 20.7). Thus, recently organic/amino acids and alkali have been recovered as two separate streams from direct EDBPM treatment of fermentation broth. The H^+ and OH^- produced at the interface of BPM are transported through anion and cation exchange layers, respectively (Pourcelly, 2002). EDBPM application on Na_2SO_4 solution will produce NaOH and H_2SO_4.

The presence of two oppositely charged polar layers with a hygroscopic interface (~10 nm) increases the membrane's overall resistance to ion transport and renders a need for higher applied potential. Water present in the interface experiences tremendous electric potential due to the narrow gap between two oppositely charged membranes and facilitates ~50 million times higher electro-splitting, relative to the normal electrolysis of water. The imbalance between the rate of H^+ and OH^- release through polar layers and the diffusion of water from bulk to the interface results in osmotic imbalance, bulging of the interface and drop in performance (Krol et al., 1998). Major challenges with EDBPM are (i) energy consumption, (ii) ion selectivity, (iii) stability (thermal/pH) of the membrane and (iv) capital investment.

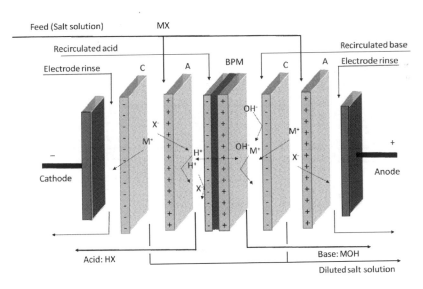

FIGURE 20.7
Cell arrangement for generation of acid and base from salt solution using bipolar membrane. (From Pourcelly, G., *Russian Journal of Electrochemistry*, 38 (8): 1026–1033, 2002.)

Several studies (Yuan et al., 2010; Rajesh et al., 2011; Xue et al., 2006) addressed some of the above challenges to using BPM by three means: (i) by modifying the base polymer used for polar membranes, (ii) by changing the hygroscopic interface layer (Ag doping with gelatin) or (iii) by modification of contact surfaces between the interface layers and polar membranes, e.g., incorporation of micron size resin particles to achieve extra surface and facilitate water splitting. In spite of a series of attempts on membrane modifications, very few of these have seen the light of the day as far as commercialization is concerned. There are only two companies worldwide (FuMA-Tech DE GmbH and Tokuyama Co. Japan) who have been able to commercialize their bipolar membranes (Germandi et al., 2009).

20.14 Bipolar Membrane Preparation and Characterization

BPMs need to experience two contrasting pH conditions on their two surfaces and still maintain specific physical and chemical properties. Therefore, the base polymer needs to be sufficiently stable to accommodate wide pH variation without much deterioration in its performance. Commonly used base polymers are PE, PES and PSF, whereas PVC and PVDF are chemically and mechanically unstable. AEM poses the biggest threat to stability due to slow hydrolysis (at higher pH) of quaternary amine groups that present as charged moieties in this category of membranes. In Table 20.2 lists popular cation and anion exchange layers, ion exchange group and synthesis techniques followed (Germandi et al., 2009).

A series of methods (Bauer et al., 1988) are available for the synthesis of bipolar membranes. In general, synthesis steps involve three main components: (i) synthesis of monopolar layer, (ii) casting of interface layer and (iii) casting of the remaining monopolar layer. Sata (2004) and Strathmann (2004) listed out the principal steps of BPM synthesis:

i. Laminated bipolar membrane: two separate monopolar layers are glued together with an intermediate catalyst layer placed over any of the monopolar layers.

TABLE 20.2

Classification of IEM with Respect to Base Materials and Synthesis Techniques Adopted

Polymer	Ion Exchange Group	Synthesis Technique
Anion Exchange Layer		
Polysulfone	Diamine	Chloro-methylation
Brominated poly (2, 6-dimethyl-1, 4 phenylene oxide)	Quaternary amines	Bromination and quaternary amination instead of chlorination
Blend of PS and PVDF	Diamine	Chloromethylation
PVDF	Tertiary amine	Plasma induced grafting, co-polymerization of 2-methacrylic acid 3-(biscarboxymethylamino)
Cation Exchange Layer		
Sulfonated polysulfone	Sulfonic acid	Dispersed in cation resin powder
Poly(ether ether ketone)	Sulfonic acid	
Blend of sulfonated poly(ether ether ketone) and poly(ether sulfone)	Sulfonic acid	
Poly(vinylidene fluoride)	Sulfonic acid	

Source: Germandi, A., S. Abdu, and E. Van de Ven, *Advanced Bipolar Membrane Processes for Highly Saline Waste Streams,* 2009.

ii. Multilayer bipolar membrane: casting of cation exchange polymer solution over already catalyzed anion exchange membrane layer.

iii. Single film bipolar membrane: two surfaces of a single membrane layer are separately and differently charged in order to achieve adequate polarity of the surfaces.

For large-scale industrial application of the BPM multilayer, solution casting is preferred because the process is reproducible, robust and cost-effective and gives high mechanical strength, selectivity and low electrical resistance. BPM are characterized using voltammetry, chronopotentiometry and impedance measurement.

20.15 Bipolar Membrane Applications

The unique property of this membrane to produce acid, alkali and pure water simultaneously in a once through process has made it economically attractive for a wide range of applications. Since 1977, when commercial production of this membrane started (Franken 2000; Kemperman 2000) a large number of potential applications have been identified. In recent years ED with bipolar membranes has been widely used for production of ultrapure deionized water used in semiconductor (Hernon et al., 2010), pharmaceuticals (Huang et al., 2007; Zhang et al., 2011) and for deep cleaning of organic and inorganic substances in chemical and biochemical industries, production of organic and inorganic acids and bases from respective salts in food industries (Bazinet et al., 1998), removal of salts from whey proteins (Huang & Xu, 2006) and the simultaneous production of acid and alkali (Trivedi et al., 1999).

20.16 Capacitive Deionization

Capacitive deionization is an electrolyte separation technique where no chemical is added to the feed water and potential is applied with the help of highly porous graphite electrodes. The purpose of using graphite electrodes is to obtain maximum surface area such that a large quantity of cations and anions get adsorbed over both cathode and anode. Elimination of electric field will desorb all ions and electrodes get regenerated (Oren, 2008; Suss et al., 2015). This is undoubtedly a low cost electrolyte removal technique, and its effectiveness is purely dependent on the nature and interstitial surfaces of the electrode materials used. A wide range of electrode materials with different interstitial configurations that can enhance the performance of electrolyte removal has been reported in the literature. Modified carbon aerogels have been found to be the best electrode material for their high electrical conductivity, high specific surface area and easily controllable pore size (Landon et al., 2014). An excellent review report by Suss et al. (2015) presents a wide spectrum of capacitive electrode materials. There is a huge scope for construction of electrode materials that can have an interstitial surface of higher magnitude. The performance and stability of this technique are intrinsically bonded to the electrode material, which until recently was exclusively carbon. Following the development of the hybrid CDI (the desalination battery), the strict limitation of carbon materials has ended, with preliminary trials shedding light on the prospect of using heteroatom carbons or carbon hybrid materials. This field is widely unexplored and the community will only be able to benefit from novel and possibly exotic electrode materials if more research is done. This CDI technique is very competitive as long as the electrolyte concentration remains below 3000 mg/L (Landon et al., 2014). Application of this technology will produce low cost water, free from all soluble electrolytes.

20.17 Energy Generation through Reverse Electrodialysis

Energy generation by utilization of chemical potential of two streams of high and low concentration electrolytes is known as reverse electrodialysis (RED). RED is one of the most promising methods to extract energy by separating seawater and river water using a polar membrane. The method is certainly sustainable, green and clean. Theoretically, the energy that can be generated by mixing 1 m³ of river water is 2.5 MJ when mixed with a large quantity of seawater (Veerman et al., 2009). Assuming an ideal process, there is no overall energy effect and the process is balanced by cooling the effluent by 0.2° or so. There are various techniques of recovering energy from seawater: (i) RED, (ii) pressure retarded osmosis, (iii) utilization of vapor pressure difference, (iv) mechanical methods and (v) utilization of cryoscopic (freezing point) techniques. Out of these techniques, RED is the most promising if operational challenges can be overcome.

The cell used for generation of energy using RED is very similar to an ordinary ED apparatus (Figure 20.8). Usually, iron electrodes are chosen while cation and anion exchange

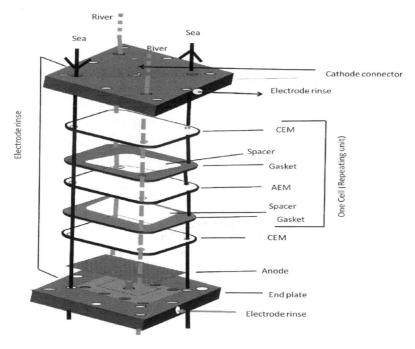

FIGURE 20.8
A reverse electrodialysis stack with single cell. (Adapted from Veerman, J., M. Saakes, S.J. Metz, and G.J. Harmsen, *Journal of Membrane Science*, 327 (1–2): 136–144, doi:10.1016/j.memsci.2008.11.015, 2009.)

membranes are placed alternately. The technological challenge is to scale-up the apparatus from 100 mW to 100 MW. Major process challenges experienced by RED are the same as in ED, and only economic feasibility needs to be explored. The list below shows some of the limitations needing serious attention:

i. Unwanted electrode reactions consuming energy.
ii. Leakage current interrupting the effectiveness in ion transport.
iii. Minimization of membrane fouling.
iv. Reduction of total cell resistance to facilitate ion transport.
v. Modification of flow spacer geometries to improve flow hydrodynamics and minimization of concentration polarization.
vi. Minimization of pressure drop and cost of pumping.

Veerman et al. (2009) carried out multistage experiments using a RED cell containing 50 cell pairs (10 cm × 10 cm) with synthetic river water (1g NaCl/L) and synthetic seawater (30g NaCl/L). Experiments were performed at maximal power condition (current density of about 30A/m^2 and flow rate of 300 mL/min). Under these conditions, the theoretical energy efficiency was maximum at 50% but the actual energy efficiency of a single stack was only 9%. Effluent streams were introduced into another stack and so on. With 3 stages of RED operation, the overall efficiency only reached a meager 18%. Incorporation of a fourth stage showed no improvement in efficiency. Power density of the 3rd stage was found to be 50% of the value noted in the first stage.

Very recently W. Li et al. (2013) have tried to incorporate RED technology in a hybrid process scheme to desalinate water by combining it with a RO unit. RED will harvest energy out of the concentration difference between seawater and brackish water while the RO assembly will use it to desalinate partially into dilute seawater. The treated seawater from the RED unit will have lower salinity and act as feed to the RO unit and minimize pump work. The concentrated RO brine provides the RED unit with a higher solute concentration. The authors explored different configurations of the hybrid RED-RO process and concluded that an appropriate arrangement of this hybrid process can substantially lower specific energy consumption and provide a better control on discharge brine concentration, compared to conventional seawater desalination by RO (W. Li et al., 2013). Cell design (spacer geometry, electrode surface, flow hydrodynamics, etc.) appears to be the most crucial parameter explaining the below par performance of this technology. RED technique needs tremendous improvement to make it worthy of industrial application. To date, very limited literature is available on RED technique.

20.18 Conclusions

In this chapter, we discussed the reason to choosing electrodialysis as a route to water purification. Based on different process stream conditions, electrodialysis has been diversified into other electro-deionization techniques, such as desalination, bipolar membrane electrodialysis and capacitive deionization. Common terminologies and fundamental concepts (*I-V* characteristics, Ohmic-Plateau-over limiting regions, concentration polarization, limiting current density and its determination) that are essential for understanding ion transport phenomena in electrodialysis were discussed. Factors affecting ion transport and probable remedies to overcome those challenges were addressed. Four important applications of electrodialysis, (i) nitrate and hardness removal from ground water, (ii) heavy metal ions removal from mining/tannery wastewater, (iii) recovery of organic acids from fermentation broths and (iv) electrodeionization for generation of ultrapure water for pharmaceutical industries, were discussed to show the wide range of its applicability.

Problems experienced with separation of organic acid salts were effectively dealt with bipolar membrane electrodialysis. We also presented a short discussion on bipolar membranes, their synthesis routes, the common base materials used and choice of application. Low cost deionization of wastewater using capacitive deionization was highlighted with a brief discussion of electrode types and porosity that influence its effectiveness.

Finally, we took up the application of an electrodialysis cell into an energy generation device via reverse electrodialysis. Although practical application of this technology is still forthcoming, we discussed its theoretical feasibility along with directions for further research in this context.

References

Arar, Ö., Ü. Yüksel, N. Kabay, and M. Yüksel. 2014. Various applications of electrodeionization (edi) method for water treatment—A short review. *Desalination* 342: 16–22.

Balster, J. 2006. "Membrane Module and Process Development for Monopolar and Bipolar Membrane Electrodialysis." University of Twente, The Netherlands.

Balster, J., I. Punt, D. Stamatialis, and M. Wessling. 2006. Multi-layer spacer geometries with improved mass transport. *Journal of Membrane Science* 282 (1–2): 351–361. doi:10.1016/j.memsci.2006.05.039.

Balster, J., M.H. Yildirim, D.F. Stamatialis, R. Ibanez, R.G.H. Lammertink, V. Jordan, and M. Wessling. 2007. Morphology and microtopology of cation-exchange polymers and the origin of the overlimiting current. *The Journal of Physical Chemistry. B* 111 (9): 2152–2165. doi:10.1021/jp068474t.

Bauer, B., F.J. Gerner, and H. Strathmann. 1988. Development of bipolar membranes. *Desalination* 68: 279–292.

Bazinet, L., F. Lamarche, D. Ippersiel, and St Hyacinthe Que. 1998. Bipolar-membrane electrodialysis: Applications of electrodialysis in the food industry. *Trends in Food Science & Technology* 9: 107–113.

Chandra, A., J. Ganesh Dattatreya Tadimeti, and S. Chattopadhyay. 2017. Transport hindrances with electrodialytic recovery of citric acid from solution of strong electrolytes. *Chinese Journal of Chemical Engineering.* doi:10.1016/j.cjche.2017.05.010.

Dubrawski, M. M. Czaplicka, and J. Mrozowski. 2014. The application of electrodialysis to the treatment of industrial copper and zinc electrolytes. *Desalination and Water Treatment* 55 (May 2015). Taylor & Francis: 1–12. doi:10.1080/19443994.2014.913995.

Elyanow, D. and J. Persechino. 2005. Advances in nitrate removal. https://www.suezwatertechnologies.com/kcpguest/documents/Technical Papers_Cust/Americas/English/TP1033EN.pdf.

Ferrer, J.S. Jaime, S. Laborie, G. Durand, and M. Rakib. 2006. Formic acid regeneration by electromembrane processes. *Journal of Membrane Science* 280 (1–2): 509–516. doi:10.1016/j.memsci.2006.02.012.

Fimbres-Weihs, G.A., and D.E. Wiley. 2010. Review of 3D CFD modeling of flow and mass transfer in narrow spacer-filled channels in membrane modules. *Chemical Engineering and Processing: Process Intensification* 49 (7). Elsevier B.V.: 759–781. doi:10.1016/j.cep.2010.01.007.

Franken, T. 2000. Bipolar membrane technology and its applications. *Membrane Technology*, no. 125: 8–11. doi:10.1016/S0958-2118(00)80212-8.

Geraldes, V. and M. Diná Afonso. 2010. Limiting current density in the electrodialysis of multi-ionic solutions. *Journal of Membrane Science* 360 (1–2): 499–508. doi:10.1016/j.memsci.2010.05.054.

Germandi, A., S. Abdu, and E. Van de Ven. 2009. *Advanced Bipolar Membrane Processes for Highly Saline Waste Streams.*

Hernon, B.P., L. Zhang, L.R. Siwak, and E.J. Schoepke. 2010. "Progress Report: Application of Electro-Deionization in Ultrapure Water Production."

Hosseini, S.M., S.S. Madaeni, A.R. Heidari, and A. Amirimehr. 2012. Preparation and characterization of ion-selective polyvinyl chloride based heterogeneous cation exchange membrane modified by magnetic iron–nickel oxide nanoparticles. *Desalination* 284 (January). Elsevier B.V.: 191–199. doi:10.1016/j.desal.2011.08.057.

Huang, C. and T. Xu. 2006. Electrodialysis with bipolar membranes for sustainable development. *Environmental Science & Technology* 40 (17): 5233–5243.

Huang, C., T. Xu, Y. Zhang, Y. Xue, and G. Chen. 2007. Application of electrodialysis to the production of organic acids: State-of-the-art and recent developments. *Journal of Membrane Science* 288: 1–12. doi:10.1016/j.memsci.2006.11.026.

Kabay, N., O. Ipek, H. Kahveci, and M. Yüksel. 2006. Effect of salt combination on separation of monovalent and divalent salts by electrodialysis. *Desalination* 198(1–3): 84–91. doi:10.1016/j.desal.2006.09.013.

Kannan, R., H.N. Kagalwala, H.D. Chaudhari, U.K. Kharul, S. Kurungot, and V.K. Pillai. 2011. Improved performance of phosphonated carbon nanotube–polybenzimidazole composite membranes in proton exchange membrane fuel cells. *Journal of Materials Chemistry* 21 (20): 7223–7231. doi:10.1039/c0jm04265j.

Kemperman, A.J.B. 2000. *Handbook on Bipolar Membrane Technology.* Twente University Press. https://research.utwente.nl/en/publications/handbook-bipolar-membrane-technology.

Kim, J., and B. Van Der Bruggen. 2010. The use of nanoparticles in polymeric and ceramic membrane structures: Review of manufacturing procedures and performance improvement for water treatment. *Environmental Pollution* 158 (7). Elsevier Ltd: 2335–2349. doi:10.1016/j.envpol.2010.03.024.

Klaysom, C., R. Marschall, L. Wang, P. Ladewig, and G.Q. Max Lu. 2010. Synthesis of composite ion-exchange membranes and their electrochemical properties for desalination applications. *Journal of Materials Chemistry* 20: 4669–4674. doi:10.1039/b925357b.

Krol, J.J., M. Jansink, M. Wessling, and H. Strathmann. 1998. Behaviour of bipolar membranes at high current density water diffusion limitation, *Separation and Purification Technology* 14: 41–52.

Krol, J.J. 1997. "Monopolar and Bipolar Ion Exchange Membranes Mass Transport Limitations." University of Twente, The Netherlands.

Kumar, M., B.P. Tripathi, and V.K. Shahi. 2009. Ionic transport phenomenon across sol-gel derived organic-inorganic composite mono-valent cation selective membranes. *Journal of Membrane Science* 340 (1–2): 52–61.

Laktionov, E., E. Dejean, J. Sandeaux, R. Sandeaux, C. Gavach, and G. Pourcelly. 1999. Production of high resistivity water by electrodialysis. Influence of ion-exchange textiles as conducting spacers. *Separation Science and Technology* 34 (1): 69–84. doi:10.1081/SS-100100637.

Landon, J., X. Gao, A. Omosebi, and K. Liu. 2014. "Integrated Capacitive Deionization for Reduced Cost of Wastewater Treatment." *2014 Eastern Unconventional Oil and Gas Symposium - November 5–7, 2014 in Lexington, Kentucky, USA*, 1–27.

Lee, H.-J., H. Strathmann, and S.-H. Moon. 2006. Determination of the limiting current density in electrodialysis desalination as an empirical function of linear velocity. *Desalination* 190 (1–3): 43–50. doi:10.1016/j.desal.2005.08.004.

Li, F., W. Meindersma, A.B. de Haan, and T. Reith. 2002. Optimization of commercial net spacers in spiral wound membrane modules. *Journal of Membrane Science* 208 (1–2): 289–302. doi:10.1016/S0376-7388(02)00307-1.

Li, F., W. Meindersma, A.B. de Haan, and T. Reith. 2005. Novel spacers for mass transfer enhancement in membrane separations. *Journal of Membrane Science* 253 (1–2): 1–12. doi:10.1016/j.memsci.2004.12.019.

Li, Q., C. Huang, and T. Xu. 2009. Bipolar membrane electrodialysis in an organic medium: Production of methyl methoxyacetate. *Journal of Membrane Science* 339 (1–2): 28–32. doi:10.1016/j.memsci.2009.04.026.

Li, W., W.B. Krantz, E.R. Cornelissen, J.W. Post, A.R.D. Verliefde, and C.Y. Tang. 2013. A novel hybrid process of reverse electrodialysis and reverse osmosis for low energy seawater desalination and brine management. *Applied Energy* 104: 592–602. doi:10.1016/j.apenergy.2012.11.064.

Mishchuk, N.A. 1998. Perspectives of the electrodialysis intensification. *Desalination* 117: 283–295. doi:10.1016/S0011-9164(98)00120-9.

Mishra, A.K. and P.K. Bhattacharya. 1984. Alkaline black liquour treatment by batch electrodialysis. *Can. J. Chem. Eng.* 62: 723–727. doi:10.1002/cjce.5450620520.

Mishra, A.K. and P.K. Bhattacharya. 1987. Alkaline black liquour treatment by continuous electrodialysis. *Journal of Membrane Science* 33: 83–95. doi:10.1016/S0376-7388(00)80053-8.

Ng, L.Y., A.W. Mohammad, C.P. Leo, and N. Hilal. 2013. Polymeric membranes incorporated with metal/metal oxide nanoparticles: A comprehensive review. *Desalination* 308. Elsevier B.V.: 15–33. doi:10.1016/j.desal.2010.11.033.

Oren, Y. 2008. Capacitive deionization (cdi) for desalination and water treatment—Past, present and future (a review). *Desalination* 228 (1–3): 10–29. doi:10.1016/j.desal.2007.08.005.

Ortiz, J.M., J.A. Sotoca, E. Expósito, F. Gallud, V. García-García, V. Montiel, and A. Aldaz. 2005. Brackish water desalination by electrodialysis: Batch recirculation operation modeling. *Journal of Membrane Science* 252 (1–2): 65–75. doi:10.1016/j.memsci.2004.11.021.

Parulekar, S.J. 1998. Optimal current and voltage trajectories for minimum energy consumption in batch electrodialysis. *Journal of Membrane Science* 148: 91–103. doi:10.1016/S0376-7388(98)00148-3.

Pourcelly, G. 2002. Electrodialysis with bipolar membranes: Principles, optimization, and applications. *Russian Journal of Electrochemistry* 38 (8): 919–926. doi:10.1023/A:1016882216287.

Rajesh, A.M., M. Kumar, and V.K. Shahi. 2011. Functionalized biopolymer based bipolar membrane with poly ethylene glycol interfacial layer for improved water splitting. *Journal of Membrane Science* 372 (1–2). Elsevier B.V.: 249–257. doi:10.1016/j.memsci.2011.02.009.

Rohman, F.S., M.R. Othman, and N. Aziz. 2010. Modeling of batch electrodialysis for hydrochloric acid recovery. *Chemical Engineering Journal* 162 (2). Elsevier B.V.: 466–479. doi:10.1016/j.cej.2010.05.030.

Sata, T. 2004. Ion Exchange Membranes: Preparation, Characterization, Modification and Application. *Royal Society of Chemistry* doi:10.1039/9781847551177.

Shaposhnik, V.A., O.V. Grigorchuk, E.N. Korzhov, V.I. Vasil'eva, and V. YaKlimov. 1998. The effect of ion-conducting spacers on mass transfer—Numerical analysis and concentration field visualization by means of laser interferometry. *Journal of Membrane Science* 139: 85–96. doi:10.1016/S0376-7388(97)00247-0.

Strathmann, H. 2004. *Ion-Exchange Membrane Separation Processes. Membrane Science and Technology Series, 9.*

Strathmann, H. 2010. Electrodialysis, a mature technology with a multitude of new applications. *Desalination* 264 (3). Elsevier B.V.: 268–288. doi:10.1016/j.desal.2010.04.069.

Suss, M.E., S. Porada, X. Sun, P.M. Biesheuvel, J. Yoon, and V. Presser. 2015. Water desalination via capacitive deionization: What is it and what can we expect from it? *Energy Environ Sci* 8 (8). Royal Society of Chemistry: 2296–2319. doi:10.1039/C5EE00519A.

Tadimeti, J.G.D. and S. Chattopadhyay. 2016a. Uninterrupted swirling motion facilitating ion transport in electrodialysis. *Desalination* 392: 54–62. doi:10.1016/j.desal.2016.04.007.

Tadimeti, J.G.D. and S. Chattopadhyay. 2016b. Physico-chemical local equilibrium influencing cation transport in electrodialysis of multi-ionic solutions. *Desalination* 385: 93–105. doi:10.1016/j.desal.2016.02.016.

Tadimeti, J.G.D., V. Kurian, A. Chandra, and S. Chattopadhyay. 2016. Corrugated membrane surfaces for effective ion transport in electrodialysis. *Journal of Membrane Science* 499. Elsevier: 418–428. doi:10.1016/j.memsci.2015.11.001.

Tanaka, Y. 2002. Current density distribution, limiting current density and saturation current density in an ion-exchange membrane electrodialyzer. *Journal of Membrane Science* 210: 65–75.

Trivedi, G.S., B.G. Shah, S.K. Adhikary, and R. Rangarajan. 1999. Studies on bipolar membranes part III: Conversion of sodium phosphate to phosphoric acid and sodium hydroxide. *Reactive and Functional Polymers* 39: 91–97. doi:10.1016/S1381-5148(97)00159-4.

Vadthya, P., A. Kumari, C. Sumana, and S. Sridhar. 2015. Electrodialysis aided desalination of crude glycerol in the production of biodiesel from oil feed stock. *Desalination* 362: 133–140. doi:10.1016/j.desal.2015.02.001.

Vasil'eva, V.I., V.A. Shaposhnik, and O.V. Grigorchuk. 2001. Local mass transfer during electrodialysis with ion-exchange membranes and spacers. *Russian Journal of Electrochemistry* 37 (11): 1164–1171.

Veerman, J., M. Saakes, S.J. Metz, and G.J. Harmsen. 2009. Reverse electrodialysis: Performance of a stack with 50 cells on the mixing of sea and river water. *Journal of Membrane Science* 327 (1–2): 136–144. doi:10.1016/j.memsci.2008.11.015.

Wang, Z., Y. Luo, and P. Yu. 2006. Recovery of organic acids from waste salt solutions derived from the manufacture of cyclohexanone by electrodialysis. *Journal of Membrane Science* 280 (1–2): 134–137. doi:10.1016/j.memsci.2006.01.015.

Xu, T. 2005. Ion exchange membranes: State of their development and perspective. *Journal of Membrane Science* 263 (1–2): 1–29. doi:10.1016/j.memsci.2005.05.002.

Xue, Y.H., R.Q. Fu, Y.X. Fu, and T.W. Xu. 2006. Fundamental studies on the intermediate layer of a bipolar membrane. v. Effect of silver halide and its dope in gelatin on water dissociation at the interface of a bipolar membrane. *Journal of Colloid and Interface Science* 298 (1): 313–320. doi:10.1016/j.jcis.2005.11.049.

Yuan, X.-Z., C. Song, H. Wang, and J. Zhang. 2010. *Electrochemical Impedance Spectroscopy in PEM Fuel Cells.* Springer. doi:10.1007/978-1-84882-846-9.

Zhang, K., M. Wang, and C. Gao. 2011. Tartaric acid production by ion exchange resin-filling electrometathesis and its process economics. *Journal of Membrane Science* 366 (1–2). Elsevier B.V.: 266–271. doi:10.1016/j.memsci.2010.10.013.

21

Adsorption-Membrane Filtration Hybrid Process in Wastewater Treatment

Kulbhushan Samal, Chandan Das, and Kaustubha Mohanty

CONTENTS

21.1 Introduction

Clean water is a primary need of all living beings, but access to clean water for millions of people worldwide has dramatically become restricted and has induced water scarcity in several regions. Increased industrial and other human activities due to rapid population growth generate a huge amount of wastewater containing a variety of pollutants that are hazardous to our environment. This has imposed an adverse effect on living things, with widespread illness in developing countries occurring mainly due to consumption of unsafe water. The toxic pollutant in a water body can also affect aquatic life. Therefore, cost-effective treatment of wastewater has been a major environmental challenge worldwide.

As a result, significant efforts have been made and continuous research is underway to develop cost-effective wastewater treatment techniques and strategies. Various wastewater treatment methods, such as coagulation and flocculation, chemical precipitation, sedimentation, chemical oxidation, ozonation, ion exchange and electrochemical treatment have been examined for the removal of various pollutants from wastewater (Bhatnagar & Sillanpaa, 2010; Samal et al., 2017). These processes have limited applications due to their limitations, which include causing secondary pollution, incomplete removal of pollutants, longer time duration, the requirement of expensive equipment and chemicals. Also, the complex nature of wastewater makes it difficult to achieve complete removal of pollutants by using a single separation process. Among all available methods, membrane separation process has emerged as the most efficient wastewater treatment technique. Membrane process has considerable advantages due to high throughput, feasibility of scale-up, easier and less energy intensive operation, lower footprint and convenience of inline integration. However, membrane fouling due to scale deposition and pore blocking by pollutants is a major concern in these processes. Fouling of the membrane lowers the permeate flux and demands increased energy requirement, which leads to high operating costs.

In order to overcome the limitations of the individual conventional process, membrane filtration process was coupled with conventional methods in many studies. Membrane hybrid process utilizes the advantage of the individual processes and defeats their flaws. The hybrid processes have shown advantages such as higher removal efficiency, high quantity of treated wastewater, low energy consumption, higher membrane life and lower backwashing time. Among the conventional process, adsorption technique is one of the widely used wastewater treatment processes due to its high separation efficiency and low operating cost. Membrane processes coupled with adsorption process are used in a number of studies for removal of various pollutants since it promises greater economical and efficiency. Different types of adsorbents, such as activated carbon or granular carbon, polymeric resins, mineral oxides and biomass-derived adsorbent were tested as adsorbents for removal of various pollutants (Samal et al., 2016; Song et al., 2009; Ipek et al., 2012). The pollutant removal efficiency of membrane filtration–adsorption hybrid system depends on various operating parameters, such as adsorbent dose, uptake capacity, critical flux, wastewater characteristics, membrane configuration and operational modes.

21.2 Adsorption Process

Adsorption is the accumulation of molecules or substances (adsorbate) on a solid surface (adsorbent). The adsorbents have various functional groups present on it as active sites, and the pollutants have an affinity toward the active sites such that pollutants are attracted to the surface of adsorbent or pore walls and bind. The adsorption process has been categorized into two categories, physical adsorption and chemisorption, depending on the force of attraction between adsorbent and adsorbate. The adsorption process in which the force of attraction is due to weak Van der Waals forces is called physisorption or physical adsorption (Figure 21.1). If the force of attraction is very strong due to chemical bonding between adsorbent and adsorbate molecule, it is termed as chemisorption (Figure 21.1) (Dabrowski, 2001).

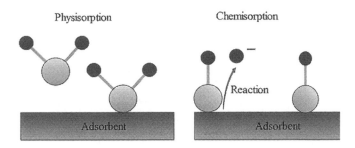

FIGURE 21.1
Schematic diagram of on physical adsorption (physisorption) and chemical adsorption (chemisorption).

Uptake and percentage sorption: The uptake and % sorption can be calculated according to Equations 21.1 and 21.2, respectively:

$$q = \frac{(C_o - C_e)V}{W} \tag{21.1}$$

$$\text{Sorption}\% = \frac{(C_o - C_e)}{C_o} \times 100 \tag{21.2}$$

Where q is equilibrium uptake capacity in mg g^{-1}, C_o is initial adsorbate concentration in mg L^{-1}, C_e is equilibrium adsorbate concentration in mg L^{-1}, V is the volume of the solution in L and W is the mass of adsorbent in g. Adsorption process has been tested as one of the efficient water treatment techniques that uses activated carbon for the removal of a variety of pollutants from water. The main drawback of using activated carbon is its high cost and requirement of a regeneration step. A lot of work has been done to develop low-cost adsorbents utilizing waste biomass. It is interesting to use waste materials as low-cost adsorbents because it also contributes to safe disposal of waste and environmental protection. A wide range of low-cost adsorbents are discussed in literature along with their adsorption capacities for different pollutants. The low-cost adsorbents derived from bio-waste have shown some promising results for the removal of various pollutants from the aqueous system (Bhatnagar & Sillanpaa, 2010).

The success of adsorption process depends considerably on the adsorbent used in the process. Therefore, selection of the adsorbent is one of the key issues in making the process economical and efficient. Identification of an adequate adsorbent to attain maximum removal of a specific type of pollutant depends upon the characteristics of the adsorbent as well as the adsorbate. The physiochemical characteristics of an adsorbent such as BET surface area, porosity, pore size, particle size, bulk density, moisture content and point of zero charge plays an important role in uptake capacity of the adsorbent. The physicochemical properties of different types of adsorbent reported in the literature are presented in Table 21.1. The efficiency of the adsorption process depends not only on the physicochemical properties of the adsorbent, but also on various operating parameters of the process, such as adsorbent dose, initial concentration of adsorbate, contact time, pH, temperature, ionic strength, speed of rotation and the existence of competing compounds in solution.

TABLE 21.1

Different Types of Adsorbents and Their Physicochemical Characteristics

Adsorbent	Average Particle Size (μm)	Average Pore Size (nm)	Pore Volume (mL g⁻¹)	Surface Area (m²g⁻¹)	Reference
Activated carbon	4–5	60	40-85	1100	Kim et al. (2009)
Silica gel	5	6	–	640–800	Backhaus et al. (2001)
Polymeric resins	535 ± 85	0.15–80	1–1.1	700–900	Ipek et al. (2012)
Biomass-derived activated carbon	150–200	–	0.542	712	Mohanty et al. (2006)
Biosorbent (*P. Hysterophorus L.*)	< 125	–	0.0105	20.79	Samal et al. (2017)

The following points must be considered while selecting an adsorbent for any adsorption process:

1. The uptake of pollutants by the adsorbent should be high. To achieve a high uptake value of a pollutant, the surface modification of adsorbent should be optimized along with process conditions.

2. Adsorption kinetics and isotherms of the process must be examined to identify the correct sorption mechanism, which is important in design and simulation of the process.

3. The recovery or disposal of the adsorbate, as well as adsorbent, is an important concern of the process. Therefore, the reusability of adsorbent after regeneration should be economically feasible, or its disposal must be environmentally safe.

4. The efficiency of the adsorbent to remove multicomponent pollutants should be satisfactory.

5. Last but not the least, the economy of the process should not be ignored. Low cost of adsorbent preparation and higher removal efficiency of adsorbents are desirable.

21.2.1 Adsorption Kinetics

The kinetics of an adsorption process was mainly tested using pseudo-first-order model (Lagergren, 1898), pseudo-second-order model (Ho & McKay, 2000) and Weber-Morris model (Weber & Morris, 1963). The pseudo-first-order model is a one-parameter model, expressed as Equation 21.3:

$$\log\left(q_e - q\right) = \log q_e - \left(\frac{K'}{2.303}\right) \times t \tag{21.3}$$

Where q_e and q are the mass of adsorbate uptake (mg g⁻¹) at equilibrium, respectively, t is time (min) and K' is the first-order reaction rate constant of adsorption (min⁻¹). The pseudo-second-order model is a two-parameter model, which is expressed as Equation 21.4:

$$\frac{t}{q} = \left(\frac{1}{K''q_e^2}\right) + \frac{t}{q_e} \tag{21.4}$$

Where q_e and q are the mass adsorbed (mg g^{-1}) at equilibrium and time t (min) and K'' is the pseudo-second-order rate constant of adsorption (mg g^{-1} min^{-1}).

The intra-particle diffusion can be determined by using the equation described by Weber-Morris, expressed as Equation 21.5:

$$K_p = \frac{q}{t^{0.5}} \tag{21.5}$$

Where q (mg g^{-1}) is the amount adsorbed at time (t) (min) and K_p is the intra-particle rate constant (mg g^{-1} min$^{-0.5}$).

21.2.2 Adsorption Isotherms

Various isotherms models such as Langmuir isotherm model (Langmuir, 1916), Freundlich isotherm model (Freundlich, 1906) and Dubinin-Radushkevich (D-R) isotherm model (Hobson, 1969) were widely employed to understand the adsorption equilibrium behavior of the system and to model the equilibrium sorption system. The Langmuir model is the simplest model capable of representing such behavior with some assumptions: adsorption occurs on a homogeneous surface by monolayer adsorption, there is no interaction between adsorbed ions and the energy is same for all active sites of adsorption (Langmuir, 1916). The nonlinear form of Langmuir equation is represented as Equation 21.6:

$$q_e = \frac{q_m K_L C_e}{1 + K_L C_e} \tag{21.6}$$

Where q_e is the equilibrium adsorbate concentration on the adsorbent (mg g^{-1}), q_m is monolayer adsorption capacity of the adsorbent (mg g^{-1}), C_e is the equilibrium adsorbate concentration in the solution (mg L^{-1}) and K_L is the Langmuir adsorption constant (L mg^{-1}). Separation factor R_L is a dimensionless constant of Langmuir isotherm that indicates the feasibility of the adsorption process (Ertugay & Bayhan, 2008). Separation factor R_L can be explained by Equation 21.6:

$$R_L = \frac{1}{1 + K_L C_o} \tag{21.6}$$

Where C_o (mg L^{-1}) is the initial concentration of pollutant. The value of separation factor R_L gives an idea about the feasibility of adsorption process whether adsorption is favorable, unfavorable, linear or irreversible. The magnitude of R_L is a positive number that determines the feasibility of adsorption process.

The Freundlich isotherm model is empirical in nature and assumes that adsorption is taking place on a heterogeneous surface by multilayer adsorption. Furthermore, this model assumes that stronger binding sites are occupied first and that the binding strength decreases with increasing degree of site occupation. It also assumes that the amount of adsorbate uptake increases infinitely with an increase in concentration (Freundlich, 1906). The nonlinear relation of Freundlich isotherm model can be expressed as Equation 21.7:

$$q_e = K_F C_e^{1/n} \tag{21.7}$$

Where q_e is the equilibrium adsorbate concentration on the adsorbent (mg g^{-1}), C_e is the equilibrium adsorbate concentration in the solution (mg L^{-1}), K_F is Freundlich model constant related to the adsorption capacity and $1/n$ is Freundlich model constant related to the intensity of adsorption.

The adsorption on microporous materials has been also described by the Dubinin-Radushkevich (DR) model equation. The DR equation is semi-empirical in nature and is based on the assumptions of a change in the potential energy between the bulk and adsorbed phases and a characteristic energy of adsorbent surface. This model equation gives an idea of macroscopic behavior of adsorbent loading. The Dubinin-Radushkevich model (Hobson, 1969) is represented by Equation 21.8:

$$q_e = q_m \exp\left(-Be^2\right) \tag{21.8}$$

$$e = RT \ln\left(1 + \frac{1}{C_e}\right) \tag{21.9}$$

$$E = \frac{1}{\sqrt{2B}} \tag{21.10}$$

Where q_m is the theoretical saturation uptake capacity (mg g^{-1}), B is a constant related to adsorption energy (mol^2 kJ^{-2}), e is the Polanyi potential, which can be calculated by using Equation 21.9, R is the gas constant (kJ mol^{-1} K^{-1}) and T is the temperature (K). The constant B is related to the energy of adsorption (E, kJ mol^{-1}) per molecule of adsorbate and can be calculated by using Equation 21.10. The value of mean free energy of adsorption gives an idea about adsorption mechanism, whether the adsorption that occurred is physical or chemical (Srividya & Mohanty, 2009). The value of E in the range 8 to 16 kJ mol^{-1} signifies that the adsorption process takes place chemically and the value of E < 8 kJ mol^{-1} signifies that adsorption takes place physically.

21.3 Membrane Process

Membrane process is a rate governed separation process, gaining importance in several purification processes as compared to equilibrium governed distillation, adsorption, etc. The membrane is defined as a selective barrier between two phases that selectively allows some species to pass through it and restricts remaining chemical species of the mixture from passing through one phase to another. The driving forces for separation can be a pressure gradient, concentration gradient or electrochemical potential gradient. Membrane processes have been categorized depending on their pore size of the membrane into Microfiltration (> 1000 Å), Ultrafiltration (20–1000 Å), Nanofiltration (5–20 Å) and Reverse Osmosis (2–10 Å) (Dutta, 2009). The membrane process can be operated in both batch and continuous mode. Various modules have been developed for membrane filtration such as dead-end filtration, cross-flow filtration, spiral wound membrane filtration, hollow fiber membrane, rotating disc and spinning basket module. Concentration polarization and irreversible membrane fouling, which cause a decline in flux, are the main drawbacks of the process (Sen et al., 2010). The most practiced way to reduce concentration polarization is to create turbulence over membrane surface through stirring the feed solution or to allow the feed to flow tangentially over membrane surface (Venkataganesh et al., 2012). The other possible technique to reduce membrane fouling is a coupling of a conventional process with the membrane process. Permeate flux (J) of a membrane process is determined by using Equation 21.11:

$$J = \frac{v}{At} \tag{21.11}$$

Where v is permeate volume in m^3, t is time in sec, A is membrane area in m^2.

In order to achieve high rejection of solute, the operating conditions of membrane process need to be optimized. The most influencing parameters of membrane process are molecular weight cut-off of the membrane (MWCO), trans-membrane pressure drop, solute concentration in the feed, cross-flow velocity in tubular, spiral wound and hollow fiber modules besides the rotation speed in rotating disc and spinning basket modules. MWCO is defined as the solute size, which is at least 90% rejected by the membrane. The MWCO of the membrane must be selected in a manner that ensures maximum retention of the solute as well as high permeate flux. Trans-membrane pressure is an important parameter since it is directly related to the energy requirement of the process. Increased trans-membrane pressure allows higher permeate flux due to the enhancement in the driving force. The optimum trans-membrane pressure must be determined in such a way that it provides for high permeate flux and solute rejection, low energy consumption and long membrane life. Higher permeate flux value can be achieved with cross-flow velocity in the turbulent regime as compared to laminar regime due to decreasing concentration polarization (Das et al., 2006). Rotation of membrane plate is allowed in the spinning basket module and the rotating disk module to minimize concentration polarization. The module must be operated at an optimum rotational speed to achieve efficient and economical separation.

21.4 Adsorption-Membrane Hybrid Process

In many studies, adsorption process was coupled with membrane filtration to improve the performance of the overall process. Adsorption process has been coupled with different membrane filtration processes in various schemes, such as adsorption in submersed membrane filtration, adsorption followed by membrane filtration and membrane filtration followed by adsorption. The efficiency and economy of the hybrid process depend on individual processes as well as the hybridization scheme.

21.4.1 Adsorption-Membrane Filtration Hybridization Scheme

21.4.1.1 First Scheme: Adsorption-Submersed Membrane Filtration

Hollow fiber membranes have been frequently applied in submersed membrane filtration process especially, known as submersed membrane reactor (SMR). U-shaped membrane modules are generally used in SMR type of systems. The schematic diagram of a submersed membrane reactor is represented as Figure 21.2, which was used by Lee et al. (2009) for

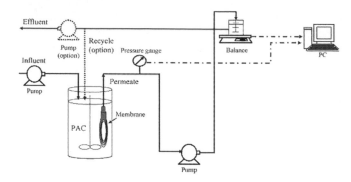

FIGURE 21.2
Schematic diagram of a submersed adsorption-membrane filtration hybrid process. (From Lee, S., J.W. Lee, S. Kim, P.K. Park, J.H. Kim, and C.H. Lee, *Journal of Membrane Science*, 326(1): 84–91, doi:10.1016/j.memsci.2008.09.031, 2009.)

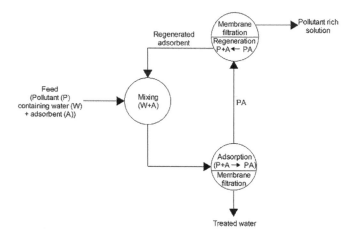

FIGURE 21.3
Flow diagram of the adsorption-ultrafiltration hybrid system. (From Ipek, I.Y., N. Kabay, M. Yüksel, D. Yapici, and Ü. Yüksel, *Desalination*, 306: 24–28, doi:10.1016/j.desal.2012.08.033, 2012.)

removal of 17β-estradiol by a powdered activated carbon (PAC) based adsorption–microfiltration hybrid process. The hollow-fiber membrane module was immersed in the reactor containing wastewater, after which the adsorbent was added to the reactor. Permeate was continuously withdrawn with the help of a peristaltic pump. Since membrane filtration was operated in a continuous mode, the feed was also continuously fed into the reactor. HRT can be controlled by recycling permeate into the reactor in a certain proportion or by altering permeate flux. The separation mechanism of an integrated system includes separation due to adsorption as well as membrane process. The flow diagram of hybrid process presented in Figure 21.3 shows that the compound (*P*) binds onto adsorbent (*A*) and forms a complex (*PA*), which is retained by an appropriate membrane. In the second stage, the adsorbent (A) is regenerated and separated out from the compound (P), then reused.

21.4.1.2 Second Scheme: Coupling of Adsorption Process with Membrane Filtration

In another hybridization scheme, the influence of adsorption on the hybrid process was investigated by coupling adsorption with membrane filtration process. Samal et al. (2016)

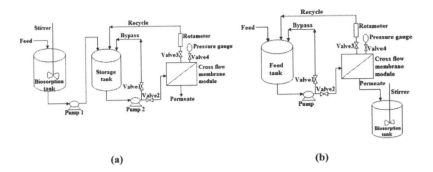

FIGURE 21.4

(a) Biosorption process followed by membrane filtration (b) membrane filtration followed by biosorption process. (From Samal, K., C. Das, and K. Mohanty, *Journal of Water Process Engineering*, 10(April): 30–38, doi:10.1016/j .jwpe.2016.01.013, 2016.)

used biosorbent derived from *T. rossica* leaves for adsorption process that was coupled with membrane filtration (NF/RO) in two different schemes as shown in Figure 21.4 (a) and Figure 21.4 (b). In the first hybrid process (Figure 21.4 (a)), the Pb ion contaminated wastewater was fed into a biosorption tank and treated in batch mode. Then the biosorption treated aqueous solution was fed to the membrane. In this type of hybridization scheme, the pollutant was partially removed through biosorption and remnants through membrane filtration. The addition of adsorption process before the membrane process enhanced the percentage removal percentage of pollutant. In the second hybrid process (Figure 21.4 (b)), the contaminated wastewater was fed into the cross-flow membrane cell and then permeate of the membrane was treated by biosorption. The pollutant concentration in the aqueous solution was first lowered by membrane filtration and then higher removal was achieved by adsorption. Since the wastewater was fed into the membrane cell without any pretreatment, a lower flux value was attained due to the concentration polarization. Additionally, the pollutant removal efficiency and flux in this process was slightly lower than the first hybrid process. However, the removal efficiency of the hybrid process was still better than the individual processes. From an economic point of view, biosorption followed by membrane process (Figure 21.4 (a)) was better compared to membrane filtration followed by adsorption (Figure 21.4 (b)). In another study, Koltuniewicz et al. (2004) have calculated and compared the total operating cost of adsorption process in column adsorption and adsorption-membrane hybrid process. It has also been reported that the total operating cost of sorption in adsorption-membrane hybrid process was around 98% lower than the adsorption in column process.

21.4.1.3 *Third Scheme: PAC Dynamic Membrane*

The dynamic membrane of PAC is another option that utilizes the adsorption characteristics of PAC adsorbent and simultaneous filtration by the dynamically formed membrane. Wu et al. (2017) have studied the dynamic membrane of PAC for removal of microbes and organic matter from seawater. The main steps of this process work to form a gel layer of PAC over the membrane surface during the microfiltration of PAC suspended solution. The gel layer of PAC particle forms a porous structure on the membrane surface that acts like a membrane itself. After the formation of PAC layer the feed is switched to wastewater. Since the MF membrane has larger pore size, it is difficult to remove the microbes, especially viruses, and organic matter from permeate by MF filtration alone (Figure 21.5 (a)).

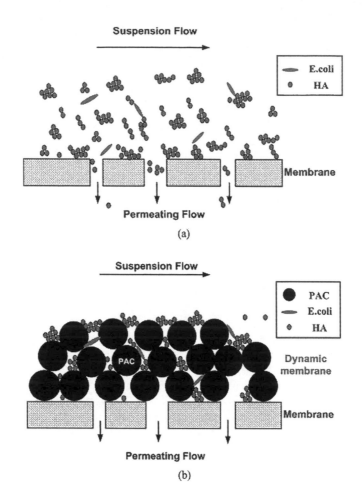

FIGURE 21.5
Schematic diagram of (a) direct filtration and (b) dynamic membrane filtration of seawater in cross-flow mode. (From Wu, S-E., K-J. Hwang, T-W. Cheng, Y-C. Lin and K-L. Tung, *Journal of Membrane Science*, 541: 189–197, doi org/10.1016/j.memsci.2017.07.006, 2017.)

On the other hand, the gel layer of PAC on the MF membrane surface adsorbs the pollutant and behaves like a membrane to remove the pollutants (Figure 21.5 (b)). Wu et al. (2017) have observed that the thickness of a dynamic membrane was mainly influenced by applied trans-membrane pressure and cross-flow velocity. The higher operating trans-membrane pressure and lower cross-flow velocity can cause an increase in the thickness of the dynamic membrane. The PAC gel layer thickness (L_c) could be calculated theoretically by using Equation 21.12 (Wu et al., 2017):

$$L_c = 9.56 \times 10^{-4} \left(\frac{F_t}{F_n} \right)^{0.15} \tag{21.12}$$

Where F_t/F_n is the ratio of tangential to normal drag forces exerted on the PAC particles present at the membrane surface.

The flux of the PAC dynamic filtration could be calculated using the Equation 21.13 (Wu et al., 2017):

$$J = \frac{\Delta P}{\mu\left(R_m + R_c + R_{cf}\right)} \tag{21.13}$$

Where Δp is the trans-membrane pressure difference, μ is the viscosity and R_m, R_c, and R_{cf} are the resistances caused by the membrane, cake, and cake clogging respectively. The clogging resistance (R_{cf}) can be determined using Equation 21.14:

$$R_{cf} = au_s^b \Delta P^c \tag{21.14}$$

Where a, b and c are the coefficients of the equation, u_s is the cross-flow velocity and ΔP is the trans-membrane pressure difference.

21.4.2 Effect of Adsorption Process on Hybrid Process

The performance and economy of the adsorption-membrane filtration hybrid process depends on various factors of the adsorption phenomenon, including configuration, mode of operation, adsorbent dose, sorption characteristics and operating conditions. Therefore, understanding the mechanisms of adsorption process in the hybrid system is essential. The deposition of adsorbent on membrane surface is common in the first type of hybrid process (Figure 21.2). Therefore, mode and extent of deposition of the adsorbent on the membrane surface must be considered in the adsorption-membrane hybrid process because it may decrease the effective adsorbent concentration in the bulk phase. As a result, the deposition of the adsorbent may reduce the effective sorption capacity of smaller adsorbent doses that are intended to help avoid plugging and to enable greater removal of pollutant. Also, at a particular cross-flow velocity, mixtures of the adsorbent in solutions can facilitate small adsorbent dose, which is appropriate to achieve higher rate of adsorption as compare to large adsorbent doses. However, use of lower adsorbent dose may result in early saturation of adsorbent active sites. The residence time plays an important role in adsorption process: if residence time is too short, incomplete adsorption may occur but on the other hand, if it is too long, adsorbent gets saturated. The high dose of the adsorbent with adequate residence time results in a high rate of sorption. The optimum adsorbent dose in the hybrid process is subject to sorption capacity of adsorbent as well as the operating time of the process. The particle size of the adsorbent also plays a significant role in the efficiency of the hybrid process. Adsorbents of smaller particle size have a high surface area, which results in higher uptake of the adsorbate, thus increasing the overall efficiency. Koltuniewicz et al. (2004) have compared the operation costs of sorption in the adsorption-membrane hybrid system and conventional adsorption process using adsorbents of different particle sizes. The adsorption-membrane hybrid process is more economical for fine adsorbent particles than the conventional adsorption process (Koltuniewicz et al., 2004). It is recommended to use adsorbents of smaller particle size to achieve higher removal efficiency and make the process economically viable.

21.4.3 Effect of Membrane Process on Hybrid Process

Membrane filtration is a key unit of the adsorption-membrane hybrid process; therefore, it is obvious that the efficiency of the hybrid process significantly depends on the membrane process. Process parameters, such as trans-membrane pressure, cross-flow velocity, sorbent concentration on the membrane surface and permeate flux exert a significant effect on the efficiency of the hybrid process since they affect the pollutant rejection efficiency. Trans-membrane pressure has a direct effect on flux and separation efficiency. With increasing trans-membrane pressure the permeate flux increases but allows lower contact time between sorbent and pollutant. High trans-membrane pressure also causes higher sorbent layer thickness on the membrane surface, which affects separation. The cross-flow velocity is another important parameter of the membrane process since it decreases the adsorbent layer thickness on the membrane surface and enhances permeate flux in a proportional manner. However, higher cross-flow velocity also reduces the contact time of the pollutants and adsorbent. To select an appropriate cross-flow velocity, the residence time of the adsorbent on the membrane surface must be optimum to attain the highest separation efficiency in the hybrid process. Membrane processes have the advantage of high throughput due to high flux values, but high throughput is not desirable in all cases. In adsorption-membrane hybrid system, higher flux value results in higher adsorbent concentration in the bulk phase and increased adsorbent layer thickness on the membrane surface. Moreover, higher flux lowers separation efficiency by reducing the contact time of the pollutant with the adsorbent. Therefore, the flux value must be optimum so that contact time between pollutant and adsorbent is adequate for high separation efficiency.

21.4.4 Application of Adsorption-Membrane Filtration Hybrid Process

A review of the literature shows that adsorption-membrane hybrid process has been used frequently for removal of different pollutants. The hybrid processes were applied and tested at different operating conditions to investigate their effect on pollutant removal efficiency. A few of the adsorption-membrane hybrid processes and their applications are summarized in Table 21.2. The adsorption-membrane hybrid process was mainly applied in wastewater treatment. The PAC adsorbent was used dominantly as the adsorbent process in the hybrid system due to its high uptake capacity. However, a few low-cost adsorbents derived from biomasses were also tested. The adsorption-membrane hybrid processes have shown an advantage over individual adsorption and membrane processes.

21.4.5 Extension of Adsorption-Membrane Hybrid Process

In order to improve the efficiency of the hybrid process, different pretreatment techniques such as coagulation, flocculation, ozonation and ultrasound radiation have been used in hybrid processes (Yu et al., 2014; Guo et al., 2004; Fan et al., 2014; Secondes et al., 2014). Yu et al. (2014) have investigated the role of coagulation and adsorption process on membrane fouling by organic matter. They observed that the addition of PAC during alum coagulation produced a greater removal of DOM compared to alum coagulation without PAC (Yu et al., 2014). The application of PAC adsorbent at lower dose along with alum coagulation helped to lower the irreversible and reversible fouling of UF membrane, for operating periods lasting fewer than 20 days. After 20 days of operation, increased membrane fouling and TMP rates were observed due to the substantial development of microorganisms and associated EPS. Guo et al. (2004) examined the flocculation process coupled with adsorption-membrane filtration hybrid process for the removal of organics and phosphorus. They studied the effect of flocculation and adsorption process on the critical flux of the membrane process. They observed that there was only a 33% flux increase with pre-flocculation, but when pre-flocculation

TABLE 21.2

Application of Adsorption-Membrane Hybrid Processes Reported in the Literature

S. No.	Adsorbent	Membrane	Pollutant	Remarks	Reference
1	PAA, PEI, PSSA, Aquatreat AR-4, Alcosperse AS-104	MF (Nylon: 1.2 µm, PP: 0.2 µm, and Ceramic: 0.8 µm), UF (PES: 10 kDa)	Ground water (metal ion), 1–50 ppm	75–80% rejection of Ca (10–30 ppm).	Volchek et al. (1993)
2	PAC (Picachem 8P, Pulsorb RD 90)	HF (PP: 0.2 µm, PES: 0.1 µm)	–	PAC 100 times bigger than the membrane pore size helps to prevent membrane blocking.	Florencia et al. (2006)
3	Clay (Kaolin and montmorillonite)	MF (CA: 30 kDa) and TFC (150 and 300 Da)	Phenol and o-cresol	The clay-UF process enhances removal efficiency of phenols by 80%.	Lin et al. (2006)
4	Boron selective resins (Diaion CRB 02 and Dowex XUS 43594.00)	SWRO (Toray™, UTC-80-AB and Filmtec™, SW30HR)	Boron (seawater)	98–99% salt rejections at seawater pH of 8.2, and 95.5–97% at pH of 10.5.	Kabay et al., (2008)
5	Boron selective resin (Dowex XUS 43594.00)	MF (PP: 0.4 µm)	Boron (seawater)	Boron concentration reduced from 2 mg L^{-1} to the WHO recommended level.	Bryjak et al. (2009)
6	Adsorption	RO (PA and PES)	Phenol	No significant differences in phenol rejection coefficients for different feed condition.	Bódalo et al. (2008)
7	PAC (Norit SA-UF)	Submersed (PES: 0.1 µm)	Pharmaceuticals	99.8% rejection of pollutant and 60% removal of DOC.	Saravia and Frimmel (2008)
8	PAC (Norit SA Super)	MF (PE: 0.4 µm)	E2 and NOM	The removal of E2 was approximately 92%.	Song et al. (2009)
9	GAC (Norit GAC 1240)	MF (Polyolefin: 0.22 µm)	NOM	The removal efficiency of UV_{260} was 60% compared to 30% by MF alone.	Kim et al. (2009)
10	PAC (Norit SA Super)	Submersed MF	E2	PAC deposition on the membrane should be maintained as low as possible.	Lee et al. (2009)

(Continued)

TABLE 21.2 (CONTINUED)

Application of Adsorption-Membrane Hybrid Processes Reported in the Literature

S. No.	Adsorbent	Membrane	Pollutant	Remarks	Reference
11	Polymer adsorbents	UF (0.04 μm)	Phenol	Phenol removal of 90% by the hybrid process.	Ipek et al. (2012)
12	PAC	MF-RO	Petrochemicals	A high overall recovery (91%) of reverse osmosis (RO).	Zhao et al. (2013)
13	Biosorbent (*T. rossica*)	NF/RO (PES: 400Da)	Lead	>99% removal of lead ion with hybrid process.	Samal et al. (2016)

Abbreviations: CA: Cellulose acetate, E2:17β-estradiol, HF: Hollow fiber, GAC: Granular activated carbon, MF: Microfiltration, NF: Nanofiltration, NOM: Natural organic matter, PA: Polyamide, PAA: Polyacrylic acid, PAC: Powdered activated carbon, PE: Polyethylene, PEI: Polyethylenimine, PES: Polyethersulphone, PP: Polypropylene, PSSA: Polystyrene sulfonic acid, RO: Reverse osmosis, SWRO: Spiral wound reverse osmosis, TFC: Thin film composite.

was combined with PAC adsorption, it resulted in nine times higher critical flux (Guo et al., 2004). In another study, Fan et al. (2014) studied the effect of coagulation and ozonation process on adsorption-membrane filtration hybrid process in river water treatment. The separation efficiency of coagulation-ozonation-adsorption (PAC)-membrane filtration (UF) was found to be 99% turbidity, 99.9% particle counts, 100% coliform bacteria, 64% DOC, 73% THMFPs, 75% HAAFPs, 98% ammonia, 96% geosmin, 88% 2-MIB, 98% EDCs and 98% PPCPs. The effect of ultrasound (US) radiation on adsorption-membrane hybrid process was tested by Secondes et al. (2014) for the removal of emerging contaminants. They studied the influence of PAC dose and US frequency on the removal of pollutants by the hybrid process. They reported > 99% removal of the emerging contaminants in the ultrasound-irradiation-adsorption (PAC)-membrane filtration (UF) process (Secondes et al., 2014).

21.5 Conclusions and Future Scope

The adsorption–membrane hybrid process has been tested as an efficient process for removal of a wide range of pollutants from wastewater. The incorporation of adsorption process with membrane process resulted in higher removal efficiency of pollutants and reduced flux decline. The hybrid processes have exhibited rapid removal of pollutants, lower pressure requirements, and ease of adsorbent regeneration. The main advantage of this process is that it can utilize waste biomass, scrap and waste organic material as efficient sorbents, which reduces the overall cost of the process. At the same time, it reduces the burden of safe disposal of these waste materials. However, the overall mechanism of the process is not well understood and recycling of pollutant-containing waste is a matter of concern. Further research on the adsorption-membrane hybrid process on a larger scale could help establish this process as a commercially viable alternative for wastewater treatment in industries.

Nomenclature

A	Membrane area in m^2
B	Constant related to adsorption energy E in mol^2 kJ^{-2}
C_e	Equilibrium pollutant concentration in mg L^{-1}
C_o	Initial pollutant concentration in mg L^{-1}
E	Free energy of adsorption in kJ mol^{-1}
e	Polanyi potential
J	Calculated permeate flux in $m^3 m^{-2} s^{-1}$
K'	Pseudo-second-order rate constant of adsorption in mg g^{-1} min^{-1}
K_d	First-order kinetics rate constant of adsorption in min^{-1}
K_F	Freundlich model constants related to adsorption capacity
K_L	Langmuir adsorption constant in L mg^{-1} relating the free energy of adsorption
Kp	Intra-particle diffusion rate constant in mg g^{-1} $min^{-0.5}$
1/n	Freundlich model constants related to intensity of adsorption
q	Uptake capacity in mg g^{-1}
q_e	Equilibrium uptake capacity in mg g^{-1}
q_m	Monolayer adsorption capacity in mg g^{-1}
q_t	Uptake capacity at time t in mg g^{-1}
R	Gas constant in kJ $mol^{-1} K^{-1}$
R_L	Dimensionless constant separation factor
t	Time in seconds
T	Temperature in K
v	Permeate volume in m^3
V	Volume of the solution in L
W	Mass of adsorbent in g

References

Backhaus, W.K., E. Klumpp, H-D. Narres, and M.J. Schwuger. 2001. Adsorption of 2,4-dichlorophenol on montmorillonite and silica: Influence of nonionic surfactants. *Journal of Colloid and Interface Science* 242(1): 6–13. doi:10.1006/jcis.2001.7781.

Bhatnagar, A., and M. Sillanpaa. 2010. Utilization of agro-industrial and municipal waste materials as potential adsorbents for water treatment—A review. *Chemical Engineering Journal* 157(2–3): 277–296. doi:10.1016/j.cej.2010.01.007.

Bódalo, A., J.L. Gómez, M. Gómez, G. León, M. Hidalgo, and M. Ruíz. 2008. Phenol removal from water by hybrid processes: Study of the membrane process step. *Desalination* 223(1–3): 323–329. doi:10.1016/j.desal.2007.01.219.

Bryjak, M., J. Wolska, I. Soroko, and N. Kabay. 2009. Adsorption-membrane filtration process in boron removal from first stage seawater ro permeate. *Desalination* 241(1–3): 127–132. doi:10.1016/j.desal.2008.01.062.

Dabrowski, A. 2001. Adsorption—From theory to practice. *Advances in Colloid and Interface Science* 93(1–3): 135–224. doi:10.1016/S0001-8686(00)00082-8.

Das, C., P. Patel, S. De, and S. DasGupta. 2006. Treatment of tanning effluent using nanofiltration followed by reverse osmosis. *Separation and Purification Technology* 50(3): 291–299. doi:10.1016/j.seppur.2005.11.034.

Ertugay, N. and Y.K. Bayhan. 2008. Biosorption of Cr (VI) from aqueous solutions by biomass of agaricus bisporus. *Journal of Hazardous Materials* 154(1–3): 432–439.

Fan, X., Y. Tao, L. Wang, X. Zhang, Y. Lei, Z. Wang, and H. Noguchi. 2014. performance of an integrated process combining ozonation with ceramic membrane ultra-filtration for advanced treatment of drinking water. *Desalination* 335(1): 47–54. doi:10.1016/j.desal.2013.12.014.

Freundlich, H. 1906. Adsorption in Solids. *Zeitschrift für Physikalische Chemie* 57: 385–470.

Guo, W.S., S. Vigneswaran, H.H. Ngo, and H. Chapman. 2004. Experimental investigation of adsorption-flocculation-microfiltration hybrid system in wastewater reuse. *Journal of Membrane Science* 242(1–2): 27–35. doi:10.1016/j.memsci.2003.06.006.

Ho, Y.S. and G. McKay. 2000. The kinetics of sorption of divalent metal ions onto sphagnum moss peat. *Water Research* 34(3): 735–742. doi:10.1016/S0043-1354(99)00232-8.

Hobson, J.P. 1969. Physical adsorption isotherms extending from ultrahigh vacuum to vapor pressure. *Journal of Phyical Chemistry* 73(8): 2720–2727. doi:10.1021/j100842a045.

Ipek, I.Y., N. Kabay, M. Yüksel, D. Yapici, and Ü. Yüksel. 2012. Application of adsorption-ultrafiltration hybrid method for removal of phenol from water by hypercrosslinked polymer adsorbents. *Desalination* 306: 24–28. doi:10.1016/j.desal.2012.08.033.

Kabay, N., M. Bryjak, S. Schlosser, M. Kitis, S. Avlonitis, Z. Matejka, I. Al-Mutaz, and M. Yuksel. 2008. Adsorption-membrane filtration (AMF) hybrid process for boron removal from seawater: An overview. *Desalination* 223(1–3): 38–48. doi:10.1016/j.desal.2007.01.196.

Kim, K.Y., H.S. Kim, J. Kim, J.W. Nam, J.M. Kim, and S. Son. 2009. A hybrid microfiltration-granular activated carbon system for water purification and wastewater reclamation/reuse. *Desalination* 243(1–3): 132–144. doi:10.1016/j.desal.2008.04.020.

Koltuniewicz, A.B., A. Witek, and K. Bezak. 2004. Efficiency of membrane-sorption integrated processes. *Journal of Membrane Science* 239(1): 129–141. doi:10.1016/j.memsci.2004.02.037.

Lagergren, S. 1898. About the theory of so-called adsorption of soluble substances. *Kungl. Svenska vetenskapsakademiens handlingar* 24(4): 1–39.

Langmuir, I. 1916. Constitution and fundamental properties of solids and liquids. I. Solids. *Journal of the American Chemical Society* 38: 2221–2295. doi:10.1021/ja02268a002.

Lee, S., J.W. Lee, S. Kim, P.K. Park, J.H. Kim, and C.H. Lee. 2009. Removal of 17β-Estradiol by powdered activated carbon-microfiltraion hybrid process: The effect of PAC deposition on membrane surface. *Journal of Membrane Science* 326(1): 84–91. doi:10.1016/j.memsci.2008.09.031.

Lin, S.H., R.C. Hsiao, and R.S. Juang. 2006. Removal of soluble organics from water by a hybrid process of clay adsorption and membrane filtration. *Journal of Hazardous Materials* 135(1–3): 134–140. doi:10.1016/j.jhazmat.2005.11.030.

Mohanty, K., D. Das, and M.N. Biswas. 2006. Preparation and characterization of activated carbons from sterculia alata nutshell by chemical activation with zinc chloride to remove phenol from wastewater. *Adsorption* 12(2): 119–132. doi:10.1007/s10450-006-0374-2.

Samal, K., C. Das, and K. Mohanty. 2016. Development of hybrid membrane process for pb bearing wastewater treatment. *Journal of Water Process Engineering* 10(April): 30–38. doi:10.1016/j.jwpe.2016.01.013.

Samal, K., K. Mohanty, and C. Das. 2017. treatment of pb ion contaminated wastewater using hazardous parthenium (*P. Hysterophorus* L.) weed. *Water Science and Technology* 75(2): 427–438. doi:10.2166/wst.2016.536.

Saravia, F. and F.H. Frimmel. 2008. Role of NOM in the performance of adsorption-membrane hybrid systems applied for the removal of pharmaceuticals. *Desalination* 224(1–3): 168–171. doi:10.1016/j.desal.2007.02.089.

Saravia, F., P. Naab, and F.H. Frimmel. 2006. Influence of particle size and particle size distribution on membrane-adsorption hybrid systems. *Desalination* 200(1–3): 446–448. doi:10.1016/j.desal.2006.03.370.

Secondes, M.F.N., V. Naddeo, V. Belgiorno, and F. Ballesteros. 2014. Removal of emerging contaminants by simultaneous application of membrane ultrafiltration, activated carbon adsorption, and ultrasound irradiation. *Journal of Hazardous Materials* 264: 342–349. doi:10.1016/j.jhazmat.2013.11.039.

Sen, M., A. Manna, and P. Pal. 2010. Removal of arsenic from contaminated groundwater by membrane-integrated hybrid treatment system. *Journal of Membrane Science* 354(1–2): 108–113. doi:10.1016/j.memsci.2010.02.063.

Song, K.Y., P.K. Park, J.H. Kim, C.H. Lee, and S. Lee. 2009. Coupling effect of 17β-estradiol and natural organic matter on the performance of a pac adsorption/membrane filtration hybrid system. *Desalination* 237(1–3): 392–399. doi:10.1016/j.desal.2008.11.004.

Srividya, K., and K. Mohanty. 2009. Biosorption of hexavalent chromium from aqueous solutions by catla catla scales: Equilibrium and kinetics studies. *Chemical Engineering Journal* 155(3): 666–673. doi:10.1016/j.cej.2009.08.024.

Venkataganesh, B., A. Maiti, S. Bhattacharjee, and S. De. 2012. Electric field assisted cross flow micellar enhanced ultrafiltration for removal of naphthenic acid. *Separation & Purification Technology* 98: 36–45. doi:10.1016/j.seppur.2012.06.017.

Volchek, K., L. Keller, D. Velicogna, and H. Whittaker. 1993. Selective removal of metal ions from ground water by polymeric binding and microfiltration. *Desalination* 89(3): 247–262. doi:10.1016/0011-9164(93)80140-I.

Weber, W.J., and J.C. Morris. 1963. Kinetics of adsorption on carbon from solution. *Journal of the Sanitary Engineering Division* 89(2): 31–60.

Wu, S-E., K-J. Hwang, T-W. Cheng, Y-C. Lin and K-L. Tung. 2017. Dynamic membranes of powder-activated carbon for removing microbes and organic matter from seawater. *Journal of Membrane Science* 541: 189–197. doi.org/10.1016/j.memsci.2017.07.006.

Yu, W., L. Xu, J. Qu, and N. Graham. 2014. Investigation of pre-coagulation and powder activate carbon adsorption on ultrafiltration membrane fouling. *Journal of Membrane Science* 459: 157–168. doi:10.1016/j.memsci.2014.02.005.

Zhao, C., P. Gu, and G. Zhang. 2013. A hybrid process of powdered activated carbon countercurrent two-stage adsorption and microfiltration for petrochemical ro concentrate treatment. *Desalination* 330: 9–15. doi:10.1016/j.desal.2013.09.010.

22

Layer-by-Layer (Lbl) Coated Multilayer Membranes in Dye House Effluent Treatment

Usha K. Aravind, Subha Sasi, Mary Lidiya Mathew, and Charuvila T. Aravindakumar

CONTENTS

22.1 Introduction

Clean energy and water are the key challenges of the twenty-first century. Sustainable technologies and renewable materials are necessary to face these two interrelated resources. There are plenty of technological developments related to these two aspects, with a certain amount of limitations. Fresh water resources are overexploited and at higher risk of contamination. Run off from domestic, industrial and agricultural sources continue to pollute water resources and the entry of new kinds of compounds is on the rise. The protection of fresh water resources is vital and hence more comprehensive technologies are needed for the treatment of resultant wastewater. Existing technologies have to be addressed from this angle to make it more sustainable. It is also necessary to give a serious outlook to industrial effluent treatment to assure zero waste discharge. Among industries, textile continues to be the biggest consumer of fresh water (Sungur et al., 2015). A strong scientific understanding of the contributing factors to textile wastewater is therefore necessary. Along with this, current treatment methods and advancement in their evolution are also important.

Adsorption, advanced oxidation processes (AOPs) and membranes play a prominent role in water treatment technologies. Various AOPs are successful in dye degradation and decolorization (Rayaroth et al., 2015). Certain AOPs can degrade only a particular class of dyes and the chemical oxygen demand (COD) remains high. Though there is lot of advancement in each technique, membrane technology continues to dominate in water purification. The entry of new membrane materials and environmentally benign surface coating methods has created new heights in pressure-driven membrane processes. Surface modification of commercially available membrane, via the layer-by-layer (LbL) method, is well established as an efficient method to achieve desired characteristics into the membranes (Decher, 1997). Novel composite membranes with highly selective separation skin at nanoscale precision can be obtained. The layers of alternately charged polyelectrolytes can be constructed on porous membrane substrates from water. A few bilayers change the surface characteristics dramatically. The fact that their properties can be well tuned with respect to surface charge, thickness and porosity makes multilayered membrane an ideal tool for analytical separation. With suitable support materials, low pressure membrane separation processes can replace high pressure processes in the future. On a small scale, they have been used in protein purification, drug delivery, ion separation, gas separation, pervaporation and fuel cells. Multilayer assembly already shows industrial presence in packaging and biomedical field. In water treatment, the skin formed from weak polyelectrolytes' combination is worth a special mention due to their flexibility and adaptability. The separation ability of a polyelectrolyte pair such as chitosan and polyacrylic acid (CHI/PAA) assembled on polyamide microfiltration membrane is illustrated in this chapter. The working of this membrane under adsorption and filtration mode is also discussed. The response of CHI/PAA LbL assembly to external stimuli enables the reusability of the membrane and recovery of concentrated dye from the medium.

22.2 Water Consumption in the Textile Sector

The textile industry contributes significantly to the overall economic growth of developing countries by providing employment opportunities, industrial production and exports (Tanange, 2010). At the same time, the negative impact to the ecosystem of the discharged pollutants from textile factories cannot be overlooked (Rajaram & Das, 2008). The textile industry is in the forefront in fresh water consumption where more than 90% of the water is utilized as processing water, mainly for manufacture of chemicals and rinsing. The major difference with other industries is that only a small percentage is used as cooling water. The amount of water required varies depending on the type of fabric and method of processing. For instance, wool and felted fabrics consume more water compared to synthetic fibers. The statistics on water consumption for an average-sized textile mill shows about 1.6 million liters of water for the production of about 8000 kg of fabric per day (Kant, 2012).

Pre-treatment is a major step that utilizes water for different purposes including sizing/desizing, scouring, bleaching and mercerizing. About 45% of the total wastewater runs down from this process. Dyeing and printing also demand large volumes of water. After dyeing, the textile materials are subjected to washing. Water is also required for cleaning the printing machines. The textile industry on average generates about 21–377 m^3 of wastewater per ton of manufactured product (Buscio et al., 2015).

TABLE 22.1

Major Physico-Chemical Parameters of Textile Effluents
and Their Permissible Limits

Parameter	BIS Permissible Limits
Color	—
pH	5.5–9
Electrical conductivity (EC) (μmho/cm)	600
Total suspended solids (mg/l)	100
Total dissolved solids (mg/l)	2100
BOD	30
COD	250

22.3 Overview of Textile Processing and Major Pollutants

Textile processing essentially involves various stages with each stage contributing to color, TDS, TSS, COD and BOD of the effluent. An overview of these processes is summarized in the block diagram given below (Figure 22.1).

The permissible limits for major physico-chemical parameters in textile wastewater are presented in Table 22.1.

22.4 Natural vs. Synthetic Dyes

The uses of natural dyes dates back to the Neolithic period, and colorants derived from natural materials remained in practice until the 1850s. Natural dyes find their origin mainly from plant materials including roots, berries, barks, leaves and seeds. Some mollusks and minerals also serve as a source of natural color (Yusuf et al., 2017). These dyes are mostly derived from renewable resources and have no adverse effects on humans or the environment. The waste produced in certain cases—for instance harda, indigo etc.— formed ideal fertilizers and thus added value. Certain constituents in natural dye stuffs also possess anti-allergic, moth-proofing potential that makes them suitable colorants for children's garments. In addition, the range of colors produced from the natural materials is pleasant. The emission of greenhouse gases can also be reduced by the replacement of synthetic dyes. The era of synthetic dyes started with the accidental synthesis of a violet dyestuff called "mauve" by William Henry Perkin in 1856. The subsequent production and commercialization of aniline-based dyes triggered new research in the area and resulted in the advent of coal tar–based synthetic dyes. They became popular owing to easy availability, widespread application, low cost and higher reproducibility of shades. Due to the increased demand of dyes and new innovations in dye chemistry, the synthetic dye industry flourished rapidly. But over the years health hazards associated with synthetic dyes became evident. In the modern era, some countries have banned a few dyes.

Over the last few years natural dyes have become popular again. This is because of the increased environmental consciousness and awareness of health hazards. Significant

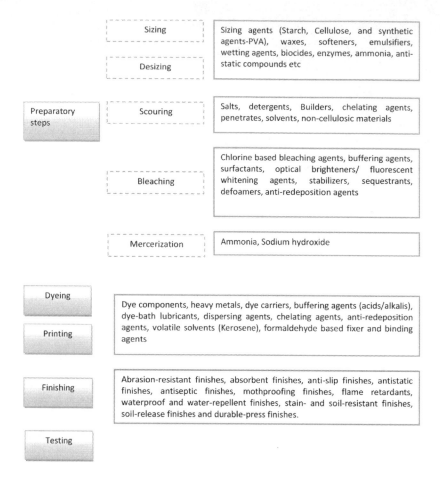

FIGURE 22.1
Stages of textile processing.

research improved the limitations of natural dyes especially regarding color yield, reproducibility of individual shades and complexity of the dyeing process. Steps have also been taken to replace harmful mordants like chromium with eco-friendly compounds (Yusuf et al., 2017). There is a constant effort to derive colorants from bio-resources. Most countries now promote green colorants, which is actually paving the way for the re-emergence of bio-colorants (Saxena & Raja, 2014).

22.5 Treatment Methods for Textile Wastewater

Today, various methods are in practice for the treatment of textile industry wastewater. Commonly employed methods are adsorption, AOP and membrane filtration. An overview is presented in the following sections. To give a clear picture, the overview is summarized in Figure 22.2.

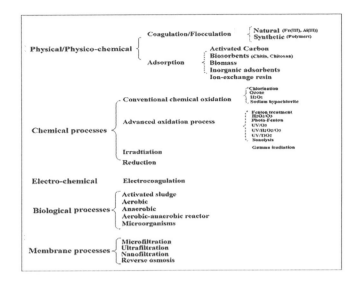

FIGURE 22.2
General treatment protocols for textile wastewater.

22.6 Membrane Separation Processes

Membrane separation, the most popular and sustainable technique, provides a highly promising platform for analytical separation. The major membrane processes such as reverse osmosis (RO), nanofiltration (NF), ultrafiltration (UF), microfiltration (MF), dialysis, electrodialysis (ED) and pervaporation (PV) cover a wide range of particle/molecular size separation. Membrane filtration is an attractive separation process, as it is usually performed under gentle conditions. Separation of different components is achieved by a combination of sieving, or hindered transport through the narrow membrane pores and specific interactions between the components and the membrane material (such as adsorption, electrical interactions). Pressure driven processes, where the feed side is subjected to hydraulic pressure is of prime interest, in the present context. They are generally classified on the basis of size of the solutes that the membranes reject. In its definition, RO retains all components other than the solvent (e.g., water) itself, NF stops higher valent ions, while UF retains only macromolecules or particles larger than about 10-2000A. MF, on the other hand, is designed to retain particles in the micron range, that is, suspended particles in the range of 0.1 µm to about 5 µm. NF is relatively a new process that uses charged membranes with pores that are larger than RO membranes, but too small to allow permeation of many organic compounds such as sugars. Though these processes are driven by an externally applied pressure difference, advent of concentration difference and electrical field (for charged compounds) cannot be ruled out, which strongly influences the filtration process.

22.6.1 Microfiltration in Textile Effluent Treatment

In the textile industry, membrane processes are utilized for various applications amongst which effluent treatment is of utmost importance. The textile wastewater is highly complex in nature in terms of acidity/alkalinity, nature and type of dyes, presence of auxiliary chemicals, solvents and temperature. Thus, the membranes employed in the textile industry are designed to have high mechanical, chemical and thermal stability. In addition, the anti-fouling potential of the membrane also plays a crucial role. These innate properties of MF membranes make them least feasible for textile wastewater treatment. However, the technique is effectively employed as a pre-treatment step especially in hybrid systems. The application of MF followed by UF was found effective in the reclamation of wash water (Uzal et al., 2009). The high rejection for colloids and particles makes the process more apt, compared to other conventional pre-treatment steps including coagulation, flocculation etc. (Tahri et al., 2013). Moreover, MF reduces the extent of fouling to facilitate longer life of UF, NF or RO membranes. The successful application of MF as a pre-treatment process has been demonstrated for textile effluents carrying indigo blue (Couto et al., 2017), acid yellow 36 (AY-36) and kaolin (Juang et al., 2013). Combined electro-oxidation with MF is another option, where COD and color removal was highly satisfactory on a laboratory scale. Recent studies show that MF is also explored as a single step for textile wastewater treatment with tubular carbon (Tahri et al., 2013) and organoclay/chitosan nanocomposite MF membrane (Daraei et al., 2013) being examples in this category.

22.6.2 Membrane Fouling

The major constraint towards the extensive use of membranes for water treatment is fouling, a phenomenon that can severely affect overall process performance. The extent of fouling always depends on the membrane properties, feed characteristics and operating conditions of the system (Acero et al., 2010).

A novel method employed to overcome the problem, without compromising the efficiency of the membrane, is the preparation of antifouling membranes. Generally, the materials involved in the synthesis of membranes are hydrophobic in nature and prone to fouling by different kinds of solutes. The key tactic involved in the production of antifouling membranes is the generation of fouling resistant surfaces that are characterized by hydrophilic or super hydrophilic nature. Due to the great diversity of foulants, a single antifouling mechanism may not be fruitful, thus inducing a real challenge to developing strategies based on multiple antifouling mechanisms. Surface coating methods can be adopted for conferring antifouling potential. Two techniques, namely "coating-to" and "coating-from" are employed for surface coating. In "coating-from" technique, an antifouling surface is generated on the membrane "in situ" through a specific pre-treatment (Zhao et al., 2015). In "coating-to," a post-modification process is carried out with suitable antifouling materials (polymers and inorganic nano-materials) either by spin coating or by dip coating. The surface of the membrane is modified to meet the targets without altering its structure. Hydrophilization is mostly done through grafting, pulsed laser ablation and chemical vapor deposition which are tedious and limit wider application.

Advancements in surface coating have lately opened up the possibility of new materials with enhanced separation properties. Strategies for the fabrication of ordered monolayers are mainly Langmuir–Blodgett (LB) (Blodgett, 1935; Langmuir & Schaefer, 1937) and self-assembled monolayers (SAMs) (Prime & Whitesides, 1991). LB is one of the elegant ways to build multilayer structures. In LB, the monomolecular films of lipids are transferred from air/water interface onto a solid substrate. The control of layer thickness and ordering of molecules is possible with molecular precision. The requirement of amphiphilic materials and long processing time restricts the popularity of this technique. LB multilayers have limited stability against solvents and thermal treatments. SAM can be applied to a wide range of materials but requires strong interaction between the monolayer and the substrate. For this, pre-deposition of compounds like thiols are necessary. Hence both LB and SAM are complicated to implement on macroscopic surfaces. For practical applications, LbL has emerged as an alternative for multilayer assembly where nanostructured components with distinct structure and composition can be combined.

22.7 Layer-by-Layer (LbL) Assembly

The LbL assembly technique was first mentioned by Iler in 1966, for the buildup of inorganic films (Iler, 1966). This assembly was based on negatively charged silica particles and positively charged alumina. In 1991, Decher and Hong expanded this technique using alternately charged bolaform molecules. Further, the assemblies of alternately charged polyelectrolytes were introduced (Decheret al., 1992). Since then, this technology has remained in the forefront of materials science and engineering. The LbL assembly technique involves alternate dipping of a substrate or a solid support in two oppositely charged polyelectrolytes along with intermittent water rinse steps to remove any extra material that is loosely bound to the surface. Thus, one can build as many layers as needed, depending on the requirement. The LbL assembly can be realized by dip coating, spin coating, spray coating and de-wetting techniques. Ultrathin films could be built offering high flux and selectivity for membrane separations, filtrations and purification applications. Thicker membranes could also be built offering hydrophobic coatings, hydrophilic coatings or anti-corrosive coatings. Multilayers reported in the literature had thickness that ranged between 10 nm and 10 μm.

The multilayer buildup is primarily attributed to electrostatic interaction. But it is also driven by hydrogen bond, coordination bond, π-π stacking, hydrophobic interaction and molecular recognition. Many applications of LbL assembly can be cited but the theoretical knowledge of these primary and secondary interactions on the buildup is rather limited. It is perhaps more dependent on the polyelectrolyte (building unit) character which means that in case of strong polyelectrolytes, the electrostatic interactions have the upper edge. Alternatively, the interactions involved in weak polyelectrolytes are of different kinds depending on the degree of ionization.

LbL technique is a very robust, reproducible, durable, cost-effective and environmental friendly method. It is done at room temperature, so there is no need for any vacuum equipment or special instrumentation. It can be used to assemble various types of materials, polymers, composites, clay, proteins, dyes, carbon nanotubes or nanoparticles. Also, it can be coated on various kinds of substrates such as silicon, gold, platinum, plastics, glass, quartz, stainless steel, clay, nanoparticles, blood cells and colloidal particles. The whole process can be automated making the coating procedure less time consuming and applicable for commercial purposes where productivity and labor are major problems. Above all, this technique offers very precise control over thickness and unprecedented uniformity of the coating down to the subnanometer scale. The LbL procedure can be applied for the fabrication of multicomposite films, i.e., nanoscopic assembly of hundreds of different materials in a single device using environmental friendly, ultra-low–cost techniques.

Several recent studies suggest that polyelectrolyte multilayer (PEM) films prepared by the alternate deposition of cationic and anionic polyelectrolytes on suitable substrates are promising candidates for "skin" layers in composite membranes for NF and RO processes (Stanton et al., 2003; Miller & Bruening 2004). NF is used for applications such as water softening, brackish water reclamation, and dye-salt separations. LbL deposition of anionic and cationic polyelectrolytes readily converts polymeric ultrafiltration membranes into materials capable of nanofiltration (Malaisamy & Bruening, 2005). Depending on the polyelectrolytes employed, PEM membranes can remove salt from sugar solutions, separate proteins, or allow size-selective passage of specific sugars. A polyelectrolyte is a macromolecular species which dissociates into a highly charged polymeric molecule in water or other ionizing solvents. These macromolecules carry covalently bound charged groups (cationic or anionic), and low molecular counter ions (Decher, 1997). However, every neutral polymer can be transformed into a polyelectrolyte by covalently attaching an appropriate number of ionic groups.

22.8 Materials in LbL Assembly

Given the large set of materials that are easily incorporated into multilayer films, layer-by-layer deposition is a rather general approach for the fabrication of complex surface coatings. It is possible to coat almost any solvent-accessible surface starting with submicron objects up to the inside of tubing or even objects with a surface of several square meters. Like a chemical reaction, the precise structure of each layer depends on a set of control parameters such as concentration, adsorption times, ionic strength, pH, temperature, nature and concentration of added salt, rinsing time, humidity of the surrounding air, drying etc. While the LbL technique generally works very well due to the fact that the processing window is rather large, it is highly recommended to keep the deposition conditions as consistent as possible in order to get highly reproducible results.

Recent experiments on polyelectrolyte multilayers formed by consecutive absorption of negatively charged and positively charged polyelectrolytes have shown that, upon addition of a new layer, the number of charges carried by the incorporated polymer is large enough to neutralize the charge of the previous layer and even to invert the sign of the zeta potential (Dubas & Schlenoff, 1999, 2001). Neutron experiments also clearly indicate that the newly incorporated layer strongly interpenetrates the previous layer and even the one before (Steitz et al., 2000). But there is no complete mixing between the layers, each layer thus keeping its identity. Often it is said that polyelectrolyte multilayer films are independent of the underlying substrate. Since polyanion and polycation adsorption is repeated consecutively, after a few layers, the structure and properties of each layer are governed by the choice of polyanion/polycation pair and deposition conditions, which means that the influence of the substrate is typically lost after a few deposition cycles. Cellulose acetate, polysulfone, polyether sulfone, polyamides, polycarbonate, nylon, polyacrylonitrile, polyvinylidene fluoride, polystyrene etc. are polymers employed for membrane manufacture. However, despite the vast number of materials that have been studied as possible membrane materials, only a few of these have succeeded commercially. Figure 22.3 summarizes the polymers and polyelectrolytes generally employed in LbL assembly.

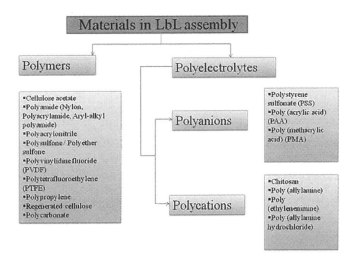

FIGURE 22.3
Substrates and coating materials in LbL assembly.

22.9 LbL Assembled MF Membranes for Textile Dye Removal

Polyelectrolyte multilayered (PEM) membranes are an ideal tool for the separation of dye house effluents (Aravind et al., 2010; Baburaj et al., 2012; Gopalakrishnan et al., 2015). The most attractive feature is that it is possible to tune their electrostatic nature and overall properties. The properties of each layer can be controlled at nanoscale. The nature of the polyelectro-lyte pair, in terms of a weak or strong functional group, decides its response factor towards external solution condition, such as pH or ionic strength. Weak polyelectrolytes respond to pH changes and the degree of ionization can be controlled by altering the deposition pH; often multilayers are incorporated with non-ionized chain segments. For such multilayers, swelling extent will be high in water with higher permeation rates of ions. On the other hand, strong polyelectrolytes respond to ionic strength and multilayers accommodate more charged segments. A combination of weak PE is more appropriate for the removal/recovery of dyes as it permits the PEM to behave as a pH responsive shell, i.e., it can switch between completely permeable (open) to impermeable (closed) state depending on pH. This enables high recovery of dyes and is a potential recovery option providing a considerable dye/salt separation window.

Dye loading and release behavior of PEM (PAA/CHI and PAA/PEI on polyamide MF membrane) is illustrated taking effluents containing two different model dyes, Coomassie Brillent Blue (CBB), Methylene Blue (MB). The uptake of cationic dye, MB is different for the two combinations. PAA/CHI was more efficient in the loading of MB which is reflected in the UV-vis spectra (Figure 22.4). The loading is layer dependent and with higher bilayers, dye aggregation is observed. The aggregated state itself indicates heavy adsorption. The difference in the nanoskin architecture is clear as more number of bilayers are needed to bring in considerable loading by PEI/PAA. With PEI as the cationic layer, the dye uptake is less and exists mostly in monomeric and dimeric forms.

The character of the polyelctrolyte, the functional groups and the range of interactions within the multilayer is a decisive factor in dye loading. The adsorption and desorption behavior at the PEM depends on the molecular architecture of the multilayers. The architecture in turn depends on the number of deposited layers, pH of deposition, feed pH

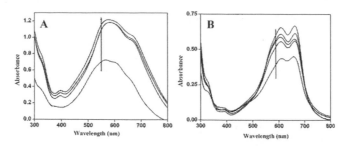

FIGURE 22.4

UV-vis spectra of MB adsorbed PAA/CHI membranes. Number of bilayers 3.5, 4.5, 6.5 and 9.5 respectively in the increasing order. Deposition pH of PAA and CHI, 4 and 2. (Adapted from Baburaj, M.S., C.T. Aravindakumar, S. Sreedhanya, A.P. Thomas, and U.K. Aravind, *Desalination*, 288: 72–79, doi: http://dx.doi.org/10.1016/j.desal.2011.12.015, 2012.)

etc. The topmost layer, its overall charge, charge of the multilayer and charge on the dye molecule are also key parameters. The analysis of the deposition pH of PE give a firsthand information about the difference in dye loading that further substantiates the domination of one of the polyelectrolyte and the involvement of other forces. The assembly pH of PAA/CHI is 4/2 and that of PEI/PAA is 6/2. The pH selection allows a combination of partially ionized (PAA)/fully ionized (CHI) polyelectrolyte, thus ensuring non stochiometric pairing of chain segments to yield thick and interpenetrating bilayers. In addition, CHI, the natural polysaccharide has certain other advantages. In addition to electrostatic interaction, hydrogen bonding and van der Waals forces support adsorption.

Generally, in multilayer systems with PAA as the anionic polyelctroyle, the main contributor to thickness will be that of PAA itself. Hence, when the dye solution has pH above the pKa of PAA especially towards alkaline pH, there will be excess negative charge enhancing the loading of cationic dye. Also in cases with CHI based multilayers, the dye will diffuse into inner layers. But for PEI, adsorption is mostly restricted to the surface layer. The deposition pH of 4/6 allowed a less favored combination with respect to the degree of ionization. The architecture itself is different in PEI/PAA with a high degree of crosslinking. Swelling is less when the film is placed in contact with aqueous solution and hence, less permeable. The loading of dye was pH dependent with alkaline pH being the most beneficial in terms of dye removal. At alkaline pH, the swelling is more pronounced as the carboxylic group deprotonate with the increase in the surface charge density of the polymer strands. The pH dependent swelling is schematically depicted in Figure 22.5.

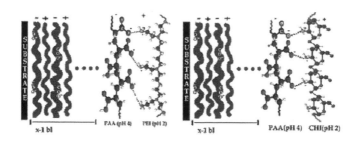

FIGURE 22.5

A schematic representation composite membrane. (a) Self-assembled bilayers of PAA/CHI and (b) Self assembled bilayers of PAA/PEI at effluent pH 10.5. (Adapted from Baburaj, M.S., C.T. Aravindakumar, S. Sreedhanya, A.P. Thomas, and U.K. Aravind, *Desalination*, 288: 72–79, doi: http://dx.doi.org/10.1016/j.desal.2011.12.015, 2012.).

The anionic dye (CBB) loading required a larger number of bilayers. The difference is mainly due to the molecular size. Alkaline pH was also unsuitable and less loading occurred for CBB. The presence of PEI in the multilayer induced floc formation with dye sedimentation. The performance of polyelectrolyte skinned membranes in the removal of dye from model effluent is summarized in Table 22.2. Further, this study proved the reusability of membranes through desorption studies. Acidic pH could trigger complete release of MB whereas CBB was released at neutral pH.

The ability of LbL to enhance the surface properties of polymer membranes to enhance its capacity in dye removal is proved beyond a doubt. The combination of weak/strong polyelectrolyte (CHI/PSS) is found to be effective for the reduction of COD and color from textile and paper mill effluents. Here too, the degree of ionization of CHI is the deciding factor.

The ratio of the polyelectrolyte within the multilayer, its thickness and interpretation are decided by the degree of ionization. For weak polyelectrolytes, this is again controlled by the buildup pH. The degree of ionization can be quite different in the deposited state as the pKa value of poly electrolyte is sensitive to the environment. In order to overcome such discrepancies, CHI and PAA paired at identical pHs on polyamide MF membranes is evaluated for dye removal. The industrial presence of membranes for dye house effluent treatment is mostly based on pressure driven processes. The molecular size of the dye indicates the need of high pressure filtration, which means NF or RO would be ideal. Again, cost considerations motivate the design of membranes suited to low pressure operation. The proper choice of the buildup condition of PEM allows low pressure salt/dye separation on a laboratory scale that is realized through CHI/PAA multilayered polyamide microfiltration membranes in NaCl/RB-5 separation. CHI/PAA skin of 50 nm exhibited a separation window greater than 8000 (NaCl/RB-5) with a flux of 7 $m^3m^{-2}day^{-1}$. This study also illustrated linear and nonlinear growth regions of multilayers and their share in dye removal in real time filtration. The multilayers in LbL assembly grow in two patterns. In linear growth, an equal amount of polyelectrolyte is adsorbed at each step with exact charge compensation. In nonlinear growth, one or both of the polyelectrolytes

TABLE 22.2

Dependence of Dye Adsorption with Number of Bilayers. Membranes Used are PAA/CHI and PAA/PEI (Initial Concentration of CBB and MB are 14.75 mg L^{-1} and 14.40 mg L^{-1}). The Value in Parenthesis Represents the number of Bilayers Used for MB, where PAA/CHI and PAA/PEI Terminated with Anionic layer. Deposition pH of PAA, PEI and CHI was 4, 6 and 2, Respectively

Number of Bilayers	Concentration of CBB After Treatment (mg L^{-1})		Concentration of MB After Treatment (mg L^{-1})	
	PAA/CHI	**PAA/PEI**	**PAA/CHI**	**PAA/PEI**
3 (3.5)	10.85	2.75	9.26	11.25
5 (4.5)	10.65	2.12	7.90	10.62
7 (6.5)	10.5	2.10	7.90	10.75
9 (9.5)	9.35	2.08	6.62	–
15 (15.5)	7.85	2.0	–	10.00
20 (19.5)	6.75	1.5	2.6	6.75

Source: Adapted from Baburaj, M.S., C.T. Aravindakumar, S. Sreedhanya, A.P. Thomas, and U.K. Aravind, *Desalination*, 288: 72–79, doi: http://dx.doi.org/10.1016/j.desal.2011 .12.015, 2012.

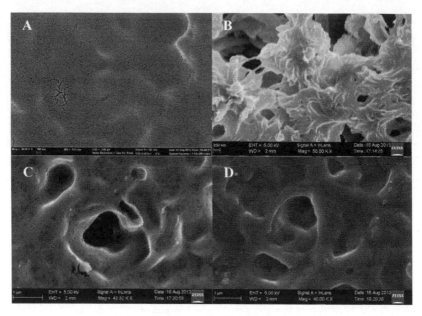

FIGURE 22.6
FESEM images of (a) CHI/PAA 5bl loaded RB5 active side, (b) CHI/PAA 5bl loaded RB5 rare side, (c) CHI/PAA 5bl loaded RB5 with 10000 ppm NaCl in the feed and (d) CHI/PAA 5bl after desorption.

constituting the multilayer inter diffuse, causing large mass gain. Buildup conditions such as pH, ionic strength, type of polyelectrolyte etc. influence growth pattern. The occurrence of two growth patterns in CHI/PAA multilayers and its share in the sustainability of multilayers in response to large salt and dye is discussed in this article (Gopalakrishnan et al., 2015). The FESEM view of CHI/PAA membrane after filtration and desorption are collectively shown in Figure 22.6. The FESEM view of dye adsorbed membrane clearly shows the effectiveness of nanoskinned surface. The performance of the membrane processes are affected by fouling that cause lower mass transfer rate, decreased life and increased use of cleaning chemicals. The multilayered nanoskin confers fouling resistant surfaces. The response of these layers to external stimuli (pH or ionic strength) is expected to make back flushing quite successful. High fouling resistance is attained at the cost of compromising the high initial flux. Several optimization strategies in the deposition medium can potentially increase the flux.

Dye molecule loaded multilayer combinations are utilized for various purposes (Jiang et al., 2009). They are used instead of one polyelectrolyte counterpart to explore the multilayer growth patterns. Another interesting aspect is the possibility of using multilayer as a platform for controlled drug delivery or as a reservoir for sustained release. Thermal and photochemical stabilities is yet another field of study. Dye molecule also enabled the existence of various interactive forces within the multilayer and at the interfaces. Such forces are rich in weak polyelectrolyte combinations and hence they are the best candidates for multilayered skins performing filtration. Strong/weak polyelectrolyte combinations also provides separation skin to improve the performance under nanofiltration condition (Hong et al., 2006).

Nanomaterials are gaining significance in the scenario of dye effluent treatment. The introduction of nanoparticles in the membrane matrix generally improves its mechanical stability and antimicrobial properties. Cellulose nanocrystals (CNC) embedded polyethersulfone membrane matrix shows large improvement in dye-salt rejection and flux recovery (Daraei et al., 2017). The nanocystals imparted high hydrophiliciy and dense skin. CNC in chitosan matrix enhances the removal efficiency of positively charged dyes. The latest trend is to form diamond based nano-composite membranes via LbL. Controlled preparation of diamond based organic/inorganic composite material is highly promising due to its thermal, mechanical properties and high surface area (Zhao et al., 2017). Carboxyl terminated microdiamond coated with PEI/PAA has shown good selectivity toward MB and rhodamine B. The nanomaterial can also be separated well from wastewater for reuse. Nanocomposite membrane materials exhibit good adsorption capacity.

22.10 Conclusions and Future Prospects

The introduction of LbL assembled polyelectrolyte on commercial membrane substrate enables fabrication of materials with a wide range of properties. Such materials can give new dimensions to membrane based filtrations bringing down the cost. LbL can impart NF and UF characteristics to MF membranes. Existing membrane modules can be coated in LbL fashion to enhance life, rejection properties and flux. Low pressure filtrations with LbL coated membranes respond not only to macromolecules or particulates but to low molecular weight solutes and multivalent ions. Most importantly, in textile wastewater purification, the dye-salt separation window is quite large. This ensures the much-desired dye recovery and water recycling. This composite membrane material is also promising for its antifouling nature. Moreover, its longevity is further demonstrated because the deterioration, if any, to the multilayered skin gets rectified on its own.

References

Acero, J.L., F.J. Benitez, A.I. Leal, F.J. Real, and F. Teva. 2010. Membrane filtration technologies applied to municipal secondary effluents for potential reuse. *Journal of Hazardous Materials* 177(1–3): 390–398. doi: http://dx.doi.org/10.1016/j.jhazmat.2009.12.045.

Aravind, U.K., B. George, M.S. Baburaj, S. Thomas, A.P. Thomas, and C.T. Aravindakumar. 2010. Treatment of industrial effluents using polyelectrolyte membranes. *Desalination* 252(1): 27–32. doi: http://dx.doi.org/10.1016/j.desal.2009.11.006.

Baburaj, M.S., C.T. Aravindakumar, S. Sreedhanya, A.P. Thomas, and U.K. Aravind. 2012. Treatment of model textile effluents with PAA/CHI and PAA/PEI composite membranes. *Desalination* 288: 72–79. doi: http://dx.doi.org/10.1016/j.desal.2011.12.015.

Blodgett, K.B. 1935. Films built by depositing successive monomolecular layers on a solid surface. *Journal of the American Chemical Society* 57(6): 1007–1022. doi: 10.1021/ja01309a011.

Buscio, V., M.J. Marín, M. Crespi, and C. Gutiérrez-Bouzán. 2015. Reuse of textile wastewater after homogenization–decantation treatment coupled to PVDF ultrafiltration membranes. *Chemical Engineering Journal* 265: 122–128. doi: http://dx.doi.org/10.1016/j.cej.2014.12.057.

Couto, C.F., L.S. Marques, M.C.S. Amaral, and W.G. Moravia. 2017. Coupling of nanofiltration with microfiltration and membrane bioreactor for textile effluent reclamation. *Separation Science and Technology*: 1–11. doi: 10.1080/01496395.2017.1321670.

Daraei, P., N. Ghaemi, and H.S. Ghari. 2017. An ultra-antifouling polyethersulfone membrane embedded with cellulose nanocrystals for improved dye and salt removal from water. *Cellulose* 24(2): 915–929. doi: 10.1007/s10570-016-1135-3.

Daraei, P., S.S. Madaeni, E Salehi, N. Ghaemi, H.S. Ghari, M.A. Khadivi, and E. Rostami. 2013. Novel thin film composite membrane fabricated by mixed matrix nanoclay/chitosan on PVDF micro-filtration support: Preparation, characterization and performance in dye removal. *Journal of Membrane Science* 436: 97–108. doi: http://dx.doi.org/10.1016/j.memsci.2013.02.031.

Decher, G., J.D. Hong, and J. Schmitt. 1992. Buildup of ultrathin multilayer films by a self-assembly process: III. Consecutively alternating adsorption of anionic and cationic poly-electrolytes on charged surfaces. *Thin Solid Films* 210: 831–835. doi: http://dx.doi.org/10.1016/0040-6090(92)90417-A.

Decher, G. 1997. Fuzzy nanoassemblies: Toward layered polymeric multicomposites. *Science* 277(5330): 1232–1237. doi: 10.1126/science.277.5330.1232.

Dubas, S.T., and J.B. Schlenoff. 1999. Factors controlling the growth of polyelectrolyte multilayers. *Macromolecules* 32(24): 8153–8160. doi: 10.1021/ma981927a.

Dubas, S.T. and J.B. Schlenoff. 2001. Swelling and smoothing of polyelectrolyte multilayers by salt. *Langmuir* 17(25): 7725–7727. doi: 10.1021/la0112099.

Gopalakrishnan, A., M.L. Mathew, J. Chandran, J. Winglee, A.R. Badireddy, M. Wiesner, C.T. Aravindakumar, and U.K. Aravind. 2015. Sustainable polyelectrolyte multilayer surfaces: Possible matrix for salt/dye separation. *ACS Applied Materials & Interfaces* 7(6): 3699–3707. doi: 10.1021/am508298d.

Hong, S.U., M.D. Miller, and M.L. Bruening. 2006. Removal of dyes, sugars, and amino acids from nacl solutions using multilayer polyelectrolyte nanofiltration membranes. *Industrial & Engineering Chemistry Research* 45(18): 6284–6288. doi: 10.1021/ie060239+.

Iler, R.K. 1966. Multilayers of colloidal particles. *Journal of Colloid and Interface Science* 21(6): 569–594. doi: http://dx.doi.org/10.1016/0095-8522(66)90018-3.

Jiang, B., J.B. Barnett, and B. Li. 2009. Advances in polyelectrolyte multilayer nanofilms as tunable drug delivery systems. *Nanotechnology, Science and Applications* 2: 21–27.

Juang, Y., E. Nurhayati, C-P. Huang, J.R. Pan, and S. Huang. 2013. A hybrid electrochemical advanced oxidation/microfiltration system using BDD/Ti anode for acid yellow 36 dye wastewater treatment. *Separation and Purification Technology* 120: 289–295.

Kant, R. 2012. Textile dyeing industry an environmental hazard. *Natural Science* 04(01): 5. doi: 10.4236/ns.2012.41004.

Langmuir, I. and V.J. Schaefer. 1937. Monolayers and multilayers of chlorophyll. *Journal of the American Chemical Society* 59(10): 2075–2076. doi: 10.1021/ja01289a506.

Lee, S., J. Cho, and M. Elimelech. 2005. Combined influence of natural organic matter (NOM) and colloidal particles on nanofiltration membrane fouling. *Journal of Membrane Science* 262(1–2): 27–41. doi: http://dx.doi.org/10.1016/j.memsci.2005.03.043.

Malaisamy, R. and M.L. Bruening. 2005. High-flux nanofiltration membranes prepared by adsorption of multilayer polyelectrolyte membranes on polymeric supports. *Langmuir* 21(23): 10587–10592. doi: 10.1021/la051669s.

Miller, M.D. and M.L. Bruening. 2004. Controlling the nanofiltration properties of multilayer poly-electrolyte membranes through variation of film composition. *Langmuir* 20(26): 11545–11551. doi: 10.1021/la0479859.

Prime, K.L. and G.M. Whitesides. 1991. Self-assembled organic monolayers: Model systems for studying adsorption of proteins at surfaces. *Science* 252(5009): 1164–1167. doi: 10.1126/science.252.5009.1164.

Rajaram, T. and A. Das. 2008. Water pollution by industrial effluents in India: Discharge scenarios and case for participatory ecosystem specific local regulation. *Futures* 40(1): 56–69. doi: http://dx.doi.org/10.1016/j.futures.2007.06.002.

Rayaroth, M.P., U.K. Aravind, and C.T. Aravindakumar. 2015. Sonochemical degradation of Coomassie Brilliant Blue: Effect of frequency, power density, pH and various additives. *Chemosphere* 119: 848–855. doi: http://dx.doi.org/10.1016/j.chemosphere.2014.08.037.

Saxena, S. and A.S.M. Raja. 2014. Natural dyes: Sources, chemistry, application and sustainability issues. In S.S. Muthu (Ed.), *Roadmap to Sustainable Textiles and Clothing: Eco-friendly Raw Materials, Technologies, and Processing Methods* (37–80). Singapore: Springer Singapore.

Stanton, B.W., J.J. Harris, M.D. Miller, and M.L. Bruening. 2003. Ultrathin, multilayered polyelectrolyte films as nanofiltration membranes. *Langmuir* 19(17): 7038–7042. doi: 10.1021/la034603a.

Steitz, R., V. Leiner, R. Siebrecht, and R. v. Klitzing. 2000. Influence of the ionic strength on the structure of polyelectrolyte films at the solid/liquid interface. *Colloids and Surfaces A: Physicochemical and Engineering Aspects* 163(1): 63–70. doi: http://dx.doi.org/10.1016/S0927-7757(99)00431-8.

Sungur, S. and F. Gülmez. 2015. Determination of metal contents of various fibers used in textile industry by MP-AES. *Journal of Spectroscopy* 2015:5. doi: 10.1155/2015/640271.

Tahri, N., I. Jedidi, S. Cerneaux, M. Cretin, and R.B. Amar. 2013. Development of an asymmetric carbon microfiltration membrane: Application to the treatment of industrial textile wastewater. *Separation and Purification Technology* 118: 179–187. doi: http://dx.doi.org/10.1016/j.seppur.2013.06.042.

Tanange, K.R. 2010. Indian textile industry: Growth and current scenario. *Cyber Literature: The International Online Journal - Literature, Humanities & Communication Technologies* 3(2): 24–30.

Uzal, N., L. Yilmaz, and U. Yetis. 2009. Microfiltration/ultrafiltration as pretreatment for reclamation of rinsing waters of indigo dyeing. *Desalination* 240(1): 198–208. doi: http://dx.doi.org/10.1016/j.desal.2007.10.092.

Yusuf, M., M. Shabbir, and F. Mohammad. 2017. Natural Colorants: Historical, Processing and Sustainable Prospects. *Natural Products and Bioprospecting* 7(1): 123–145. doi: 10.1007/s13659-017-0119-9.

Zhao, X., K. Ma, T. Jiao, R. Xing, X. Ma, J. Hu, H. Huang, L. Zhang, and X. Yan. 2017. Fabrication of hierarchical layer-by-layer assembled diamond-based core-shell nanocomposites as highly efficient dye absorbents for wastewater treatment. *Scientific Reports* 7: 44076. doi: 10.1038/srep44076.

Zhao, X., Y. Su, H. Dai, Y. Li, R. Zhang, and Z. Jiang. 2015. Coordination-enabled synergistic surface segregation for fabrication of multi-defense mechanism membranes. *Journal of Materials Chemistry A* 3(7): 3325–3331. doi: 10.1039/c4ta06179a.

23

Membrane Technology—A Sustainable Approach for Environmental Protection

Ranjana Das, Arijit Mondal, and Chiranjib Bhattacharjee

CONTENTS

23.1 Introduction

Increasing population, rapidly expanding urbanization, economic growth and industrial revolution and have led to increased global consumption of various feedstock in different sectors like minerals, metals, food, plastics, wood products, automobiles etc. It is believed that the trend of increasing consumption will continue to rise in the next several decades. As a result, increasing waste generation has dramatically affected the environment (Hurk et al., 2017). The urgent requirement is to improve old technologies or adopt new technologies to reduce industrial and domestic emissions into the environment. A series of policies and measures have been issued globally to guide industrial transformation and realize green development. The main focus in these environmental protection policies is to deploy effective technologies for emission management in every operation of waste process industries.

Pollutants in water, air and soil have contaminated the ecosystem. "Hazardous Metals" also termed as "Trace Elements" are a concern because these chemical entities are found in nature in very low concentrations. Industries such as mining, battery processing, paint, electroplating, chemical manufacturing, viscous-rayon, copper picking, galvanizing, rubber processing, thermal power plant, solid waste dumping and some agricultural practices are the main sources of heavy metals as well as gaseous pollutants. Copper, lead, arsenic, mercury, cadmium and nickel are primary toxic metals. River water contamination by these toxic heavy metals prevents the reuse of water for industrial and drinking purposes. Arsenic and lead are the main culprits in drinking water because these toxic metals damage human body chromosomes (Nidheesha, & Singh, 2017). Trace concentrations of mercury, nickel, lead and zinc can damage the growth of aquatic algae (Kovacika et al., 2017). Many fresh water bodies like lakes, ponds or rivers—as well as ground water resources—contain high metal concentrations from geochemical origin because of gradual solubilization of these heavy metals from the soil with the help of active pH value. Air pollutants such as CO_2, SO_X, NO_X, O_3, particles settling on water, acid rain and soluble metallic dusts are also harmful to our ecosystem. Metals may be bound to the soil in sediments or may be precipitated as hydroxides. Metals can also exist as colloidal suspensions or suspended particles, or they could be absorbed by biota. Water pollution by toxic heavy metals like arsenic, chromium, zinc, mercury, lead, cadmium, copper and iron originate from industrial wastes, burning of fossil fuels, discharge of domestic sewage and land run-off to a common collection point. To aquatic life and humans, these elements demonstrate different environmental characteristics and toxicity. Further investigation is needed to measure the impact of toxic heavy metals in the aquatic environment.

Several conventional technologies are adopted by different industries to overcome this critical situation. A few advanced technologies are trying to retrofit to overcome barriers faced by conventional technologies. In this study, we have made an attempt to highlight the advantages of membrane-based processes among other advanced technologies. Membrane separation has become widely popular for domestic as well as industrial applications including wastewater treatment, drinking water purification, hemodialysis in medical industry, filtration in food industry and materials' recovery for environmental protection (Bernardo et al., 2009). Membrane technologies have advantages over other conventional technologies because they maintain high purity while also having lower energy demand, and light weight and compact modular designs with less impact on the environment. The availability of new types of stable membrane materials paves the way for future growth and process solutions to industrial problems faced by developing countries. In general, a membrane is defined as a semi permeable barrier placed between two fluid streams, which selectively separates one or multiple compounds from the feed fluid stream. The driving force for the separation is a chemical potential gradient. Most membranes work on the principle of pore dimensions for selective transport of smaller sized species that are forced through the membrane matrix under pressure. Separation driven by hydraulic pressure can be categorized into microfiltration (MF), ultrafiltration (UF), nanofiltration (NF) and reverse osmosis (RO) based on pore size distribution in the matrix. Components are selectively separated, concentrated or purified from the feed stock for a wide range of particles of varying sizes like bacteria and pyrogens, proteins, amino acids, sugars, inorganic acids, salts etc. Membrane technology does not require a phase change and ancillary equipment such as heat generators, evaporators or condensers. As a result, large volumes can be treated with remarkable energy efficiency.

23.2 Brief about Membrane Technology

Membrane separation is kinetically controlled separation of a fluid mixture through a semipermeable membrane. The membrane process is seldom spontaneous and requires some driving force for desired separation. Different forms of energies act as the driving force for desired separation like mechanical potential gradient, electrical potential gradient, chemical potential gradient (osmotic pressure and vapor pressure). Membrane processes operate without application of heat and require less energy than conventional thermal separation processes (distillation, sublimation and crystallization). It is typically a physical separation that rarely requires use of any chemicals, thus making it especially suitable for separation of azeotropic liquids and separation of solutes that form isomorphic crystals. Significant technical application of membrane technology involves drinking water production by reverse osmosis (worldwide approximately 7 million cubic meters annually). In the wastewater treatment sector, membrane technology has a strong foothold and its impact is growing. Through ultrafiltration and microfiltration processes, removal of suspended particles, colloids and macromolecules with the aim of disinfecting and clarifying water—especially those assigned for water-sports and recreation—can be attained.

The use of membrane separation processes for environment protection and conservation is well known over the past 30 years. Membranes are increasingly becoming important as smart energy recovery techniques, such as in fuel cells and osmotic power plants. Membrane processes are widely used for treatment of various process streams because of a unique combination of properties like flexibility, separation efficiency and low-energy consumption. The membrane market is categorized based on the type of material used for formation: polymeric or ceramic or a special hybrid type. Among these types, the polymeric section reportedly has the largest market share since 2015 due to the easily obtainable nature and cost-effectiveness of polymers that are suitable for applications such as water and wastewater treatment, medical and pharmaceutical, and industrial processing. The ceramic division is projected to grow at the highest compounded annual growth rate (CAGR) during the forecast period, owing to advantageous membrane characteristics of high porosity, defined pore size, high permeability and good mechanical properties. According to a statistical report, the Asia-Pacific region is the fastest growing market for membranes that is projected to grow from an estimated US\$ 7.32 billion in 2016 to US\$ 11.95 billion by 2021, at a CAGR of 10.3%. The scope of membranes in environmental protection includes surface water purification, ground water recovery, wastewater treatment, purification of landfill leachates, remediation of contaminated soils, flue gas cleaning and biogas conditioning. Membrane technology has been reported to have wide application spectrum for treatment of emerging contaminants from water (Rodriguez-Narvaez et al., 2017). Ultrafiltration process has been used for the removal of a diverse group of emerging contaminants because of smaller pore size (in the range of 0.001–0.01 µm). Removal efficiency can vary widely depending on the membrane type and closely related to the contaminant type. Nanofiltration and reverse osmosis process have also been reported for selective removal of the toxins and contaminants from waste streams (Setiawan et al., 2011). Environmental protection is a broad subject area. Within the scope of this review, only a few areas have been covered in detail like municipal wastewater treatment, pharmaceutical effluent treatment, heavy metal removal from ground water, gas separation to prevent air pollution, tannery effluent treatment besides processing of dairy, and paper and pulp industry wastewater.

23.3 Application of Membrane Technology in Environmental Protection

This section is an intense review of the typical applications of membrane technology in different industrial and agro-industrial sections aiming at environmental protection and waste management for recovery of valuable components.

23.3.1 Municipal Wastewater Treatment

Membrane processes have received global recognition as a reliable and proven technology in municipal wastewater treatment (Madaeni et al., 2015). Figure 23.1 presents the global market status on municipal waste stream treatment as reported by Frost and Sulivan in 2013, who have also compiled a detailed statistical report on the global as well as Indian market for water and wastewater treatment and desalination (Frost & Sullivan, 2009). In 2012, the global market for water and wastewater solutions and services were worth US$ 180 billion, while the share of treatment equipment in this market was valued at only US$ 34.59 billion. At the same time, the market for global water and wastewater filtration systems was valued at US$ 6,117 million (Royan, 2012). Because the technology offers savings and reduction in water and energy footprints, it has been recognized for its sustainable approach. Investigations carried out by different authors related to wastewater treatment by membrane technology have been summarized in Table 23.1. Membrane technology for wastewater treatment was first applied around 30 years ago, and though formerly considered uneconomical, the development of new generation membrane bioreactor (MBR) units working with low negative pressure (out-to-in permeate suction) and membrane aeration with reduced fouling have increased the stakes for utilizing membranes in treating real time effluents. MBR systems are typical immersed filtration systems that involve lower costs of installation and operation than previous methodologies with the additional benefits of a small footprint, flexible design for the upgradation of old treatment plants, complete solids' removal, effluent disinfection and no problem of sludge bulking in comparison to conventional processes (Mutamim et al., 2013).

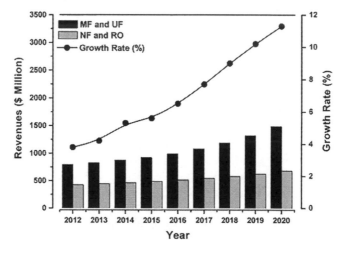

FIGURE 23.1
Global market status of membrane technology.

TABLE 23.1

Typical Investigations Carried out by Various Researchers on wastewater/Sewage Treatment

Investigation	Interpretation	Reference
Development of new materials and methods for fabricating and modifying polymeric membranes for municipal wastewater treatment	Study revealed potential for treatment of municipal water with considerable water reclamation and solid sludge concentration for easier disposal	S.S. Madaeni et al. (2015)
Assessment of the quality of effluents obtained from physicochemical-UV and microfiltration–ultrafiltration as tertiary treatments. A physicochemical-UV urban wastewater treatment system was designed for the study.	The membrane system remains less viable than physico-chemical UV treatment, particularly owing to the high cost of installation and membrane replacement	M. Gómeza et al. (2007)
Municipal wastewater was treated by biological route followed by nanofiltration for wastewater reuse	Authors concluded that NF could be a potential method to reuse biologically treated municipal sewage	Bunani et al. (2013)
Authors used ultrafiltration and microfiltration as polishing steps after biological treatment of municipal wastewater	Authors concluded the process as energy efficient and almost achieving zero-discharge	Rautenbach et al. (1996)
Treatment of secondary and tertiary municipal wastewater by membrane filtration has been investigated by hollow-fiber polysulfone ultrafiltration membranes.	Authors highlighted the importance of membrane structure, membrane polymer, membrane cartridge and operating procedures of the MF-UF system for wastewater treatment	Tchobanoglous et al. (1998)
The performance of a UF pilot plant was investigated under two opposite operating conditions ("stressed operating condition" versus "conventional operating condition").	Results indicate that for both conditions, the reclaimed effluent complied with the Italian regulations for unrestricted wastewater reuse	Falsanisi et al. (2009)

23.3.2 Pharmaceutical Waste Treatment

Conventional municipal wastewater treatment plants (MWWTPs) are not able to completely remove micro-pollutants at ng L^{-1} to mg L^{-1} levels, for instance they cannot remove pharmaceuticals, personal care products, pesticides, detergents and various industrial additives. As a consequence, these undesirable compounds accumulate in the environment and cause a horde of risks to all living organisms like bacterial resistance, feminization of aquatic organisms, neurotoxicity, endocrine disruption and cancer. Pharmaceuticals and pesticide residues are labelled as emerging water pollutants having adverse effects on the endocrine systems of human beings and wildlife. According to world pollution control regulations, WWTPs must reduce at least 80% of micro-pollutants. Major removal of micro-pollutants in conventional MWWTPs occurs during secondary treatment in activated sludge systems and membrane bioreactors (Verlicchi et al., 2012). Ganiyu et al. (2015) and other research publications have presented in detail the diverse application of membrane technology for the treatment of pharmaceutical effluents (Table 23.2). Since conventional WWTPs are designed to remove a large spectrum of pollutants from industrial, domestic and farming wastewater, they are not efficient in removal of pharmaceutically active components. Owing to their low concentrations in the ecosystem and diverse physicochemical properties, these compounds are not completely eliminated during treatment processes and ultimately are discharged into river bodies. Additionally, the concentrations of certain pharmaceuticals are reported to increase during treatment as a consequence

TABLE 23.2

Summary of the Investigations Related to Pharmaceutical Effluent Treatment by Membrane Process

Investigation	Interpretation	Reference
Studies on the application of NF and RO for removal of endocrine disrupting chemical from effluent.	More than 73% removal of the targeted components was reported.	Verliefde et al. (2015)
Nanofiltration process efficiency was studied on nine selected pharmaceuticals and five endocrine disruptors.	Significant removal efficiency has been observed and removal was observed, depending on molecular characteristics of the targeted compounds	Quintanilla et al. (2009)
Treatment of triclosan contaminated water with NF and RO	Significantly high removal efficiency was observed depending on membrane characteristics.	Nghiem and Coleman (2008)
Treatment of antibiotic contaminated veterinary effluent with NF and RO	98% removal of the targeted compound was observed	Košutic et al. (2007)
Treatment of wastewater contaminated with carbamazepine	About 60% reduction of carbamazepine was observed with significant fouling effect on membrane surface.	Mahlangu et al. (2014)
Treatment of seven different pharmaceutically active compounds (PhACs) using ceramic ultrafiltration membrane	Efficient separation with high rejection results (> 70%) were observed for diclofenac (75.9%), diazepam (72.6%), erythromycin (85.4%), and triclosan (72.9%) during filtration experiments using ceramic membrane of 1 kDa MWCO at pH 8.	Garcia-Ivars et al. (2017)

of their transformation into conjugates. Such complexities have led the researchers to explore new technological alternatives like pressure-driven ion processes such as RO, NF, UF and MF for pilot- and full-scale installations. Membrane processes have been successfully implemented either distinctly or in combination with other conventional techniques of wastewater reclamation to achieve high quality product water by efficient removal of dissolved solids, proteins, sugars, inorganic ions, bacteria, viruses and organic micropollutants (Taheran et al., 2016). Several publications have reported the efficiency of NF and RO techniques in efficient removal of pharmaceutically active component and pesticides from raw wastewaters and natural effluents using polymeric membranes (Ganiyu et al., 2015). Studies have also been reported on application of ceramic membranes and their pros and cons in treatment of real effluents contaminated with pharmaceutically active components (Wang et al., 2015).

23.3.3 Heavy Metal Removal from Ground Water

Ground water pollution due to release of heavy metals into the ecosystem has been causing worldwide concern. Density and atomic weights of heavy metals normally lie in the range of 5 gm/cm^3 and 63.5–200.6, respectively (Srivastava et al., 2008). Wastewaters from different chemical industries like metal plating facilities, battery manufacturing, fertilizer, mining, paper and pesticides, metallurgical, mining, fossil fuel and tannery are the main sources of heavy metals. Industrial growth during the last decade has contributed to heavy metals' increased release into the environment. The toxic effects of heavy metals such as cadmium, arsenic, mercury, zinc, chromium and lead on human health have been assessed broadly. The possible symptoms of toxic heavy metal poisoning include high blood pressure, sleep disabilities, speech disorders, fatigue, poor concentration, irritability, memory loss, depression, increased allergic reactions, vascular occlusion, and autoimmune diseases (Qu et al., 2013). Cadmium, lead, arsenic and mercury are directly harmful to the human body. Though our bodies need some heavy metals like manganese, iron, chromium, copper and zinc at desirably low concentrations, their presence in large amounts may be extremely dangerous (Qu et al., 2013). A number of methods have been utilized for the removal of heavy metals from ground water. The selection of specific technology is mainly based on its economics and initial concentration levels of heavy metals. From ground water, the heavy metals are usually removed by chemical or physical processes such as oxidation, reduction, bioremediation, electrochemical treatment, lime softening, coagulation, precipitation, adsorption and ion exchange (Peric et al., 2004). Nevertheless, all these methods have their own limitations and capabilities especially regarding efficiencies and costs. Among membrane processes, NF and RO are capable of removing heavy metals from ground water. However, they require higher transmembrane pressure and expensive membranes in comparison to low pressure processes such as MF and UF. However, microfiltration and ultrafiltration membranes are not capable of removing heavy metals, mainly due to the large pore sizes. To overcome this limitation, special kinds of MF and UF membranes have to be developed to remove small-sized pollutants based on their morphology, synthesis procedure and materials. For this purpose, the concept of reactive mixed matrix membrane was introduced that are a kind of MF/UF membranes developed by incorporation of inorganic adsorptive materials into the polymeric matrix. The separation mechanism of reactive mixed matrix membranes involves a combination of adsorption as well as convenience of filtration. The inorganic dispersed phase provides selectivity due to its specific chemical structure and polymer matrix grants mechanical support. Recently, several reactive mixed matrix membranes have been developed for efficient removal of pollutants such as arsenic, fluoride, lead, sulphur and phenolic compounds from water (Peric et al., 2004).

23.3.4 CO_2 Separation

Global warming, a subject of increasing emphasis, is believed to be caused by the emission of greenhouse gases (GHGs) which traps heat radiating from the surface of the earth when solar energy is being transmitted through the atmosphere and thus increases the global surface temperature. As estimated by researchers, the global average surface temperature has increased between 0.6 and 1.0° C during the last 150 years. As a result of global warming, melting icebergs in polar oceans causes the sea level to rise globally. Other alarming consequences of global warming include droughts, expansion of desert regions, heat waves, disrupted ecosystems, increasingly severe weather and loss of agricultural productivity. CO_2 sequestration attracts more interest for its large greenhouse forcing of the climate, its substantial projected future forcing and its lingering persistence in the atmosphere. The concentration of CO_2 in the atmosphere has been increased from 280 ppmv in the preindustrial era to about 400 ppmv in 2014 (Ledley et al., 1999). Conventional CO_2 capture technology is not relevant for the entire range of separation. For post-combustion technology, absorption of CO_2 in aqueous alkanolamine solutions or its adsorption onto a solid surface such as zeolite or activated carbon are the most common techniques for CO_2 capture. However, the scientific community is actively involved in search for alternate technologies due to the high energy requirement involved in amine absorption and the regeneration processes in addition to the corrosive nature of the solvents as well as costly equipment. Membranes have received considerable attention in recent years for CO_2

separation, since membrane units are frequently used in gas separation due to their light weight and space efficiency arising from compact modular design. Also, separating agents are not required and hence the regeneration step is absent (Zhao et al., 2012). Membrane systems discussed here can be classified into six categories: Zeolite Membranes, Carbon Molecular Sieves, Silica Membranes, Mixed-matrix Membranes, Supported Ionic Liquid Membranes and Polymeric Membranes.

Well-defined pore structure can be found in zeolites which are crystalline alumino silicates. A thin zeolite layer deposited on different types of support (porous α-alumina or stainless steel) constitutes a zeolite membrane which has numerous advantages in the field of gas separation compared to traditional polymeric membranes due to excellent thermal, mechanical and chemical stabilities. Gas separation through zeolite membranes is governed by molecular sieving accompanied by surface diffusion. For CO_2 separation at low temperature, zeolite membranes are very useful due to their preferential adsorption (Shekhawat et al., 2003), which helps to get both high permeability and selectivity. At high temperatures, selectivity of zeolite membranes starts decreasing because the selective adsorption of CO_2 decreases. Additionally, zeolite membranes are very expensive, and difficult to process and handle.

Carbon membranes are prepared by pyrolysis of thermosetting polymers under controlled conditions that provides a porous random network throughout the membrane. The pore size of carbon membranes varies from 0.35 to 1 nm depending on preparation conditions (Hagg, 2009). Two types of carbon membranes are normally available (i.e., supported or unsupported). Unsupported carbon membranes are brittle in nature. A few limitations of carbon membranes are poor selectivity due to frequent defects in the thin active layer, expensive nature, toughness in processing and difficulty in handling.

Two types of silica membranes normally reported according to their morphology are mesoporous and microporous membranes for gaseous separations. Different techniques are available to prepare silica membranes like hydrothermal synthesis, sol-gel method etc. There are certain issues associated with the preparation of silica membranes like defect formation during preparation and thermal cracking during heat treatment that affect membrane reproducibility. Another type of problem associated with the silica membrane is hydrothermal instability at elevated temperatures. Permeance and selectivity are drastically decreased when the membranes are exposed to water vapor.

Composite organic-inorganic membranes consist of inorganic particles incorporated into a polymer matrix to form mixed matrix membranes. Gas molecules pass through both polymer and inorganic phases. There are plenty of inorganic materials like zeolites, silica particles, metal organic frameworks and carbon nanotubes that can be incorporated into the polymer matrix. At low loading of the inorganic phase in mixed-matrix membranes, gas transport is primarily governed by the polymeric phase. Overall, the mixed matrix membranes cannot show a significant improvement in gas permeance or selectivity over conventional polymeric membranes.

In case of supported ionic liquid membranes (SILMs), ionic liquids are impregnated into polymeric or inorganic supports that help to transport gas molecules through the membrane by solution-diffusion mechanism or facilitated transport mechanism. Ionic liquids, normally treated as organic salts, are nonflammable and thermally stable liquids at room temperature (Jiang et al., 2007). Traditional SILMs suffer from stability problems under pressurized conditions or vacuum which disrupts performance. The instability is possibly caused by carrier washout and carrier degradation (Zhao et al., 2012).

Organic polymers are the most widely used materials in membrane based gas separation (Abetz et al., 2006). Conventional polymers are classified into two different categories, one is rubbery polymer (used above their glass transition temperature (T_g)) and the other is glassy polymers (used below their T_g). The sorption of gases in rubbery polymers and glassy polymers follows Henry's law and complex sorption isotherms, respectively. Commercial organic membranes contain thin selective asymmetric structure supported on a porous layer. A thin selective layer allows gas separation through the membrane while the thick support layer provides mechanical strength. The main limitation of existing conventional polymeric membranes is low CO_2 permeability as well as low CO_2/N_2 selectivity.

23.3.5 Tannery and Dye Waste Treatment

Leather industry effluent is ranked high amongst contemporary industrial pollutants that contribute to high foreign exchange as well as severe environmental hazards. Leather tanning is described as a production process related to the transformation of animal hides and skins by using water, chemicals and mechanical processing. Waste streams from leather tanning process are reported to contain a high concentration of pollutants, characteristics of which depends on the type of production process, including the source of the tanning (UNEP, 2010). According to the study, the process involves approximately 15–20 m^3 of water for each ton of raw skin produced and the effluents are characterized by high levels of COD, BOD, salts, suspended solids and conductivity values. Tannery waste is one of the most threatening effluent streams that has severe adverse effects on aquatic system and ground water quality because of toxic heavy metal contamination by Cr (VI). The deadly impact of this pollutant on human health and the ecosystem has been recognized by several authors: it acutely affects the skin, the respiratory tract, liver function, while also causing severe pulmonary congestions, vomiting, diarrhoea and genetic deformations by damaging DNA structure (Dasgupta et al., 2015). These authors have presented an integrated scheme based on nanofiltration process to enhance the removal of toxic Cr (VI) species from tannery effluents. Mohammed and Sahu (2015) have published a study on an integrated bioadsorption and membrane separation process for the recovery of chromium from tannery waste. The authors have used different membranes (UF, NF and RO) for recovery of chromium among which RO is reported to exhibit 99.9% separation at optimum process conditions (pH 6.8, flow rate 0.72 m^3/h, feed pressure of 40 bar). MBR is another advanced approach for tannery effluent treatment, with different configurations attempted in treating tannery effluent for recovery of Cr (VI) with an eco-friendly process discharge. Suganthi et al. (2015) have presented a hybrid MBR for treatment of a tannery waste stream that is designed by combining three different processes; coagulation, biological treatment (activated sludge process) followed by microfiltration. A report on quality assessment of the final discharge from MF has shown about 90.2% reduction of COD and 92.75% reduction of color in hybrid MBR, which fulfils the condition of discharge effluent quality and provides better fallouts than conventional MBR, which has shown only 72.69% and 75.82% of COD and color removal, respectively. Angelucci et al. (2017) have suggested an advanced membrane separation technique for treating a simulated tannery effluent to achieve effective removal of the organic load and the toxic heavy metal, chromium. This group of authors have developed a continuous two-phase partitioning bioreactor that prevents direct contact between the toxic wastewater and the microorganisms themselves and hence, microbial activity remains unaffected by the influent wastewater composition. Munz et al. (2008) have done a detailed study on the effectiveness of the MBR system and comparison of its process efficiency with a conventional activated sludge process. Their work was based on development of a pilot-scale MBR and evaluation was done in terms of degradation and removal of organic and nitrogen compounds.

The designed MBR exhibited comparatively higher COD removal and a more stable and complete nitrification compared to conventional processes. Use of the MBR system in tannery waste treatment not only has improved process efficiency but also has lightened up a new route of value addition to the waste treatment process for a generation of alternative fuel source (biogas). The process involves treatment of raw tannery wastewater (with high suspended solids) using a flat sheet submerged anaerobic membrane (0.4 μm) bioreactor (SAMBR) with hypersaline anaerobic seed sludge for recovering biogas. The SAMBR system demonstrated higher COD removal efficiency and biogas yield, affirming its feasibility for treating raw tannery wastewater loaded with high suspended solids with minimized chemical use and a lower number of unit operations (clarifloculation and chemical sludge disposal). The authors have developed a novel ultrafiltration membrane (hydrophilic) based on nanochitosan, cellulose acetate, and polyethylene glycol using a phase inversion technique and claimed efficiency in reducing typical physicochemical parameters of BOD, COD, TDS, TSS, salinity, turbidity and electrical conductivity. Miceller enhanced ultrafiltration (MEUF) technique is another technological advancement for recovery of heavy metals from tannery effluent. The concept of miceller enhanced ultrafiltration was introduced by Leung (1979) as a promising method for the removal of toxic heavy metal ions (low concentration) and organic compounds from industrial effluents. Several other authors have reported the effectiveness of MEUF and NF in removal of various metal ions from diluted effluent and real effluent samples (Abbasi-Garravand et al., 2014).

23.3.6 Application in Paper and Pulp Industries

Membrane technology has been used in the pulp and paper industry since the late 1960s. The primary application at that time was the treatment of dilute process streams by RO with the primary objective of reducing water consumption in paper mills. Later, installations were created for reclaiming "white water" from colored effluents by ultrafiltration and nanofiltration. Fractionation of spent cooking liquor (black liquor) in the kraft chemical pulping process is another example of the efficient utilization of membrane technology. All the segments of membrane processes like MF, UF, NF, and RO have distinct roles to play in the pulp and paper industry. Another possibility of employing membranes in paper and pulp industry is the separation of black liquor into organic compounds (lignin) and inorganic chemicals like sodium and sulphur. Lignin is a polydispersed group of phenolic compounds and its fractionation to purer products is highly dependent on the molecular cut-off of the membrane in use. Wallberg et al. (2003) have studied UF process efficiency in fractionation and concentration of kraft black liquor. In this study, the lignin retention was reported as 80%, 67% and 45% for membranes with cut-off specifications of 4,000, 8,000 and 20,000 Da, respectively with insignificant retention of retention of sodium and sulphur. The authors have concluded that the membrane cut-off influences the recovery of lignin, its molar mass distribution and purity value. Researchers have established the application potential of membrane technology in the purification of the pulp and paper industry–generated waste stream. Their study aimed to evaluate different segments of membrane methods (MF, UF, NF) to generate permeated streams of acceptable quality with energy efficient process control. This group of authors designed a pilot plant equipped with different membrane modules based on a ceramic tubular membrane for high percentage water recovery with low energy consumption. An integrated membrane process consisting of MBR, continuous membrane filtration (CMF) unit and RO was used to treat the paper mill waste stream. The recovery of water in RO system was reported to be over 65%, which meets the high quality standards of process water from paper mill similar to distilled water (conductivity less than 200 mS/cm, COD lower than 15 mg/L, chroma less than 15 PCU and turbidity less than 0.1 NTU) (Zhang et al., 2009). A hybrid technique of electrochemical separation and membrane has also been reported for treatment of the paper and pulp industry effluent at ambient temperature. The results confirmed that the process can facilitate the removal of lignin and other organic and inorganic compounds, in terms of reduced color, COD, BOD, TSS and TDS. Nanofiltration process was explored for selective separation of the contaminants from paper mill effluent. The process involved biological treatment of newsprint mill effluent followed by an NF step to produce water suitable as feed to RO. Several researchers have also attempted methodologies with modified membranes and advanced module geometries besides ceramic membranes and hybrid processes for selective separation of the components from waste streams to enable water reclamation and eco-friendly discharge (Negaresha et al., 2012).

23.3.7 Dairy Wastewater Treatment

Membrane technology has modernized the dairy sector in all segments and is practiced in dairy industries for various purposes like prolonging the shelf life of milk without heat treatment, isolating the major components of milk for making new products, improving yield and quality of the dairy products, as well as concentration, fractionation and purification of milk components in their natural state. Dairy industry wastewater is recognized as a potential source of environmental pollution having a high BOD and COD load. The organic compounds in the waste streams, mainly casein, after discharge to surface water, lead to the formation of heavy black sludges and strong butyric acid odor. Drainage of whey causes a huge loss of valuable nutrients and creates environmental hazards. Hence, the discharge of process water from dairy industries necessitates cost-effective treatment technologies. Membranes must isolate valuable organic components to lower the BOD and COD load of the discharged stream, which comes under whey processing. Kumar et al. (2013) have presented a review on the utilization of the membrane process in the dairy industry. MF is used to remove suspended solids as the primary clarification step. UF with membranes of desired MWCO is used to produce whey protein concentrate and defatted whey concentrate. NF is meant for desalting of whey and recovery of lactose. RO is used to reduce the volume of water and recovery of total solids. The process scheme is based on a two-stage UF + NF process for the treatment of simulated dairy wastewater to recycle nutrients and water, and facilitate bioenergy production. The initial UF step involves isolation of whey protein and lipids in retentate while the permeate is further subjected to NF for lactose recovery. The retentate is subjected to algal fermentation to produce biodiesel (trans-esterified fat) and biofuel (bioethanol). The NF retentate (lactose fraction) is utilized to produce biogas under anaerobic digestion and the permeate is reusable quality water. Some authors have also used the lactose fraction for production of valuable nutraceuticals (β-galactosyl oligosaccharides) (Das et al., 2011). Another aspect for proliferating membrane technology in the dairy industry is the production of high quality effluent compared to other conventional treatment processes to tackle water crisis. To cope with the increasing scarcity of fresh water, membranes play a vital role in the production of water for subsequent use (i.e., water reclamation) and help in providing safe, clean and affordable water for industries. Studies have demonstrated the reutilization aspects of treating dairy effluent by an integrated MBR + NF approach. The outcome of experimental work projects MBR as an efficient process in removing organic matter and color from the effluent, while high concentration of dissolved solids in permeate necessitates the use of NF as a polishing step. The authors also confirmed the utility of treated effluent in the industry for cooling, steam generation and cleaning of external areas.

23.3.7.1 Case Study on Application of Membrane Technology in Dairy Effluent Treatment

Membrane processes find widespread application in dairy industries that include the production of milk protein concentrates and isolates (containing up to 90% protein on a dry basis) by UF and MF to reduce microbial load ("cold sterilization"), selective fractionation of dairy components and enhanced functional properties of dairy ingredients, production of cheese using ultrafiltered milk; both soft (e.g., cottage) cheese and hard (e.g., cheddar) cheese production from UF milk retentate, primarily with a view to understand the factors that govern the textural, ultra-structural and organoleptic properties of the UF-processed cheese with process modifications to optimize these properties. Application of RO saves energy during the manufacturing of concentrated and dried milk product from milk with reduced bulk transportation and refrigeration cost. In dairy industries, well planned use of MF/UF membranes, along with other advanced membrane techniques could lead to complete utilization of all ingredients contained in whey (Figure 23.2), thus leading to an effective environmental management solution, along with the production of value-added components, which makes the process sustainable in both environmental and economical points of view (Das et al., 2011(a, b); Sen et al., 2014; Nath et al., 2015). The authors of this article have studied in detail the application of membrane technology to develop a fully integrated effluent treatment process conforming to "zero effluent concept." The work has been carried out to assess the efficacy of the membrane-based process to produce nutraceuticals from dairy effluent (Figure 23.3), which could effectively lead to zero-effluent implementation in dairy industries. The authors have also published an article on separation of casein whey specifically with an aim to obtain a relative separation of β-lactoglobulin (β-LG) from whey protein concentrate by fractionation of proteins using a two-stage ultrafiltration (UF) with eco-friendly discharge (Figure 23.4). Experimental results are summarised in Table 23.3 which illustrates selective separation of valuable protein components to high purity levels.

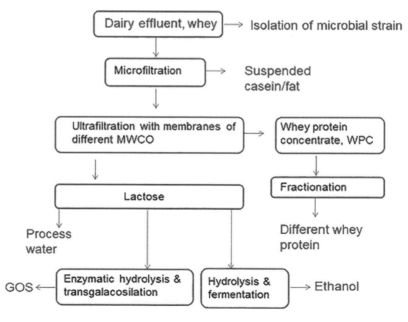

FIGURE 23.2
Schematic representation of zero-effluent implementation with dairy waste.

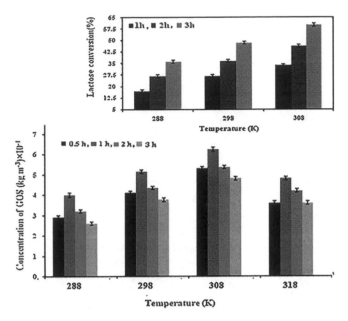

FIGURE 23.3
Effect of operating conditions on galactosyl oligosaccharide (Sen et al., 2014). Reproduced with permission.

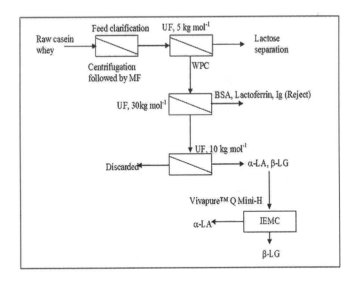

FIGURE 23.4
Schematic diagram of the proposed process flow-diagram (Bhattacharjee et al., 2006). Reproduced with permission.

TABLE 23.3

Compositions of Pretreated Casein Whey, UF Permeate (30 kg mol⁻¹), UF Permeate (10 kg mol⁻¹) and UF Retentate (10 kg mol⁻¹) at Different *pH*, under Fixed Operating Conditions (Bhattacharjee et al., 2006) Reproduced with permission

Compound	Concentration in Raw Casein Whey (kg m⁻³)	Concentration in Pretreated Casein Whey (kg m⁻³)	pH	Concentration (kg m⁻³)			
				UF Permeate (30 kg mol⁻¹)	UF Permeate (10 kg mol⁻¹)	UF Retentate (10 kg mol⁻¹)	IEMC* Permeate (buffer pH = 5.0)
α-LA	1.26 ± 0.0466	0.93 ± 0.032	2.8	0.81 ± 0.028	0.32 ± 0.011	1.02 ± 0.035	0.031 ± 0.001
			5.6	0.80 ± 0.025	0.28 ± 0.009	0.99 ± 0.034	0.028 ± 0.0008
β-LG	3.46 ± 0.13	3.23 ± 0.12	2.8	2.73 ± 0.1	0.43 ± 0.013	3.71 ± 0.14	0.539 ± 0.02
			5.6	2.29 ± 0.09	0.21 ± 0.007	3.05 ± 0.11	0.437 ± 0.017
TPC	7.66 ± 0.29	6.16 ± 0.23	2.8	3.87 ± 0.15	0.83 ± 0.03	5.11 ± 0.19	0.615 ± 0.024
			5.6	3.51 ± 0.14	0.71 ± 0.024	4.55 ± 0.17	0.502 ± 0.018
TS	645 ± 25	13.89 ± 0.5	2.8	11.50 ± 0.41	7.80 ± 0.31	15.20 ± 0.61	2. 050 ± 0.077
			5.6	11.15 ± 0.39	7.66 ± 0.31	14.64 ± 0.54	1.975 ± 0.098
%-Purity of β-LG w.r.t. TPC	45.2 ± 2.1	52.4 ± 2.6	2.8	70.6 ± 4.2	52. 4 ± 3.1	72.7 ± 4.5	87.6 ± 5.25
			5.6	65.1 ± 3.9	30.4 ± 1.9	67.0 ± 4.6	87.0 ± 5.1
%-Purity of β-LG w.r.t. TS	0.54 ± 0.032	23.2 ± 1.4	2.8	23.7 ± 1.64	5. 6 ± 0.38	24.4 ± 1.46	26.3 ± 1.55
			5.6	20.5 ± 1.22	2. 8 ± 0.18	20.8 ± 1.45	22.1 ±1.32

23.4 Advanced Membrane Separation Process for Treatment of Different Waste Streams

MBR is categorised as an advanced separation process for wastewater treatment. MBR is termed as "advanced integrated recycle and reuse solution." Several authors have covered several aspects of MBR, with a comprehensive overview of its operational performances with real effluents with the objective of water reclamation and environmental protection. Quite a lot of reports relate to modifications in configuration and hydraulics of MBR systems for betterment of its performance with respect to removal of organic contaminants present at trace concentrations (Radjenović et al., 2008). MBR technology combines the operation of the biological activated sludge processes and membranes to approach "process intensification" and overcome disadvantages of conventional activated sludge processes that cannot cope with varying compositions of wastewater or fluctuations in the wastewater flow rate.

References

Abbasi-Garravand, E. and C.N. Mulligan. 2014. Using micellar enhanced ultrafiltration and reduction techniques for *removal* of Cr(VI) and Cr(III) from water. *Separation and Purification Technology* 132: 505–512.

Abetz, V., T. Brinkmann, M. Dijkstra, K. Ebert, D. Fritsch, and K. Ohlrogge. 2006. Developments in membrane research: From material via process design to industrial application. *Advanced Engineering Materials* 8: 328–358.

Angelucci, D.M., V. Stazi, A.J. Daugulis, and M.C. Tomei. 2017. Treatment of synthetic tannery wastewater in a continuous two-phase partitioning bioreactor: Biodegradation of the organic fraction and chromium separation. *Journal of Cleaner Production* 152: 321–329.

Bernardo, P., E. Drioli, and G. Golemme. 2009. Membrane gas separation: A review/state of the art. *Industrial & Engineering Chemistry Research* 48: 4638–4663.

Bhattacharjee, S., C. Bhattacharjee, and S. Datta. 2006. Studies on the fractionation of β-lactoglobulin from casein whey using ultrafiltration and ion-exchange membrane chromatograph. *Journal of Membrane Science* 275: 141–150.

Bunani, S., E. Yörükoğlu, G. Sert, U. Yüksel, M. Yüksel, and N. Kabay. 2013. Application of nanofiltration for reuse of municipal wastewater and quality analysis of product water. *Desalination* 315: 33–36.

Das, R., D. Sen, A. Sarkar, and C. Bhattacharjee. 2011b. Study on the effect of membrane speed on enzymatic synthesis of galacto-oligosaccharides using immobilized β-Galactosidase in rotating disk membrane reactor (RDMR). *Journal of Institution of Engineers (India)* 91: 3–10.

Das, R., D. Sen, A. Sarkar, S. Bhattacharyya, and C. Bhattacharjee. 2011a. A comparative study on the production of galacto-oligosaccharide from whey permeate in recycle membrane reactor and in enzymatic batch reactor. *Industrial & Engineering Chemistry Resource* 50: 806–816.

Dasgupta, J., D. Mondal, S. Chakraborty, J. Sikder, S. Curcio, and H.A. Arafat. 2015. Nano filtration based water reclamation from tannery effluent following coagulation pretreatment. *Ecotoxicology and Enviromental Safety* 121: 22–30.

Falsanisi, D., L. Liberti, and M. Notarnicola. 2009. Ultrafiltration (UF) pilot plant for municipal wastewater reuse in agriculture: Impact of the operation mode on process performance. *Water* 1: 872–885.

Frost & Sullivan. 2009. Assessment of Indian Desalination Market.

Frost & Sullivan. 2013. Global Water and Wastewater Filtration Systems Market, Mountain View.

Ganiyu, S.O., E.D. Van Hullebusch, M. Cretin, G. Esposito, and M.A. Oturan. 2015. Coupling of membrane filtration and advanced oxidation processes for removal of pharmaceutical residues: A critical review. *Separation and Purification Technology* 156: 891–914.

Garcia-Ivars, J., J. Durá-María, C. Moscardó-Carreño, C. Carbonell-Alcaina, M.I. Alcaina-Miranda, and M.I. Iborra-Clar. 2017. Rejection of trace pharmaceutically active compounds present in municipal wastewaters using ceramic fine ultrafiltration membranes: Effect of feed solution pH and fouling phenomena. *Separation and Purification Technology* 175: 58–71.

Gómeza, M., F. Plazab, G. Garralónb, J. Péreza, and M.A. Gómeza. 2007. A comparative study of tertiary wastewater treatment by physico-chemical-UV process and macrofiltration-ultrafiltration technologies. *Desalination* 202: 369–376.

Hagg, M. 2009. Membranes in gas separation. In A.K. Pabby, S.S.H. Rizvi, and A.M. Sastre (Eds.), *Handbook of membrane separations: Chemical, pharmaceutical, food, and biotechnological applications* (65–105). Boca Raton, FL: CRC Press.

Hurk, P.V.D., L.E. Gerzel, P. Calomiris, and D.C. Haney. 2017. Phylogenetic signals in detoxification pathways in Cyprinid and Centrarchid species in relation to sensitivity to environmental pollutants. *Aquatic Toxicology* 188: 20–25.

Jiang, Y.Y., Z. Zhou, Z. Jiao, L. Li, Y.T. Wu, and Z.B. Zhang. 2007. SO_2 gas separation using supported ionic liquid membranes. *Journal of Physical Chemistry B* 111: 5058–5061.

Košutic, K., D. Dolar, D. Asperger, and B. Kunst. 2007. Removal of antibiotics from a model wastewater by RO/NF membranes. *Separation and Purification Technology* 53: 244–249.

Kovacika, J., P. Babulab, and J. Hedbavnyd. 2017. Comparison of vascular and non-vascular aquatic plant as indicators of cadmium toxicity. *Chemosphere* 180: 86–92.

Kumar, P., N. Sharma, R. Rajan, S. Kumar, and Z.F. Bhat. 2013. Dong keeJeong, Perspective of membrane technology in dairy industry: A review. *Asian-Australas Journal of Animal Sciences* 26: 1347–1358.

Ledley, T.S., E.T. Sundquist, S.E. Schwartz, D.K. Hall, J.D. Fellows, and T.L. Killeen. 1999. Climate change and greenhouse gases. *EOS-Transactions American Geophysical Union* 80: 453–458.

Leung, P.S. 1979. Surfactant micelle enhanced ultrafiltration. In A.R. Cooper (Ed.), *Ultrafiltration Membranes and Applications* (415–421). New York, NY: Plenum Press.

Madaeni, S.S., N. Ghaemi, H. Rajabi. 2015. Advances in polymeric membranes for water treatment. In *Advances in Membrane Technologies for ater Treatment, Materials, Processes and Applications, A volume in Woodhead Publishing Series in Energy* (3–41). A. Basile, A. Cassano, and N.K., Rastogi (Eds.). Woodhead Publishing.

Mahlangu, T.O., E.M.V. Hoek, B.B. Mamba, and A.R.D. Verlifede. 2014. Influence of organic, colloidal and combined fouling on NF rejection of NaCl salt and carbamazepine: Role of solute-foulant-membrane interactions and cake enhanced concentration polarization. *Journal of Membrane Science* 471: 35–46.

Mohammed, K. and O. Sahu. 2015. Bioadsorption and membrane technology for reduction and recovery of chromium from tannery industry wastewater. *Environmental Technology & Innovation* 4: 150–158.

Munz, G., M. Gualtiero, L. Salvadori, B. Claudia, and L. Claudio. 2008. Process efficiency and microbial monitoring in MBR (membrane bioreactor) and CASP (conventional activated sludge process) treatment of tannery wastewater. *Bioresource Technology* 99: 8559–8564.

Mutamim, N.S.A., Z.Z. Noor, M.A.A. Hassan, A. Yuniarto, and G. Olsson. 2013. Membrane bioreactor: Applications and limitations in treating high strength industrial wastewater. *Chemical Engineering Journal* 225: 109–119.

Nath, A., S. Chakraborty, C. Bhattacharjee, and R. Chowdhury. 2015. Studies on the separation of proteins and lactose from casein whey by cross-flow ultrafiltration. *Desalination and Water Treatment* 54(2): 481–501.

Negaresha, E., A. Antony, M. Bassandeh, D.E. Richardson, and G. Leslie. 2012. Selective separation of contaminants from paper mill effluent using nanofiltration. *Chemical Engineering Research & Design* 90: 576–583.

Nghiem, L.D. and P.J. Coleman. 2008. NF/RO filtration of the hydrophobic ionogenic compound triclosan: Transport mechanisms and the influence of membrane fouling. *Separation and Purification Technology* 62: 709–716.

Nidheesha, P.V. and T.S. Anantha Singh. 2017. Arsenic removal by electrocoagulation process: Recent trends and removal mechanism. *Chemosphere* 181: 418–432.

Peric, J., M. Trgo, and N.V. Medvidovic. 2004. Removal of zinc, copper and lead by natural zeolite-a comparison of adsorption isotherms. *Water Research* 38: 1893–1899.

Qu, X., P.J.J. Alvarez, and Q. Li. 2013. Applications of nanotechnology in water and wastewater treatment. *Water Research* 47: 3931–3946.

Radjenović, J., M. Matošić, I. Mijatović, M. Petrović, and D. Barceló. 2008. Membrane Bioreactor (MBR) as an advanced wastewater treatment technology. *Handbook of Enrvionmental Chemistry* 5: 37–101.

Rautenbach, R., K. Vossenkaul, T. Linn, and T. Katz, T. 1996. Wastewater treatment by membrane processes - New development in ultrafiltration, nanofiltration and reverse osmosis. *Desalination* 108: 247–253.

Rodriguez-Narvaez, O.M., J.M. Peralta-Hernandez, A. Goonetilleke, and E.R. Bandala. 2017. Treatment technologies for emerging contaminants in water: A review. *Chemical Engineering Journal* 323: 361–380.

Royan, F. 2012. Sustainable water treatment technologies in the 2020 global water market.

Sen, P., A. Nath, C. Bhattacharjee, R. Chowdhury, and P. Bhattacharya. 2014. Process engineering studies of free and micro-encapsulated β-Galactosidase in batch and packed bed bioreactors for production of galacto oligosaccharides. *Biochemical Engineering Journal* 90: 59–72.

Setiawan, L., R. Wang, K. Li, and A.G. Fane. 2011. Fabrication of novel poly(amide-imide) forward osmosis hollow fiber membranes with a positively charged nanofiltration-like selective layer. *Journal of Membrane Science* 369: 196–205.

Shekhawat, D., D.R. Luebke, and H.W. Pennline. 2003. A review of carbon dioxide selective membranes-A topical report. National Energy Technology Laboratory, U.S. Department of Energy.

Srivastava, N.K. and C.B. Majumder. 2008. Novel biofiltration methods for the treatment of heavy metals from industrial wastewater. *Journal of Hazardous Materials* 151: 1–8.

Taheran, M., S.K. Brar, M. Verma, R.Y. Surampalli, T.C. Zhang, and J.R. Valero. 2016. Membrane processes for removal of pharmaceutically active compounds (PhACs) from water and wastewaters. *Science of the Total Environment* 547: 60–77.

Tchobanoglous, G., J. Darby, K. Bourgeous, J. McArdle, P. Genest, and M. Tylla. 1998. Ultrafiltration as an advanced tertiary treatment process for municipal wastewater. *Desalination* 119: 315–322.

UNEP (United Nation Education Programme). 2010. The central role of wastewater management in sustainable development.

Verlicchi, P., M. Al-Aukidy, and E. Zambello. 2012. Occurrence of pharmaceutical compounds in urban wastewater: Removal, mass load and environmental risk after a secondary treatment-A review. *Science of the Total Environment* 429: 123–155.

Verliefde, A.R.D., E.R. Cornelissen, S.G.J. Heijman, J.Q.J.C. Verberk, G.L. Amy, B. van der Bruggen, and J.C. Van Dijk. 2008. The role of electrostatic interactions on the rejection of organic solutes in aqueous solutions with nanofiltration. *Journal of Membrane Science* 322: 52–66.

Wallberg, O., A.S. Jönsson, and R. Wimmerstedt. 2003. Fractionation and concentration of kraft black liquor lignin with ultrafiltration. *Desalination* 154: 187–199.

Wang, Y., J. Zhu, H. Huang, and H.H. Cho. 2015. Carbon nanotube composite membranes for microfiltration of pharmaceuticals and personal care products: Capabilities and potential mechanisms. *Journal of Membrane Science* 479: 165–174.

Yangali-Quintanilla, V., A. Sadmani, M. McConville, M. Kennedy, and G. Amy. 2009. Rejection of pharmaceutically active compounds and endocrine disrupting compounds by clean and fouled nanofiltration membranes. *Water Research* 43: 2349–2362.

Zhang, Y., C. Ma, F. Ye, Y. Kong, and H. Li. 2009. The treatment of wastewater of paper mill with integrated membrane process. *Desalination* 236: 349–356.

Zhao, Y. and W.S.W. Ho. 2012. Steric hindrance effect on amine demonstrated in solid polymer membranes for CO_2 transport. *Journal of Membrane Science* 415 416: 132–138.

24

Processing of Dairy Industrial Effluent and Kitchen Wastewater by Integration of Microbial Action with Membrane Processes

S.S. Chandrasekhar, Nivedita Sahu, and Sundergopal Sridhar

CONTENTS

24.1 Introduction

Increasing demands for clean water supplies continue to drive the hunt for new and sustainable water sources throughout the world. Wastewater reclamation and reuse are indispensable to avoid water shortage and degradation of the environment. Sustainable and low-energy demanding technologies are especially important to achieve these goals. Industrial effluents contain highly variable levels of Biological Oxygen Demand (BOD) and Chemical Oxygen Demand (COD) that are directly released into the environment, causing hazardous pollution. In order to reduce the effect of these pollutants/effluents, membrane processes and their applications are one of the possible solutions for tackling problems in wastewater treatment and reuse (Ashe and Paul, 2010; Singh et al., 2014). Membrane process/ filtration technology is a promising and proven method mainly used in industrial process steps meant for production or separation of specific component or product. Most companies use membranes to improve bioprocess efficiency, lower their energy costs and increase product recovery. Membranes are physical barriers that are used in different processes for separation of liquid-liquid and solid-liquid mixtures. Membrane filtration has been extensively used in many manufacturing industries for years (Amin et al., 2016). Wastewater treatment was first started in the late 19th century and became a standard method by the 1930s. Additionally, both aerobic and anaerobic biological wastewater treatment technologies can be used for kitchen wastewater and industrial effluents (Shah, 2016; Visvanathan et al., 2000).

24.2 Membrane Filtration Technology

The word "membrane" is derived from Latin word "membrana." Membrane filtration is a separation process in which a membrane acts as a semi-permeable thin barrier that separates permeate and retentate from a feed when driving force is applied due to the availability of fine pores.

24.2.1 Membrane Characteristics

Most membranes are made up of polymers or inorganic materials, and they consist of either a dense nonporous structure or a lot of microscopic tiny pores. The focus of this study is on 'porous membranes' that control the passage of feed components from one side to the other based on size exclusion though fine pores that allow smaller particles and water to penetrate, resulting in a permeate containing mostly water, while larger particles are retained. The nature, molecular weight and amount of retained particles depend upon the pore diameter and pore size distribution of the membrane. Classification of membranes is based on the type of material, structure, morphology and driving force. Organic membranes are generally based on polyethersulfone, polyacrylonitrile, polyvinylidene

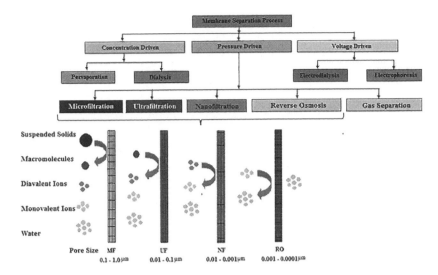

FIGURE 24.1
Classification of pressure driven membrane filtration processes.

fluoride, polyetherimide and polyamide. Inorganic membranes are based on zeolites, ceramics and reinforced carbon material.

24.2.2 Classification of Membranes

Membrane processes are broadly classified according to the driving force that brings about separation as shown in Figure 24.1. Pressure difference, concentration gradient or electrical potential between the two sides of the membrane acts as the driving force. Separation is achieved due to the difference in the rates of transport or diffusivities of different species though the membrane, and is therefore kinetically driven.

24.3 Potential of Membrane Filtration When Combined with Biological Process

The membrane filtration process is a potentially viable, efficient and cost-effective process with a wide range of applications in many branches in industrial sectors, such as water, wastewater, food processing, pharmaceutical, chemicals, biotechnology, medicines and more (Praneeth et al., 2014). Membrane filtration is technically and economically more advantageous than conventional activated sludge process. It is easy to operate, simple to install, requires low process costs and is eco-friendly (Rafik et al., 2015). For the past two decades, this technology has provided major solutions in addressing water-related problems. To overcome the disadvantage of conventional technologies, biological treatment technologies have been integrated with membrane methods. In the last few years, membrane bioreactors (MBR) have emerged as a key treatment method for the separation of contaminants from wastewater (Santasmasas et al., 2014). A membrane bioreactor (MBR) can be defined as a combination of biological activated sludge or anaerobic digestion with membrane filtration technology that is used to treat wastewater, and it has risen in popularity in the context of wastewater reclamation and reuse. The technology is quite effective in removing organic biological entities and inorganic contaminants from wastewater (Venkata Narayana et al., 2016). MBRs use membrane modules to remove of turbidity, microbes and colloidal impurities like silica, while biological reactions account for organic carbon and nitrogen removal. Compared to conventional biological processes (e.g., an activated sludge process), MBRs have a smaller footprint and produce a higher quality effluent with less sludge. MBR is considered an advanced treatment process over a variety of activated sludge processes and is also an alternative to conventional activated sludge (CAS) systems due to its high treatment efficiency. MBR is used in both municipal and industrial wastewater treatment to aid in water reclamation and reuse (Lin et al., 2012; Chu et al., 2006). The advantages of MBR compared to other methods include good control of biological activity, production of high quality water that is free of bacteria and pathogens, smaller plant size, higher organic loading rates, less sludge production and flexibility of operation. In addition to industrial wastewater treatment, incorporation in STPs to enable water recycling and landfill leachate treatment, MBRs could be downsized and used to cater to water conservation in residential complexes for small communities by treating grey and black water (Nazim, 2002; Chen et al., 2003).

Microbial Fuel Cell (MFC) is a technology that uses exoelectrogenic biofilms on an anodic chamber to degrade organic matter and produce renewable energy in the form of electricity (Logan and Regan, 2006). In 1910, Potter demonstrated the concept of MFC. In the demonstration, electrical energy was produced from living bacterial cultures by using different electrodes that acted as both anode and cathode chambers. In the treatment of wastewater systems, membranes in the MFC play an important role. Here, an anaerobic tank acts as an anode chamber: the bacterial culture will use the organic matter in this tank to produce electrons and protons. The electrons will then transfer to an anode, subsequently flow across a conductive material containing a resistor and finally get moved to the cathode chamber. The protons are also transferred to the cathode chamber through the proton exchange membrane (PEM). Both electrons and protons eventually reach the cathode chamber. MFC is thus a bioreactor that produces a high value of energy in the form of electricity: it breaks down the organic matter present in wastewater by using a bacterial culture to constitute a new form of renewable energy (Kumar et al., 2012). Since the last decade, MFC technology has played a significant role in meeting the challenges posed by sewage and other wastewater treatment due to its capability to produce energy in the form of electricity or hydrogen. Different types of wastewater that contain biodegradable organic matter coming from sanitary, food processing, starch processing and distillery industries are well-suited for generating electricity or hydrogen gas by using MFC technology (Zhang et al., 2012).

24.4 Challenges Facing MBR and MFC Technologies

MBR and MFC processes have emerged rapidly as process intensification strategies for the treatment of different types of wastewater by combining membrane separation with a biological process to reduce equipment size (smaller footprint), lower running costs, enhance the process and increase environmental safety (Melin et al., 2006; Bixio et al., 2008). The key advantage of these treatment methods is water reclamation and the reuse of the treated water for ground water recharge, flushing toilets, irrigation, gardening and industrial utility, especially when compared to instead of previously when the untreated discharge entered into the ecosystem and caused environmental hazards (Tadkaew et al., 2007; Pham et al., 2006). MFCs have exhibited over 90% of charge efficiency in some cases and as much as 80% COD removal. One of the major disadvantages of MFC is the thick layer of biofilm that forms on the anode chamber, which contains a community of bacterial and microbial organic substances. Due to the formation of the cake-like layer, electron and proton transport rates will be less in the reactor (Rabaey and Verstraete, 2005).

24.5 Types of Membrane Bioreactors

MBR systems can be configured in two ways. When membrane filtration occurs inside the bioreactor it is known as submerge, internal or immersed MBR; in contrast, filtration occurring outside the bioreactor is called side-stream or external MBR.

24.5.1 Submerged/Immersed MBR

In the submerged or immersed configuration of MBR, the membrane is directly immersed in the bioreactor shown in Figure 24.2 (a). The membranes can be configured either

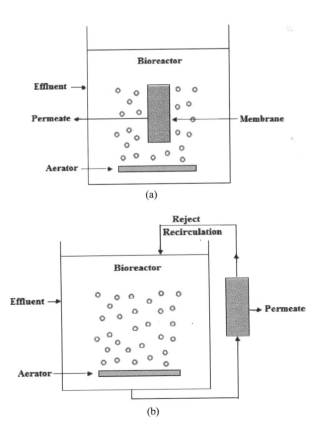

(a)

(b)

FIGURE 24.2
Schematic representations of (a) submerged/internal and (b) side-stream/external membrane bioreactors.

horizontally or vertically and can be made of spirally wound or plate and frame geometry, although hollow fibers are more commonly used.

24.5.2 Side-Stream/External MBR

In the side-stream type of MBR configuration, the membrane module is located outside the reactor as shown in Figure 24.2 (b). The mixed liquor suspended solids (MLSS) are also circulated though the external module along with water.

24.5.3 Advancements in MBR Process

There have been progressive studies from 1992 on MBR process for treating different effluents as shown in Table 24.1.

TABLE 24.1

Performance of Various Membrane Process Configurations for the Treatment of Different Industrial Effluents

S. No	Membrane Configuration	Type of Effluent	Reduction Parameter	Reference
1	Ultrafiltration Side stream	Food industry	97% of COD	Ross et al. (1992)
2	Ultrafiltration Side stream	Different industries	97% of COD	Krauth and Staab (1993)
3	Ultrafiltration Side stream	Automotive industry	94% of COD	Knoblock et al. (1994)
4	Microfiltration Side stream	Sludge digestion	Not available	Pillay et al. (1994)
5	Ultrafiltration Side stream tubular	Kitchen and municipal	94% of COD	Fan et al. (1996)
6	Ultrafiltration Side stream flat plate	Industrial	Not available	Choo and Lee (1996)
7	Microfiltration Submerged tubular	Kitchen	Not available	Shimizu et al. (1996)
8	Microfiltration Submerged hollow fiber	Kitchen	Not available	Ueda et al. (1997)
9	Ultrafiltration submerged	Municipal	96% of COD	Buisson et al. (1998)
10	Microfiltration Submerged flat plate	Synthetic	Not available	Nagaoka et al. (1998)
11	Ultrafiltration Submerged Tubular	Kitchen and municipal	99% of COD	Cicek et al. (1998)
12	Ultrafiltration Submerged hollow fiber	Kitchen and municipal	96% of COD	Cote et al. (1998)
13	Microfiltration Side stream tubular	Kitchen	Not available	Tardieu et al. (1998)
14	Microfiltration Side stream tubular	Synthetic	Not available	Elmaleh and Abdelmoummi (1998)
15	Microfiltration Submerged hollow fiber	Kitchen	Not available	Parameshwaran et al. (1999)
16	Microfiltration Submerged hollow fiber	Kitchen	Not available	Shin et al. (1999)

(Continued)

TABLE 24.1 (CONTINUED)

Performance of Various Membrane Process Configurations for the Treatment of Different Industrial Effluents

S. No	Membrane Configuration	Type of Effluent	Reduction Parameter	Reference
17	Microfiltration Submerged tubular	Kitchen	Not available	Ueda and Horan (2000)
18	Microfiltration Side stream tubular	Kitchen and industrial	Not available	Defrance et al. (2000)
19	Ultrafiltration Side stream flat plate	Synthetic	Not available	Chang et al. (2001)
20	Ultrafiltration Submerged plate and frame	Kitchen and municipal	91% of COD	Van de Roest et al. (2002)
21	Ultrafiltration Submerged hollow fiber	Kitchen and municipal	95% of COD	Lorenz et al. (2002)
22	Ultrafiltration Submerged hollow fiber	Kitchen and municipal	95% of COD	Rosenberger et al. (2002)
23	Ultrafiltration	Metal plating, chemical and textile industry	90% of COD and 50% of heavy metals	Katsou et al. (2012)
24	Nanofiltration	Winery industry	95% of polyphenols and polysaccharides	Alexandre et al. (2013)
25	Ultrafiltration Flat sheet	Municipal and various industries	70% of COD and 50% of heavy metals	Majid et al. (2014)
26	Ultrafiltration	Dairy	95% of TSS	Devendra and Rohit (2014)
27	Ultrafiltration	Textile	90% of COD and color	Devendra and Rohit (2014)
28	Ultrafiltration Hollow fiber	Pharmaceutical	90% of COD	Devendra and Rohit (2014)
29	Ultrafiltration Hollow fiber	Refinery	60% of COD	Devendra and Rohit (2014)
30	Ultrafiltration Flat sheet	Oil	90% of COD	Devendra and Rohit (2014)
31	Microfiltration Flat sheet	Crude oil	90% of COD	Jiang et al. (2015)
32	Reverse osmosis	Reject water from different industries	85% of TDS and ions removal	Virapan et al. (2017)

Sources: Chang, I.S., J.S. Kim, and C.H. Lee, *Proceeding of MBR3 Conference*, 16th of May, Cranfield: 19–28, 2001; Cicek, N., J.P. Franco, M.T. Suidan, and V. Urbain, *Journal American Water Works Association*, 90(11): 105–113, 1998.

24.5.4 Microbial Consortia Used in MBR

Various microbes exhibit effective reduction in different parameters as depicted in Table 24.2.

TABLE 24.2

Treatment of Various Effluents Using Mixed Microbial Consortia

S. No	Type of Microbial Consortia	Type of Effluent	Reference
1	Lactic acid bacteria	Ice cream wastewater	Scott and Smith (1996)
2	Mixed group of bacteria	Oil wastewater	Soltani et al. (2010)
3	Pseudomonas species - LZ-Q	Soil at Petrochemical	Jiang et al. (2015)
4	Pseudomonas stutzeri - ZP2	Soil at oil refinery	Janbandhu and Fulekar (2011)

24.6 Types of MFCs

Generally, two types of MFCs are available, either single chambered MFC or multiple chambered MFC.

24.6.1 Design and Working of Microbial Fuel Cell (MFC)

The design of prototype MFC model plays a major role in influencing the performance of the system in terms of power generation, positive and negative charge efficiency and stability. A prototypical model of MFC consists of an anode chamber and a cathode chamber separated by a PEM as shown in Figure 24.3 (a).

(a)

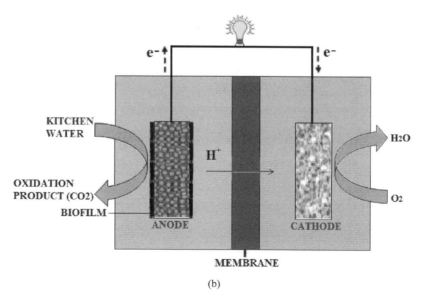

(b)

FIGURE 24.3

(a) Prototype model of MFC separated by a PEM and (b) Schematic diagram of double chambered MFC.

24.6.2 Advantages in MFC Process (Table 24.3)

TABLE 24.3

Efficiency of Different Chambered Fuel Cells in Treating Industrial Wastewater

S. No	Type of Microbial Fuel Cell	Type of Effluent	Reduction Parameter	Power/Current	Reference
1	Two-chamber	Synthetic waste	COD 70%	2.37 mA	Liu et al. (2004)
2	Single-chamber	Food processing (Cereal)	COD 95%	371 ± 10 mW/m^2	Oh and Logan (2005)
3	Air cathode	Municipal	COD 70%	10^3 mW/m^2	You et al. (2006)
4	Two-chamber	Chocolate industry	COD 75%	3.02 mA	Zhang et al. (2008)
5	Air cathode	Starch processing	COD 95%, Ammonia-nitrogen 90%	239.4 mW/m^2	Zhang et al. (2011)
6	Single-chamber	Dairy	COD 91%	20.2 W/m^2	Mahdi Mardanpour et al. (2012)
7	Two-chamber	Dairy	COD 90%, 99% turbidity	14.92 mA	Wang et al. (2013)
8	Single-chamber	Agriculture	COD 70%	55 mW	Santoro et al. (2013)
9	Single-chamber	Municipal	COD 60%, BOD 65%	Not available	Sciarria et al. (2013)
10	Two-chamber	Dairy industry wastewater	COD 90%, BOD 80%, TSS 70%	621.13 mW/m^2	Mansoorian et al. (2016)
11	Up-flow anaerobic sludge blanket reactor	Beet-sugar	COD 50%, Sulfate 50%, Color 40%	1410.2 mW/m^2	Cheng et al. (2016)

Source: Li, X., X. Hu, and T. Cai, *Langmuir*, 33: 4477–4489, 2017.

24.6.3 Microbial Consortia Used in MFC (Table 24.4)

TABLE 24.4

Evaluation of the Performance of Mixed Microbial Consortia for Effluent Treatment

S. No	Type of Microbial Consortia	Type of Effluent	Reference
1	*Comamonas testosterone* (Gram-negative bacteria)	Artificial wastewater	Der-Fong Juang (2012)
2	*Candida albicans, Bacillus subtilis*	Dairy effluent	Mostafa (2013)
3	Anaerobic micro organisms	Sludge from wastewater plant	Passos et al. (2015)
4	Aerobic and anaerobic micro organisms	Dairy effluent	Mansoorian et al. (2016)

24.7 Experimental Case Study with Kitchen Wastewater

24.7.1 Collection of Kitchen Wastewater

30 L of kitchen wastewater was collected from the CSIR-IICT canteen, Hyderabad, and characterized for various parameters like pH, TSS, TDS, turbidity, conductivity, COD and BOD (Ashish, 2014).

24.7.2 Preparation of Inoculums

The mixed microbial consortia were isolated from the yogurt of buttermilk for treatment of kitchen wastewater by using nutrient agar plates. The inoculation was done by serial dilution method and incubated at 30°C. Isolated colonies on nutrient agar plates were inoculated in 2000 mL of nutrient broth and kept for incubation at 30°C for 24 h (overnight incubation) (Sreemoyee and Priti, 2013).

24.7.3 Description of Submerged/Internal MBR (SMBR)

The schematic experimental model of SMBR with 30 L feed capacity is described in Figure 24.4 (a). The reactor was submerged with a hydrophilized ultrafiltration (HF-UF) membrane module of 10 kDa molecular weight cut off (MWCO), and the flow line was connected to a booster pump's suction end for drawing the permeate as the final treated water. The kitchen wastewater was taken into a reactor in which 5% of mixed microbial consortia were added and oxygen was supplied for aerobic digestion by an aerator. The filtration unit was operated in batch mode at room temperature (30°C). The wastewater level in the bioreactor was maintained constant at 20 L, and the hydraulic retention time (HRT) was maintained at 21 days. Continuous coarse bubble aeration by the air diffuser was applied to promote local cross flow velocity along the membrane surface and simultaneously to produce enhanced dissolved oxygen (DO). 15 L was collected as permeate after 21 days for analysis.

24.8 Experimental Case Study on Dairy Industrial Effluent

24.8.1 Collection of Dairy Industrial Effluent

30 L of dairy industrial effluent was collected from the nearby Vijaya Dairy, Hyderabad, and analyzed for various parameters like pH, TSS, TDS, turbidity, conductivity, COD and BOD.

24.8.2 Preparation of Inoculums

The mixed microbial consortia were isolated from the dairy effluent and cultured on nutrient agar plates. The inoculation was done by serial dilution method and incubated at 30°C. Isolated colonies on nutrient agar plates were inoculated in 3000 mL of nutrient broth and kept for incubation at 30°C for 24 h (overnight incubation).

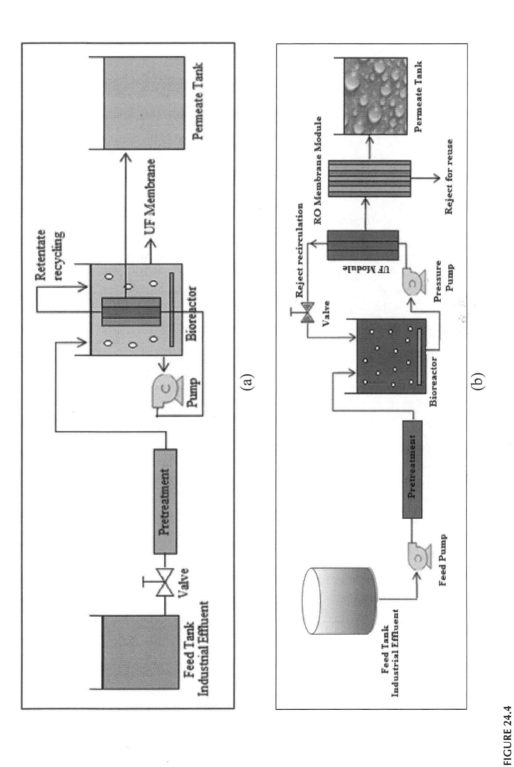

FIGURE 24.4

(a) Schematic diagram of (a) Aerobic SMBR and (b) Aerobic integrated SSMBR+RO.

24.8.3 Description of Side-Stream/External MBR (SSMBR)

The experimental model of SSMBR of 40 L feed capacity is presented in Figure 24.4 (b). The reactor was side-stream/external with hydrophilized ultrafiltration (HF-UF) membrane module of 10 kDa molecular weight cut off (MWCO), and the permeate tube was connected to the suction side of a booster pump for drawing the permeate as the final treated effluent. The dairy effluent was taken into the reactor to which 8% of mixed microbial consortia was added and oxygen was supplied for aerobic digestion by the aerator. The filtration unit was operated in batch mode at room temperature (30°C). The effluent level in the bioreactor was maintained constant at 30 L while HRT was again at 21 days following which was 20 L was collected as permeate. Continuous coarse bubble aeration by air diffuser was applied to promote local cross flow velocity along the membrane surface and simultaneously produce enhanced dissolved oxygen (DO).

24.8.4 Description of Laboratory MFC Unit for Treatment of Kitchen Wastewater

The experiment was done using kitchen wastewater collected from the canteen of CSIR-IICT, Hyderabad. The lab scale MFC setup contained two identical compartments of cubical shape, fabricated from Plexiglass material, as seen in Figure 24.3 (b). Both the compartments had equal volume of 0.6 L, between which the indigenously prepared PEM was placed. Graphite was used as electrodes with a circular area of 19.6 cm^2 for both the anode and cathode compartments. The anode compartment where microorganisms grew and the biofilms formed were fed with 100 mL of culture isolated from cow manure and kept under anaerobic condition. To maintain the anaerobic conditions in the anode compartment, it was purged with N_2, whereas potassium permanganate ($KMnO_4$) at 0.2 g L^{-1} and pH 7.0 ($\pm$ 0.2) was used in the cathode side as an effective electron acceptor. The environment friendly oxidant nature of $KMNO_4$ helps in the enhancement of MFC performance. The cathode chamber was aerated with proper mixing, whereas continuous stirring was provided in the anode chamber for complete circulation and uniform mixing of the feed solution. The copper wire present at the top of the both compartments connected the electrodes to different external resistors of load ranging from 1 Ω – 12 kΩ. Furthermore, a digital multi-meter was used to record voltage and current values. The operating conditions for MFC were ambient temperature (30 $\pm$ 5°C) and atmospheric pressure.

In order to measure the cell voltage, the following equation was used:

$$V = I \times R \tag{24.1}$$

Where I is the current (ampere), V is the voltage (volt) and R is the resistance (ohms). After acquiring stable current reading, the power density was calculated by the following equation:

$$P = I \times V / A \tag{24.2}$$

Where A is the projected area of the membrane (cm^2).

24.9 Sampling and Analytical Methods

The performance of the MBR and MFC systems was assessed by monitoring both water quantity and quality under various operating conditions. Permeate and retentate samples were analyzed at regular intervals for total dissolved solids (TDS), total suspended solids (TSS) and COD according to APHA methods. Turbidity was evaluated by using a HACH Make Colorimeter (DR/890). The conductivity was determined by using a digital conductivity meter (Model DCM-900, Global Electronics, Hyderabad, India) and pH of the samples were determined using a pH meter (Model DPH-504, Global Electronics, Hyderabad, India). COD was determined by closed reflux titrimetric method and modified Winkler's method. The total suspended solids (TSS) concentration was assessed by weighing a sample after passing it through a Whatman filter paper and drying for 1 h at 105°C. Approximately 2 ml of samples were refluxed in a COD digester for a period of 3 h at 150°C. COD was then calculated by using the following formula:

$$COD = (kb - ks) \times DF \times M \times 8000 \qquad (24.3)$$

Where *kb* and *ks* are the amount of ferrous ammonium sulfate (FAS) consumed for blank and actual sample, respectively. *DF* is dilution factor of the sample, and *M* is the molarity of the FAS solution. The percentage COD reduction was calculated by the following equation:

$$COD \text{ reduction } \% = (CODi - CODf) / CODi \times 100 \qquad (24.4)$$

Where *CODi* and *CODf* are the initial and final COD values in ppm (before and after treatment), respectively.

24.10 Membrane Fouling and Its Prevention

In general, fouling of the membrane is caused by suspended solids, inorganic salts, microbes, organic compounds, oil and grease and metal hydroxides (Fe, Mn, Al) present in the feed water that accumulate either on the membrane surface or within the pores. Fouling of the membrane results in higher operating costs due to low water flux and reduced process efficiency, and can necessitate an early replacement of the membrane. The membrane modules in the case study were dismantled from the reactor and cleaned at regular intervals by chemical cleaning followed by water wash at 2 kg/cm^2 pressure. The chemical cleaning was performed by treating with 1-2% citric acid for 15 min followed by water wash for 10 min. The membrane was further washed by a mixture of 1% NaOH + 0.5% EDTA + 0.1% sodium lauryl sulphate (surfactant) for 15 min. and again washed with water for 10 min. Following the above washes, the scales formed were completely removed (Chang et al., 2002; Judd, 2005).

24.11 Results and Discussion

24.11.1 Treatment of Kitchen Wastewater and Dairy Effluent

For kitchen wastewater, the initial turbidity, COD, TDS, TSS, conductivity and pH were recorded to be 71 FAU, 3,967 ppm, 1,382 ppm, 450 ppm, 1.89 mS/cm and 8.35, respectively (Table 24.5), accompanied by bad odor and a dark yellow color. Microbial culture isolated from curd was found to actively grow in the SMBR after the third day of the experiment. Similarly, initial parameters for dairy industry effluent was recorded as 32 FAU, 6,000 ppm, 4,220 ppm, 2,875 ppm, 11 mS/cm and 8.66, respectively (Table 24.6), and accompanied by bad odor and black color. The culture isolated from curd was not found effective in reducing the feed values. Bacteria isolated from the dairy effluent itself were found to rapidly multiply in the SSMBR after the third day of the experiment. Subsequently, a reduction in impurity levels in dairy effluent was found (Table 24.6) (Chandrasekhar et al., 2017).

In both cases, the MBR system was operated for 21 days at a temperature of 30°C. After 21 days of experimentation, neither the SMBR nor the SSMBR showed a reduction in the impurity levels. The reduction trends for turbidity, COD, TDS, TSS, conductivity, and pH in kitchen wastewater and dairy effluent are shown in Figures 24.5 (a–d) and 24.6 (a–b), respectively.

TABLE 24.5

Characteristics of Kitchen Wastewater before and after Treatment by Submerged MBR

S. No	Quality Parameter	Units	Feed Value	Permeate Value
1	pH	–	8.35	6.90
2	TDS	ppm	1382	1000
3	TSS	ppm	450	120
4	Turbidity	FAU	71	6
5	COD	ppm	3967	120
6	Conductivity	mS/cm	1.89	1.40

TABLE 24.6

Characteristics of Dairy Effluent before and after Treatment with Side-Stream MBR

S. No	Quality Parameter	Units	Feed Value	Permeate Value
1	pH	–	8.66	8.19
2	TDS	ppm	4220	1590
3	TSS	ppm	2875	1750
4	Turbidity	FAU	32	3
5	COD	ppm	6000	280
6	Conductivity	mS/cm	11	2.7

Source: Chandrasekhar, S.S., D. Srinath, S. Nivedita, and S. Sridhar, *International Journal of Pure and Applied Bioscience*, 5: 71–79, 2017.

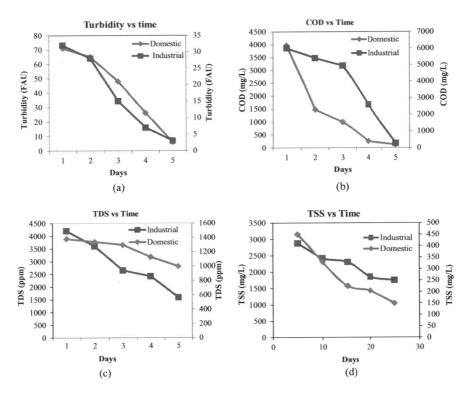

FIGURE 24.5
(a) Rejection of turbidity, (b) Efficiency of COD removal, (c) Reduction in TDS and (d) Decrease in TSS, at a pressure 4 kg cm^{-2}.

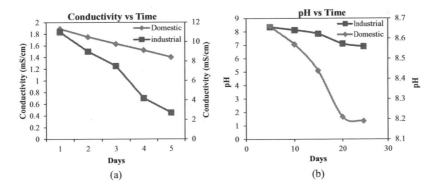

FIGURE 24.6
(a) Reduction in conductivity and (b) Change in pH at a pressure of 4 kg cm^{-2}.

24.11.2 Case Study on Kitchen Wastewater

A significant reduction in turbidity and COD was recorded with the HF-UF membrane in aerobic mode of SMBR after 21 days of experiment. A sharp drop in turbidity from 71 FAU in the feed to 6 FAU in permeate was recorded, which corresponds to 92% turbidity removal efficiency (Figure 24.5 (a)).

Similarly, 97% COD reduction was noted with a drop from 3967 ppm to 120 ppm as shown in Figure 24.5 (b). Decreases in TDS from 1382 ppm to 1000 ppm and TSS from 450 ppm to 120 ppm were also observed and presented in Figure 24.5 (c) and 24.5 (d), respectively. Marginal reductions in conductivity and pH from 1.89 mS/cm to 1.4 mS/cm and from 8.35 to 6.9, respectively, are presented in Figure 24.6 (a, b). TSS rejection of 73% and a 30% decrease in TDS, along with reduced turbidity and COD, were quite encouraging. The SMBR set-up with microbial cultures isolated from curd was found to have potential for treating kitchen wastewater.

24.11.3 Case Study on Dairy Effluent

The HF-UF membrane was operated in aerobic mode of SSMBR for 21 days, as shown in Figure 24.5 (a). Considerable reduction in turbidity from 32 FAU to 3 FAU was observed, corresponding to 91% turbidity removal. Subsequently, a 95% decrease in COD was noted (from 6,000 ppm in feed to 280 ppm in permeate), as shown in Figure 24.5 (b) (Chandrasekhar et al., 2017). The COD concentration in permeate reached an approximately constant level of 280 ppm after 21 days of commencement of the treatment. TDS of the effluent was found to be reduced from 4220 ppm to 1590 ppm with 62% rejection, as displayed in Figure 24.5 (c) (Chandrasekhar et al., 2017). Similarly, reductions in TSS, conductivity and pH were observed, from 2875 ppm to 1750 ppm, 11 mS/cm to 2.7 mS/cm and 8.66 to 8.19, respectively (Chandrasekhar et al., 2017). The results are illustrated in Figure 24.5 (d), Figure 24.6 (a) and Figure 24.6 (b).

Results obtained from this study indicate the efficiency of microorganism and the MBR process. The microbes isolated from the effluent were found to be highly effective in reducing contaminants. The organic load in the effluent created a favorable environment that allowed for the growth of the microbes. The treated permeate still had bad odor and color.

24.11.4 Color Removal from Kitchen Wastewater and Dairy Effluent

Color removal from kitchen wastewater and dairy effluent is a critical aspect. The filtrate/permeate showed a decrease from yellow to pale yellow for kitchen wastewater and black color to light color in the case of dairy effluent after 21 days in SMBR and SSMBR using mixed microbial consortia. Reduction in color was observed in permeate after passing through RO as seen in Figure 24.7 (a, b).

FIGURE 24.7
Color intensities of feed and permeate samples.

24.11.5 Treatment of Kitchen Wastewater by MFC

Electrochemical performance of biofilm grown in the closed circuit was recorded. The presence of bacteria acts as a biocatalyst in the anodic cell to oxidize the substrate, generating electrons and protons. These electrons get transported though an external circuit from the anode side to the cathode side, while the protons pass though the membrane and enter the aerated cathode cell where they combine with oxygen and produce clean water as a byproduct. The potential of MFC was studied under a 12 kΩ external load as a function of time. The maximum OCV after 7 days was observed to be 920 ± 5 mV. This indicates the efficiency of membrane for MFC application. After achieving stable OCV, different resistors were connected to obtain a power density of 190 mW m^{-2}. The high current density with low voltage drop indicates good performance arising from high conductivity, optimum water uptake and low internal resistance. This preliminary result also confirms the potential of the membrane to generate energy using kitchen wastewater as a substrate. At low concentrations of organic matter, cell voltage decreased. A power density of 190 mW m^{-2} was achieved along with significant COD removal efficiency of 65% for kitchen wastewater. The encouraging result of MFC creates the need for an elaborate study of the nature and morphology of membrane, as well as an assessment of its thermal and mechanical stability and detailed characterization by adopting various methods like FTIR, XRD, SEM, TGA, and UTM.

24.12 Conclusions

The treatment of kitchen and industrial wastewaters was investigated using membrane bioreactor (MBR) and microbial fuel cell (MFC) technologies. MBR and MFC are good devices for water reclamation and power generation during advanced wastewater treatment. It is most likely that the high quality of the effluent produced by the MBR technology will promote future projects for the reuse and recycling of several industrial effluents containing recalcitrant compounds. The present study focuses on design and implementation of the MBR technology for wastewater treatment on a small scale. To address the design issues for simultaneous treatment and valorization of wastewater, the development of integrated MBR is required. The interesting and attractive advantages in characteristics of integrated MBR have great potential to play an important role in wastewater treatment process or sustainable development. MBR shows good performance in the reduction of organic pollutants making it an attractive process for wastewater recycle and reuse, whereas MFC shows good energy generation efficiency. The flexibility of both SMBR and SSMBR configurations allows their operation under both favorable and challenging conditions, which represents a significant development for extending their implementation to remote areas and small towns. The performance of the synthesized membrane in the MBR system for treatment of kitchen wastewater and dairy effluent was assessed on the basis of COD and turbidity rejection values. The removal of organic pollutants in terms of COD was considerably high and good quality permeate was achieved during long-term operation. The model developed helped to provide information on the variation of flux with trans-membrane pressure. Membrane flux decreased over time after filtering the mixed liquor, although chemical cleaning appeared to be successful in order to maintain the membrane's permeability. Most of the membrane fouling was found to come from organic sources, and to a lesser extent, inorganic compounds. The present investigation concludes that an SMBR and SSMBR with an indigenously developed spiral wound hydrophilized ultrafiltration (HF-UF) membrane of 10 kDa MWCO and mixed microbial consortia isolated from buttermilk or dairy effluent could be used for the treatment of kitchen and dairy effluents on a pilot scale. The results of the present study indicate that both the MBR configurations can achieve high impurity removal efficiencies in wastewater treatment and that MBR permeate is suitable for urban utility, agricultural and recreational reuse, according to quality criteria. Permeate could be further processed though reverse osmosis (RO) to obtain high quality water for reuse.

References

Alexandre, G., A. Moura Bernardes, and P. de Maria Norberta. 2013. Nanofiltration for the recovery of low molecular weight polysaccharides and polyphenols from winery effluents. *Separation Science and Technology* 48(17): 2524–2530.

Amin, S.K., H.A.M. Abdallah, M.H. Roushdy, and S.A. El-Sherbiny. 2016. An overview of production and development of ceramic membranes. *International Journal of Applied Engineering Research* 11(12): 7708–7721.

Ashe, B. and S. Paul. 2010. Isolation and characterization of lactic acid bacteria from dairy effluents. *Journal of Environmental Research and Development* 4(4): 983–991.

Ashish, T. 2014. Study of characteristics and treatments of dairy industry wastewater. *Journal of Applied and Environmental Microbiology* 2(1): 16–22.

Bixio, D., C. Thoeye, and T. Wintgens. 2008. Water reclamation and reuse: Implementation and management issues. *Desalination* 218: 13–23.

Buisson, H., P. Cote, M. Praderie, and H. Paillard. 1998. The use of immersed membranes for upgrading wastewater treatment plants. *Water Science and Technology* 37(9): 89–95.

Chandrasekhar, S.S., D. Srinath, S. Nivedita, and S. Sridhar. 2017. Treatment of dairy industry effluent using membrane bioreactor. *International Journal of Pure and Applied Bioscience* 5: 71–79.

Chang, I.S., J.S. Kim, and C.H. Lee. 2001. The effects of EPS on membrane fouling in a MBR process. *Proceeding of MBR3 Conference*, 16th of May, Cranfield: 19–28.

Chang, I.S., P. Le-Clech, B. Jefferson, and S. Judd. 2002. Membrane fouling in membrane bioreactors for wastewater treatment. *Journal of Environmental Engineering* 128(11): 1018–1029.

Chen, T.K., J.N. Chen, C.H. Ni, G.T. Lin, and C.Y. Chang. 2003. Application of a membrane bioreactor system for opto-electronic industrial wastewater treatment–a pilot study. *Water Science and Technology* 48: 195–202.

Cheng, X., L. He, H.W. Lu, Y.W. Chen, and L.X. Ren. 2016. Optimal water resources management and system benefit for the Marcellus shale-gas reservoir in Pennsylvania and West Virginia. *Journal of Hydrology* 540: 412–22.

Choo, K.H. and C.H. Lee. 1996. Membrane fouling mechanisms in the membrane coupled anaerobic bioreactor. *Water Research* 30: 1771–1780.

Chu, L., X. Zhang, F. Yang, and X. Li. 2006. Treatment of kitchen wastewater by using a microaerobic membrane bioreactor. *Desalination* 189: 181–192.

Cicek, N., J.P. Franco, M.T. Suidan, and V. Urbain. 1998. Using a membrane bioreactor to reclaim wastewater. *Journal American Water Works Association* 90(11): 105–113.

Cote, P., H. Buisson, and M. Praderie. 1998. Immersed membranes activated sludge process applied to the treatment of municipal wastewater. *Water Science and Technology* 38: 437–442.

Defrance, L., M.Y. Jaffrin, B. Gupta, P. Paullier, and V. Geaugey. 2000. Contribution of various constituents of activated sludge to membrane bioreactor fouling. *Bioresource Technology* 73: 105–112.

Der-Fong, J. 2012. Organic removal efficiencies and power production capabilities of microbial fuel cells with pure cultures and mixed culture. *Journal of Asia-Pacific Chemical, Biological & Environmental Engineering Society* 1: 2–7.

Devendra, D. and T. Rohit. 2014. A Review on membrane bioreactors: An emerging technology for industrial wastewater treatment. *International Journal of Emerging Technology and Advanced Engineering* 4(12): 226–236.

Elmaleh, S. and L. Abdelmoummi. 1998. Experimental test to evaluate performance of an anaerobic reactor provided with an external membrane unit. *Water Science Technology* 38: 385–392.

Fan, X.J., V. Urbain, Y. Qian, and J. Manem. 1996. Nitrification and mass balance with a membrane bioreactor for municipal wastewater treatment. *Water Science and Technology* 34: 129–136.

Janbandhu, A. and M. Fulekar. 2011. Biodegradation of phenanthene using adapted microbial consortium isolated from petrochemical contaminated environment. *Journal of Hazard Material* 187: 333–340.

Jiang, Y., H. Huang, M. Wu, X. Yu, Y. Chen, P. Liu, and X. Li. 2015. Pseudomonas species LZ-Q continuously degrades phenanthene under hypersaline and hyperalkaline condition in a membrane bioreactor system. *Biophysics Reports* 1(3): 156–167.

Judd, S. 2005. Fouling control in submerged membrane bioreactors. *Water Science and Technology* 51: 27–34.

Katsou, E., S. Malamis, T. Kosanovic, K. Souma, and K.J. Haralambo. 2012. Application of adsorption and ultrafiltration processes for the pre-treatment of several industrial wastewater streams. *Water Air Soil and Pollution* 223: 5519–5534.

Knoblock, M.D., P.M. Sutton, P.M. Misha, K. Gupta, and A. Janson. 1994. Membrane biological reactor system for treatment of oily wastewaters. *Water Environment Research* 66(2): 133–139.

Krauth, K.H. and K.F. Staab. 1993. Pressurized bioreactor with membrane filtration for wastewater treatment. *Water Research* 27(3): 405–411.

Kumar, A.K., M.V. Reddy, K. Chandrasekhar, S. Srikanth, and S.V. Mohan. 2012. Endocrine disruptive estrogens role in electron transfer: Bio-electrochemical remediation with microbial mediated electro genesis. *Bioresource Technology* 104: 547–556.

Li, H., P. Du, Y. Chen, H. Lu, X. Cheng, B. Chang, and Z. Wang. 2017. Advances in microbial fuel cells for wastewater treatment. *Renewable and Sustainable Energy Reviews* 71: 388–403.

Lin, H., W. Gao, F. Meng, B.Q. Liao, K.T. Leung, L. Zhao, J. Chen, and H. Hong. 2012. Membrane bioreactors for industrial wastewater treatment: A critical review. *Critical Reviews in Environmental Science and Technology* 42(7): 677–740.

Liu, H., R. Ramanarayanan, and B.E. Logan. 2004. Production of electricity during wastewater treatment using a single chamber microbial fuel cell. *Environmental Science and Technology* 38(14): 2281–2285.

Logan, B.E. and J.M. Regan. 2006. Electricity producing bacterial communities in microbial fuel cells. *Trends in Microbiology* 14(12): 512–518.

Lorenz, W., T. Cunningham, and J.P. Penny. 2002. "Phosphorus removal in a membrane bioreactor system. A full scale wastewater demonstration study." Paper presented at WEFTEC, Chicago. *Water Environment Federation*, VA, U.S.

Mahdi Mardanpour, M.D., M. Nasr Esfahany, T. Behzad, and R. Sedaqatvand. 2012. Single chamber microbial fuel cell with spiral anode for dairy wastewater treatment. *Biosensors and Bioelectronics* 38: 264–269.

Majid, H., G.N. Bidhendi, A. Torabian, and N. Mehrdadi. 2014. A study on membrane bioreactor for water reuse from the effluent of industrial town wastewater treatment plant. *Iranian Journal of Toxicology* 8(24): 983–990.

Mansoorian, J.H., A.H. Mahvi, A.J. Jafari, and N. Khanjani. 2016. Evaluation of dairy industry wastewater treatment and simultaneous bioelectricity generation in a catalyst-less and mediator-less membrane microbial fuel cell. *Journal of Saudi Chemical Society* 20: 88–100.

Melin, T., B. Jefferson, and D. Bixio. 2006. Membrane bioreactor technology for wastewater treatment and reuse. *Desalination* 187: 271–282.

Mostafa, A.A. 2013. Treatment of cheese processing wastewater by physicochemical and biological methods. *International Journal of Microbiological Research* 4(3): 321–332.

Nagaoka, H., S. Yamanishi, and A. Miya. 1998. Modeling of biofouling by extracelluar polymers in a membrane separation activated sludge. *Water Science Technology* 38: 497–504.

Nazim, C. 2002. "Membrane bioreactors in the treatment of wastewater generated from agricultural industries and activities." AIC 2002 Meeting, CSAR/SCGR Program, Saskatoon, Saskatchewan.

Oh, S.E. and B.E. Logan. 2005. Hydrogen and electricity production from a food processing wastewater using fermentation and microbial fuel cell technologies. *Water Research* 39: 4673–4682.

Parameshwaran, K., C. Visvanathan, and R. Ben Aim. 1999. Membrane as solid/liquid separator and air diffuser in bioreactor. *Journal of Environmental Engineering* 125(9): 825–834.

Passos, V.F., S.A. Neto, N. Sidney, A.R. de Andrade, and V. Reginatto. 2015. Energy generation in a microbial fuel cell using anaerobic sludge from a wastewater treatment plant. *Journal of Scientia Agricola* 73: 424–428.

Pham, T.H., K. Rabaey, P. Aelterman, P. Clauwaert, L. De Schamphelaire, N. Boon, and W. Verstraete. 2006. Microbial fuel cells in relation to conventional anaerobic digestion technology. *Engineering Life Science* 6: 285–292.

Pillay, V.L., B. Townsend, and C.A. Buckley. 1994. Improving the performance of anaerobic digesters at wastewater treatment works: The coupled cross-flow microfiltration/digester process. *Water Science and Technology* 30(12): 329–337.

Praneeth, K., S. Moulik, V. Pavani, K. Bhargava Suresh, T. James, and S. Sridhar. 2014. Performance assessment and hydrodynamic analysis of a submerged membrane bioreactor for treating industrial effluent. *Journal of Hazardous Materials* 274: 300–313.

Rabaey, K. and W. Verstraete. 2005. Microbial fuel cells: Novel biotechnology for energy generation. *Trends Biotechnology* 23: 291–298.

Rafik, M., H. Qabli, S. Belhamidi, F. Elhannouni, A. Elkhedmaoui, and A. Elmidaoui. 2015. Membrane separation in the sugar industry. *Journal of Chemical and Pharmaceutical Research* 7(9): 653–658.

Rosenberger, S., U. Kruger, R. Witzig, W. Manz, U. Szewzyk, and M. Kraume. 2002. Performance of a bioreactor with submerged membranes for aerobic treatment of municipal wastewater. *Water Research* 36(2): 413–420.

Ross, W.R., J.P. Barnard, N.K.H. Strohwald, C.J. Grobler, and J. Sanetra. 1992. Practical application of the ADUF process to the full-scale treatment of maize processing effluent. *Water Science and Technology* 25(10): 27–39.

Santasmasas, C., M. Rovira, F. Clarens, and C. Valderrama. 2014. Wastewater treatment by MBR pilot plant: Flat sheet and hollow fiber case studies. *Desalination and Water Treatment* 51: 2423–2430.

Santoro, C., I. Ieropoulos, J. Greenman, P. Cristiani, T. Vadas, and A. Mackay. 2013. Power generation and contaminant removal in single chamber microbial fuel cells (SCMFCs) treating human urine. *International Journal of Hydrogen Energy* 38(26): 11543–11551.

Sciarria, T.P., A. Tenca, A.D. Epifanio, B. Mecheri, G. Merlino, and M. Barbato. 2013. Using olive mill wastewater to improve performance in producing electricity from kitchen wastewater by using single-chamber microbial fuel cell. *Bioresource Technology* 147: 246–253.

Scott, J.A. and K.L. Smith. 1996. A bioreactor coupled to a membrane to provide aeration and filtration in ice cream factory wastewater. *Water Resources* 31: 1–69.

Shah, M.P. 2016. Industrial wastewater treatment: A challenging task in the industrial waste management. *Advances in Recycling & Waste Management* 2(1): 115.

Shimizu, Y., K. Uryu, Y.I. Okuno, and A. Watanabe. 1996. Cross flow microfiltration of activated sludge using submerged membrane with air bubbling. *Journal of Fermentation and Bioengineering* 81: 55–60.

Shin, H.S., H. An, S.T. Kang, K.H. Choi, and K.S. Jun. 1999. "Fouling characteristics in pilot scale submerged membrane bioreactor." Presented at *Proceedings 1st WEFTEC*, New Orleans, LA.

Singh, N.B., S. Ruchi, and M.D. Manzer Imam. 2014. Waste water management in dairy industry: Pollution abatement and preventive attitudes. *International Journal of Science, Environment and Technology* 3(2): 672–683.

Soltani, S., D. Mowla, M. Vossoughi, and M. Hesampour. 2010. Experimental investigation of oily water treatment by membrane bioreactor. *Desalination* 250: 598–600.

Sreemoyee, C. and P. Priti. 2013. Assessment of physico-chemical parameters of dairy wastewater and isolation and characterization of bacterial strains in terms of COD reduction. *International Journal of Science, Environment and Technology* 2(3): 395-400.

Tadkaew, N., M. Sivakumar, and L.D. Nghiem. 2007. Membrane bioreactor technology for decentralised wastewater treatment and reuse. *International Journal of Water* 3(4): 368–380.

Tardieu, E., A. Grasmick, V. Geaugey, and J. Manem. 1998. Hydrodynamic control of bioparticle deposition in a MBR applied to wastewater treatment. *Journal of Membrane Science* 147: 1–12.

Ueda, T., T. Hata, Y. Kikuoka, and O. Seino. 1997. Effects of aeration on suction pressure in a submerged membrane bioreactor. *Water Research* 31: 489–494.

Ueda, T. and N. Horan. 2000. Fate of indigenous bacteriophage in a membrane bioreactor. *Water Research* 34: 2151–2159.

Van de Roest, H.F., D.P. Lawrence, and A.G.N. Van Bentem. 2002. *Membrane Bioreactors for Municipal Wastewater Treatment. Strowa Report* (150). London, UK: IWA Publishing.

Venkata Narayana, A., B. Sumalatha, K. Kiran Kumar, D. John Babu, T.C. Venkateswarulu. 2016. Membrane bioreactors for waste water treatment. *Journal of Chemical and Pharmaceutical Research* 8(4): 258–260.

Virapan, Saravanane, R. and V. Murugaiyan. 2017. Treatment of reverse osmosis reject water from industries. *International Journal of Applied Environmental Sciences* 12(3): 489–503.

Visvanathan, C., R. Ben Aim, and K. Parameshwaran. 2000. Membrane separation bioreactors for wastewater treatment. *Critical Reviews in Environmental Science and Technology* 30: 1–48.

Wang, Y.K., G.P. Sheng, B.J. Shi, W.W. Li, and H.Q. Yu. 2013. A novel electrochemical membrane bioreactor as a potential net energy producer for sustainable wastewater treatment. *Science Reports* 3: 1864.

You, S.J., Q.L. Zhao, J.Q. Jiang, and J.N. Zhang. 2006. Treatment of kitchen wastewater with simultaneous electricity generation in microbial fuel cell under continuous operation. *Chemical and Biochemical Engineering* 20: 407–412.

Zhang, G., Q. Zhao, Y. Jiao, D.J. Lee, and N. Ren. 2012. Efficient electricity generation from sewage sludge using biocathode microbial fuel cell. *Water Research* 46: 43–52.

Zhang, J.N., Q.L. Zhao, S.J. You, J.Q. Jiang, and N.Q. Ren. 2008. Continuous electricity production from leachate in a novel upflow air-cathode membrane-free microbial fuel cell. *Water Science Technology* 57: 1017–1021.

Zhang, Y.F., J.S. Noori, and I. Angelidaki. 2011. Simultaneous organic carbon, nutrients removal and energy production in a photo-microbial fuel cell (PFC). *Energy Environmental Science* 4: 4340–4346.

List of Abbreviations

BOD	Biological oxygen demand
COD	Chemical oxygen demand
DO	Dissolved oxygen
EDTA	Ethylene diamine tetra acetic acid
FAU	Ferroin attenuated unit
HF-UF	Hydrophilized Ultrafiltration
HT	Hydraulic retention time
MBR	Membrane bioreactor
MCWO	Molecular Weight Cut-Off

MF   Microfiltration
NA   Nutrient agar
NaOH   Sodium hydroxide
NB   Nutrient broth
NF   Nanofiltration
PAN   Polyacrylonitrile
PE   Polyethylene
PES   Polyethersulfone
PET   Polyethylene terephthalate
PFS   Polymeric ferric sulfate
PP   Polypropylene
PS   Polysulfone
PTFE   Polytetrafluoroethylene
PVB   Polyvinyl butyral
PVDF   Polyvinylidene fluoride
RO   Reverse osmosis
SMBR   Submerged membrane bioreactor
SSMBR   Side-stream membrane bioreactor
TSS   Total suspended solids
TDS   Total dissolved solids
UF   Ultrafiltration

Index

Page numbers followed by f and t indicate figures and tables, respectively.

Taylor & Francis Group
an **informa** business

Taylor & Francis eBooks

www.taylorfrancis.com

A single destination for eBooks from Taylor & Francis
with increased functionality and an improved user
experience to meet the needs of our customers.

90,000+ eBooks of award-winning academic content in
Humanities, Social Science, Science, Technology, Engineering,
and Medical written by a global network of editors and authors.

TAYLOR & FRANCIS EBOOKS OFFERS:

A streamlined
experience for
our library
customers

A single point
of discovery
for all of our
eBook content

Improved
search and
discovery of
content at both
book and
chapter level

REQUEST A FREE TRIAL
support@taylorfrancis.com

Routledge
Taylor & Francis Group

CRC Press
Taylor & Francis Group